8TH EDITION

Automatic Control Systems

BENJAMIN C. KUO
University of Illinois at Urbana-Champaign

FARID GOLNARAGHI
University of Waterloo

JOHN WILEY & SONS, INC.

Executive Editor   *Bill Zobrist*
Senior Marketing Manager   *Katherine Hepburn*
Senior Production Editor   *Caroline Sieg*
Senior Designer   *Karin Kincheloe*
Illustration Coordinator   *Anna Melhorn*

About the Cover:
The cover depicts images from the emerging field of Micro-Electro-Mechanical Systems (MEMS) and a screenshot of the ACSYS controls programming tool used in the text. This MEMS photo shows a modern application of a sophisticated control system. In this case, researchers at the University of Wisconsin developed an ultrasonic actuation method in which ultrasonic shaking at 1–4 Mhz reduces the static friction at the hinges and above ambient temperature (~ 80 C) on the substrate leads to a thermokinetic force to lift the flaps simultaneously! In turn, this will enable massively parallel assembly of micromachines. The ACSYS programming tool depicts the advancement in software technologies allowing for the ease in which controls can be simulated and learned by students today.

Photo courtesy of Amit Lal and Ville Kaajakari, SonicMEMS Laboratory, University of Wisconsin-Madison.

This book was set in 10/12 Times Roman by Techbooks and printed and bound by RR Donnelley/Willard. The cover was printed by Phoenix Color.

This book is printed on acid free paper.   ∞

ISBN 0-471-13476-7

Printed in the United States of America

10 9 8 7 6 5 4 3

Dedications

Benjamin C. Kuo

To my family and Pugsley, Baobei, Buppy, and Tuskers

M. Farid Golnaraghi

To my wife, Mitra, for standing by me and for showing me the meaning of true love, and to baby Sophia, the joy of my life.

Preface (Readme)

This is the first time I have written a book for John Wiley & Sons, although it is a revision of an old edition. In 2000 Simon and Schuster was sold to Pearson, and the U.S. Justice Department stipulated the condition of merger was to divest a list of titles of Simon and Schuster. Apparently, *Automatic Control Systems*, 7th Edition, was among the titles. Maybe there are experts at the Justice Department who understood control systems. So, this is a brief history on how *Automatic Control Systems*, after being with Simon Schuster for nearly 40 years suddenly ended up as a book at John Wiley & Sons. However, we couldn't be happier as Wiley authors, and the forced transition has turned out to be a blessing.

In order to bring on new ideas and current material the 8th Edition has brought on a new coauthor, Professor Farid Golnaraghi from the University of Waterloo Ontario, Canada.

What we attempted to do for the revision is to make the book more streamlined while retaining the essential material. We have added more computer-aided tools for students and teachers. The prepublication manuscript has been reviewed by many professors, and most of the relevant suggestions have been adopted.

In the 8th Edition, the following material has been moved into appendices on the CD-ROM. These are

Appendix A: Complex Variable Theory

Appendix B: Differential and Difference Equations

Appendix C: Elementary Matrix Theory and Algebra

Appendix D: Laplace Transform Table

Appendix E: Operational Amplifier

Appendix F: Properties and Construction of the Root Loci

Appendix G: Frequency-Domain Plots

Appendix H: General Nyquist Criterion

Appendix I: Discrete-Data Control Systems

Appendix J: *z*-Transform Table

Appendix K: **ACSYS** 2002: Description of the Software

Answers to Selected Problems

In addition, the CD-ROM contains the MATLAB^{TM} files for **ACSYS**, which are software tools for solving control-system problems, and Powerpoint files for the illustrations in the text. We have pulled all the material on discrete-data control systems in each chapter and placed it in Appendix I.

The following paragraphs are aimed at three groups: professors who have adopted the book or who we hope will select it as their text; practicing engineers looking for answers to solve their day-to-day design problems; and finally, students who are going to live with the book because it has been assigned for the control-systems course they are taking.

To the Professor: The material assembled in this book is an outgrowth of senior-level control-system courses taught by the authors at their universities throughout their teaching career. The first seven editions have been adopted by hundreds of universities in the United States and around the world, and have been translated into at least six languages.

---

^{TM}MATLAB is a registered trademark of MathWorks Inc.

Practically all the design topics presented in the 7th Edition have been retained. One of the significant changes is that the subject of discrete-data control systems is now in Appendix I on the CD-ROM. We would prefer to teach discrete-data systems as extensions of their analog counterparts. However, realistically, it would be difficult to cover both analog and discrete control systems in a one-semester course.

The software added to this edition is very different from the software accompanying any other control book. Here, through extensive use of MATLAB GUI programming, we have created software that is easy to use. As a result, students will need to focus only on learning control problems, not programming! We also have added two very new applications: SIMLab and Virtual Lab, where students work on realistic problems and conduct speed and position control labs in software environment. In SIMLab, students have access to the system parameters and can alter them (as in any simulation). In Virtual Lab, we have introduced a black-box approach, where the students have no access to the plant parameters and have to use some sort of system identification technique to find them. Through Virtual Lab we have essentially provided students with a realistic online lab with all the problems they would encounter in a real speed- or position-control lab, for example, amplifier saturation, noise, and nonlinearity. We welcome your ideas for the future editions of this book.

Finally, a sample section-by-section a one-semester course is given in the *Instructor's Manual*, which is available from the publisher to qualified instructors. The *Manual* also contains detailed solutions to all the problems in the book.

To Practicing Engineers: This book was written with the readers in mind and is very suitable for self-study. Our objective was to treat subjects clearly and thoroughly. The book does not use the theorem–proof–Q.E.D. style and is without heavy mathematics. The authors have consulted extensively for wide sectors of the industry for many years, and have participated in solving numerous control-systems problems, from aerospace systems to industrial controls, automotive controls, and control of computer peripherals. Although it is difficult to adopt all the details and realism of practical problems in a textbook at this level, some examples and problems reflect simplified versions of real-life systems.

To Students: You have had it now that you have signed up for this course and your professor has assigned this book! You had no say about the choice, although you can form and express your opinion on the book after reading it. Worse yet, one of the reasons that your professor made the selection is because he or she intends to make you work hard. But please don't misunderstand us: what we really mean is that although this is an easy book to study (in our opinion), it is a no-nonsense book. It doesn't have cartoons or nice-looking photographs to amuse you. From here on, it is all business and hard work. You should have had the prerequisites on subjects found in a typical linear-systems course, such as how to solve linear ordinary differential equations, Laplace transform and appplications, and time-response and frequency-domain analysis of linear systems. In this book you will not find too much new mathematics to which you have not been exposed before. What is interesting and challenging is that you are going to learn how to apply some of the mathematics that you have acquired during the last two or three years of study in college. In case you need to review some of the mathematical foundations, you can find them in the appendices on the CD-ROM that accompanies this text. The CD-ROM also contains lots of other goodies, including the **ACSYS** software, which is GUI software that uses MATLAB-based programs for solving linear control systems problems. You will also find the Simulink-based SimLab and Virtual Lab, which will help you to gain understanding of real-world control systems.

This book has numerous illustrative examples. Some of these are deliberately simple for the purpose of illustrating new ideas and subject matter. Some examples are more elaborate, in order to bring the practical world closer to you. Furthermore, the objective of this book is to present a complex subject in a clear and thorough way. One of the important learning strategies for you as a student is not to rely strictly on the textbook assigned. When studying a certain subject, go to the library and check out a few similar texts to see how other authors treat the same subject. You may gain new perspectives on the subject and discover that one author may treat the material with more care and thoroughness than the others. Do not be distracted by written-down coverage with oversimplified examples. The minute you step into the real world, you will face the design of control systems with nonlinearities and/or time-varying elements as well as orders that can boggle your mind. It may be discouraging to tell you now that strictly linear and first-order systems do not exist in the real world.

Some advanced engineering students in college do not believe that the material they learn in the classroom is ever going to be applied directly in industry. Some of our students come back from field and interview trips totally surprised to find that the material they learned in courses on control systems is actually being used in industry today. They are surprised to find that this book is also a popular reference for practicing engineers. Unfortunately, these fact-finding, eye-opening, and self-motivating trips usually occur near the end of their college days, which is often too late for students to get motivated.

There are many learning aids available to you: The MATLAB-based **ACSYS** software will assist you in solving all kinds of control-systems problems. The SIMLab and Virtual Lab software can be used for simulation of virtual experimental systems. These are all found on the CD-ROM accompanying this text. In addition, the Review Questions and Summary at the end of each chapter should all be useful to you. You should also visit the Web site dedicated to the book, where you will find the errata and other supplemental material.

We hope that you will enjoy this book. It will represent another major textbook acquisition (investment) in your college career. Our advice to you is not to sell it back to the bookstore at the end of the semester. If you do so, but find out later in your professional career that you need to refer to a control systems book, you will have to buy it back at a higher price again.

An Important Note Regarding the ACSYS Software: At the time of publication of this book, there is an issue of compatibility between MATLAB version 6.0 (R12), the student version of MATLAB (R12) and MATLAB version 6.1 (R12.1), and Windows XP. Upon our request Mathworks Inc. issued the following statement:

"The Student Version of MATLAB 6.0 (R12) is not officially supported under Windows XP. For more information on the system requirements for the Student Version of MAT-LAB 6.0 (R12), please see the following URL:

http://www.mathworks.com/products/studentversion/sys_req.shtml

Currently, there are no plans to officially support the Student Version of MATLAB 6.0 (12) on Windows XP, though our development staff may readdress support for Windows XP on the Student Version in the future."
Further in a statement in the following URL they suggest:

http://www.mathworks.com/support/solutions/data/30479.shtml

"MATLAB 6.1 (R12.1) was released before Windows XP was finalized and thus was not validated under Windows XP. Windows XP will be officially supported in our next release of MATLAB. The system requirements for MATLAB 6.1 (R12.1) can be found at the following URL:

http://www.mathworks.com/products/system.shtml/Windows

In the minimal testing that we have done, we have experienced some incompatibilities with MATLAB 6.1 (R12.1) and Windows XP. There are two possible workarounds that you can do to address these issues:
1. You can use Windows XP in the "Windows Classic Style" mode and/or
2. You can download a new file, hg.dll ...

PLEASE NOTE: The new hg.dll file is not meant for use with the Student Version of MATLAB 6.0 (R12) on Windows XP."

Based on the previous statements, it is our understanding that MATLAB R12 is not compatible with Windows XP and MATLAB version 6.1 (R12.1) may be used with Windows XP with some possible problems.

Further, at the final stages of publication of this book the pre-release version of MATLAB 6.5 R13 became available. The ACSYS software was successfully tested on the pre-release of MATLAB 6.5 R13, and has worked properly using all Microsoft Windows operating systems. It is expected that the student and full versions of MATLAB 6.5 R13 are available in 2003.

As a result, we have decided to release three versions of the ACSYS software, which accompanies this book:

1. **ACSYS** 2002 (R12) is supported by all Microsoft Operating Systems except for Windows XP. The users of the student version of MATLAB (R12) must use this version.
2. **ACSYS** 2002 (R12.1) is supported by all Microsoft Operating Systems, and appear to work fine with the Windows XP. Although we have not observed any problems running MATLAB 6.1 under Windows XP Operating System, Windows XP users may expect to encounter some problems.
3. **ACSYS** 2002 (Pre-release R13) is supported by all Microsoft Operating Systems. This version should work properly with the student and full versions of MATLAB R13 once they are available.

Special Acknowledgments: The authors wish to thank the following reviewers, who have given permission to have their names mentioned. The prepublication reviews have had great impact on the revision project.

Professor B. Ross Barmish, University of Wisconsin
Professor Peter Reischi, San Jose State University
Hy D. Tran, University of New Mexico
Hoss Cyrus Ahmadi, University of British Columbia
Professor Peng-Yung Woo, Northern Illinois University
Professor John L. Crassidis, University at Buffalo
Professor Horacio J. Marquez, University of Alberta
Professor Joseph F. Horn, The Pennsylvania State University
Professor Semyon M. Meerkov, University of Michigan
Professor Bill Diong, University of Texas at El Paso
Professor Swapan Kumar Mukherjee, Regional Institute of Technology
Professor L. F. Yeung, City University of Hong Kong

The authors are indebted to Professor Duane Hanselman at the University of Maine for his valuable suggestions on the revision.

The authors thank Tony Kim, Peter Won and Hamid Karbasi, graduate students at University of Waterloo, for their help with the software development, and Professor Jan Huissoon of the University of Waterloo for his help in creating the SIMLab and Virtual Lab experiments. Farid Golnaraghi also wishes to thank Professor Benjamin Kuo for sharing the pleasure of writing this wonderful book, and for his teachings, patience, and support throughout this experience.

B. C. Kuo; Champaign, Illinois U.S.A.

M. F. Golnaraghi; Waterloo, Ontario, Canada

2002, Year of the Horse

Table of Contents

CHAPTER 1

Introduction

▷ 1-1 INTRODUCTION

The objective of this chapter is to familiarize the reader with the following subjects:

1. What a control system is.
2. Why control systems are important.
3. What the basic components of a control system are.
4. Some examples of control system applications.
5. Why feedback is incorporated into most control systems.
6. Types of control systems.

One of the most commonly asked questions by a novice on control system is: What is a control system? To answer the question, we can cite that in our daily life there are numerous "objectives" that need to be accomplished. For instance, in the domestic domain, we need to regulate the temperature and humidity of homes and buildings for comfortable living. For transportation, we need to control the automobile and airplane to go from one point to another accurately and safely. Industrially, manufacturing processes contain numerous objectives for products that will satisfy the precision and cost-effectiveness requirements. A human being is capable of performing a wide range of tasks, including decision making. Some of these tasks, such as picking up objects and walking from one point to another, are commonly carried out in a routine fashion. Under certain conditions, some of these tasks are to be performed in the best possible way. For instance, an athlete running a 100-yard dash has the objective of running that distance in the shortest possible time. A marathon runner, on the other hand, not only must run the distance as quickly as possible, but in doing so, he or she must control the consumption of energy and devise the best strategy for the race. The means of achieving these "objectives" usually involve the use of control systems that implement certain control strategies.

• Control systems are in abundance in modern civilization.

In recent years, control systems have assumed an increasingly important role in the development and advancement of modern civilization and technology. Practically every aspect of our day-to-day activities is affected by some type of control systems. Control systems are found in abundance in all sectors of industry, such as quality control of manufactured products, automatic assembly line, machine-tool control, space technology and weapon systems, computer control, transportation systems, power systems, robotics, MicroElectroMechanical Systems (MEMS), nanotechnology, and many others. Even the control of inventory and social and economic systems may be approached from the theory of automatic control.

1-1-1 Basic Components of a Control System

The basic ingredients of a control system can be described by:

1. Objectives of control.
2. Control-system components.
3. Results or outputs.

The basic relationship among these three components is illustrated in Fig. 1-1. In more technical terms, the **objectives** can be identified with **inputs**, or **actuating signals**, u, and the results are also called **outputs**, or **controlled variables**, y. In general, the objective of the control system is to control the outputs in some prescribed manner by the inputs through the elements of the control system.

1-1-2 Examples of Control-System Applications

Smart Transportation Systems

Perhaps one of the most innovative fields in controls used in consumer products is that of the automotive industry. We have grown to desire cars that are "intelligent" and provide maximum levels of comfort, safety, and fuel efficiency. Examples of intelligent systems in cars include climate control, cruise control, antilock brake systems (ABSs), active suspensions that reduce vehicle vibration over rough terrain, air springs that self-level the vehicle in high-G turns (in addition to providing better ride), integrated vehicle dynamics that provide yaw control when the vehicle is either over- or understeering (by selectively activating the brakes to regain vehicle control), traction control systems to prevent spinning of wheels during acceleration, active sway bars to provide "controlled" rolling of vehicle.

Intelligent Systems

Applications of control systems have significantly increased through the development of new materials, which provide unique opportunities for highly efficient actuation and sensing, thereby reducing energy losses and environmental impacts. State-of-the-art actuators and sensors may be implemented in virtually any system, including biological propulsion; locomotion; robotics, material handling; biomedical, surgical, and endoscopic; aeronautics; marine; and the defense and space industries. Potential applications of control of these systems may benefit the following areas:

- *Machine tools:* Improve precision and increase productivity by controlling chatter.
- *Flexible robotics:* Enable faster motion with greater accuracy.
- *Photolithography:* Enable the manufacture of smaller microelectronic circuits by controlling vibration in the photolithography circuit-printing process.
- *Biomechanical and biomedical:* Artificial muscles, drug delivery systems, and other assistive technologies.
- *Process control:* For example, on/off shape control of solar reflectors or aerodynamic surfaces.

Control in Virtual Prototyping and Hardware in the Loop

The concept of virtual prototyping has become a widely used phenomenon in the automotive, aerospace, defense, and space industries. In all these areas, pressure to cut costs

Figure 1-1 Basic components of a control system.

has forced manufacturers to design and test an entire system in a computer environment before a physical prototype is made. Design tools such as **MATLAB** and **Simulink** enable companies to design and test controllers for different components (e.g., suspension, ABS, steering, engines, flight control mechanisms, landing gear, and specialized devices) within the system and examine the behavior of the control system on the virtual prototype in real time. This allows the designers to change or adjust controller parameters online before the actual hardware is developed. Hardware in the loop terminology is a new approach of testing individual components by attaching them to the virtual and controller prototypes. Here the physical controller hardware is interfaced with the computer and replaces its mathematical model within the computer!

Steering Control of Automobile

As a simple example of the control system, as shown in Fig. 1-1, consider the steering control of an automobile. The direction of the two front wheels can be regarded as the controlled variable, or the output, y; the direction of the steering wheel is the actuating signal, or the input, u. The control system, or process in this case, is composed of the steering mechanism and the dynamics of the entire automobile. However, if the objective is to control the speed of the automobile, then the amount of pressure exerted on the accelerator is the actuating signal, and the vehicle speed is the controlled variable. As a whole, we can regard the simplified automobile control system as one with two inputs (steering and accelerator) and two outputs (heading and speed). In this case, the two controls and two outputs are independent of each other, but, there are systems for which the controls are coupled. Systems with more than one input and one output are called **multivariable systems**.

Idle-Speed Control of Automobile

As another example of a control system, we consider the idle-speed control of an automobile engine. The objective of such a control system is to maintain the engine idle speed at a relatively low value (for fuel economy) regardless of the applied engine loads (e.g., transmission, power steering, air conditioning). Without the idle-speed control, any sudden engine-load application would cause a drop in engine speed that might cause the engine to stall. Thus the main objectives of the idle-speed control system are (1) to eliminate or minimize the speed droop when engine loading is applied, and (2) to maintain the engine idle speed at a desired value. Figure 1-2 shows the block diagram of the idle-speed control system from the standpoint of inputs–system–outputs. In this case, the throttle angle α and the load torque T_L (due to the application of air conditioning, power steering,

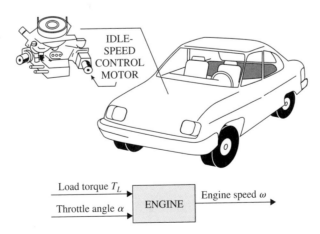

Figure 1-2 Idle-speed control system.

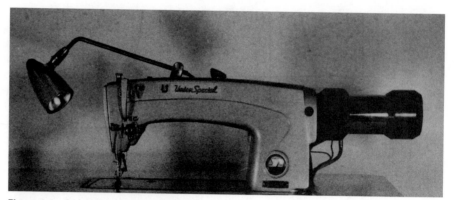

Figure 1-3 Industrial sewing machine.

transmission, or power brakes, etc.) are the inputs, and the engine speed ω is the output. The engine is the controlled process of the system.

Industrial Sewing Machine

Sewing, as the basic joining operation in a garment-making process, is in principle a rather complicated and laborious operation. For low cost and high productivity, the sewing industry has to rely on sophisticated sewing machines to increase the speed and accuracy of the sewing operations. Figure 1-3 shows a photograph of a typical industrial sewing machine, which, compared to household machines, is strictly a single-purpose, high-precision device. It can produce only one type of stitch, but is extremely fast, with a typical rate of over 100 stitches per second. One stitch corresponds to one revolution of the machine main shaft, which translates to top speeds of as high as 8000 rpm. An ideal velocity profile of one start-stop cycle of the machine is shown in Fig. 1-4. Typically, there

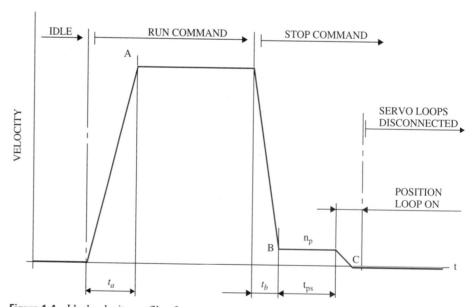

Figure 1-4 Ideal velocity profile of one start-stop cycle of an industrial sewing machine.

Figure 1-5 Solar collector field.

should be no velocity overshoot at point A and no undershoot at point B. Acceleration time t_a, deceleration time t_b, and position search time t_{ps} should be as short as possible. When the machine reaches the stopping point, C, there should be zero or negligible oscillations. To achieve these performance objectives, the control system in the machine should be designed with stringent requirements.

Sun-Tracking Control of Solar Collectors

To achieve the goal of development of economically feasible non-fossil-fuel electrical power, the U.S. government has sponsored many organizations in research and development of solar power conversion methods, including the solar-cell conversion techniques. In most of these systems, the need for high efficiencies dictates the use of devices for sun tracking. Figure 1-5 shows a solar collector field. Figure 1-6 shows a conceptual method of efficient water extraction using solar power. During the hours of daylight, the solar collector would produce electricity to pump water from the underground water table to a reservoir (perhaps on a nearby mountain or hill), and in the early morning hours, the water would be released into the irrigation system.

One of the most important features of the solar collector is that the collector dish must track the sun accurately. Therefore, the movement of the collector dish must be controlled by sophisticated control systems. The block diagram of Fig. 1-7 describes the general philosophy of the sun-tracking system together with some of the most important components. The basic philosophy of the control system is that a predetermined desired rate

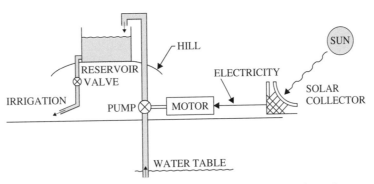

Figure 1-6 Conceptual method of efficient water extraction using solar power.

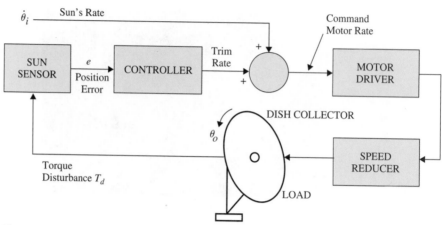

Figure 1-7 Important components of the sun-tracking control system.

is modified or trimmed by actual position errors determined by the sun sensor. The controller ensures that the tracking collector is pointed toward the sun in the morning and sends a "start track" command. The controller constantly calculates the sun's rate for the two axes (azimuth and elevation) of control during the day. The controller uses the sun rate and sun sensor information as inputs to generate proper motor commands to slew the collector.

1-1-3 Open-Loop Control Systems (Nonfeedback Systems)

The idle-speed control system illustrated in Figs. 1-2 is rather unsophisticated and is called an **open-loop control system**. It is not difficult to see that the system as shown would not satisfactorily fulfill critical performance requirements. For instance, if the throttle angle α is set at a certain initial value that corresponds to a certain engine speed, then when a load torque T_L is then applied, there is no way to prevent a drop in the engine speed. The only way to make the system work is to have a means of adjusting α in response to a change in the load torque in order to maintain ω at the desired level. The conventional electric washing machine is another example of an open-loop control system because, typically, the amount of machine wash time is entirely determined by the judgment and estimation of the human operator.

• Open-loop systems are economical but usually inaccurate.

The elements of an open-loop control system can usually be divided into two parts: the **controller** and the **controlled process**, as shown by the block diagram of Fig. 1-8. An input signal or command r is applied to the controller, whose output acts as the actuating signal u; the actuating signal then controls the controlled process so that the controlled variable y will perform according to some prescribed standards. In simple cases, the controller

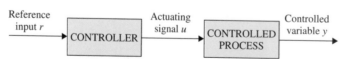

Figure 1-8 Elements of an open-loop control system.

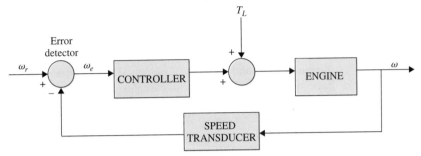

Figure 1-9 Block diagram of a closed-loop idle-speed control system.

can be an amplifier, mechanical linkage, filter, or other control elements, depending on the nature of the system. In more sophisticated cases, the controller can be a computer such as a microprocessor. Because of the simplicity and economy of open-loop control systems, we find this type of system in many noncritical applications.

1-1-4 Closed-loop Control Systems (Feedback Control Systems)

What is missing in the open-loop control system for more accurate and more adaptive control is a link or feedback from the output to the input of the system. To obtain more accurate control, the controlled signal y should be fed back and compared with the reference input, and an actuating signal proportional to the difference of the input and the output must be sent through the system to correct the error. A system with one or more feedback paths such as that just described is called a **closed-loop system**.

A closed-loop idle-speed control system is shown in Fig. 1-9. The reference input ω_r sets the desired idling speed. The engine speed at idle should agree with the reference value ω_r, and any difference such as the load torque T_L is sensed by the speed transducer and the error detector. The controller will operate on the difference and provide a signal to adjust the throttle angle α to correct the error. Figure 1-10 compares the typical performances of open-loop and closed-loop idle-speed control systems. In Fig. 1-10(a), the idle speed of the open-loop system will drop and settle at a lower value after a load torque is applied. In Fig. 1-10(b), the idle speed of the closed-loop system is shown to recover quickly to the preset value after the application of T_L.

- Closed-loop systems have many advantages over open-loop systems.

The objective of the idle-speed control system illustrated, also known as a **regulator system**, is to maintain the system output at a prescribed level.

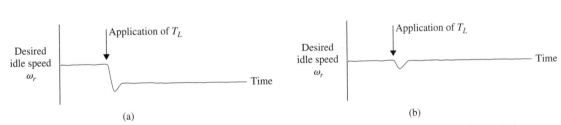

(a)   (b)

Figure 1-10 (a) Typical response of the open-loop idle-speed control system. (b) Typical response of the closed-loop idle-speed control system.

▶ 1-2 WHAT IS FEEDBACK AND WHAT ARE ITS EFFECTS?

The motivation for using feedback, as illustrated by the examples in Section 1-1, is somewhat oversimplified. In these examples, feedback is used to reduce the error between the reference input and the system output. However, the significance of the effects of feedback in control systems is more complex than is demonstrated by these simple examples. The reduction of system error is merely one of the many important effects that feedback may have upon a system. We show in the following sections that feedback also has effects on such system performance characteristics as **stability**, **bandwidth**, **overall gain**, **impedance**, and **sensitivity**.

• Feedback exists whenever there is a closed sequence of cause-and-effect relationships.

To understand the effects of feedback on a control system, it is essential to examine this phenomenon in a broad sense. When feedback is deliberately introduced for the purpose of control, its existence is easily identified. However, there are numerous situations wherein a physical system that we recognize as an inherently nonfeedback system turns out to have feedback when it is observed in a certain manner. In general, we can state that whenever a closed sequence of **cause-and-effect relationships** exists among the variables of a system, feedback is said to exist. This viewpoint will inevitably admit feedback in a large number of systems that ordinarily would be identified as nonfeedback systems. However, control-system theory allows numerous systems, with or without physical feedback, to be studied in a systematic way once the existence of feedback in the sense mentioned previously is established.

We shall now investigate the effects of feedback on the various aspects of system performance. Without the necessary mathematical foundation of linear-system theory, at this point we can rely only on simple static-system notation for our discussion. Let us consider the simple feedback system configuration shown in Fig. 1-11, where r is the input signal; y, the output signal; e, the error; and b, the feedback signal. The parameters G and H may be considered as constant gains. By simple algebraic manipulations, it is simple to show that the input-output relation of the system is

$$M = \frac{y}{r} = \frac{G}{1 + GH} \tag{1-1}$$

Using this basic relationship of the feedback system structure, we can uncover some of the significant effects of feedback.

1-2-1 Effect of Feedback on Overall Gain

• Feedback may increase the gain of a system in one frequency range but decrease it in another.

As seen from Eq. (1-1), feedback affects the gain G of a nonfeedback system by a factor $1 + GH$. The system of Fig. 1-11 is said to have **negative feedback**, since a minus sign is assigned to the feedback signal. The quantity GH may itself include a minus sign, so the *general effect of feedback is that it may increase or decrease the gain G*.

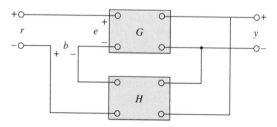

Figure 1-11 Feedback system.

In a practical control system, G and H are functions of frequency, so the magnitude of $1 + GH$ may be greater than 1 in one frequency range but less than 1 in another. Therefore, *feedback could increase the gain of system in one frequency range but decrease it in another.*

1-2-2 Effect of Feedback on Stability

• A system is unstable if its output is out of control.

Stability is a notion that describes whether the system will be able to follow the input command, that is, be useful in general. In a nonrigorous manner, *a system is said to be unstable if its output is out of control.* To investigate the effect of feedback on stability, we can again refer to the expression in Eq. (1-1). If $GH = -1$, the output of the system is infinite for any finite input, and the system is said to be unstable. Therefore, we may state that *feedback can cause a system that is originally stable to become unstable.* Certainly, feedback is a double-edged sword; when it is improperly used, it can be harmful. It should be pointed out, however, that we are only dealing with the static case here, and, in general, $GH = -1$ is not the only condition for instability. The subject of system stability will be treated formally in Chapter 6.

It can be demonstrated that one of the advantages of incorporating feedback is that it can stabilize an unstable system. Let us assume that the feedback system in Fig. 1-11 is unstable because $GH = -1$. If we introduce another feedback loop through a negative feedback gain of F, as shown in Fig. 1-12, the input-output relation of the overall system is

$$\frac{y}{r} = \frac{G}{1 + GH + GF} \tag{1-2}$$

• Feedback can improve stability or be harmful to stability.

It is apparent that although the properties of G and H are such that the inner-loop feedback system is unstable, because $GH = -1$, the overall system can be stable by properly selecting the outer-loop feedback gain F. In practice, GH is a function of frequency, and the stability condition of the closed-loop system depends on the **magnitude** and **phase** of GH. The bottom line is that *feedback can improve stability or be harmful to stability if it is not properly applied.*

Sensitivity considerations often are important in the design of control systems. Since all physical elements have properties that change with environment and age, we cannot always consider that the parameters of a control system to be completely stationary over

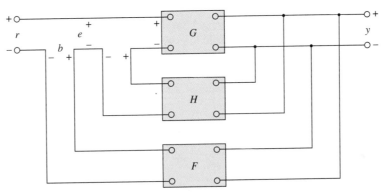

Figure 1-12 Feedback system with two feedback loops.

the entire operating life of the system. For instance, the winding resistance of an electric motor changes as the temperature of the motor rises during operation. Control systems with electric components may not operate normally when first turned on, because of the still-changing system parameters during warmup. This phenomenon is sometimes called "morning sickness." Most duplicating machines have a warmup period during which time operation is blocked out when first turned on.

In general, a good control system should be very insensitive to parameter variations but sensitive to the input commands. We shall investigate what effect feedback has on sensitivity to parameter variations. Referring to the system in Fig. 1-11, we consider G to be a gain parameter that may vary. The sensitivity of the gain of the overall system M to the variation in G is defined as

$$S_G^M = \frac{\partial M/M}{\partial G/G} = \frac{\text{percentage change in } M}{\text{percentage change in } G} \qquad (1\text{-}3)$$

• Note: Feedback can increase or decrease the sensitivity of a system.

where ∂M denotes the incremental change in M due to the incremental change in G, or ∂G. By using Eq. (1-1), the sensitivity function is written

$$S_G^M = \frac{\partial M}{\partial G} \frac{G}{M} = \frac{1}{1 + GH} \qquad (1\text{-}4)$$

This relation shows that if GH is a positive constant, the magnitude of the sensitivity function can be made arbitrarily small by increasing GH, providing that the system remains stable. It is apparent that in an open-loop system, the gain of the system will respond in a one-to-one fashion to the variation in G (i.e., $S_G^M = 1$). Again, in practice, GH is a function of frequency; the magnitude of $1 + GH$ may be less than unity over some frequency ranges, so feedback could be harmful to the sensitivity to parameter variations in certain cases. In general, the sensitivity of the system gain of a feedback system to parameter variations depends on where the parameter is located. The reader can derive the sensitivity of the system in Fig. 1-11 due to the variation of H.

1-2-3 Effect of Feedback on External Disturbance or Noise

All physical systems are subject to some types of extraneous signals or noise during operation. Examples of these signals are thermal-noise voltage in electronic circuits and brush or commutator noise in electric motors. External disturbances, such as wind gusts acting on an antenna, are also quite common in control systems. Therefore, control systems should be designed so that they are insensitive to noise and disturbances and sensitive to input commands.

• Feedback can reduce the effect of noise.

The effect of feedback on noise and disturbance depends greatly on where these extraneous signals occur in the system. No general conclusions can be reached, but in many situations, *feedback can reduce the effect of noise and disturbance on system performance.* Let us refer to the system shown in Fig. 1-13, in which r denotes the command signal and n is the noise signal. In the absence of feedback, that is, $H = 0$, the output y due to n acting alone is

$$y = G_2 n \qquad (1\text{-}5)$$

With the presence of feedback, the system output due to n acting alone is

$$y = \frac{G_2}{1 + G_1 G_2 H} n \qquad (1\text{-}6)$$

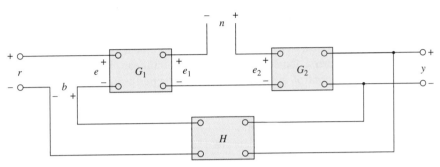

Figure 1-13 Feedback system with a noise signal.

• Feedback also can affect bandwidth, impedance, transient and frequency responses.

Comparing Eq. (1-6) with Eq. (1-5) shows that the noise component in the output of Eq. (1-6) is reduced by the factor $1 + G_1G_2H$ if the latter is greater than unity and the system is kept stable.

In Chapter 10 the feedforward and forward controller configurations are used along with feedback to reduce the effects of disturbance and noise inputs. In general, feedback also has effects on such performance characteristics as bandwidth, impedance, transient response, and frequency response. These effects will become known as we continue.

▶ 1-3 TYPES OF FEEDBACK CONTROL SYSTEMS

Feedback control systems may be classified in a number of ways, depending upon the purpose of the classification. For instance, according to the method of analysis and design, control systems are classified as **linear** or **nonlinear**, and **time-varying** or **time-invariant**. According to the types of signal found in the system, reference is often made to **continuous-data** or **discrete-data** systems, and **modulated** or **unmodulated** systems. Control systems are often classified according to the main purpose of the system. For instance, a **position-control system** and a **velocity-control system** control the output variables just as the names imply. In Chapter 7, the **type** of control system is defined according to the form of the open-loop transfer function. In general, there are many other ways of identifying control systems according to some special features of the system. It is important to know some of the more common ways of classifying control systems before embarking on the analysis and design of these systems.

1-3-1 **Linear versus Nonlinear Control Systems**

• Most real-life control systems have nonlinear characteristics to some extent.

This classification is made according to the methods of analysis and design. Strictly speaking, linear systems do not exist in practice, since all physical systems are nonlinear to some extent. Linear feedback control systems are idealized models fabricated by the analyst purely for the simplicity of analysis and design. When the magnitudes of signals in a control system are limited to ranges in which system components exhibit linear characteristics (i.e., the principle of superposition applies), the system is essentially linear. But when the magnitudes of signals are extended beyond the range of the linear operation, depending on the severity of the nonlinearity, the system should no longer be considered linear. For instance, amplifiers used in control systems often exhibit a saturation effect when their input signals become large; the magnetic field of a motor usually has saturation properties. Other common nonlinear effects found in control systems are the backlash or dead play between coupled gear members, non-linear spring characteristics, nonlinear friction force or torque between moving members, and so on. Quite often, nonlinear characteristics are intentionally introduced in a control system to improve its performance or provide more effective control. For instance, to achieve minimum-time control, an on-off (bang-bang or relay) type controller is used in many missile or spacecraft control systems. Typically in these systems, jets are mounted on sides of the vehicle to provide reaction torque for attitude control. These jets are often controlled in a full-on or full-off fashion, so a fixed amount of air is applied from a given jet for a certain time period to control the attitude of the space vehicle.

• There are no general methods for solving a wide class of nonlinear systems.

For linear systems, a wealth of analytical and graphical techniques is available for design and analysis purposes. A majority of the material in this text is devoted to the analysis and design of linear systems. Nonlinear systems, on the other hand, are usually difficult to treat mathematically, and there are no general methods available for solving a wide class of nonlinear systems. It is practical to first design the controller based on the linear-system model by neglecting the nonlinearities of the system. The designed controller is then applied to the nonlinear system model for evaluation or redesign by computer simulation. The Virtual Lab introduced in Chapter 11 is mainly used to model the characteristics of practical systems with realistic physical components.

1-3-2 Time-Invariant versus Time-Varying Systems

When the parameters of a control system are stationary with respect to time during the operation of the system, the system is called a **time-invariant system**. In practice, most physical systems contain elements that drift or vary with time. For example, the winding resistance of an electric motor will vary when the motor is being first excited and its temperature is rising. Another example of a time-varying system is a guided-missile control system in which the mass of the missile decreases as the fuel on board is being consumed during flight. Although a time-varying system without nonlinearity is still a linear system, the analysis and design of this class of systems are usually much more complex than that of the linear **time-invariant systems**.

Continuous-Data Control Systems

A **continuous-data system** is one in which the signals at various parts of the system are all functions of the continuous time variable t. The signals in continuous-data systems may be further classified as ac or dc. Unlike the general definitions of ac and dc signals used in electrical engineering, **ac** and **dc control systems** carry special significance in control systems terminology. When one refers to an ac control system, it usually means that the signals in the system are *modulated* by some form of modulation scheme. On the other hand, when a dc control system is referred to, it does not mean that all the signals in the system are unidirectional; then there would be no corrective control movement. A **dc control system** simply implies that the signals are *unmodulated*, but they are still ac signals according to the conventional definition. The schematic diagram of a closed-loop dc control system is shown in Fig. 1-14. Typical waveforms of the signals in response to a step-function input are shown in the figure. Typical components of a dc control system are potentiometers, dc amplifiers, dc motors, dc tachometers, and so on.

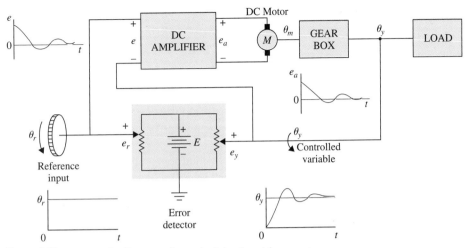

Figure 1-14 Schematic diagram of a typical dc closed-loop system.

Figure 1-15 shows the schematic diagram of a typical **ac control system** that performs essentially the same task as the dc system in Fig. 1-14. In this case, the signals in the system are modulated; that is, the information is transmitted by an ac carrier signal. Notice that the output controlled variable still behaves similarly to that of the dc system. In this case, the modulated signals are demodulated by the low-pass characteristics of the ac motor. Ac control systems are used extensively in aircraft and missile control systems in which noise and disturbance often create problems. By using modulated ac control systems with carrier frequencies of 400 Hz or higher, the system will be less susceptible to low-frequency noise. Typical components of an ac

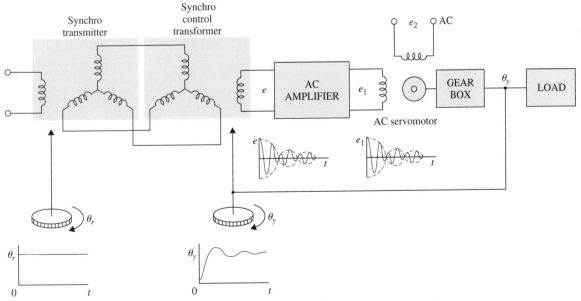

Figure 1-15 Schematic diagram of a typical ac closed-loop control system.

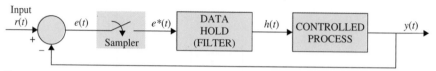

Figure 1-16 Block diagram of a sampled-data control system.

control system are synchros, ac amplifiers, ac motors, gyroscopes, accelerometers, and so on.

In practice, not all control systems are strictly of the ac or dc type. A system may incorporate a mixture of ac and dc components, using modulators and demodulators to match the signals at various points in the system.

Discrete-Data Control Systems

Discrete-data control systems differ from the continuous-data systems in that the signals at one or more points of the system are in the form of either a pulse train or a digital code. Usually, discrete-data control systems are subdivided into **sampled-data** and **digital control systems**. Sampled-data control systems refer to a more general class of discrete-data systems in which the signals are in the form of pulse data. A digital control system refers to the use of a digital computer or controller in the system, so that the signals are digitally coded, such as in binary code.

In general, a sampled-data system receives data or information only intermittently at specific instants of time. For example, the error signal in a control system can be supplied only in the form of pulses, in which case the control system receives no information about the error signal during the periods between two consecutive pulses. Strictly, a sampled-data system can also be classified as an ac system, since the signal of the system is pulse modulated.

Figure 1-16 illustrates how a typical sampled-data system operates. A continuous-data input signal $r(t)$ is applied to the system. The error signal $e(t)$ is sampled by a sampling device, the **sampler**, and the output of the sampler is a sequence of pulses. The sampling rate of the sampler may or may not be uniform. There are many advantages to incorporating sampling into a control system. One important advantage is that expensive equipment used in the system may be time-shared among several control channels. Another advantage is that pulse data are usually less susceptible to noise.

• Digital control systems are usually less susceptible to noise.

Because digital computers provide many advantages in size and flexibility, computer control has become increasingly popular in recent years. Many airborne systems contain digital controllers that can pack thousands of discrete elements into a space no larger than the size of this book. Figure 1-17 shows the basic elements of a digital autopilot for guided-missile control.

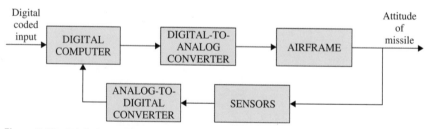

Figure 1-17 Digital autopilot system for a guided missile.

► 1-4 SUMMARY

In this chapter, we introduced some of the basic concepts of what a control system is and what it is supposed to accomplish. The basic components of a control system were described. By demonstrating the effects of feedback in a rudimentary way, the question of why most control systems are closed-loop systems was also clarified. Most important, it was pointed out that feedback is a double-edged sword—it can benefit as well as harm the system to be controlled. This is part of the challenging task of designing a control system, which involves consideration of such performance criteria as stability, sensitivity, bandwidth, and accuracy. Finally, various types of control systems were categorized according to the system signals, linearity, and control objectives. Several typical control-system examples were given to illustrate the analysis and design of control systems. Most systems encountered in real life are nonlinear and time-varying to some extent. The concentration on the studies of linear systems is due primarily to the availability of unified and simple-to-understand analytical methods in the analysis and design of linear systems.

► REVIEW QUESTIONS

1. List the advantages and disadvantages of an open-loop system.

2. List the advantages and the disadvantages of a closed-loop system.

3. Give the definitions of ac and dc control systems.

4. Give the advantages of a digital control system over a continuous-data control system.

5. A closed-loop control system is usually more accurate than an open-loop system.   **(T)**   **(F)**

6. Feedback is sometimes used to improve the sensitivity of a control system.   **(T)**   **(F)**

7. If an open-loop system is unstable, then applying feedback will always improve its stability.   **(T)**   **(F)**

8. Feedback can increase the gain of a system in one frequency range but decrease it in another.   **(T)**   **(F)**

9. Nonlinear elements are sometimes intentionally introduced to a control system to improve its performance.   **(T)**   **(F)**

10. Discrete-data control systems are more susceptible to noise due to the nature of its signals.   **(T)**   **(F)**

CHAPTER 2

Mathematical Foundation

▷ 2-1 INTRODUCTION

One of the most important tasks in the analysis and design of control systems is the mathematical modeling of the systems. The control systems engineer often has the task of determining not only how to accurately describe a system mathematically, but more important, how to make proper assumptions and approximations, whenever necessary, so that the system may be realistically characterized by a mathematical model, linear if possible. It is not difficult to understand that the analysis and computer simulation of any system are only as good as the model used to describe it.

The studies of control systems rely to a great extent on applied mathematics. One of the major purposes of control-system studies is to develop a set of analytical tools so that the designer can arrive at reasonably predictable and reliable designs without depending solely on the drudgery of experimentation or extensive computer simulation.

For the study of classical control theory, which represents a good portion of this text, the required mathematical background includes such subjects as **complex-variable theory**, **differential** and **difference equations**, **Laplace transformation** and z**-transformation**, and so on. Modern control theory, on the other hand, requires considerably more intensive mathematical background, such as **matrix theory**, **set theory**, **linear algebra** and **transformation**, **variational calculus**, **mathematical programming**, **probability theory**, and other advanced mathematics.

In this chapter, we present the theory and applications of the Laplace transform. A review on the complex variable concept is given in Appendix A. Appendix B covers a review on differential and difference equations with emphasis on applications for modeling control systems. Elementary matrix algebra is covered in Appendix C. Because of space limitations, as well as the fact that most subjects are considered as review material for the reader, the treatment of these mathematical subjects is not exhaustive. The reader who wishes to conduct an in-depth study of any of these subjects should refer to books that are devoted to them.

At the end of the chapter we present a MATLAB tool to solve most problems addressed in this chapter. The reader is encouraged to apply this tool to all the problems identified by the MATLAB Toolbox in the left margin of the text throughout this chapter. (See example of toolbox in the margin of this page.) We have developed an easy-to-use and fully graphics-based package to eliminate the users' need to write computer code. This software is based on MATLAB R12 and prerelease R13 (optimized for MATLAB 6.1[1]).

---

[1] Visit *www.mathworks.com.*

The main objectives of this chapter are

1. To introduce the fundamentals of Laplace transforms.
2. To demonstrate the applications of Laplace transform to solve linear ordinary differential equations.
3. To introduce the concept of transfer functions and how to apply them to the modeling of linear time-invariant systems.
4. To demonstrate the MATLAB tools using case studies.

▶ 2-2 LAPLACE TRANSFORM

The Laplace transform is one of the mathematical tools used to solve linear ordinary differential equations. In contrast with the classical method of solving linear differential equations, the Laplace transform method has the following two features:

1. The homogeneous equation and the particular integral of the solution of the differential equation are obtained in one operation.
2. The Laplace transform converts the differential equation into an algebraic equation in s. It is then possible to manipulate the algebraic equation by simple algebraic rules to obtain the solution in the s-domain. The final solution is obtained by taking the inverse Laplace transform.

2-2-1 Definition of the Laplace Transform

Given the real function $f(t)$ that satisfies the condition

$$\int_0^\infty \left| f(t)e^{-\sigma t} \right| dt < \infty \tag{2-1}$$

for some finite, real σ, the Laplace transform of $f(t)$ is defined as

$$F(s) = \int_0^\infty f(t)e^{-st}\, dt \tag{2-2}$$

or

$$F(s) = \text{Laplace transform of } f(t) = \mathcal{L}[f(t)] \tag{2-3}$$

• The response of a causal system does not precede the input.

The variable s is referred to as the **Laplace operator**, which is a complex variable; that is, $s = \sigma + j\omega$, where σ is the real component and ω is the imaginary component. The defining equation in Eq. (2-2) is also known as the **one-sided Laplace transform**, as the integration is evaluated from $t = 0$ to ∞. This simply means that all information contained in $f(t)$ prior to $t = 0$ is ignored or considered to be zero. This assumption does not impose any limitation on the applications of the Laplace transform to linear systems, since in the usual time-domain studies, time reference is often chosen at $t = 0$. Furthermore, for a physical system when an input is applied at $t = 0$, the response of the system does not start sooner than $t = 0$; that is, response does not precede excitation. Such a system is also known as being **causal** or simply **physically realizable**.

Strictly, the one-sided Laplace transform should be defined from $t = 0^-$ to $t = \infty$. The symbol $t = 0^-$ implies the limit of $t \to 0$ is taken from the left side of $t = 0$. This limiting process will take care of situations under which the function $f(t)$ has a jump discontinuity or an impulse at $t = 0$. For the subjects treated in this text, the defining equation of the Laplace transform in Eq. (2-2) is almost never used in problem solving, since the transform expressions encountered are either given or can be found from the Laplace transform table, such as the one given in Appendix F. Thus, the fine point of using 0^- or 0^+ never needs to be addressed. For simplicity, we shall simply use $t = 0$ or $t = t_0$ (≥ 0) as the initial time in all subsequent discussions.

The following examples illustrate how Eq. (2-2) is used for the evaluation of the Laplace transform of $f(t)$.

▶ **EXAMPLE 2-1**   Let $f(t)$ be a unit-step function that is defined as

$$f(t) = u_s(t) = 1 \qquad t > 0$$
$$= 0 \qquad t < 0 \tag{2-4}$$

The Laplace transform of $f(t)$ is obtained as

$$F(s) = \mathcal{L}[u_s(t)] = \int_0^\infty u_s(t)e^{-st}\, dt = -\frac{1}{s}e^{-st}\Big|_0^\infty = \frac{1}{s} \tag{2-5}$$

Equation (2-5) is valid if

$$\int_0^\infty \left| u_s(t)e^{-\sigma t} \right| dt = \int_0^\infty \left| e^{-\sigma t} \right| dt < \infty \tag{2-6}$$

which means that the real part of s, σ, must be greater than zero. In practice, we simply refer to the Laplace transform of the unit-step function as $1/s$, and rarely do we have to be concerned with the region in the s-plane in which the transform integral converges absolutely.   ◀

▶ **EXAMPLE 2-2**   Consider the exponential function

$$f(t) = e^{-\alpha t} \qquad t \geq 0 \tag{2-7}$$

where α is a real constant. The Laplace transform of $f(t)$ is written

$$F(s) = \int_0^\infty e^{-\alpha t} e^{-st}\, dt = \frac{e^{-(s+\alpha)t}}{s + \alpha}\Big|_0^\infty = \frac{1}{s + \alpha} \tag{2-8}$$

◀

2-2-2   Inverse Laplace Transformation

Given the Laplace transform $F(s)$, the operation of obtaining $f(t)$ is termed the **inverse Laplace transformation,** and is denoted by

$$f(t) = \text{Inverse Laplace transform of } F(s) = \mathcal{L}^{-1}[F(s)] \tag{2-9}$$

The inverse Laplace transform integral is given as

$$f(t) = \frac{1}{2\pi j} \int_{c-j\infty}^{c+j\infty} F(s)e^{st}\, ds \tag{2-10}$$

where c is a real constant that is greater than the real parts of all the singularities of $F(s)$. Equation (2-10) represents a line integral that is to be evaluated in the s-plane. For simple functions, the inverse Laplace transform operation can be carried out simply by referring to the Laplace transform table, such as the one given in Appendix D and on the inside back cover. For complex functions, the inverse Laplace transform can be carried out by first performing a partial-fraction expansion (Section 2-3) on $F(s)$ and then use the transform table. Our transfer-function analysis **MATLAB** tool (TFtool) can also be used for partial-fraction expansion/inverse Laplace transformation. More details on partial-fraction expansion will be given in Section 2-3.

2-2-3 Important Theorems of the Laplace Transform

The applications of the Laplace transform in many instances are simplified by utilization of the properties of the transform. These properties are presented by the following theorems, for which no proofs are given here.

▣ **Theorem 1.** *Multiplication by a Constant*
Let k be a constant, and $F(s)$ be the Laplace transform of $f(t)$. Then

$$\mathcal{L}[kf(t)] = kF(s) \tag{2-11}$$

▣ **Theorem 2.** *Sum and Difference*
Let $F_1(s)$ and $F_2(s)$ be the Laplace transform of $f_1(t)$ and $f_2(t)$, respectively. Then

$$\mathcal{L}[f_1(t) \pm f_2(t)] = F_1(s) \pm F_2(s) \tag{2-12}$$

▣ **Theorem 3.** *Differentiation*
Let $F(s)$ be the Laplace transform of $f(t)$, and $f(0)$ is the limit of $f(t)$ as t approaches 0. The Laplace transform of the time derivative of $f(t)$ is

$$\mathcal{L}\left[\frac{df(t)}{dt}\right] = sF(s) - \lim_{t \to 0} f(t) = sF(s) - f(0) \tag{2-13}$$

In general, for higher-order derivatives of $f(t)$,

$$\mathcal{L}\left[\frac{d^n f(t)}{dt^n}\right] = s^n F(s) - \lim_{t \to 0}\left[s^{n-1} f(t) + s^{n-2}\frac{df(t)}{dt} + \cdots + \frac{d^{n-1} f(t)}{dt^{n-1}}\right]$$
$$= s^n F(s) - s^{n-1} f(0) - s^{n-2} f^{(1)}(0) - \cdots - f^{(n-1)}(0) \tag{2-14}$$

where $f^{(i)}(0)$ denotes the ith-order derivative of $f(t)$ with respect to t, evaluated at $t = 0$.

▣ **Theorem 4.** *Integration*
The Laplace transform of the first integral of $f(t)$ with respect to t is the Laplace transform of $f(t)$ divided by s; that is,

$$\mathcal{L}\left[\int_0^t f(\tau)d\tau\right] = \frac{F(s)}{s} \tag{2-15}$$

For nth-order integration,

$$\mathcal{L}\left[\int_0^{t_1}\int_0^{t_2}\cdots\int_0^{t_n} f(t)d\tau dt_1 dt_2 \cdots dt_{n-1}\right] = \frac{F(s)}{s^n} \tag{2-16}$$

▣ **Theorem 5.** *Shift in Time*
The Laplace transform of $f(t)$ delayed by time T is equal to the Laplace transform $f(t)$ multiplied by e^{-Ts}; that is

$$\mathcal{L}[f(t - T)u_s(t - T)] = e^{-Ts} F(s) \tag{2-17}$$

where $u_s(t - T)$ denotes the unit-step function that is shifted in time to the right by T.

▣ **Theorem 6.** *Initial-Value Theorem*
If the Laplace transform of $f(t)$ is $F(s)$, then

$$\lim_{t \to 0} f(t) = \lim_{s \to \infty} sF(s) \tag{2-18}$$

if the limit exists.

• The final-value theorem is valid only if $sF(s)$ does not have any poles on the $j\omega$ axis and in the right half of the s-plane.

■ **Theorem 7.** *Final-Value Theorem*

If the Laplace transform of $f(t)$ is $F(s)$, and if $sF(s)$ is analytic (see Appendix A on the definition of an analytic function) on the imaginary axis and in the right half of the s-plane, then

$$\lim_{t\to\infty} f(t) = \lim_{s\to 0} sF(s) \qquad (2\text{-}19)$$

The final-value theorem is very useful for the analysis and design of control systems, since it gives the final value of a time function by knowing the behavior of its Laplace transform at $s = 0$. The final-value theorem is *not* valid if $sF(s)$ contains any pole whose real part is zero or positive, which is equivalent to the analytic requirement of $sF(s)$ in the right-half s-plane, as stated in the theorem. The following examples illustrate the care that must be taken in applying the theorem.

► **EXAMPLE 2-3** Consider the function

$$F(s) = \frac{5}{s(s^2 + s + 2)} \qquad (2\text{-}20)$$

Since $sF(s)$ is analytic on the imaginary axis and in the right-half s-plane, the final-value theorem may be applied. Using Eq. (2-19), we have

$$\lim_{t\to\infty} f(t) = \lim_{t\to 0} sF(s) = \lim_{s\to 0} \frac{5}{s^2 + s + 2} = \frac{5}{2} \qquad (2\text{-}21)$$

◄

► **EXAMPLE 2-4** Consider the function

$$F(s) = \frac{\omega}{s^2 + \omega^2} \qquad (2\text{-}22)$$

which is the Laplace transform of $f(t) = \sin \omega t$. Since the function $sF(s)$ has two poles on the imaginary axis of the s-plane, the final-value theorem *cannot* be applied in this case. In other words, although the final-value theorem would yield a value of zero as the final value of $f(t)$, the result is erroneous.

◄

■ **Theorem 8.** *Complex Shifting*

The Laplace transform of $f(t)$ multiplied by $e^{\mp \alpha t}$, where α is a constant, is equal to the Laplace transform $F(s)$, with s replaced by $s \pm \alpha$; *that is,*

$$\mathcal{L}[e^{\mp \alpha t} f(t)] = F(s \pm \alpha) \qquad (2\text{-}23)$$

■ **Theorem 9.** *Real Convolution (Complex Multiplication)*

Let $F_1(s)$ and $F_2(s)$ be the Laplace transforms of $f_1(t)$ and $f_2(t)$, respectively, and $f_1(t) = 0$, $f_2(t) = 0$, for $t < 0$, then

$$F_1(s)F_2(s) = \mathcal{L}[f_1(t) * f_2(t)]$$

$$= \mathcal{L}\left[\int_0^t f_1(\tau)f_2(t - \tau)d\tau\right] = \mathcal{L}\left[\int_0^t f_2(\tau)f_1(t - \tau)d\tau\right] \qquad (2\text{-}24)$$

where the symbol * denotes **convolution** in the time domain.

Equation (2-24) shows that multiplication of two transformed functions in the complex s-domain is equivalent to the convolution of two corresponding real functions of t in the t-domain. An important fact to remember is that *the inverse Laplace transform of the product of two functions in the s-domain is **not** equal to the product of the two corresponding real functions in the t-domain*; that is, in general,

$$\mathcal{L}^{-1}[F_1(s)F_2(s)] \neq f_1(t)f_2(t) \qquad (2\text{-}25)$$

TABLE 2-1   Theorems of Laplace Transforms

Multiplication by a constant	$\mathcal{L}[kf(t)] = kF(s)$
Sum and difference	$\mathcal{L}[f_1(t) \pm f_2(t)] = F_1(s) \pm F_2(s)$
Differentiation	$\mathcal{L}\left[\dfrac{df(t)}{dt}\right] = sF(s) - f(0)$
	$\mathcal{L}\left[\dfrac{d^n f(t)}{dt^n}\right] = s^n F(s) - s^{n-1} f(0) - s^{n-2} f^{(1)}(0)$
	$\qquad\qquad\qquad - \cdots - sf^{(n-2)}(0) - f^{(n-1)}(0)$
	where
	$f^{(k)}(0) = \dfrac{d^k f(t)}{dt^k}\bigg|_{t=0}$
Integration	$\mathcal{L}\left[\displaystyle\int_0^t f(\tau)d\tau\right] = \dfrac{F(s)}{s}$
	$\mathcal{L}\left[\displaystyle\int_0^{t_1}\int_0^{t_2}\cdots\int_0^{t_n} f(t)d\tau dt_1 dt_2 \cdots dt_{n-1}\right] = \dfrac{F(s)}{s^n}$
Shift in time	$\mathcal{L}[f(t - T)u_s(t - T)] = e^{-Ts}F(s)$
Initial-value theorem	$\lim\limits_{t\to 0} f(t) = \lim\limits_{t\to\infty} sF(s)$
Final-value theorem	$\lim\limits_{t\to\infty} f(t) = \lim\limits_{s\to 0} sF(s)$ if $sF(s)$ does not have poles on or to the right of the imaginary axis in the s-plane.
Complex shifting	$\mathcal{L}[e^{\mp\alpha t}f(t)] = F(s \pm \alpha)$
Real convolution	$F_1(s)F_2(s) = \mathcal{L}\left[\displaystyle\int_0^t f_1(\tau)f_2(t - \tau)d\tau\right]$
	$\qquad\qquad = \mathcal{L}\left[\displaystyle\int_0^t f_2(\tau)f_1(t - \tau)d\tau\right] = \mathcal{L}[f_1(t) * f_2(t)]$

There is also a dual relation to the real convolution theorem, called the **complex convolution**, or **real multiplication**. Essentially, the theorem states that multiplication in the real t-domain is equivalent to convolution in the complex s-domain; that is,

$$\mathcal{L}[f_1(t)f_2(t)] = F_1(s) * F_2(s) \qquad\qquad (2\text{-}26)$$

where * denotes complex convolution in this case. Details of the complex convolution formula are not given here.

Table 2-1 summarizes the theorems of the Laplace transforms represented.

▶ 2-3 INVERSE LAPLACE TRANSFORM BY PARTIAL-FRACTION EXPANSION

In a majority of problems in control systems, the evaluation of the inverse Laplace transform does not rely on the use of the inversion integral of Eq. (2-10). Rather, the inverse Laplace transform operation involving rational functions can be carried out using a Laplace transform table and partial-fraction expansion, both of which can also be done by computer programs.

2-3-1 Partial-Fraction Expansion

When the Laplace transform solution of a differential equation is a rational function in s, it can be written as

$$G(s) = \frac{Q(s)}{P(s)} \tag{2-27}$$

where $P(s)$ and $Q(s)$ are polynomials of s. It is assumed that *the order of $P(s)$ in s is greater than that of $Q(s)$*. The polynomial $P(s)$ may be written

$$P(s) = s^n + a_{n-1}s^{n-1} + \cdots + a_1 s + a_0 \tag{2-28}$$

where $a_0, a_1, \ldots, a_{n-1}$ are real coefficients. The methods of partial-fraction expansion will now be given for the cases of simple poles, multiple-order poles, and complex-conjugate poles of $G(s)$.

$G(s)$ Has Simple Poles[2] If all the poles of $G(s)$ are simple and real, Eq. (2-27) can be written as

$$G(s) = \frac{Q(s)}{P(s)} = \frac{Q(s)}{(s + s_1)(s + s_2)\cdots(s + s_n)} \tag{2-29}$$

where $s_1 \neq s_2 \neq \cdots \neq s_n$. Applying the partial-fraction expansion, Eq. (2-29) is written

$$G(s) = \frac{K_{s1}}{s + s_1} + \frac{K_{s2}}{s + s_2} + \cdots + \frac{K_{sn}}{s + s_n} \tag{2-30}$$

The coefficient K_{si} ($i = 1, 2, \ldots, n$) is determined by multiplying both sides of Eq. (2-29) by the factor $(s + s_i)$ and then setting s equal to $-s_i$. To find the coefficient K_{s1}, for instance, we multiply both sides of Eq. (2-29) by $(s + s_1)$ and let $s = -s_1$. Thus,

$$K_{s1} = \left[(s + s_1)\frac{Q(s)}{P(s)} \right]\Bigg|_{s=-s_1} = \frac{Q(-s_1)}{(s_2 - s_1)(s_3 - s_1)\cdots(s_n - s_1)} \tag{2-31}$$

▶ **EXAMPLE 2-5** Consider the function

$$G(s) = \frac{5s + 3}{(s + 1)(s + 2)(s + 3)} \tag{2-32}$$

which is written in the partial-fraction expanded form:

$$G(s) = \frac{K_{-1}}{s + 1} + \frac{K_{-2}}{s + 2} + \frac{K_{-3}}{s + 3} \tag{2-33}$$

The coefficients K_{-1}, K_{-2}, and K_{-3} are determined as follows:

$$K_{-1} = [(s + 1)G(s)]|_{s=-1} = \frac{5(-1)+3}{(2 - 1)(3 - 1)} = -1 \tag{2-34}$$

$$K_{-2} = [(s + 2)G(s)]|_{s=-2} = \frac{5(-2)+3}{(1 - 2)(3 - 2)} = 7 \tag{2-35}$$

$$K_{-3} = [(s + 3)G(s)]|_{s=-3} = \frac{5(-3)+3}{(1 - 3)(2 - 3)} = -6 \tag{2-36}$$

---

[2] Refer to Appendix A for the definition of a pole.

Thus, Eq. (2-32) becomes

$$G(s) = \frac{-1}{s+1} + \frac{7}{s+2} - \frac{6}{s+3} \tag{2-37}$$

◀

G(s) Has Multiple-Order Poles   If r of the n poles of $G(s)$ are identical, or we say that the pole at $s = -s_i$ is of multiplicity r, $G(s)$ is written

$$G(s) = \frac{Q(s)}{P(s)} = \frac{Q(s)}{(s+s_1)(s+s_2)\cdots(s+s_{n-r})(s+s_i)^r} \tag{2-38}$$

$(i \neq 1, 2, \ldots, n-r)$, Then $G(s)$ can be expanded as

$$G(s) = \frac{K_{s1}}{s+s_1} + \frac{K_{s2}}{s+s_2} + \cdots + \frac{K_{s(n-r)}}{s+s_{n-r}}$$
$$|\leftarrow n-r \text{ terms of simple poles} \rightarrow|$$
$$+ \frac{A_1}{s+s_i} + \frac{A_2}{(s+s_i)^2} + \cdots + \frac{A_r}{(s+s_i)^r} \tag{2-39}$$
$$|\leftarrow r \text{ terms of repeated poles} \rightarrow|$$

Then $(n-r)$ coefficients, $K_{s1}, K_{s2}, \ldots, K_{s(n-r)}$, which correspond to simple poles, may be evaluated by the method described by Eq. (2-31). The determination of the coefficients that correspond to the multiple-order poles is described as follows.

$$A_r = \left[(s+s_i)^r G(s)\right]\Big|_{s=-s_i} \tag{2-40}$$

$$A_{r-1} = \frac{d}{ds}\left[(s+s_i)^r G(s)\right]\Big|_{s=-s_i} \tag{2-41}$$

$$A_{r-2} = \frac{1}{2!}\frac{d^2}{ds^2}\left[(s+s_i)^r G(s)\right]\Big|_{s=-s_i} \tag{2-42}$$

$$\vdots$$

$$A_1 = \frac{1}{(r-1)!}\frac{d^{r-1}}{ds^{r-1}}\left[(s+s_i)^r G(s)\right]\Big|_{s=-s_i} \tag{2-43}$$

▶ **EXAMPLE 2-6**   Consider the function

$$G(s) = \frac{1}{s(s+1)^3(s+2)} \tag{2-44}$$

By using the format of Eq. (2-39), $G(s)$ is written

$$G(s) = \frac{K_0}{s} + \frac{K_{-2}}{s+2} + \frac{A_1}{s+1} + \frac{A_2}{(s+1)^2} + \frac{A_3}{(s+1)^3} \tag{2-45}$$

The coefficients corresponding to the simple poles are

$$K_0 = [sG(s)]|_{s=0} = \frac{1}{2} \tag{2-46}$$

$$K_{-2} = [(s+2)G(s)]|_{s=-2} = \frac{1}{2} \tag{2-47}$$

and those of the third-order pole are

$$A_3 = \left[(s + 1)^3 G(s)\right]\Big|_{s=-1} = -1 \tag{2-48}$$

$$A_2 = \frac{d}{ds}\left[(s + 1)^3 G(s)\right]\Big|_{s=-1} = \frac{d}{ds}\left[\frac{1}{s(s + 2)}\right]\Big|_{s=-1} = 0 \tag{2-49}$$

$$A_1 = \frac{1}{2!}\frac{d^2}{ds^2}\left[(s + 1)^3 G(s)\right]\Big|_{s=-1} = \frac{1}{2}\frac{d^2}{ds^2}\left[\frac{1}{s(s + 2)}\right]\Big|_{s=-1} = -1 \tag{2-50}$$

The completed partial-fraction expansion is

$$G(s) = \frac{1}{2s} + \frac{1}{2(s + 2)} - \frac{1}{s + 1} - \frac{1}{(s + 1)^3} \tag{2-51}$$

◀

G(s) **Has Simple Complex-Conjugate Poles**   The partial-fraction expansion of Eq. (2-30) is valid also for simple complex-conjugate poles. Since complex-conjugate poles are more difficult to handle and are of special interest in control systems studies, they deserve special treatment here.

Suppose that $G(s)$ of Eq. (2-27) contains a pair of complex poles:

$$s = -\sigma + j\omega \qquad \text{and} \qquad s = -\sigma - j\omega$$

The corresponding coefficients of these poles are found by using Eq. (2-31),

$$K_{-\sigma + j\omega} = (s + \sigma - j\omega)G(s)|_{s=-\sigma + j\omega} \tag{2-52}$$

$$K_{-\sigma - j\omega} = (s + \sigma + j\omega)G(s)|_{s=-\sigma - j\omega} \tag{2-53}$$

▶ **EXAMPLE 2-7** Consider the function

$$G(s) = \frac{\omega_n^2}{s^2 + 2\zeta\omega_n s + \omega_n^2} \tag{2-54}$$

Let us assume that the value of ζ is less than one, so that the poles of $G(s)$ are complex. Then, $G(s)$ is expanded as follows:

$$G(s) = \frac{K_{-\sigma + j\omega}}{s + \sigma - j\omega} + \frac{K_{-\sigma - j\omega}}{s + \sigma + j\omega} \tag{2-55}$$

where

$$\sigma = \zeta\omega_n \tag{2-56}$$

and

$$\omega = \omega_n\sqrt{1 - \zeta^2} \tag{2-57}$$

The coefficients in Eq. (2-55) are determined as

$$K_{-\sigma + j\omega} = (s + \sigma - j\omega)G(s)|_{s=-\sigma + j\omega} = \frac{\omega_n^2}{2j\omega} \tag{2-58}$$

$$K_{-\sigma - j\omega} = (s + \sigma + j\omega)G(s)|_{s=-\sigma - j\omega} = -\frac{\omega_n^2}{2j\omega} \tag{2-59}$$

The complete partial-fraction expansion of Eq. (2-54) is

$$G(s) = \frac{\omega_n^2}{2j\omega}\left[\frac{1}{s + \sigma - j\omega} - \frac{1}{s + \sigma + j\omega}\right] \tag{2-60}$$

Taking the inverse Laplace transform on both sides of the last equation gives

$$g(t) = \frac{\omega_n^2}{2\,j\omega}e^{-\sigma t}\left(e^{j\omega t} - e^{-j\omega t}\right) \qquad t \geq 0 \tag{2-61}$$

Or,

$$g(t) = \frac{\omega_n}{\sqrt{1-\zeta^2}}e^{-\zeta\omega_n t}\sin\omega_n\sqrt{1-\zeta^2}t \qquad t \geq 0 \tag{2-62}$$

This result shows that when a second-order function of the form of Eq. (2-54) has complex poles, the inverse Laplace transform can be written out directly in the form of Eq. (2-62), without going through the complex-conjugate manipulation of Eqs. (2-55) through (2-61) ◀

▷ 2-4 APPLICATION OF THE LAPLACE TRANSFORM TO THE SOLUTION OF LINEAR ORDINARY DIFFERENTIAL EQUATIONS

Linear ordinary differential equations can be solved by the Laplace transform method with the aid of the theorems on Laplace transform given in Section 2-2, the partial-fraction expansion, and a table of Laplace transforms. The procedure is outlined as follows:

1. Transform the differential equation to the s-domain by Laplace transform using the Laplace transform table.
2. Manipulate the transformed algebraic equation and solve for the output variable.
3. Perform partial-fraction expansion to the transformed algebraic equation.
4. Obtain the inverse Laplace transform from the Laplace transform table.

The following examples illustrate this method.

▶ **EXAMPLE 2-8** Consider the differential equation

$$\frac{d^2y(t)}{dt^2} + 3\frac{dy(t)}{dt} + 2y(t) = 5u_s(t) \tag{2-63}$$

where $u_s(t)$ is the unit-step function. The initial conditions are $y(0) = -1$ and $y^{(1)}(0) = dy(t)/dt|_{t=0} = 2$. To solve the differential equation, we first take the Laplace transform on both sides of Eq. (2-63):

$$s^2Y(s) - sy(0) - y^{(1)}(0) + 3sY(s) - 3y(0) + 2Y(s) = 5/s \tag{2-64}$$

Substituting the values of the initial conditions into the last equation and solving for $Y(s)$, we get

$$Y(s) = \frac{-s^2 - s + 5}{s(s^2 + 3s + 2)} = \frac{-s^2 - s + 5}{s(s+1)(s+2)} \tag{2-65}$$

Equation (2-65) is expanded by partial-fraction expansion to give

$$Y(s) = \frac{5}{2s} - \frac{5}{s+1} + \frac{3}{2(s+2)} \tag{2-66}$$

Taking the inverse Laplace transform of Eq. (2-66), we get the complete solution as

$$y(t) = \frac{5}{2} - 5e^{-t} + \frac{3}{2}e^{-2t} \qquad t \geq 0 \tag{2-67}$$

The first term in Eq. (2-67) is the steady-state solution or the particular integral; the last two terms represent the transient, or homogeneous, solution. Unlike the classical method, which requires separate steps to give the transient and the steady-state solutions, the Laplace transform method gives the entire solution in one operation.

If only the magnitude of the steady-state solution of $y(t)$ is of interest, the final-value theorem of Eq. (2-19) may be applied. Thus,

$$\lim_{t \to \infty} y(t) = \lim_{s \to 0} sY(s) = \lim_{s \to 0} \frac{-s^2 - s + 5}{s^2 + 3s + 2} = \frac{5}{2} \tag{2-68}$$

where we have first checked and found that the function $sY(s)$ has poles only in the left-half s-plane, so that the final-value theorem is valid. ◀

▶ **EXAMPLE 2-9** Consider the linear differential equation

$$\frac{d^2y(t)}{dt^2} + 34.5 \frac{dy(t)}{dt} + 1000y(t) = 1000u_s(t) \tag{2-69}$$

The initial values of $y(t)$ and $dy(t)/dt$ are zero. Taking the Laplace transform on both sides of Eq. (2-69), and solving for $Y(s)$, we have

$$Y(s) = \frac{1000}{s(s^2 + 34.5s + 1000)} = \frac{\omega_n^2}{s(s^2 + 2\zeta\omega_n s + \omega_n^2)} \tag{2-70}$$

where $\zeta = 0.5455$, and $\omega_n = 31.62$. The inverse Laplace transform of Eq. (2-70) can be executed in a number of ways. The Laplace transform table in Appendix F gives the transform pair of the expression in Eq. (2-70) directly. The result is

$$y(t) = 1 - \frac{e^{-\zeta\omega_n t}}{\sqrt{1 - \zeta^2}} \sin(\omega_n \sqrt{1 - \zeta^2}\, t + \theta) \qquad t \geq 0 \tag{2-71}$$

where

$$\theta = \cos^{-1}\zeta = 56.94° \tag{2-72}$$

Thus,

$$y(t) = 1 - 1.193e^{-17.25t} \sin(26.5t + 56.94°) \qquad t \geq 0 \tag{2-73}$$

Equation (2-73) can be derived by performing the partial-fraction expansion of Eq. (2-70) knowing that the poles are at $s = 0$, $-\sigma + j\omega$, and $-\sigma - j\omega$, where

$$\sigma = \zeta\omega_n = 17.25 \tag{2-74}$$
$$\omega = \omega_n\sqrt{1 - \zeta^2} = 26.5 \tag{2-75}$$

The partial-fraction expansion of Eq. (2-70) is written

$$Y(s) = \frac{K_0}{s} + \frac{K_{-\sigma+j\omega}}{s + \sigma - j\omega} + \frac{K_{-\sigma-j\omega}}{s + \sigma + j\omega} \tag{2-76}$$

where

$$K_0 = sY(s)\big|_{s=0} = 1 \tag{2-77}$$

$$K_{-\sigma+j\omega} = (s + \sigma - j\omega)Y(s)\big|_{s=-\sigma+j\omega} = \frac{e^{-j\phi}}{2j\sqrt{1 - \zeta^2}} \tag{2-78}$$

$$K_{-\sigma-j\omega} = (s + \sigma + j\omega)Y(s)\big|_{s=-\sigma-j\omega} = \frac{-e^{-j\phi}}{2j\sqrt{1 - \zeta^2}} \tag{2-79}$$

The angle ϕ is given by

$$\phi = 180° - \cos^{-1}\zeta \tag{2-80}$$

and is illustrated in Fig. 2-1.

The inverse Laplace transform of Eq. (2-76) is now written

$$\begin{aligned} y(t) &= 1 + \frac{1}{2j\sqrt{1 - \zeta^2}} e^{-\zeta\omega_n t}\left[e^{j(\omega t - \phi)} - e^{-j(\omega t - \phi)}\right] \\ &= 1 + \frac{1}{\sqrt{1 - \zeta^2}} e^{-\zeta\omega_n t} \sin\left[\omega_n\sqrt{1 - \zeta^2}\, t - \phi\right] \qquad t \geq 0 \end{aligned} \tag{2-81}$$

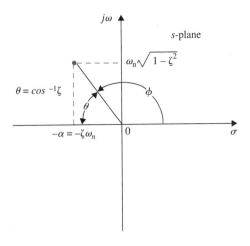

Figure 2-1   Root location in the s-plane.

Substituting Eq. (2-80) into Eq. (2-81) for ϕ, we have

$$y(t) = 1 - \frac{1}{\sqrt{1 - \zeta^2}} e^{-\zeta\omega_n t} \sin\left[\omega_n \sqrt{1 - \zeta^2}t + \cos^{-1}\zeta\right] \qquad t \geq 0 \qquad (2\text{-}82)$$

or

$$y(t) = 1 - 1.193e^{-17.25t} \sin(26.5t + 56.94°) \qquad t \geq 0 \qquad (2\text{-}83)$$

◀

▶ 2-5 IMPULSE RESPONSE AND TRANSFER FUNCTIONS OF LINEAR SYSTEMS

The classical way of modeling linear time-invariant systems is to use **transfer functions** to represent input-output relations between variables. One way to define the transfer function is to use the impulse response, which is defined as follows.

2-5-1   Impulse Response

Consider that a linear time-invariant system has the input $u(t)$ and output $y(t)$. The system can be characterized by its **impulse response** $g(t)$, which is defined as the output when the input is a unit-impulse function $\delta(t)$. Once the impulse response of a linear system is known, the output of the system $y(t)$, with any input, $u(t)$, can be found by using the **transfer function**.

2-5-2   Transfer Function (Single-Input, Single-Output Systems)

The transfer function of a linear time-invariant system is defined as the Laplace transform of the impulse response, with all the initial conditions set to zero.

Let $G(s)$ denote the transfer function of a single-input, single-output system with input $u(t)$, output $y(t)$, and impulse response $g(t)$. The transfer function $G(s)$ is defined as

$$G(s) = \mathcal{L}[g(t)] \qquad (2\text{-}84)$$

The transfer function $G(s)$ is related to the Laplace transform of the input and the output through the following relation:

$$G(s) = \frac{Y(s)}{U(s)} \qquad (2\text{-}85)$$

with all the initial conditions set to zero, and $Y(s)$ and $U(s)$ are the Laplace transforms of $y(t)$ and $u(t)$, respectively.

Although the transfer function of a linear system is defined in terms of the impulse response, in practice, the input-output relation of a linear time-invariant system with continuous-data input is often described by a differential equation, so it is more convenient to derive the transfer function directly from the differential equation. Let us consider that the input-output relation of a linear time-invariant system is described by the following nth-order differential equation with constant real coefficients:

$$\frac{d^n y(t)}{dt^n} + a_{n-1}\frac{d^{n-1} y(t)}{dt^{n-1}} + \cdots + a_1\frac{dy(t)}{dt} + a_0 y(t)$$
$$= b_m\frac{d^m u(t)}{dt^m} + b_{m-1}\frac{d^{m-1} u(t)}{dt^{m-1}} + \cdots + b_1\frac{du(t)}{dt} + b_0 u(t) \tag{2-86}$$

The coefficients $a_0, a_1, \ldots, a_{n-1}$ and $b_0, b_1, \ldots, b_m$ are real constants. Once the input $u(t)$ for $t \geq t_0$ and the initial conditions of $y(t)$ and the derivatives of $y(t)$ are specified at the initial time $t = t_0$, the output response $y(t)$ for $t \geq t_0$ is determined by solving Eq. (2-86). However, from the standpoint of linear-system analysis and design, the method of using differential equations exclusively is quite cumbersome. Thus, differential equations of the form of Eq. (2-86) are seldom used in their original form for the analysis and design of control systems. It should be pointed out that although efficient subroutines are available on digital computers for the solution of high-order differential equations, *the basic philosophy of linear control theory is that of developing analysis and design tools that will avoid the exact solution of the system differential equations*, except when computer-simulation solutions are desired for final presentation or verification. In classical control theory, even computer simulation often starts with transfer functions, rather than with differential equations.

To obtain the transfer function of the linear system that is represented by Eq. (2-86), we simply take the Laplace transform on both sides of the equation and assume **zero initial conditions**. The result is

$$(s^n + a_{n-1}s^{n-1} + \cdots + a_1 s + a_0)Y(s) = (b_m s^m + b_{m-1}s^{m-1} + \cdots + b_1 s + b_0)U(s) \tag{2-87}$$

The transfer function between $u(t)$ and $y(t)$ is given by

$$G(s) = \frac{Y(s)}{U(s)} = \frac{b_m s^m + b_{m-1}s^{m-1} + \cdots + b_1 s + b_0}{s^n + a_{n-1}s^{n-1} + \cdots + a_1 s + a_0} \tag{2-88}$$

The properties of the transfer function are summarized as follows:

1. The transfer function is defined only for a linear time-invariant system. It is not defined for nonlinear systems.

2. The transfer function between an input variable and an output variable of a system is defined as the Laplace transform of the impulse response. Alternately, the transfer function between a pair of input and output variables is the ratio of the Laplace transform of the output to the Laplace transform of the input.

3. All initial conditions of the system are set to zero.

4. The transfer function is independent of the input of the system.

5. The transfer function of a continuous-data system is expressed only as a function of the complex variable s. It is not a function of the real variable, time, or any other variable that is used as the independent variable. For discrete-data

systems modeled by difference equations, the transfer function is a function of z when the z-transform is used (refer to Appendix I).

Proper Transfer Functions: The transfer function in Eq. (2-88) is said to be **strictly proper** if the order of the denominator polynomial is greater than that of the numerator polynomial (i.e., $n > m$). If $n = m$, the transfer function is called **proper**. The transfer function is **improper** if $m > n$.

Characteristic Equation: *The **characteristic equation** of a linear system is defined the equation obtained by setting the denominator polynominal of the transfer function to zero.* Thus, from Eq. (2-88), the characteristic equation of the system described by Eq. (2-86) is

• The characteristic equation of a linear system is obtained by setting the denominator polynomial of the transfer function to zero.

$$s^n + a_{n-1}s^{n-1} + \cdots + a_1 s + a_0 = 0 \qquad (2\text{-}89)$$

Later we shall show that stability of linear, single-input, single-output systems is completely governed by the roots of the characteristic equation.

2-5-3 Transfer Function (Multivariable Systems)

The definition of transfer function is easily extended to a system with multiple inputs and outputs. A system of this type is often referred to as a **multivariable system**. In a multivariable system, a differential equation of the form of Eq. (2-86) may be used to describe the relationship between a pair of input and output variables, when all other inputs are set to zero. Since the principle of superposition is valid for linear systems, the total effect on any output due to all the inputs acting simultaneously is obtained by adding up the outputs due to each input acting alone.

Examples on multivariable systems are plentiful in practice. For example, a system in which the speed $\omega(t)$ of a motor subjected to an external disturbance torque $T_d(t)$ is determined by controlling the input voltage $v(t)$, the system is considered to have two inputs in $v(t)$ and $T_d(t)$ and one output in $\omega(t)$. Another example is the idle-speed control system of an automobile engine. In this case, the two inputs are the amounts of fuel and air intake to the engine, and the output is the idle speed of the engine. In the control of an aircraft turbopropeller engine shown in Fig. 2-2, the input variables are the fuel rate and the propeller blade angle. The output variables are the speed of rotation of the engine and the turbine-inlet temperature. In general, either one of the outputs is affected by the changes in both inputs. For instance, when the blade angle of the propeller is increased, the speed of rotation of the engine will decrease and the temperature will usually increase.

The following transfer function relations may be determined from tests performed on the system:

$$Y_1(s) = G_{11}(s)R_1(s) + G_{12}(s)R_2(s) \qquad (2\text{-}90)$$
$$Y_2(s) = G_{21}(s)R_1(s) + G_{22}(s)R_2(s) \qquad (2\text{-}91)$$

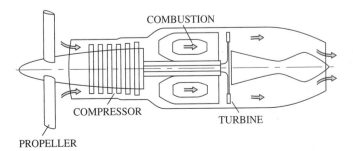

COMBUSTION

COMPRESSOR

TURBINE

PROPELLER

Figure 2-2 Aircraft turboprop engine.

where

$$Y_1(s) = \text{speed of rotation of engine}$$
$$Y_2(s) = \text{turbine-inlet temperature}$$
$$R_1(s) = \text{fuel rate}$$
$$R_2(s) = \text{propeller blade angle}$$

all in Laplace transform variables measured from some reference level. The transfer function $G_{11}(s)$ represents the transfer function between the fuel rate and the engine speed, with the propeller blade angle held at the reference value; that is, $R_2(s) = 0$. Similar definitions can be given to the other transfer functions, $G_{12}(s)$, $G_{21}(s)$, and $G_{22}(s)$.

In general, if a linear system has p inputs and q outputs, the transfer function between the jth input and the ith output is defined as

$$G_{ij}(s) = \frac{Y_i(s)}{R_j(s)} \tag{2-92}$$

with $R_k(s) = 0$, $k = 1, 2, \ldots, p$, $k \neq j$. Note that Eq. (2-92) is defined with only the jth input in effect, whereas the other inputs are set to zero. When all the p inputs are in action, the ith output transform is written

$$Y_i(s) = G_{i1}(s)R_1(s) + G_{i2}(s)R_2(s) + \cdots + G_{ip}(s)R_p(s) \tag{2-93}$$

It is convenient to express Eq. (2-93) in matrix-vector form:

$$\mathbf{Y}(s) = \mathbf{G}(s)\mathbf{R}(s) \tag{2-94}$$

where

$$\mathbf{Y}(s) = \begin{bmatrix} Y_1(s) \\ Y_2(s) \\ \vdots \\ Y_q(s) \end{bmatrix} \tag{2-95}$$

is the $q \times 1$ **transformed output vector**;

$$\mathbf{R}(s) = \begin{bmatrix} R_1(s) \\ R_2(s) \\ \vdots \\ R_p(s) \end{bmatrix} \tag{2-96}$$

is the $p \times 1$ **transformed input vector**; and

$$\mathbf{G}(s) = \begin{bmatrix} G_{11}(s) & G_{12}(s) & \cdots & G_{1p}(s) \\ G_{21}(s) & G_{22}(s) & \cdots & G_{2p}(s) \\ \cdot & \cdot & \cdots & \cdot \\ G_{q1}(s) & G_{q2}(s) & \cdots & G_{qp}(s) \end{bmatrix} \tag{2-97}$$

is the $q \times p$ **transfer-function matrix**.

▶ 2-6 MATLAB TOOLS AND CASE STUDIES

2-6-1 Description and Use of Transfer Function Tool

The Transfer Function Analysis Tool (TFtool) consists of a number of m-files (files that contain MATLAB codes) and GUIs (graphical user interfaces) for the analysis of simple control engineering transfer functions. The TFtool can be invoked either from the MAT-LAB command line by simply typing **TFtool** or from the automatic control systems launch applet (**ACSYS**) by clicking on the appropriate button. The steps involved in setting up and then solving a given problem are as follows.

- Specify the type of input preference; (i.e., polynomial or pole, zero, gain format).
- Enter the transfer function values.
- Convert the transfer function from polynomial form to poles and zeros or vice versa.
- Find the partial-fraction representation of the system.
- Use the inverse Laplace command (user must have access to the MATLAB Symbolic Toolbox for this task).
- Generate a hard copy if desired.

To better illustrate how to use TFtool, let us go through all the steps involved in solving the earlier examples in this chapter.

► **EXAMPLE 2-5**
Revisited

Consider the function

$$G(s) = \frac{5s + 3}{(s + 1)(s + 2)(s + 3)} \tag{2-98}$$

First, invoke the Transfer Function Analysis Tool by typing TFtool at the MATLAB prompt. The following window (Fig. 2-3) will appear on the computer screen.

In order to define the transfer function, click on the select input preference popup button to choose the proper input module. There are two different input mechanisms: polynomial and zero, pole, constant modules. Since the transfer function in (2-98) is in pole-zero

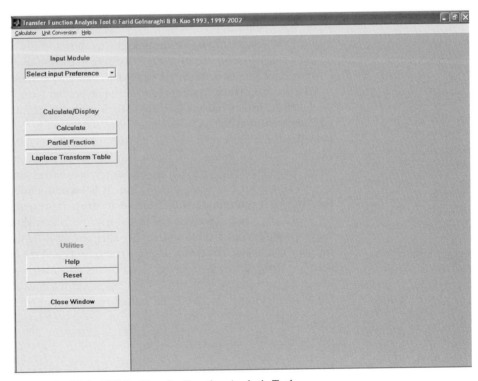

Figure 2-3 Main GUI for Transfer Function Analysis Tool.

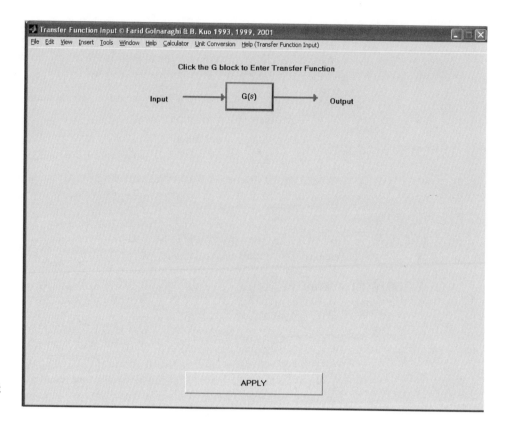

Figure 2-4 Transfer function input GUI for zero, pole, constant input mode.

format, it is easier to select the zero, pole, constant mode. Upon selecting this input mechanism, the window in Fig. 2-4 appears. Press block G to enter its properties, and follow the instructions to enter the system parameters, as shown in Fig. 2-5. Once all the values are properly entered (separated by a space), be sure to use the apply button to exit this window. Otherwise the proper transfer function values will not be transferred to the main window.

To find the system transfer function in polynomial form, press the calculate button in the calculate/display module (see Fig. 2-6). You should get the transfer function's polynomial representation and its poles and zeros as shown in Fig. 2-7. There is an alternative and more detailed representation of the transfer function available in the MATLAB command window, as shown in Fig. 2-8. Please note that at times when the coefficients in numerator or denominator polynomials are too large, the transfer function in the TFtool window (Fig. 2-7) may appear messy and unclear. **It is recommended that you always refer to the MATLAB command window for an accurate representation of transfer functions.**

Next, you need to activate the partial-fraction button shown in Fig. 2-7. The partial-fraction button is designed to appear only after you evaluate the transfer function using the calculate command. The partial-fraction coefficients and the corresponding poles will appear, as shown in Fig. 2-9. Again, as we discussed earlier, you may obtain the above information by referring to the MATLAB command window. From the values provided in Fig. 2-10, and using Eq. (2-30), the partial-fraction representation of the system is

$$G(s) = \frac{-6}{s+3} + \frac{7}{s+2} - \frac{1}{s+1} \tag{2-99}$$

Note that in case of repeated poles, Eq. (2-39) must be used.

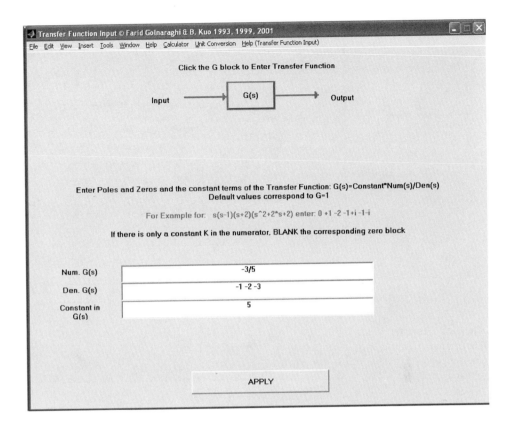

Figure 2-5 Entering transfer function parameters using the Transfer Function Input GUI.

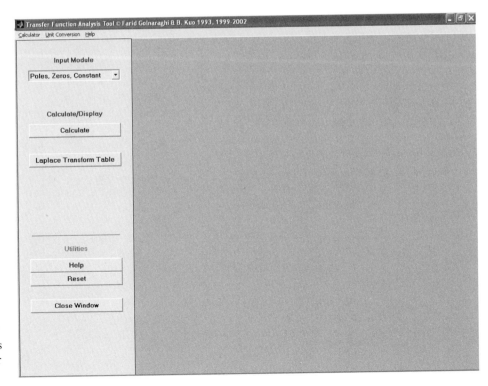

Figure 2-6 Main GUI for Transfer Function Analysis Tool after entering transfer function properties.

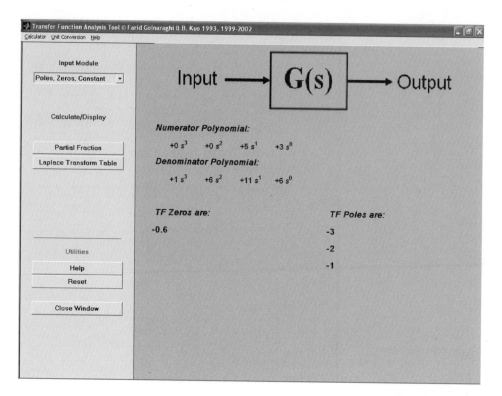

Figure 2-7 Transfer function representation in pole-zero and polynomial forms.

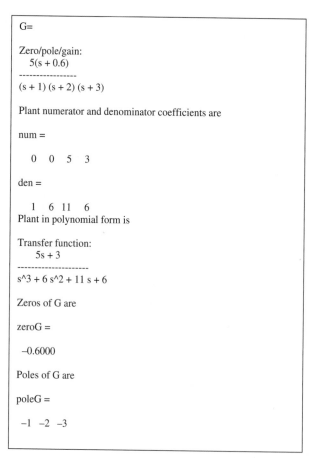

Figure 2-8 Detailed system transfer function and pole-zero values in MATLAB command window.

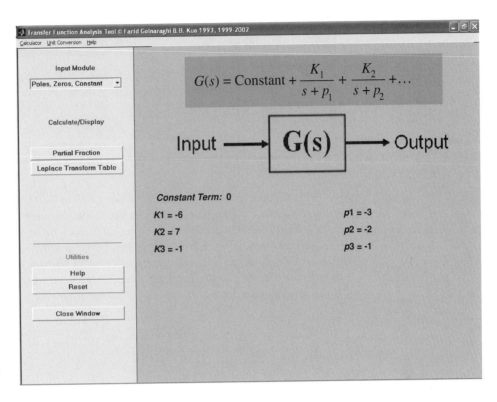

Figure 2-9 Transfer function representation in the partial fraction form.

To find the time representation of Eq. (2-98) for a given input, you may refer to the Laplace tables by activating the Laplace transforms table button or by applying the basic Laplace principles. For an impulse input, the time response of Eq. (2-98) or Eq. (2-99) is

$$g(t) = -6e^{-3t} + 7e^{-2t} - e^{-t} \qquad (2\text{-}100)$$

If you have access to the MATLAB symbolics Toolbox, you may use the **ACSYS** transfer function symbolic tool by pressing the appropriate button in the ACSYS window or by typing **TFsym** in the MATLAB command window. The symbolic tool window is shown in Fig. 2-10. Click the "Help for 1st Time User" button to see the instructions on how to use the toolbox. The instructions appear on help dialog box as shown in Fig. 2-11. As instructed, press the "Transfer Function and Inverse Laplace" button to run the program. You must run this program within the MATLAB command window. Enter the transfer function as shown in Fig. 2-12 to get the time response.

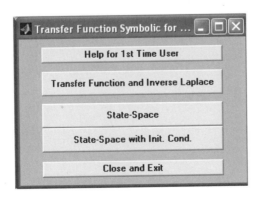

Figure 2-10 The transfer function symbolic window.

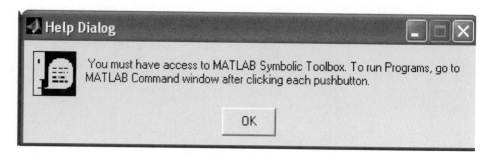

Figure 2-11 The symbolic help dialog box.

Find the inverse Laplace transform of Eq. (2-98) in the MATLAB command window by using the *ilaplace* command, as shown in Fig. 2-10. For more information on this function, refer to MATLAB help documents or type *help ilaplace* in the MATLAB command window.

To find the time representation of Eq. (2-98) for a different input function such as a step or a sinusoid, the user may combine the input transfer function (e.g., $1/s$ for a unit step input) with the transfer function in the TFtool input window. So, to obtain Eq. (2-98) time representation for a unit step input, use the following transfer function

$$G(s) = \frac{5s + 3}{s(s + 1)(s + 2)(s + 3)} \tag{2-101}$$

and repeat the previous steps.

Transfer Function Symbolic. © Kuo & Golnaraghi, 8th Edition, John Wiley & Sons. e.g., Use the following input format: (s+2)*(s^3+2*s+1)/(s*(s^2+2*s+1))

Enter G=5*(s+0.6)/((s+1)*(s+2)*(s+3))

```
      5s+3
-----------------------
   (s+1)(s+2)(s+3)
```

G in polynomial form:

Transfer function:

```
       5s+3
--------------------
s^3+6s^2+11s+6
```

G factored:

Zero/pole/gain:

```
      5(s+0.6)
--------------------
  (s+3)(s+2)(s+1)
```

Inverse Laplace transform:

Gtime =

−exp(−t)+7*exp(−2*t)−6*exp (−3*t)

Figure 2-12 The inverse Laplace transform of $G(s)$ for an impulse input in the MATLAB command window.

▶ **EXAMPLE 2-6** Consider the function
Revisited

$$G(s) = \frac{1}{s(s + 1)^3(s + 2)} \tag{2-102}$$

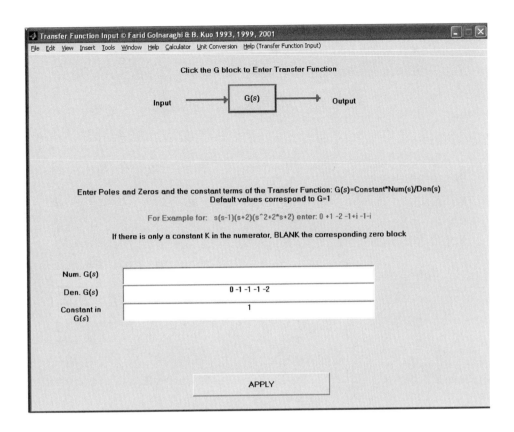

Figure 2-13 Entering transfer function parameters for Example 2-6.

To find the partial-fraction representation of this system, we first use the zero, pole, constant input option and enter the transfer function values as shown in Fig. 2-13. After clicking the partial-fraction button in the TFtool window, the partial-fraction expansion of Eq. (2-101) may be obtained by substituting the resulting coefficients and poles shown in Fig. 2-14 into Eq. (2-39) to obtain the following:

$$G(s) = \frac{1}{2s} + \frac{1}{2(s+2)} - \frac{1}{s+1} - \frac{1}{(s+1)^3} \tag{2-103}$$

Use a procedure similar to that used in the previous examples to find the inverse Laplace transform of the system.

▶ **EXAMPLE 2-8** Consider the function
Revisited

$$Y(s) = \frac{-s^2 - s + 5}{s(s^2 + 3s + 2)} = \frac{-s^2 - s + 5}{s(s+1)(s+2)} \tag{2-104}$$

Use the polynomial module in the transfer function analysis window to enter the transfer function, as in Fig. 2-15. The partial-fraction expansion in this case is

$$Y(s) = \frac{5}{2s} - \frac{5}{s+1} + \frac{3}{2(s+2)} \tag{2-105}$$

▶ **EXAMPLE 2-9** For the transfer function
Revisited

$$Y(s) = \frac{1000}{s(s^2 + 34.5s + 1000)} = \frac{\omega_n^2}{s(s^2 + 2\zeta\omega_n s + \omega_n^2)} \tag{2-106}$$

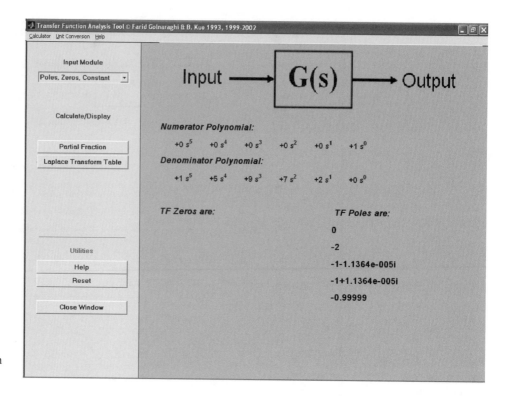

Figure 2-14 The partial-fraction expansion of Example 2-6.

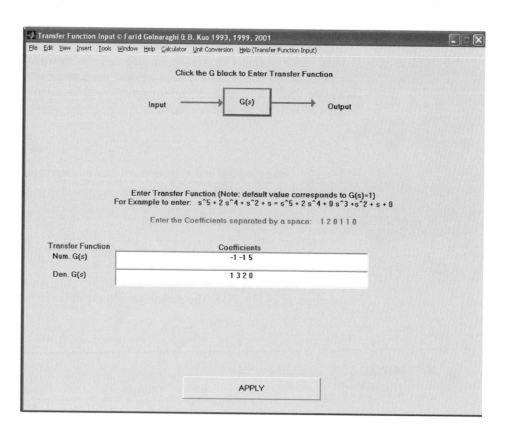

Figure 2-15 Entering transfer function parameters for Example 2-8.

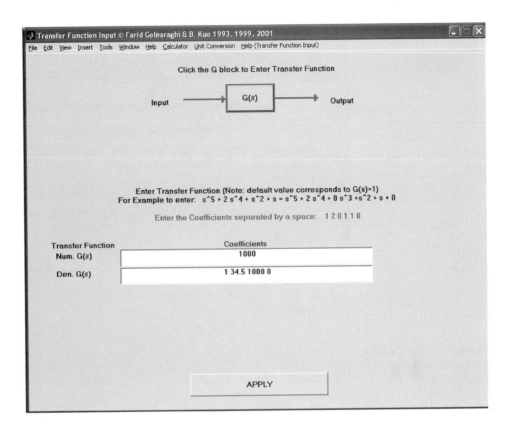

Figure 2-16 Entering transfer function parameters for Example 2-9.

we enter the transfer function parameters as in Fig. 2-16. The partial-fraction coefficients are also shown in Fig. 2-17. Using Eq. (2-30), the partial-fraction expansion in this case is

$$Y(s) = \frac{1}{s} + \frac{-0.5 + j0.325}{s + 17.25 + j26.5} + \frac{-0.5 - j0.325}{s + 17.25 - j26.5} \qquad (2\text{-}107)$$

Clearly,

$$\sigma = \zeta\omega_n = 17.25 \qquad (2\text{-}108)$$

$$\omega = \omega_n\sqrt{1 - \zeta^2} = 26.5 \qquad (2\text{-}109)$$

Using the TFsym tool, the time representation of this system is obtained as

1+(-1/2+13/40*i)*exp((-69/4-53/2*i)*t)+(1/2-13/40*i)*exp((-69/4+53/2*i)*t)

As discussed earlier, the user may also refer to the Laplace tables provided in the TFtool by clicking the Laplace tables button and entering the appropriate numbers.

In case you wish to enter the transfer function with complex poles or zeros, select the zero, pole, constant module in the Transfer Function Analysis Tool window. For our example, the pole-zero representation of Eq. (2-106) is

$$Y(s) = \frac{1000}{s(s^2 + 34.5s + 1000)} = \frac{1000}{s(s + 17.25 + j26.5)(s + 17.25 - j26.5)} \qquad (2\text{-}110)$$

Enter the transfer function values as shown in Fig. 2-18. Note that MATLAB treats both i and j as the imaginary number $\sqrt{-1}$. After you enter the transfer function values and press the APPLY button in Fig. 2-18, you can press the calculate button in the Transfer Function Analysis Tool window to obtain the screen shown in Fig. 2-18. Recall that a

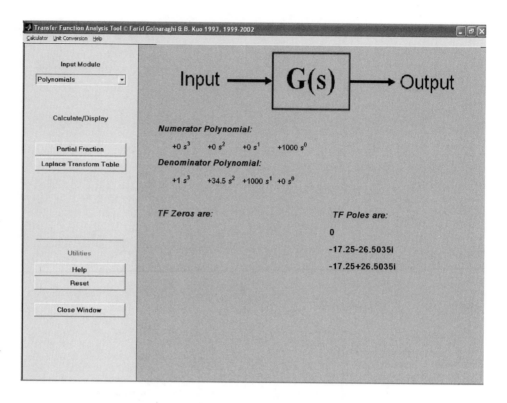

Figure 2-17 The partial-fraction expansion of Example 2-9.

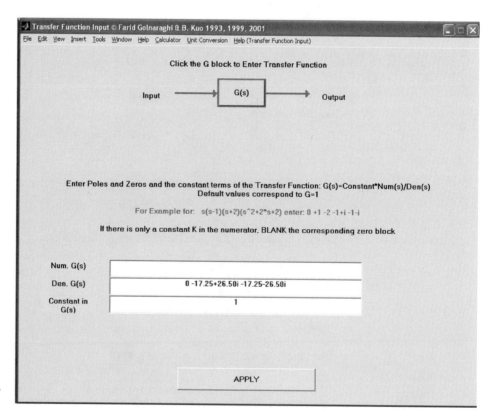

Figure 2-18 Entering the transfer function in Example 2-9 in zero, pole, constant form.

more accurate representation of the system will always appear in the MATLAB command window, upon pressing the calculate button.

In the end, the reader is encouraged to apply TFtool and TFsym to all problems identified by a MATLAB Toolbox in the left margin of the text throughout this chapter.

▶ 2-7 SUMMARY

In this chapter, we presented some fundamental mathematics required for the study of linear control systems. Specifically, the Laplace transform is used for the solution of linear ordinary differential equations. This transform method is characterized by first transforming the real-domain equations into algebraic equations in the transform domain. The solutions are first obtained in the transform domain by using the familiar methods of solving algebraic equations. The final solution in the real domain is obtained by taking the inverse transform. For engineering problems, the transform tables and the partial-fraction expansion method are recommended for the inverse transformation.

The MATLAB tools **TFtool** and **TFsym** were also introduced, and their use for the evaluation of transfer function poles and zeros, partial-fraction expansion, and inverse Laplace transforms was demonstrated.

▶ REVIEW QUESTIONS

1. Give the definitions of the poles and zeros of a function of the complex variable s.

2. What are the advantages of the Laplace-transform method of solving linear ordinary differential equations over the classical method?

3. What are state equations?

4. What is a causal system?

5. Give the defining equation of the one-sided Laplace transform.

6. Give the defining equation of the inverse Laplace transform.

7. Give the expression of the final-value theorem of the Laplace transform. What is the condition under which the theorem is valid?

8. Give the Laplace transform of the unit-step function, $u_s(t)$.

9. What is the Laplace transform of the unit-ramp function, $tu_s(t)$?

10. Give the Laplace transform of $f(t)$ shifted to the right (delayed) by T_d in terms of the Laplace transform of $f(t)$, $F(s)$.

11. If $\mathcal{L}[f_1(t)] = F_1(s)$ and $\mathcal{L}[f_2(t)] = F_2(s)$, then find $\mathcal{L}[f_1(t)]f_2(t)]$ in terms of $F_1(s)$ and $F_2(s)$.

12. Do you know how to handle the exponential term in performing the partial-fraction expansion of

$$F(s) = \frac{10}{(s + 1)(s + 2)} e^{-2s}$$

13. Do you know how to handle the partial-fraction expansion of a function whose denominator order is not greater than that of the numerator, for example,

$$F(s) = \frac{10(s^2 + 5s + 1)}{(s + 1)(s + 2)}$$

14. In trying to find the inverse Laplace transform of the following function, do you have to perform the partial-fraction expansion?

$$F(s) = \frac{1}{(s + 5)^3}$$

▶ REFERENCES

Complex Variables, Laplace Transforms

1. F. B. Hildebrand, *Methods of Applied Mathematics*, 2nd Ed., Prentice Hall, Englewood Cliffs, NJ, 1965.
2. B. C. Kuo, *Linear Networks and Systems*, McGraw-Hill Book Company, New York, 1967.
3. C. R. Wylie, Jr., *Advanced Engineering Mathematics*, 2nd Ed., McGraw-Hill Book Company, New York, 1960.

Partial-Fraction Expansion

4. C. Pottle, "On the Partial Fraction Expansion of a Rational Function with Multiple Poles by Digital Computer," *IEEE Trans. Circuit Theory*, Vol. CT-11, 161–162, Mar. 1964.
5. B. O. Watkins, "A Partial Fraction Algorithm," *IEEE Trans. Automatic Control*, Vo. AC-16, 489–491, Oct. 1971.

▶ PROBLEMS

2-1. Find the poles and zeros of the following functions (including the ones at infinity, if any). Mark the finite poles with × and the finite zeros with o in the *s*-plane.

(a) $G(s) = \dfrac{10(s + 2)}{s^2(s + 1)(s + 10)}$ (b) $G(s) = \dfrac{10s(s + 1)}{(s + 2)(s^2 + 3s + 2)}$

(c) $G(s) = \dfrac{10(s + 2)}{s(s^2 + 2s + 2)}$ (d) $G(s) = \dfrac{e^{-2s}}{10s(s + 1)(s + 2)}$

• Laplace transforms

2-2. Find the Laplace transforms of the following functions. Use the theorems on Laplace transforms if applicable.

(a) $g(t) = 5te^{-5t}u_s(t)$ (b) $g(t) = (t\sin 2t + e^{-2t})u_s(t)$

(c) $g(t) = 2e^{-2t}\sin 2t\, u_s(t)$ (d) $g(t) = \sin 2t \cos 2t\, u_s(t)$

(e) $g(t) = \displaystyle\sum_{k=0}^{\infty} e^{-5kT}\delta(t - kT)$ where $\delta(t)$ = unit-impulse function

• Laplace transforms

2-3. Find the Laplace transforms of the functions shown in Fig. 2P-3. First, write a complete expression for $g(t)$, and then take the Laplace transform. Let $g_T(t)$ be the description of the function over the basic period and then delay $g_T(t)$ appropriately to get $g(t)$. Take the Laplace transform of $g(t)$ to get

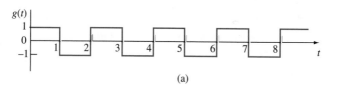

(a)

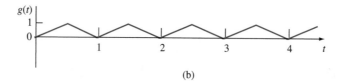

(b)

Figure 2P-3

• Laplace transforms

2-4. Find the Laplace transform of the following function.

$$g(t) = \begin{cases} t + 1 & 0 \le t < 1 \\ 0 & 1 \le t < 2 \\ 2 - t & 2 \le t < 3 \\ 0 & t \ge 3 \end{cases}$$

• Laplace transforms

2-5. Solve the following differential equations by using the Laplace transform.

(a) $\dfrac{d^2 f(t)}{dt^2} + 5\dfrac{df(t)}{dt} + 4f(t) = e^{-2t}u_s(t)$ Assume zero initial conditions.

(b) $\dfrac{dx_1(t)}{dt} = x_2(t)$

$\dfrac{dx_2(t)}{dt} = -2x_1(t) - 3x_2(t) + u_s(t)$ $x_1(0) = 1, x_2(0) = 0$

2-6. Find the inverse Laplace transforms of the following functions. Perform partial-fraction expansion on $G(s)$ first, then use the Laplace transform table. Use TFtool or any computer program that is available for the partial-fraction expansion.

(a) $G(s) = \dfrac{1}{s(s + 2)(s + 3)}$

(b) $G(s) = \dfrac{10}{(s + 1)^2(s + 3)}$

(c) $G(s) = \dfrac{100(s + 2)}{s(s^2 + 4)(s + 1)}e^{-s}$

(d) $G(s) = \dfrac{2(s + 1)}{s(s^2 + s + 2)}$

(e) $G(s) = \dfrac{1}{(s + 1)^3}$

(f) $G(s) = \dfrac{2(s^2 + s + 1)}{s(s + 1.5)(s^2 + 5s + 5)}$

2-7. Express the following set of first-order differential equations in vector-matrix form:

$$\dfrac{d\mathbf{x}(t)}{dt} = \mathbf{A}\mathbf{x}(t) + \mathbf{B}\mathbf{u}(t)$$

$$\dfrac{dx_1(t)}{dt} = -x_1(t) + 2x_2(t)$$

$$\dfrac{dx_2(t)}{dt} = -2x_2(t) + 3x_3(t) = u_1(t)$$

$$\dfrac{dx_3(t)}{dt} = -x_1(t) - 3x_2(t) - x_3(t) + u_2(t)$$

- Transfer function

2-8. The following differential equations represent linear time-invariant systems, where $r(t)$ denotes the input, and $y(t)$ the output. Find the transfer function $Y(s)/R(s)$ for each of the systems.

(a) $\dfrac{d^3y(t)}{dt^3} + 2\dfrac{d^2y(t)}{dt^2} + 5\dfrac{dy(t)}{dt} + 6y(t) = 3\dfrac{dr(t)}{dt} + r(t)$

(b) $\dfrac{d^4y(t)}{dt^4} + 10\dfrac{d^2y(t)}{dt^2} + \dfrac{dy(t)}{dt} + 5y(t) = 5r(t)$

(c) $\dfrac{d^3y(t)}{dt^3} + 10\dfrac{d^2y(t)}{dt^2} + 2\dfrac{dy(t)}{dt} + y(t) + 2\displaystyle\int_0^t y(\tau)d\tau = \dfrac{dr(t)}{dt} + 2r(t)$

(d) $2\dfrac{d^2y(t)}{dt^2} + \dfrac{dy(t)}{dt} + 5y(t) = r(t) + 2r(t - 1)$

Additional Computer Problems

The following problems are to be solved by the **TFtool**.

2-9. Perform partial-fraction expansion of the following functions.

(a) $G(s) = \dfrac{10(s + 1)}{s^2(s + 4)(s + 6)}$

(b) $G(s) = \dfrac{(s + 1)}{s(s + 2)(s^2 + 2s + 2)}$

(c) $G(s) = \dfrac{5(s + 2)}{s^2(s + 1)(s + 5)}$

(d) $G(s) = \dfrac{5e^{-2s}}{(s + 1)(s^2 + s + 1)}$

(e) $G(s) = \dfrac{100(s^2 + s + 3)}{s(s^2 + 5s + 3)}$

(f) $G(s) = \dfrac{1}{s(s^2 + 1)(s + 0.5)^2}$

2-10. Find the inverse Laplace transforms of the functions in Problem 2-9.

CHAPTER 3

Block Diagrams and Signal-Flow Graphs

▶ 3-1 BLOCK DIAGRAMS

Because of their simplicity and versatility, **block diagrams** are often used by control engineers to model all types of systems. A block diagram can be used simply to describe the composition and interconnection of a system. Or it can be used, together with transfer functions, to describe the cause-and-effect relationships throughout the system. For instance, the block diagram of Fig. 3-1(a) models an open-loop, dc-motor, speed-control system. The block diagram in this case simply shows how the system components are interconnected, and no mathematical details are given. If the mathematical and functional relationships of all the system elements are known, the block diagram can be used as a tool for the analytic or computer solution of the system. In general, block diagrams can be used to model linear as well as nonlinear systems. For example, the input-output relations of the dc-motor control system may be represented by the block diagram shown in Fig. 3-1(b). In the figure, the input voltage to the motor is the output of the power amplifier, which, realistically, has a nonlinear characteristic. If the motor is linear, or, more appropriately, if it is operated in the linear region of its characteristics, its dynamics can

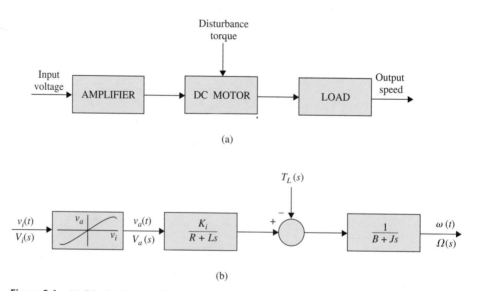

(a)

(b)

Figure 3-1   (a) Block diagram of a dc-motor control system. (b) Block diagram with transfer functions and amplifier characteristics.

be represented by transfer functions. The nonlinear amplifier gain can only be described between the time variables $v_i(t)$ and $v_a(t)$, and no transfer function exists between the Laplace transform variables $V_i(s)$ and $V_a(s)$. However, if the magnitude of $v_i(t)$ is limited to the linear range of the amplifier, then the amplifier can be regarded as linear, and the amplifier can be described by the transfer function

$$\frac{V_a(s)}{V_i(s)} = K \tag{3-1}$$

where K is a constant, which is the slope of the linear region of the amplifier characteristics.

3-1-1 Block Diagrams of Control Systems

We shall now define the block-diagram elements used frequently in control systems and the related algebra. One of the important components of a control system is the sensing device that acts as a junction point for signal comparisons. The physical components involved are the potentiometer, syncho, resolver, differential amplifier, and other signal-processing transducers. In general, sensing devices perform simple mathematical operations such as **addition** and **subtraction**. The block diagram representations of these operations are illustrated in Fig. 3-2. The addition and subtraction operations in Fig. 3-2(a) and (b) are linear, so the input and output variables of these block-diagram elements can be time-domain variables or Laplace-transform variables. Thus, in Fig. 3-2(a), the block diagram implies

$$e(t) = r(t) - y(t) \tag{3-2}$$

or

$$E(s) = R(s) - Y(s) \tag{3-3}$$

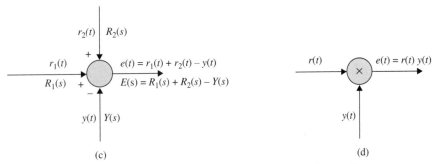

Figure 3-2 Block diagram elements of typical sensing devices of control systems. (a) Subtraction. (b) Addition. (c) Addition and Subtraction. (d) Multiplication.

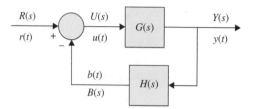

Figure 3-3 Basic block diagram of a feedback control system.

Figure 3-3 shows the block diagram of a linear feedback control system. The following terminology is defined with reference to the diagram:

$$r(t), R(s) = \text{reference input (command)}$$
$$y(t), Y(s) = \text{output (controlled variable)}$$
$$b(t), B(s) = \text{feedback signal}$$
$$u(t), U(s) = \text{actuating signal} = \text{error signal } e(t), E(s), \text{ when } H(s) = 1$$
$$H(s) = \text{feedback transfer function}$$
$$G(s)H(s) = L(s) = \text{loop transfer function}$$
$$G(s) = \text{forward-path transfer function}$$
$$M(s) = Y(s)/R(s) = \text{closed-loop transfer function or system transfer function}$$

The closed-loop transfer function $M(s)$ can be expressed as a function of $G(s)$ and $H(s)$. From Fig. 3-3, we write

$$Y(s) = G(s)U(s) \tag{3-4}$$

and

$$B(s) = H(s)Y(s) \tag{3-5}$$

The actuating signal is written

$$U(s) = R(s) - B(s) \tag{3-6}$$

Substituting Eq. (3-6) into Eq. (3-4) yields

$$Y(s) = G(s)R(s) - G(s)B(s) \tag{3-7}$$

Substituting Eq. (3-5) into Eq. (3-7) and then solving for $Y(s)/R(s)$ gives the closed-loop transfer function:

$$M(s) = \frac{Y(s)}{R(s)} = \frac{G(s)}{1 + G(s)H(s)} \tag{3-8}$$

In general, a control system may contain more than one feedback loop, and the evaluation of the transfer function from the block diagram by the algebraic method just described may be tedious. Although, in principle, the block diagram of a system with one input and one output can always be reduced to the basic single-loop form of Fig. 3-3, the algebraic steps involved in the reduction process may again be quite tedious. We show in Section 3-2 that the transfer function of any linear system can be obtained directly from its block diagram by the signal-flow graph gain formula.

3-1-2 Block Diagrams and Transfer Functions of Multivariable Systems

In this section, we shall illustrate the block diagram and matrix representations (Appendix C) of multivariable systems. Two block-diagram representations of a multivariable

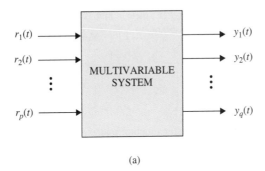

(a)

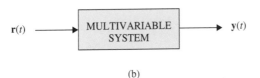

(b)

Figure 3-4 Block diagram representations of a multivariable system.

system with p inputs and q outputs are shown in Fig. 3-4(a) and (b). In Fig. 3-4(a), the individual input and output signals are designated, whereas in the block diagram of Fig. 3-4(b), the multiplicity of the inputs and outputs is denoted by vectors. The case of Fig. 3-4(b) is preferable in practice because of its simplicity.

Figure 3-5 shows the block diagram of a multivariable feedback control system. The transfer function relationships of the system are expressed in vector-matrix form (see Appendix C):

$$\mathbf{Y}(s) = \mathbf{G}(s)\mathbf{U}(s) \tag{3-9}$$

$$\mathbf{U}(s) = \mathbf{R}(s) - \mathbf{B}(s) \tag{3-10}$$

$$\mathbf{B}(s) = \mathbf{H}(s)\mathbf{Y}(s) \tag{3-11}$$

where $\mathbf{Y}(s)$ is the $q \times 1$ output vector; $\mathbf{U}(s)$, $\mathbf{R}(s)$, and $\mathbf{B}(s)$ are all $p \times 1$ vectors; and $\mathbf{G}(s)$ and $\mathbf{H}(s)$ are $q \times p$ and $p \times q$ transfer-function matrices, respectively. Substituting Eq. (3-11) into Eq. (3-10) and then from Eq. (3-10) to Eq. (3-9), we get

$$\mathbf{Y}(s) = \mathbf{G}(s)\mathbf{R}(s) - \mathbf{G}(s)\mathbf{H}(s)\mathbf{Y}(s) \tag{3-12}$$

Solving for $\mathbf{Y}(s)$ from Eq. (3-12) gives

$$\mathbf{Y}(s) = \left[\mathbf{I} + \mathbf{G}(s)\mathbf{H}(s)\right]^{-1} \mathbf{G}(s)\mathbf{R}(s) \tag{3-13}$$

provided that $\mathbf{I} + \mathbf{G}(s)\mathbf{H}(s)$ is nonsingular. The closed-loop transfer matrix is defined as

$$\mathbf{M}(s) = \left[\mathbf{I} + \mathbf{G}(s)\mathbf{H}(s)\right]^{-1} \mathbf{G}(s) \tag{3-14}$$

Then Eq. (3-13) is written

$$\mathbf{Y}(s) = \mathbf{M}(s)\mathbf{R}(s) \tag{3-15}$$

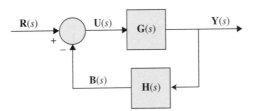

Figure 3-5 Block diagram of a multivariable feedback control system.

▶ **EXAMPLE 3-1** Consider that the forward-path transfer function matrix and the feedback-path transfer function matrix of the system shown in Fig. 3-5 are

$$\mathbf{G}(s) = \begin{bmatrix} \dfrac{1}{s+1} & -\dfrac{1}{s} \\ 2 & \dfrac{1}{s+2} \end{bmatrix} \qquad \mathbf{H}(s) = \begin{bmatrix} 1 & 0 \\ 0 & 1 \end{bmatrix} \tag{3-16}$$

respectively. The closed-loop transfer function matrix of the system is given by Eq. (3-14), and is evaluated as follows:

$$\mathbf{I} + \mathbf{G}(s)\mathbf{H}(s) = \begin{bmatrix} 1 + \dfrac{1}{s+1} & -\dfrac{1}{s} \\ 2 & 1 + \dfrac{1}{s+2} \end{bmatrix} = \begin{bmatrix} \dfrac{s+2}{s+1} & -\dfrac{1}{s} \\ 2 & \dfrac{s+3}{s+2} \end{bmatrix} \tag{3-17}$$

The closed-loop transfer function matrix is

$$\mathbf{M}(s) = [\mathbf{I} + \mathbf{G}(s)\mathbf{H}(s)]^{-1}\,\mathbf{G}(s) = \dfrac{1}{\Delta} \begin{bmatrix} \dfrac{s+3}{s+2} & \dfrac{1}{s} \\ -2 & \dfrac{s+2}{s+1} \end{bmatrix} \begin{bmatrix} \dfrac{1}{s+1} & -\dfrac{1}{s} \\ 2 & \dfrac{1}{s+2} \end{bmatrix} \tag{3-18}$$

where

$$\Delta = \dfrac{s+2}{s+1}\dfrac{s+3}{s+2} + \dfrac{2}{s} = \dfrac{s^2 + 5s + 2}{s(s+1)} \tag{3-19}$$

Thus,

$$\mathbf{M}(s) = \dfrac{s(s+1)}{s^2 + 5s + 2} \begin{bmatrix} \dfrac{3s^2 + 9s + 4}{s(s+1)(s+2)} & -\dfrac{1}{s} \\ 2 & \dfrac{3s+2}{s(s+1)} \end{bmatrix} \tag{3-20}$$

◀

▶ 3-2 SIGNAL-FLOW GRAPHS (SFGs)

A signal-flow graph (SFG) may be regarded as a simplified version of a block diagram. The SFG was introduced by S. J. Mason [2] for the cause-and-effect representation of linear systems that are modeled by algebraic equations. Besides the differences in the physical appearance of the SFG and the block diagram, the signal-flow graph is constrained by more rigid mathematical rules, whereas the block-diagram notation is more liberal. An SFG may be defined as a graphical means of portraying the input-output relationships among the variables of a set of linear algebraic equations.

Consider a linear system that is described by a set of N algebraic equations:

$$y_j = \sum_{k=1}^{N} a_{kj} y_k \qquad j = 1, 2, \ldots, N \tag{3-21}$$

It should be pointed out that these N equations are written in the form of cause-and-effect relations:

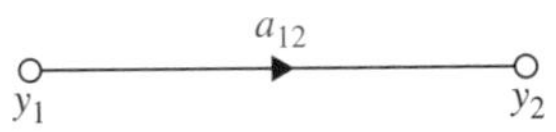

Figure 3-6 Signal flow graph of $y_2 = a_{12}y_1$.

$$\text{jth effect} = \sum_{k=1}^{N} (\text{gain from } k \text{ to } j) \times (k\text{th cause}) \qquad (3\text{-}22)$$

or simply

$$\text{Output} = \sum (\text{gain}) \times (\text{input}) \qquad (3\text{-}23)$$

This is the single most important axiom in forming the set of algebraic equations for SFGs. When the system is represented by a set of integrodifferential equations, we must first transform these into Laplace-transform equations and then rearrange the latter in the form of Eq. (3-21), or

$$Y_j(s) = \sum_{k=1}^{N} G_{kj}(s)Y_k(s) \qquad j = 1, 2, \dots, N \qquad (3\text{-}24)$$

3-2-1 Basic Elements of an SFG

• In an SFG signals can transmit through a branch only in the direction of the arrow.

When constructing an SFG, junction points, or **nodes**, are used to represent variables. The nodes are connected together by line segments called **branches**, according to the cause-and-effect equations. The branches have associated branch gains and directions. *A signal can transmit through a branch only in the direction of the arrow.* In general, given a set of equations such as Eq. (3-21) or Eq. (3-24), the construction of the SFG is basically a matter of following through the cause-and-effect relations of each variable in terms of itself and the others. For instance, consider that a linear system is represented by the simple algebraic equation

$$y_2 = a_{12}y_1 \qquad (3\text{-}25)$$

where y_1 is the input, y_2 is the output, and a_{12} is the gain, or transmittance, between the two variables. The SFG representation of Eq. (3-25) is shown in Fig. 3-6. Notice that the branch directing from node y_1 (input) to node y_2 (output) expresses the dependence of y_2 on y_1, but not the reverse. The branch between the input node and the output node should be interpreted as a unilateral amplifier with gain a_{12}, so when a signal of one unit is applied at the input y_1, a signal of strength $a_{12}y_1$ is delivered at node y_2. Although algebraically Eq. (3-25) can be written as

$$y_1 = \frac{1}{a_{12}}y_2 \qquad (3\text{-}26)$$

the SFG of Fig. 3-6 does not imply this relationship. If Eq. (3-26) is valid as a cause-and-effect equation, a new SFG should be drawn with y_2 as the input and y_1 as the output.

▶ **EXAMPLE 3-2** As an example on the construction of a SFG, consider the following set of algebraic equations:

$$\begin{aligned}
y_2 &= a_{12}y_1 + a_{32}y_3 \\
y_3 &= a_{23}y_2 + a_{43}y_4 \\
y_4 &= a_{24}y_2 + a_{34}y_3 + a_{44}y_4 \\
y_5 &= a_{25}y_2 + a_{45}y_4
\end{aligned} \qquad (3\text{-}27)$$

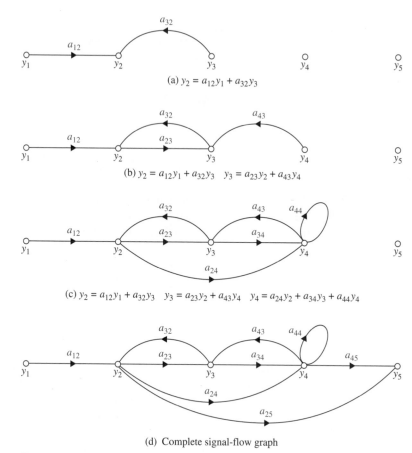

(a) $y_2 = a_{12}y_1 + a_{32}y_3$

(b) $y_2 = a_{12}y_1 + a_{32}y_3$ $y_3 = a_{23}y_2 + a_{43}y_4$

(c) $y_2 = a_{12}y_1 + a_{32}y_3$ $y_3 = a_{23}y_2 + a_{43}y_4$ $y_4 = a_{24}y_2 + a_{34}y_3 + a_{44}y_4$

(d) Complete signal-flow graph

Figure 3-7   Step-by-step construction of the signal-flow graph in Eq. (3-27).

The SFG for these equations is constructed, step by step, as shown in Fig. 3-7.   ◀

3-2-2   Summary of the Basic Properties of SFG

The important properties of the SFG that have been covered thus far are summarized as follows.

1. SFG applies only to linear systems.
2. The equations for which an SFG is drawn must be algebraic equations in the form of cause-and-effect.
3. Nodes are used to represent variables. Normally, the nodes are arranged from left to right, from the input to the output, following a succession of cause-and-effect relations through the system.
4. Signals travel along branches only in the direction described by the arrows of the branches.
5. The branch directing from node y_k to y_j represents the dependence of y_j upon y_k, but not the reverse.
6. A signal y_k traveling along a branch between y_k and y_j is multiplied by the gain of the branch a_{kj}, so a signal $a_{kj}y_k$ is delivered at y_j.

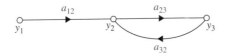

(a) Original signal-flow graph

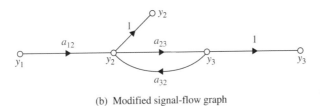

(b) Modified signal-flow graph

Figure 3-8 Modification of a signal-flow graph so that y_2 and y_3 satisfy the condition as output nodes.

3-2-3 Definitions of SFG Terms

In addition to the branches and nodes defined earlier for the SFG, the following terms are useful for the purpose of identification and executing the SFG algebra.

- An input node has only outgoing branches.

- An output node has only incoming branches.

Input Node (Source). *An* input node *is a node that has only outgoing branches.* (Example: node y_1 in Fig. 3-7.)

Output Node (Sink). *An* output node *is a node that has only incoming branches.* (Example: node y_5 in Fig. 3-7.) However, this condition is not always readily met by an output node. For instance, the SFG in Fig. 3-8(a) does not have a node that satisfies the condition of an output node. It may be necessary to regard y_2 and/or y_3 as output nodes to find the effects at these nodes due to the input. To make y_2 an output node, we simply connect a branch with unity gain from the existing node y_2 to a new node also designated as y_2, as shown in Fig. 3-8(b). The same procedure is applied to y_3. Notice that in the modified SFG of Fig. 3-8(b), the equations $y_2 = y_2$ and $y_3 = y_3$ are added to the original equations. In general, we can make any noninput node of an SFG an output by the procedure just illustrated. However, we **cannot** convert a noninput node into an input node by reversing the branch direction of the procedure described for output nodes. For instance, node y_2 of the SFG in Fig. 3-8(a) is not an input node. If we attempt to convert it into an input node by adding an incoming branch with unity gain from another identical node y_2, the SFG of Fig. 3-9 would result. The equation that portrays the relationship at node y_2 now reads

$$y_2 = y_2 + a_{12}y_1 + a_{32}y_3 \tag{3-28}$$

which is different from the original equation given in Fig. 3-8(a).

Path. *A* path *is any collection of a continuous succession of branches traversed in the same direction.* The definition of a path is entirely general, since it does not

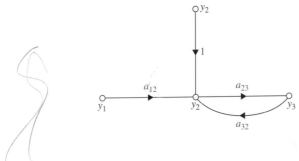

Figure 3-9 Erroneous way to make node y_2 an input node.

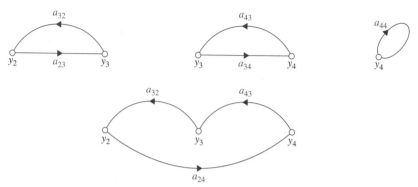

Figure 3-10 Four loops in the signal-flow graph of Fig. 3-7(d).

prevent any node from being traversed more than once. Therefore, as simple as the SFG of Fig. 3-8(a) is, it may have numerous paths just by traversing the branches a_{23} and a_{32} continuously.

Forward Path. A forward path *is a path that starts at an input node and ends at an output node, and along which no node is traversed more than once.* For example, in the SFG of Fig. 3-7(d), y_1 is the input node, and the rest of the nodes are all possible output nodes. The forward path between y_1 and y_2 is simply the connecting branch between the two nodes. There are two forward paths between y_1 and y_3: One contains the branches from y_1 to y_2 to y_3, and the other one contains the branches from y_1 to y_2 to y_4 (through the branch with gain a_{24}) and then back to y_3 (through the branch with gain a_{43}). The reader should try to determine the two forward paths between y_1 and y_4. Similarly, there are three forward paths between y_1 and y_5.

Loop. A loop *is a path that originates and terminates on the same node, and along which no other node is encountered more than once.* For example, there are four loops in the SFG of Fig. 3-7(d). These are shown in Fig. 3-10.

Path Gain. *The product of the branch gains encountered in traversing a path is called the* path gain. For example, the path gain for the path $y_1 - y_2 - y_3 - y_4$ in Fig. 3-7(d) is $a_{12}a_{23}a_{34}$.

Forward-Path Gain. *The* forward-path gain *is the path gain of a forward path.*

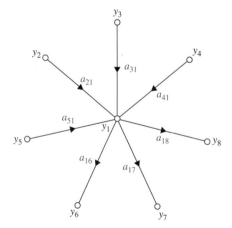

Figure 3-11 Node as a summing point and as a transmitting point.

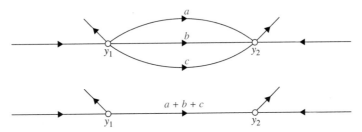

Figure 3-12 Signal-flow graph with parallel paths replaced by one with a single branch.

• Two parts of an SFG are nontouching if they do not share a common node.

Loop Gain. *The* loop gain *is the path gain of a loop.* For example, the loop gain of the loop $y_2 - y_4 - y_3 - y_2$ in Fig. 3-10 is $a_{24}a_{43}a_{32}$.

Nontouching Loops. *Two parts of an SFG are* nontouching *if they do not share a common node.* For example, the loops $y_2 - y_3 - y_2$ and $y_4 - y_4$ of the SFG in Fig. 3-7(d) are nontouching loops.

3-2-4 SFG Algebra

Based on the properties of the SFG, we can outline the following manipulation rules and algebra:

1. The value of the variable represented by a node is equal to the sum of all the signals entering the node. For the SFG of Fig. 3-11, the value of y_1 is equal to the sum of the signals transmitted through all the incoming branches; that is,

$$y_1 = a_{21}y_2 + a_{31}y_3 + a_{41}y_4 + a_{51}y_5 \qquad (3\text{-}29)$$

2. The value of the variable represented by a node is transmitted through all branches leaving the node. In the SFG of Fig. 3-11, we have

$$\begin{aligned} y_6 &= a_{16}y_1 \\ y_7 &= a_{17}y_1 \\ y_8 &= a_{18}y_1 \end{aligned} \qquad (3\text{-}30)$$

3. Parallel branches in the same direction connecting two nodes can be replaced by a single branch with gain equal to the sum of the gains of the parallel branches. An example of this case is illustrated in Fig. 3-12.

4. A series connection of unidirectional branches, as shown in Fig. 3-13, can be replaced by a single branch with gain equal to the product of the branch gains.

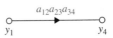

Figure 3-13 Signal-flow graph with cascade unidirectional branches replaced by a single branch.

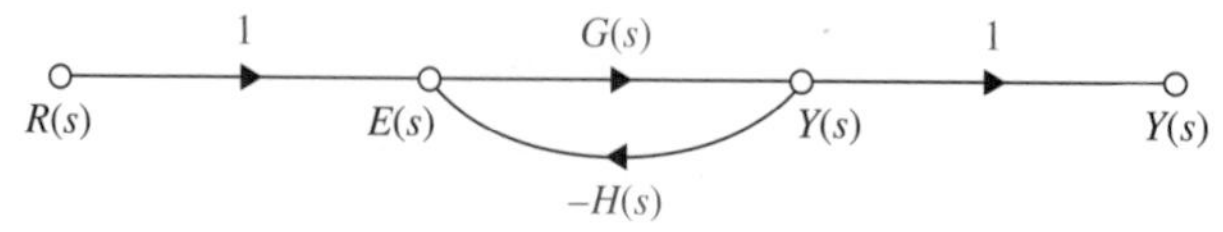

Figure 3-14 Signal-flow graph of the feedback control system shown in Fig. 3-3.

3-2-5 SFG of a Feedback Control System

The SFG of the single-loop feedback control system in Fig. 3-3 is drawn as shown in Fig. 3-14. Using the SFG algebra already outlined, the closed-loop transfer function in Eq. (3-8) can be obtained.

3-2-6 Gain Formula for SFG

Given an SFG or block diagram, the task of solving for the input-output relations by algebraic manipulation could be quite tedious. Fortunately, there is a general gain formula available that allows the determination of the input-output relations of an SFG by inspection.

Given an SFG with N forward paths and K loops, the gain between the input node y_{in} and output node y_{out} is [3]

$$M = \frac{y_{out}}{y_{in}} = \sum_{k=1}^{N} \frac{M_k \Delta_k}{\Delta} \tag{3-31}$$

where

y_{in} = input-node variable

y_{out} = output-node variable

M = gain between y_{in} and y_{out}

N = total number of forward paths between y_{in} and y_{out}

M_k = gain of the kth forward path between y_{in} and y_{out}

$$\Delta = 1 - \sum_i L_{i1} + \sum_j L_{j2} - \sum_k L_{k3} + \cdots \tag{3-32}$$

L_{mr} = gain product of the mth ($m = i, j, k, \ldots$) possible combination of r nontouching loops ($1 \le r < K$).

or

$\Delta = 1 -$ (sum of the gains of **all individual** loops) + (sum of products of gains of all possible combinations of **two** nontouching loops) − (sum of products of gains of all possible combinations of **three** nontouching loops) + $\cdots$

$$\tag{3-33}$$

Δ_k = the Δ for that part of the SFG that is nontouching with the kth forward path.

• The SFG gain formula can only be applied between an input node and an output node.

The gain formula in Eq. (3-31) may seem formidable to use at first glance. However, Δ and Δ_k are the only terms in the formula that could be complicated if the SFG has a large number of loops and nontouching loops.

Care must be taken when applying the gain formula to ensure that it is applied between an **input node** and an **output node**.

▶ **EXAMPLE 3-3** Consider that the closed-loop transfer function $Y(s)/R(s)$ of the SFG in Fig. 3-14 is to be determined by use of the gain formula, Eq. (3-31). The following results are obtained by inspection of the SFG:

1. There is only one forward path between $R(s)$ and $Y(s)$, and the forward-path gain is

$$M_1 = G(s) \tag{3-34}$$

2. There is only one loop; the loop gain is

$$L_{11} = -G(s)H(s) \tag{3-35}$$

3. There are no nontouching loops since there is only one loop. Furthermore, the forward path is in touch with the only loop. Thus, $\Delta_1 = 1$, and

$$\Delta = 1 - L_{11} = 1 + G(s)H(s) \tag{3-36}$$

Using Eq. (3-31), the closed-loop transfer function is written

$$\frac{Y(s)}{R(s)} = \frac{M_1 \Delta_1}{\Delta} = \frac{G(s)}{1 + G(s)H(s)} \tag{3-37}$$

which agrees with Eq. (3-8). ◄

► **EXAMPLE 3-4** Consider the SFG shown in Fig. 3-7(d). Let us first determine the gain between y_1 and y_5 using the gain formula.

The three forward paths between y_1 and y_5 and the forward-path gains are

$$M_1 = a_{12}a_{23}a_{34}a_{45} \qquad \text{Forward path:} \qquad y_1 - y_2 - y_3 - y_4 - y_5$$
$$M_2 = a_{12}a_{25} \qquad\qquad \text{Forward path:} \qquad y_1 - y_2 - y_5$$
$$M_3 = a_{12}a_{24}a_{45} \qquad\quad \text{Forward path:} \qquad y_1 - y_2 - y_4 - y_5$$

The four loops of the SFG are shown in Fig. 3-10. The loop gains are

$$L_{11} = a_{23}a_{32} \qquad L_{21} = a_{34}a_{43} \qquad L_{31} = a_{24}a_{43}a_{32} \qquad L_{41} = a_{44}$$

There is only one pair of nontouching loops; that is, the two loops are

$$y_2 - y_3 - y_2 \qquad \text{and} \qquad y_4 - y_4$$

Thus, the product of the gains of the two nontouching loops is

$$L_{12} = a_{23}a_{32}a_{44} \tag{3-38}$$

All the loops are in touch with forward paths M_1 and M_3. Thus, $\Delta_1 = \Delta_3 = 1$. Two of the loops are not in touch with forward path M_2. These loops are: $y_3 - y_4 - y_3$ and $y_4 - y_4$.

Thus,

$$\Delta_2 = 1 - a_{34}a_{43} - a_{44} \tag{3-39}$$

Substituting these quantities into Eq. (3-31), we have

$$\frac{y_5}{y_1} = \frac{M_1\Delta_1 + M_2\Delta_2 + M_3\Delta_3}{\Delta} = \frac{(a_{12}a_{23}a_{34}a_{45}) + (a_{12}a_{25})(1 - a_{34}a_{43} - a_{44}) + a_{12}a_{24}a_{45}}{1 - (a_{23}a_{32} + a_{34}a_{43} + a_{24}a_{32}a_{43} + a_{44}) + a_{23}a_{32}a_{44}} \tag{3-40}$$

where

$$\Delta = 1 - (L_{11} + L_{21} + L_{31} + L_{41}) + L_{12}$$
$$= 1 - (a_{23}a_{32} + a_{34}a_{43} + a_{24}a_{32}a_{43} + a_{44}) + a_{23}a_{32}a_{44} \tag{3-41}$$

• Δ is the same regardless of which output node is chosen.

The reader should verify that choosing y_2 as the output,

$$\frac{y_2}{y_1} = \frac{a_{12}(1 - a_{34}a_{43} - a_{44})}{\Delta} \tag{3-42}$$

where Δ is given in Eq. (3-41). ◄

▶ **EXAMPLE 3-5** Consider the SFG in Fig. 3-15. The following input-output relations are obtained by use of the gain formula:

$$\frac{y_2}{y_1} = \frac{1 + G_3H_2 + H_4 + G_3H_2H_4}{\Delta} \tag{3-43}$$

$$\frac{y_4}{y_1} = \frac{G_1G_2(1 + H_4)}{\Delta} \tag{3-44}$$

$$\frac{y_6}{y_1} = \frac{y_7}{y_1} = \frac{G_1G_2G_3G_4 + G_1G_5(1 + G_3H_2)}{\Delta} \tag{3-45}$$

where

$$\Delta = 1 + G_1H_1 + G_3H_2 + G_1G_2G_3H_3 + H_4 + G_1G_3H_1H_2$$
$$+ G_1H_1H_4 + G_3H_2H_4 + G_1G_2G_3H_3H_4 + G_1G_3H_1H_2H_4 \tag{3-46}$$

◀

3-2-7 Application of the Gain Formula between Output Nodes and Noninput Nodes

It was pointed out earlier that the gain formula can only be applied between a pair of input and output nodes. Often, it is of interest to find the relation between an output-node variable and a noninput-node variable. For example, in the SFG of Fig. 3-15, it may be of interest to find the relation y_7/y_2, which represents the dependence of y_7 upon y_2; the latter is not an input.

We can show that by including an input node the gain formula can still be applied to find the gain between a noninput node and an output node. Let y_{in} be an input and y_{out} be an output node of a SFG. The gain, y_{out}/y_2, where y_2 is not an input, may be written as

$$\frac{y_{\text{out}}}{y_2} = \frac{\dfrac{y_{\text{out}}}{y_{\text{in}}}}{\dfrac{y_2}{y_{\text{in}}}} = \frac{\dfrac{\sum M_k\Delta_k\big|_{\text{from } y_{\text{in}} \text{ to } y_{\text{out}}}}{\Delta}}{\dfrac{\sum M_k\Delta_k\big|_{\text{from } y_{\text{in}} \text{ to } y_2}}{\Delta}} \tag{3-47}$$

Since Δ is independent of the inputs and the outputs, the last equation is written

$$\frac{y_{\text{out}}}{y_2} = \frac{\sum M_k\Delta_k\big|_{\text{from } y_{\text{in}} \text{ to } y_{\text{out}}}}{\sum M_k\Delta_k\big|_{\text{from } y_{\text{in}} \text{ to } y_2}} \tag{3-48}$$

Notice that Δ does not appear in the last equation.

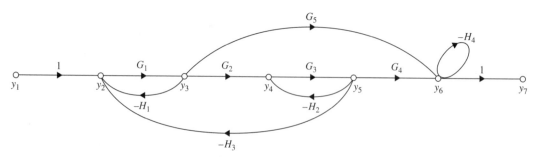

Figure 3-15 Signal-flow graph for Example 3-5.

▶ **EXAMPLE 3-6** From the SFG in Fig. 3-15, the gain between y_2 and y_7 is written

$$\frac{y_7}{y_2} = \frac{y_7/y_1}{y_2/y_1} = \frac{G_1G_2G_3G_4 + G_1G_5(1 + G_3H_2)}{1 + G_3H_2 + H_4 + G_3H_2H_4}$$

(3-49)

◀

3-2-8 Application of the Gain Formula to Block Diagrams

Because of the similarity between the block diagram and the SFG, the gain formula in Eq. (3-31) can be applied to determine the input-output gain of either. In general, the gain formula can be applied to a block diagram directly. However, in complex systems, to be able to identify all the loops and nontouching parts clearly, it may be helpful if an equivalent SFG is drawn for the block diagram first before applying the gain formula.

▶ **EXAMPLE 3-7** To illustrate how an equivalent SFG of a block diagram is constructed, and how the gain formula is applied to a block diagram, consider the block diagram shown in Fig. 3-16(a). The equivalent SFG of the system is shown in Fig. 3-16(b). Notice that since a node on the SFG is interpreted as the summing point of all incoming signals to the node, the negative feedbacks on the block diagram are represented by assigning negative gains to the feedback paths on the SFG.

The closed-loop transfer function of the system is obtained by applying Eq. (3-31) to either the block diagram or the SFG in Fig. 3-16:

$$\frac{Y(s)}{R(s)} = \frac{G_1G_2G_3 + G_1G_4}{\Delta}$$

(3-50)

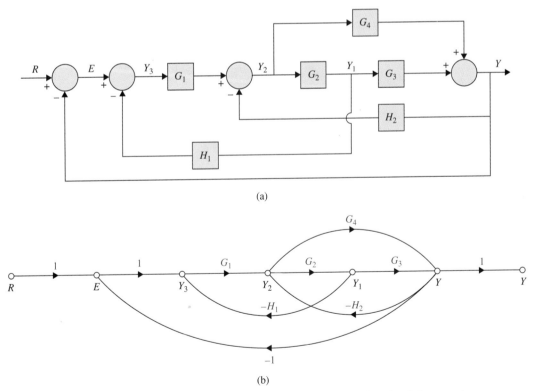

(a)

(b)

Figure 3-16 (a) Block diagram of a control system. (b) Equivalent signal-flow graph.

where

$$\Delta = 1 + G_1G_2H_1 + G_2G_3H_2 + G_1G_2G_3 + G_4H_2 + G_1G_4 \tag{3-51}$$

Similarly,

$$\frac{E(s)}{R(s)} = \frac{1 + G_1G_2H_1 + G_2G_3H_2 + G_4H_2}{\Delta} \tag{3-52}$$

$$\frac{Y(s)}{E(s)} = \frac{G_1G_2G_3 + G_1G_4}{1 + G_1G_2H_1 + G_2G_3H_2 + G_4H_2} \tag{3-53}$$

The last expression is obtained using Eq. (3-48). ◄

► 3-3 STATE DIAGRAM

In this section, we introduce the **state diagram**, which is an extension of the SFG to portray state equations and differential equations. The significance of the state diagram is that it forms a close relationship among the state equations (Appendix B and Chapter 5), computer simulation, and transfer functions. A state diagram is constructed following all the rules of the SFG using the Laplace-transformed state equations.

The basic elements of a state diagram are similar to the conventional SFG, except for the **integration** operation. Let the variables $x_1(t)$ and $x_2(t)$ be related by the first-order differentiation:

$$\frac{dx_1(t)}{dt} = x_2(t) \tag{3-54}$$

Integrating both sides of the last equation with respect to t from the initial time t_0, we get

$$x_1(t) = \int_{t_0}^{t} x_2(\tau)d\tau + x_1(t_0) \tag{3-55}$$

Since the SFG algebra does not handle integration in the time domain, we must take the Laplace transform on both sides of Eq. (3-54). We have

$$\begin{aligned}
X_1(s) &= \mathcal{L}\left[\int_{t_0}^{t} x_2(\tau)d\tau\right] + \frac{x_1(t_0)}{s} \\
&= \mathcal{L}\left[\int_{0}^{t} x_2(\tau)d\tau - \int_{0}^{t_0} x_2(\tau)d\tau\right] + \frac{x_1(t_0)}{s} \\
&= \frac{X_2(s)}{s} - \mathcal{L}\left[\int_{0}^{t_0} x_2(\tau)d\tau\right] + \frac{x_1(t_0)}{s}
\end{aligned} \tag{3-56}$$

Since the past history of the integrator is represented by $x_1(t_0)$, and the state transition is assumed to start at $\tau = t_0$, $x_2(\tau) = 0$ for $0 < \tau < t_0$. Thus, Eq. (3-56) becomes

$$X_1(s) = \frac{X_2(s)}{s} + \frac{x_1(t_0)}{s} \qquad \tau \geq t_0 \tag{3-57}$$

Equation (3-57) is now algebraic and can be represented by a SFG, as shown in Fig. 3-17. An alternative SFG with fewer elements for Eq. (3-57) is shown in Fig. 3-18. Figure 3-18 shows that *the output of the integrator is equal to s^{-1} times the input, plus the initial condition $x_1(t_0)/s$.*

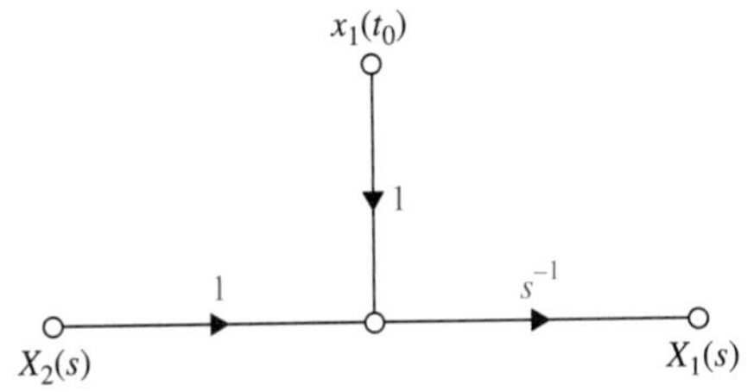

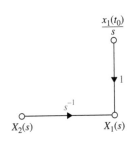

Figure 3-17 Signal-flow graph representation of $X_1(s) = [X_2(s)/s] + [x_1(t_0)/s]$.

Before embarking on several illustrative examples on the construction of state diagrams, let us point out the important uses of the state diagram.

1. A state diagram can be constructed directly from the system's differential equation. This allows the determination of the state variables and the state equations.

2. A state diagram can be constructed from the system's transfer function. This step is defined as the **decomposition** of transfer functions (Section 4-11).

3. The state diagram can be used to program the system on an analog computer or for simulation on a digital computer.

4. The state-transition equation in the Laplace transform domain may be obtained from the state diagram by using the SFG gain formula.

5. The transfer functions of a system can be determined from the state diagram.

6. The state equations and the output equations can be determined from the state diagram.

The details of these techniques will follow.

3-3-1 From Differential Equations to State Diagram

When a linear system is described by a high-order differential equation, a state diagram can be constructed from these equations, although a direct approach is not always the most convenient. Consider the following differential equation:

$$\frac{d^n y(t)}{dt^n} + a_n \frac{d^{n-1} y(t)}{dt^{n-1}} + \cdots + a_2 \frac{dy(t)}{dt} + a_1 y(t) = r(t) \tag{3-58}$$

• Notes:
The outputs of the integrators on the state diagram are usually defined as the state variables.

To construct a state diagram using this equation, we rearrange the equation as

$$\frac{d^n y(t)}{dt^n} = -a_n \frac{d^{n-1} y(t)}{dt^{n-1}} - \cdots - a_2 \frac{dy(t)}{dt} - a_1 y(t) + r(t) \tag{3-59}$$

Figure 3-18 Signal-flow graph representation of $X_1(s) = [X_2(s)/s] + [x_1(t_0)/s]$.

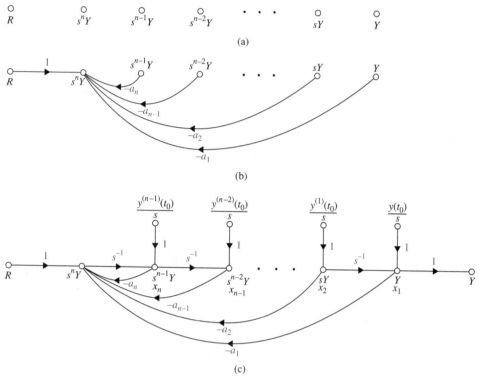

Figure 3-19 State-diagram representation of the differential equation of Eq. (3-58).

As a first step, the nodes representing $R(s)$, $s^n Y(s)$, $s^{n-1} Y(s)$, ... , $sY(s)$, and $Y(s)$ are arranged from left to right, as shown in Fig. 3-19(a). Since $s^i Y(s)$ corresponds to $d^i y(t)/dt^i$, $i = 0$, 1, 2, ... , n, in the Laplace domain, as the next step, the nodes in Fig. 3-19(a) are connected by branches to portray Eq. (3-59), resulting in Fig. 3-19(b). Finally, the integrator branches with gains of s^{-1} are inserted, and the initial conditions are added to the outputs of the integrators, according to the basic scheme in Fig. 3-19. The complete state diagram is drawn as shown in Fig. 3-19(c). *The outputs of the integrators are defined as the state variables, x_1, x_2, ... , x_n.* This is usually the natural choice of state variables once the state diagram is drawn.

When the differential equation has derivatives of the input on the right side, the problem of drawing the state diagram directly is not as straightforward as just illustrated. We will show that, in general, it is more convenient to obtain the transfer function from the differential equation first and then arrive at the state diagram through decomposition (Section 5-9).

▶ **EXAMPLE 3-8** Consider the differential equation

$$\frac{d^2 y(t)}{dt^2} + 3\frac{dy(t)}{dt} + 2y(t) = r(t) \tag{3-60}$$

Equating the highest-ordered term of the last equation to the rest of the terms, we have

$$\frac{d^2 y(t)}{dt^2} = -3\frac{dy(t)}{dt} - 2y(t) + r(t) \tag{3-61}$$

Following the procedure just outlined, the state diagram of the system is drawn as shown in Fig. 3-20. The state variables x_1 and x_2 are assigned as shown. ◀

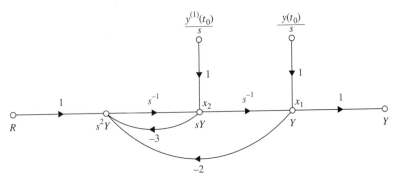

Figure 3-20 State diagram for Eq. (3-60).

3-3-2 From State Diagram to Transfer Function

The transfer function between an input and an output is obtained from the state diagram by using the gain formula and setting all other inputs and initial states to zero. The following example shows how the transfer function is obtained directly from a state diagram.

▶ **EXAMPLE 3-9** Consider the state diagram of Fig. 3-20. The transfer function between $R(s)$ and $Y(s)$ is obtained by applying the gain formula between these two nodes and setting the initial states to zero. We have

$$\frac{Y(s)}{R(s)} = \frac{1}{s^2 + 3s + 2} \tag{3-62}$$

◀

3-3-3 From State Diagram to State and Output Equations

The state equations and the output equations can be obtained directly from the state diagram by using the SFG gain formula. The general form of a state equation and the output equation for a linear system is described in Appendix B, and presented here.

State equation:

$$\frac{dx(t)}{dt} = ax(t) + br(t) \tag{3-63}$$

Output equation:

$$y(t) = cx(t) + dr(t) \tag{3-64}$$

where $x(t)$ is the state variable; $r(t)$ is the input; $y(t)$ the output; and a, b, c, and d are constant coefficients. Based on the general form of the state and output equations, the following procedure of deriving the state and output equations from the state diagram is outlined:

1. Delete the initial states and the integrator branches with gains s^{-1} from the state diagram, since the state and output equations do not contain the Laplace operator s or the initial states.

2. For the state equations, regard the nodes that represent the derivatives of the state variables as output nodes, since these variables appear on the left-hand side of the state equations. The output $y(t)$ in the output equation is naturally an output node variable.

3. Regard the state variables and the inputs as input variables on the state diagram, since these variables are found on the right-hand side of the state and output equations.

4. Apply the SFG gain formula to the state diagram.

▶ **EXAMPLE 3-10** Figure 3-21 shows the state diagram of Fig. 3-20 with the integrator branches and the initial states eliminated. Using $dx_1(t)/dt$ and $dx_2(t)/dt$ as the output nodes and $x_1(t)$, $x_2(t)$, and $r(t)$ as input nodes, and applying the gain formula between these nodes, the state equations are obtained as

$$\frac{dx_1(t)}{dt} = x_2(t)$$

$$\frac{dx_2(t)}{dt} = -2x_1(t) - 3x_2(t) + r(t) \tag{3-65}$$

Figure 3-21 State diagram of Fig. 3-20 with the initial states and the integrator branches left out.

Applying the gain formula with $x_1(t)$, $x_2(t)$, and $r(t)$ as input nodes and $y(t)$ as the output node, the output equation is written

$$y(t) = x_1(t) \tag{3-66}$$

◀

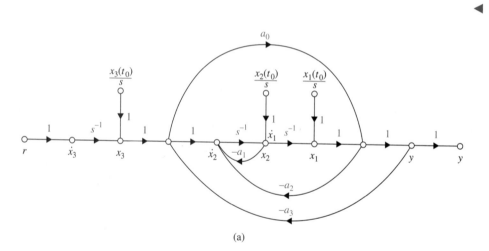

(a)

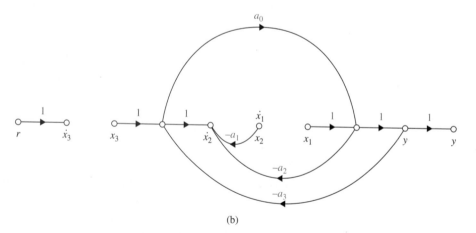

Figure 3-22 (a) State diagram. (b) State diagram in part (a) with all initial states and integrators left out.

(b)

► **EXAMPLE 3-11** As another example on the determination of the state equations from the state diagram, consider the state diagram shown in Fig. 3-22(a). This example will also emphasize the importance of applying the gain formula. Figure 3-22(b) shows the state diagram with the initial states and the integrator branches deleted. Notice that in this case, the state diagram in Fig. 3-22(b) still contains a loop. By applying the gain formula to the state diagram in Fig. 3-22(b) with $\dot{x}_1(t)$, $\dot{x}_2(t)$, and $\dot{x}_3(t)$ as output-node variables and $r(t)$, $x_1(t)$, $x_2(t)$, and $x_3(t)$ as input nodes, the state equations are obtained as follows in vector-matrix form:

$$\begin{bmatrix} \dfrac{dx_1(t)}{dt} \\[2mm] \dfrac{dx_2(t)}{dt} \\[2mm] \dfrac{dx_3(t)}{dt} \end{bmatrix} = \begin{bmatrix} 0 & 1 & 0 \\[2mm] \dfrac{-(a_2 + a_3)}{1 + a_0 a_3} & -a_1 & \dfrac{1 - a_0 a_2}{1 + a_0 a_3} \\[2mm] 0 & 0 & 0 \end{bmatrix} \begin{bmatrix} x_1(t) \\ x_2(t) \\ x_3(t) \end{bmatrix} + \begin{bmatrix} 0 \\ 0 \\ 1 \end{bmatrix} r(t) \tag{3-67}$$

The output equation is

$$y(t) = \frac{1}{1 + a_0 a_3} x_1(t) + \frac{a_0}{1 + a_0 a_3} x_3(t) \tag{3-68}$$

◄

► 3-4 MATLAB TOOLS AND CASE STUDIES

There is no specific software developed for this chapter. Although the MATLAB Controls Toolbox offers functions for finding the transfer functions from a given block diagram, students should master this subject without referring to a computer. For simple operations, however, the Transfer Function Calculator tool may be used by clicking the appropriate button in the **ACSYS** window or by typing **TFcal** in the MATLAB command window. Figure 3-23 shows the Transfer Function Calculator window. Press the "Help for 1st Time User" for instructions. The resulting MATLAB help dialog box is shown in Fig. 3-24.

As suggested in the help dialog box, press any desired button, and move to the MATLAB command window to continue with the rest of operation. The first action button is the "Transfer Function Calculator," which is used to find the transfer-function poles and

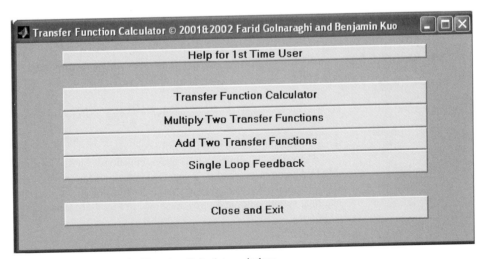

Figure 3-23 The Transfer Function Calculator window.

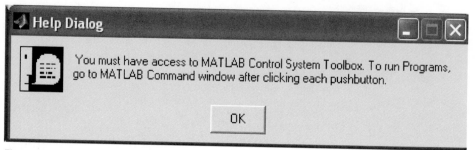

Figure 3-24 The MATLAB help dialog box for the TFcal tool.

zeros, and converts the system from transfer-function form to the state-space form (see Chapter 5 for more details). You may use the other buttons to calculate the transfer functions of simple systems as described in Fig. 3-25.

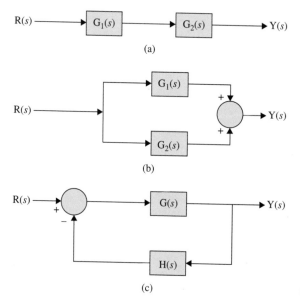

Figure 3-25 Basic block diagrams used in the TFcal tool.

▶ **EXAMPLE 3-12** Consider the following transfer functions, which correspond to the block diagrams shown in Fig. 3-25.

$$G_1(s) = \frac{1}{s + 1}, \quad G_2(s) = \frac{1}{s + 2}, \quad G(s) = \frac{1}{s(s + 1)}, \quad H(s) = 10 \tag{3-69}$$

Find the transfer function $Y(s)/R(s)$ for each case.

Press the "Multiply Two Transfer Functions" button for case (a), shown in Fig. 3-25. In MATLAB command window follow the instructions for entering $G_1(s)$ and $G_2(s)$ transfer functions. The program will generate the final transfer function in polynomial and factored forms along with the final system poles and zeros, as shown in Fig. 3-26. Hence,

$$\frac{Y(s)}{R(s)} = \frac{1}{s^2 + 3s + 2} = \frac{1}{(s + 1)(s + 2)} \tag{3-70}$$

Transfer Function Calculator. © Kuo & Golnaraghi 8th Edition, John Wiley & Sons.
e.g., Use the following input format: (s+2)*(s^3+2*s+1)/(s*(s^2+2*s+1))

Enter G1 and G2 to find G1*G2

Enter G1 = 1/(s+1)
Transfer function:
  1
----
s + 1

Enter G2 = 1/(s+2)
Transfer function:
  1
----
s + 2

G=G1*G2 is:
Transfer function:
       1
----
s^2 + 3 s + 2

polesG =
-2
-1

zerosG =
  Empty matrix: 0-by-1

G factored:

Zero/pole/gain:
      1

(s+2) (s+1)

Figure 3-26 The TFcal results for case (a) in Fig. 3-25, corresponding to Example 3-12.

Using the "Add Two Transfer Functions" and "Single Loop Feedback" options in TFcal, you can solve for cases (b) and (c) in Fig. 3-25, respectively. The results are as follows.

Case (b)
$$\frac{Y(s)}{R(s)} = \frac{2s + 3}{s^2 + 3s + 2} = \frac{2(s + 1.5)}{(s + 1)(s + 2)}$$
(3-71)

Case (c)
$$\frac{Y(s)}{R(s)} = \frac{1}{s^2 + s + 10}$$
(3-72)

◄

► 3-5 SUMMARY

This chapter was devoted to the mathematical modeling of physical systems. Transfer functions, block diagrams, and signal-flow graphs were defined. The transfer function of a linear system was defined in terms of impulse response as well as differential equations. Multivariable and single-variable systems were examined.

The block diagram representation was shown to be a versatile method of portraying linear and nonlinear systems. A powerful method of representing the interrelationships between the signals of a linear system is the signal-flow graph, or SFG. When applied properly, an SFG allows the derivation of the transfer functions between input and output variables of a linear system using the gain formula. A state diagram is an SFG that is applied to dynamic systems that are represented by differential equations.

At the end of the chapter, a brief discussion of the **ACSYS, TFcal** tool was presented, which allows the user to calculate transfer functions of simple systems.

► REVIEW
QUESTIONS

1. Define the transfer function of a linear time-invariant system in terms of its impulse response.

2. When defining the transfer function, what happens to the initial conditions of the system?

3. Define the characteristic equation of a linear system in terms of the transfer function.

4. What is referred to as a multivariable system?

5. Can signal-flow graphs (SFGs) be applied to nonlinear systems?

6. How can signal-flow graphs be applied to systems that are described by differential equations?

7. Define the input node of an SFG.

8. Define the output node of an SFG.

9. State the form to which the equations must first be conditioned before drawing the SFG.

10. What does the arrow on the branch of an SFG represent?

11. Explain how a noninput node of an SFG can be made into an output node.

12. Can the gain formula be applied between any two nodes of an SFG?

13. Explain what the nontouching loops of an SFG are.

14. Does the Δ of an SFG depend on which pair of input and output is selected?

15. List the advantages and utilities of the state diagram.

16. Given the state diagram of a linear dynamic system, how do you define the state variables?

17. Given the state diagram of a linear dynamic system, how do you find the transfer function between a pair of input and output variables?

18. Given the state diagram of a linear dynamic system, how do you write the state equations of the system?

19. The state variables of a dynamic system are not equal to the number of energy-storage elements under what condition?

► REFERENCES

Block Diagrams and Signal Flow Graphs

1. T. D. Graybeal, "Block Diagram Network Transformation," *Elec. Eng.*, Vol. 70, 985–990, 1951.
2. S. J. Mason, "Feedback Theory—Some Properties of Signal Flow Graphs," *Proc. IRE*, Vol. 41, No. 9, 1144–1156, Sept. 1953.
3. S. J. Mason, "Feedback Theory—Further Properties of Signal Flow Graphs," *Proc. IRE*, Vol. 44, No. 7, 920–926, July 1956.
4. L. P. A. Robichaud, M. Boisvert, and J. Robert, *Signal Flow Graphs and Applications*, Prentice Hall, Englewood Cliffs, NJ, 1962.
5. B. C. Kuo, *Linear Networks and Systems*, McGraw-Hill Book Company, New York, 1967.

State-Variable Analysis of Electric Networks

6. B. C. Kuo, *Linear Circuits and Systems*, McGraw-Hill Book Company, New York, 1967.

► **PROBLEMS**

• Transfer functions, final value

3-1. The block diagram of an electric train control is shown in Fig. 3P-1. The system parameters and variables are

$e_r(t)$ = voltage representing the desired train speed, V

$v(t)$ = speed of train, ft/sec

M = Mass of train = 30,000 lb/sec^2

K = amplifier gain

K_t = gain of speed indicator = 0.15 V/ft/sec

To determine the transfer function of the controller, we apply a step function of 1 volt to the input of the controller, that is, $e_c(t) = u_s(t)$. The output of the controller is measured and described by the following equation:

$$f(t) = 100(1 - 0.3e^{-6t} - 0.7e^{-10t})u_s(t)$$

(a) Find the transfer function $G_c(s)$ of the controller.

(b) Derive the forward-path transfer function $V(s)/E(s)$ of the system. The feedback path is opened in this case.

(c) Derive the closed-loop transfer function $V(s)/E_r(s)$ of the system.

(d) Assuming that K is set at a value so that the train will not run away (unstable), find the steady-state speed of the train in feet per second when the input is $e_r(t) = u_s(t)$ V.

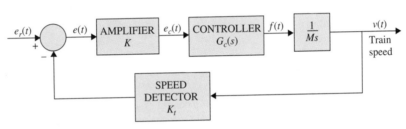

Figure 3P-1

• Transfer functions, final value

3-2. Repeat Problem 3-1 when the output of the controller is measured and described by the following expression:

$$f(t) = 100(1 - 0.3e^{-6(t-0.5)})u_s(t - 0.5)$$

when a step input of 1 V is applied to the controller.

• Multivariable system

3-3. A linear time-invariant multivariable system with inputs $r_1(t)$ and $r_2(t)$ and outputs $y_1(t)$ and $y_2(t)$ is described by the following set of differential equations.

$$\frac{d^2y_1(t)}{dt^2} + 2\frac{dy_1(t)}{dt} + 3y_2(t) = r_1(t) + r_2(t)$$

$$\frac{d^2y_2(t)}{dt^2} + 3\frac{dy_1(t)}{dt} + y_1(t) - y_2(t) = r_2(t) + \frac{dr_1(t)}{dt}$$

Find the following transfer functions:

$$\left.\frac{Y_1(s)}{R_1(s)}\right|_{R_2=0} \qquad \left.\frac{Y_2(s)}{R_1(s)}\right|_{R_2=0} \qquad \left.\frac{Y_1(s)}{R_2(s)}\right|_{R_1=0} \qquad \left.\frac{Y_2(s)}{R_2(s)}\right|_{R_1=0}$$

• Multivariable system

3-4. The aircraft turboprop engine shown in Fig. 2-2 is controlled by a closed-loop system with block diagram shown in Fig. 3P-4. The engine is modeled as a multivariable system with input vector $\mathbf{E}(s)$, which contains the fuel rate and propeller blade angle, and output vector $\mathbf{Y}(s)$, consisting of the engine speed and turbine-inlet temperature. The transfer function matrices are given as

$$\mathbf{G}(s) = \begin{bmatrix} \dfrac{2}{s(s+2)} & 10 \\ \dfrac{5}{s} & \dfrac{1}{s+1} \end{bmatrix} \qquad \mathbf{H}(s) = \begin{bmatrix} 1 & 0 \\ 0 & 1 \end{bmatrix}$$

Find the closed-loop transfer function matrix $[\mathbf{I} + \mathbf{G}(s)\mathbf{H}(s)]^{-1}\mathbf{G}(s)$.

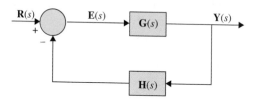

Figure 3P-4

• Signal-flow graphs

3-5. Draw signal-flow graphs for the following sets of algebraic equations. These equations should first be arranged in the form of cause-and-effect relations before SFGs can be drawn. Show that there are many possible SFGs for each set of equations.

(a)

$$x_1 = \qquad -x_2 - 3x_3 + 3$$
$$x_2 = 5x_1 - 2x_2 + x_3$$
$$x_3 = 4x_1 + x_2 - 5x_3 + 5$$

(b)

$$2x_1 + 3x_2 + x_3 = -1$$
$$x_1 - 2x_2 - x_3 = 1$$
$$3x_2 + x_3 = 0$$

• Equivalent SFG of block diagrams

3-6. The block diagram of a control system is shown in Fig. 3P-6. Draw an equivalent SFG for the system. Find the following transfer functions by applying the gain formula of the SFG directly to the block diagram. Compare the answers by applying the gain formula to the equivalent SFG.

$$\left.\frac{Y(s)}{R(s)}\right|_{N=0} \qquad \left.\frac{Y(s)}{N(s)}\right|_{R=0} \qquad \left.\frac{E(s)}{R(s)}\right|_{N=0} \qquad \left.\frac{E(s)}{N(s)}\right|_{R=0}$$

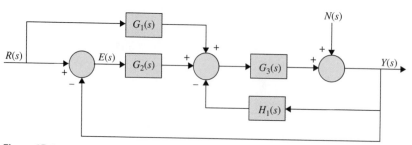

Figure 3P-6

• Transfer functions of SFG

3-7. Apply the gain formula to the SFG's shown in Fig. 3P-7 to find the following transfer functions:

$$\frac{Y_5}{Y_1} \qquad \frac{Y_4}{Y_1} \qquad \frac{Y_2}{Y_1} \qquad \frac{Y_5}{Y_2}$$

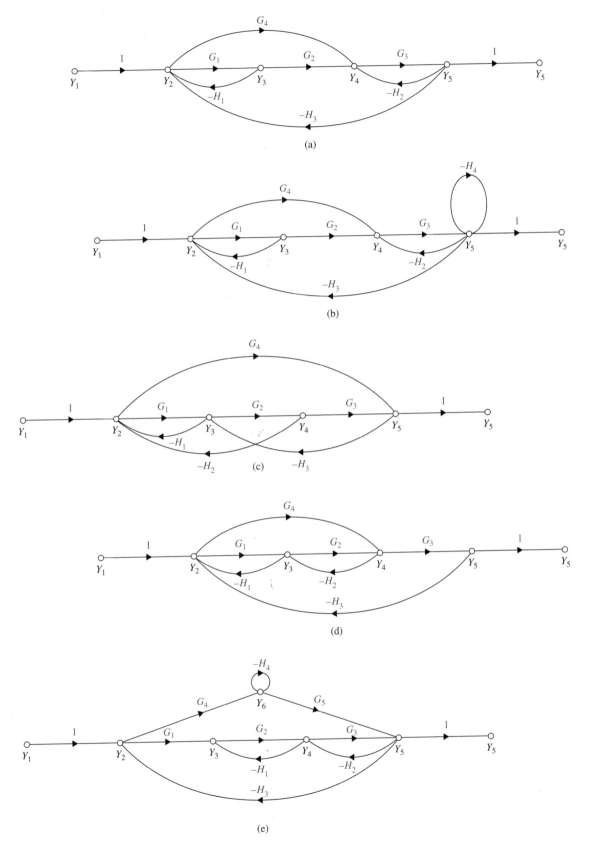

Figure 3P-7

• Transfer functions of SFG   **3-8.**   Find the transfer functions Y_7/Y_1 and Y_2/Y_1 of the SFGs shown in Fig. 3P-8.

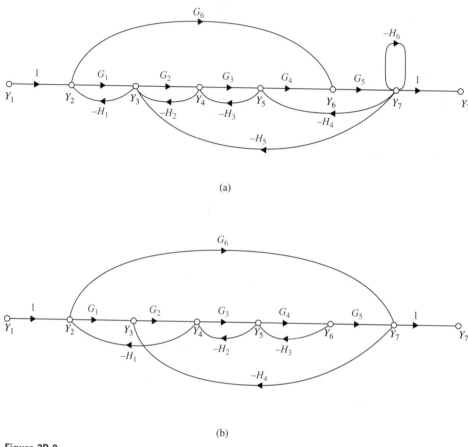

(a)

(b)

Figure 3P-8

• Application of SFG   **3-9.**   Signal-flow graphs may be used to solve a variety of electric network problems. Shown in Fig. 3P-9 is the equivalent circuit of an electronic circuit. The voltage source $e_d(t)$ represents a

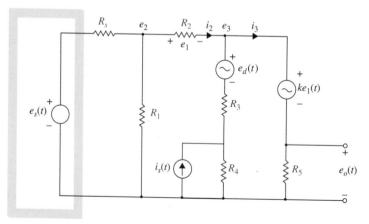

Figure 3P-9

disturbance voltage. The objective is to find the value of the constant k so that the output voltage $e_o(t)$ is not affected by $e_d(t)$. To solve the problem, it is best to first write a set of cause-and-effect equations for the network. This involves a combination of node and loop equations. Then construct an SFG using these equations. Find the gain e_o/e_d with all other inputs set to zero. For e_d not to affect e_o, set e_o/e_d to zero.

• SFG

3-10. Show that the two systems shown in Figs. 3P-10(a) and (b) are equivalent.

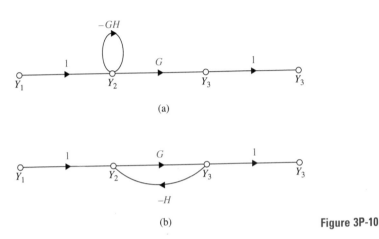

Figure 3P-10

• SFG

3-11. Show that the two systems shown in Figs. 3P-11(a) and (b) are not equivalent.

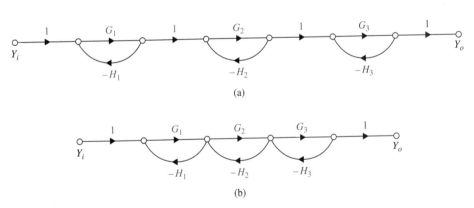

Figure 3P-11

• Transfer functions of SFG

3-12. Find the following transfer functions for the SFG shown in Fig. 3P-12.

$$\left.\frac{Y_6}{Y_1}\right|_{Y_7=0} \qquad \left.\frac{Y_6}{Y_7}\right|_{Y_1=0}$$

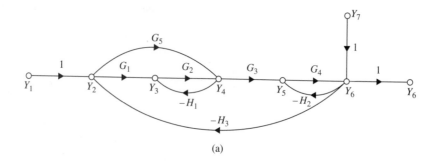

(a)

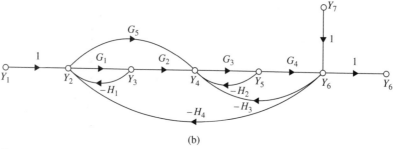

(b)

Figure 3P-12

• Transfer functions of SFG

3-13. Find the following transfer functions for the SFG shown in Fig. 3P-13. Comment on why the results for parts (c) and (d) are not the same.

(a) $\dfrac{Y_7}{Y_1}\bigg|_{Y_8=0}$ (b) $\dfrac{Y_7}{Y_8}\bigg|_{Y_1=0}$ (c) $\dfrac{Y_7}{Y_4}\bigg|_{Y_8=0}$ (d) $\dfrac{Y_7}{Y_4}\bigg|_{Y_1=0}$

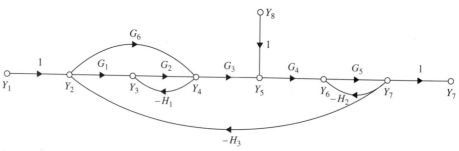

Figure 3P-13

• Transfer functions of block diagrams

3-14. The block diagram of a feedback control system is shown in Fig. 3P-14. Find the following transfer functions:

(a) $\dfrac{Y(s)}{R(s)}\bigg|_{N=0}$ (b) $\dfrac{Y(s)}{E(s)}\bigg|_{N=0}$ (c) $\dfrac{Y(s)}{N(s)}\bigg|_{R=0}$

(d) Find the output $Y(s)$ when $R(s)$ and $N(s)$ are applied simultaneously.

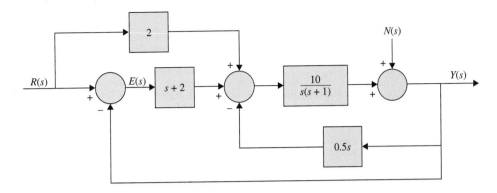

Figure 3P-14

• Transfer functions of block diagrams

3-15. The block diagram of a feedback control system is shown in Fig. 3P-15.

(a) Apply the SFG gain formula directly to the block diagram to find the transfer functions

$$\left.\frac{Y(s)}{R(s)}\right|_{N=0} \qquad \left.\frac{Y(s)}{N(s)}\right|_{R=0}$$

Express $Y(s)$ in terms of $R(s)$ and $N(s)$ when both inputs are applied simultaneously.

(b) Find the desired relation among the transfer functions $G_1(s)$, $G_2(s)$, $G_3(s)$, $G_4(s)$, $H_1(s)$, and $H_2(s)$ so that the output $Y(s)$ is not affected by the disturbance signal $N(s)$ at all.

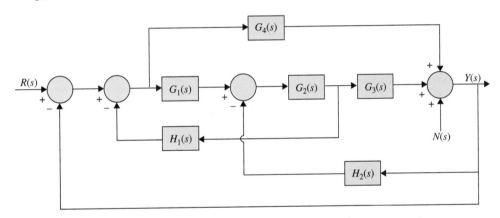

Figure 3P-15

Antenna control system

3-16. Figure 3P-16 shows the block diagram of the antenna control system of the solar-collector field shown in Fig. 1-7. The signal $N(s)$ denotes the wind gust disturbance acted on the antenna. The feedforward transfer function $G_d(s)$ is used to eliminate the effect of $N(s)$ on the output $Y(s)$. Find the transfer function $Y(s)/N(s)|_{R=0}$. Determine the expression of $G_d(s)$ so that the effect of $N(s)$ is entirely eliminated.

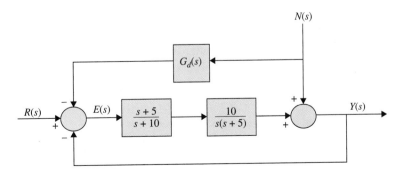

Figure 3P-16

• Transfer function of
dc-motor control system

3-17. Figure 3P-17 shows the block diagram of a dc-motor control system. The signal $N(s)$ denotes the frictional torque at the motor shaft.

(a) Find the transfer function $H(s)$ so that the output $Y(s)$ is not affected by the disturbance torque $N(s)$.

(b) With $H(s)$ as determined in part (a), find the value of K so that the steady-state value of $e(t)$ is equal to 0.1 when the input is a unit-ramp function, $r(t) = tu_s(t)$, $R(s) = 1/s^2$, and $N(s) = 0$. Apply the final-value theorem.

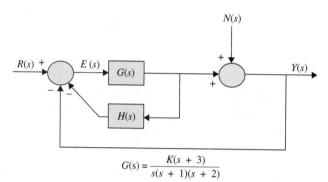

$$G(s) = \frac{K(s+3)}{s(s+1)(s+2)}$$

Figure 3P-17

• Transfer function of
position-control system

3-18. The block diagram of the position-control system of the electronic word processor is shown in Fig. 3P-18.

(a) Find the loop transfer function $\Theta_o(s)/\Theta_e(s)$ (the outer feedback path is open).

(b) Find the closed-loop transfer function $\Theta_o(s)/\Theta_r(s)$.

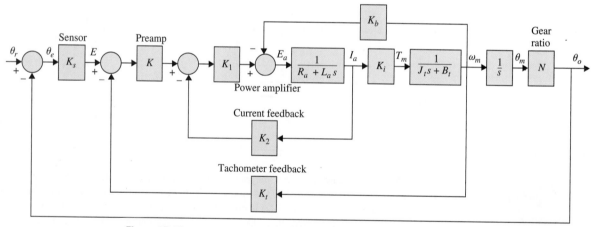

Figure 3P-18

• Turboprop engine

3-19. The coupling between the signals of the turboprop engine shown in Fig. 2-2 is shown in Fig. 3P-19. The signals are defined as

$$R_1(s) = \text{fuel rate} \qquad R_2(s) = \text{propeller blade angle}$$
$$Y_1(s) = \text{engine speed} \qquad Y_2(s) = \text{turbine inlet temperature}$$

(a) Draw an equivalent SFG for the system.

(b) Find the Δ of the system using the SFG gain formula.

(c) Find the following transfer functions:

$$\left.\frac{Y_1(s)}{R_1(s)}\right|_{R_2=0} \qquad \left.\frac{Y_1(s)}{R_2(s)}\right|_{R_1=0} \qquad \left.\frac{Y_2(s)}{R_1(s)}\right|_{R_2=0} \qquad \left.\frac{Y_2(s)}{R_2(s)}\right|_{R_1=0}$$

(d) Express the transfer functions in matrix form, $\mathbf{Y}(s) = \mathbf{G}(s)\mathbf{R}(s)$.

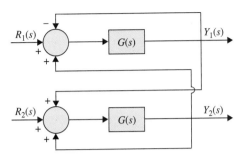

Figure 3P-19

• Transfer function of system with conditional feedback

3-20. Figure 3P-20 shows the block diagram of a control system with conditional feedback. The transfer function, $G_p(s)$, denotes the controlled process, and $G_c(s)$ and $H(s)$ are the controller transfer functions.

(a) Derive the transfer functions $Y(s)/R(s)|_{N=0}$ and $Y(s)/N(s)|_{R=0}$. Find $Y(s)/R(s)|_{N=0}$ when $G_c(s) = G_p(s)$.

(b) Let

$$G_p(s) = G_c(s) = \frac{100}{(s + 1)(s + 5)}$$

Find the output response $y(t)$ when $N(s) = 0$ and $r(t) = u_s(t)$.

(c) With $G_p(s)$ and $G_c(s)$ as given in part (b), select $H(s)$ among the following choices such that when $n(t) = u_s(t)$ and $r(t) = 0$, the steady-state value of $y(t)$ is equal to zero. (There may be more than one answer.)

$$H(s) = \frac{10}{s(s + 1)} \qquad H(s) = \frac{10}{(s + 1)(s + 2)}$$

$$H(s) = \frac{10(s + 1)}{s + 2} \qquad H(s) = \frac{K}{s^n} \quad (n = \text{positive integer}) \text{ Select } n.$$

Keep in mind that the poles of the closed-loop transfer function must all be in the left-half s-plane in order for the final-value theorem to be valid.

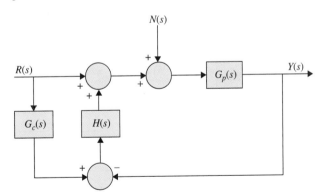

Figure 3P-20

• State diagram, characteristic equation, transfer function

3-21. **(a)** Draw a state diagram for the following state equations.

$$\frac{dx_1(t)}{dt} = -2x_1(t) + 3x_2(t)$$

$$\frac{dx_2(t)}{dt} = -5x_1(t) - 5x_2(t) + 2r(t)$$

(b) Find the characteristic equation of the system.

(c) Find the transfer functions $X_1(s)/R(s)$ and $X_2(s)/R(s)$.

• State diagram, state equations, characteristic equation, transfer function, final value

3-22. The differential equation of a linear system is

$$\frac{d^3y(t)}{dt^3} + 5\frac{d^2y(t)}{dt^2} + 6\frac{dy(t)}{dt} + 10y(t) = r(t)$$

where $y(t)$ is the output, and $r(t)$ is the input.

(a) Draw a state diagram for the system.

(b) Write the state equation from the state diagram. Define the state variables from right to left in ascending order.

(c) Find the characteristic equation and its roots. Use any computer program to find the roots.

(d) Find the transfer function $Y(s)/R(s)$.

(e) Perform a partial-fraction expansion of $Y(s)/R(s)$ and find the output $y(t)$ for $t \geq 0$ when $r(t) = u_s(t)$. Find the final value of $y(t)$ by using the final-value theorem.

• State diagram, state equations, characteristic equation, transfer function, final value

3-23. Repeat Problem 3-22 for the following differential equation.

$$\frac{d^4y(t)}{dt^4} + 4\frac{d^3y(t)}{dt^3} + 3\frac{d^2y(t)}{dt^2} + 5\frac{dy(t)}{dt} + y(t) = r(t)$$

3-24. The block diagram of a feedback control system is shown in Fig. 3P-24.

• Transfer function, characteristic equation, steady-state response

(a) Derive the following transfer functions:

$$\left.\frac{Y(s)}{R(s)}\right|_{N=0} \qquad \left.\frac{Y(s)}{N(s)}\right|_{R=0} \qquad \left.\frac{E(s)}{R(s)}\right|_{N=0}$$

(b) The controller with the transfer function $G_4(s)$ is for the reduction of the effect of the noise $N(s)$. Find $G_4(s)$ so that the output $Y(s)$ is totally independent of $N(s)$.

(c) Find the characteristic equation and its roots when $G_4(s)$ is as determined in part (b).

(d) Find the steady-state value of $e(t)$ when the input is a unit-step function. Set $N(s) = 0$.

(e) Find $y(t)$ for $t \geq 0$ when the input is a unit-step function. Use $G_4(s)$ as determined in part (b).

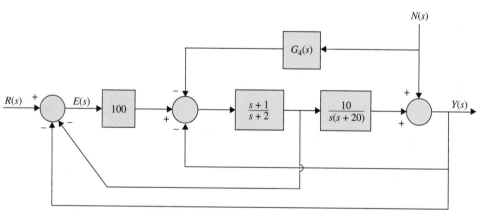

Figure 3P-24

CHAPTER 4

Modeling of Physical Systems

▶ 4-1 INTRODUCTION

One of the most important tasks in the analysis and design of control systems is mathematical modeling of the systems. The two most common methods of modeling linear systems are the transfer function method and the state-variable method. The transfer function is valid only for linear time-invariant systems, whereas the state equations can be applied to linear as well as nonlinear systems.

Although the analysis and design of linear control systems have been well developed, their counterparts for nonlinear systems are usually quite complex. Therefore, the control-systems engineer often has the task of determining not only how to accurately describe a system mathematically, but, more importantly, how to make proper assumptions and approximations, whenever necessary, so that the system may be realistically characterized by a linear mathematical model. It is not difficult to understand that the analytical and computer simulation of any system are only as good as the model used to describe it. It should also be emphasized that the modern control engineer should place special emphasis on the mathematical modeling of systems so that analysis and design problems can be conveniently solved by computers. Therefore, the main objectives of these following sections are

1. To demonstrate mathematical modeling of control systems and components.
2. To demonstrate how modeling will lead to computer solutions.

This is an introduction to the method of modeling. Since numerous types of control-system components are available, the coverage here is by no means exhaustive.

▶ 4-2 MODELING OF ELECTRICAL NETWORKS

The classical way of writing equations of electric networks is based on the loop method or the node method, both of which are formulated from the two laws of Kirchhoff. A modern method of writing network equations is the state-variable method, which is covered in Appendix B and Chapter 5. Since the electrical networks found in most control systems are rather simple, the subject is treated here only at the introductory level. The state-variable modeling of electric networks can be illustrated by the following two examples.

▶ **EXAMPLE 4-1** Let us consider the RLC network shown in Fig. 4-1(a). A practical approach is to assign the current in the inductor L, $i(t)$, and the voltage across the capacitor C, $e_c(t)$, as the state variables. The reason for this choice is because the state variables are directly related to the energy-storage element of a system. The inductor stores kinetic energy, and the capacitor stores electric potential energy.

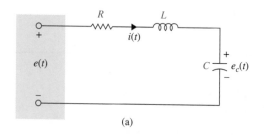

(a)

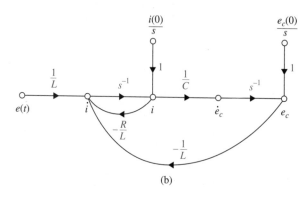

(b)

Figure 4-1 (a) RLC network.
(b) State diagram.

By assigning $i(t)$ and $e_c(t)$ as state variables, we have a complete description of the past history (via the initial states) and the present and future states of the network.

The state equations for the network in Fig. 4-1(a) are written by first equating the current in C and the voltage across L in terms of the state variables and the applied voltage $e(t)$. We have

Current in C:
$$C \frac{de_c(t)}{dt} = i(t) \tag{4-1}$$

Voltage in L:
$$L \frac{di(t)}{dt} = -e_c(t) - Ri(t) + e(t) \tag{4-2}$$

In vector-matrix form, the state equations are expressed as

$$\begin{bmatrix} \dfrac{de_c(t)}{dt} \\ \dfrac{di(t)}{dt} \end{bmatrix} = \begin{bmatrix} 0 & \dfrac{1}{C} \\ -\dfrac{1}{L} & -\dfrac{R}{L} \end{bmatrix} \begin{bmatrix} e_c(t) \\ i(t) \end{bmatrix} + \begin{bmatrix} 0 \\ \dfrac{1}{L} \end{bmatrix} e(t) \tag{4-3}$$

The state diagram of the network is shown in Fig. 4-1(b). Notice that the outputs of the integrators are defined as the state variables. The transfer functions of the system are obtained by applying the SFG gain formula to the state diagram when all the initial states are set to zero.

$$\frac{E_c(s)}{E(s)} = \frac{(1/LC)s^{-2}}{1 + (R/L)s^{-1} + (1/LC)s^{-2}} = \frac{1}{1 + RCs + LCs^2} \tag{4-4}$$

$$\frac{I(s)}{E(s)} = \frac{(1/L)s^{-1}}{1 + (R/L)s^{-1} + (1/LC)s^{-2}} = \frac{Cs}{1 + RCs + LCs^2} \tag{4-5}$$

◀

▶ **EXAMPLE 4-2** As another example of writing the state equations of an electric network, consider the network shown in Fig. 4-2(a). According to the foregoing discussion, the voltage across the capacitor, $e_c(t)$, and the currents of the inductors, $i_1(t)$ and $i_2(t)$, are assigned as state variables, as shown in Fig. 4-2(a). The state equations of the network are obtained by writing the voltages across the inductors and the

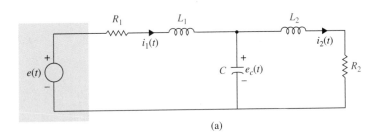

(a)

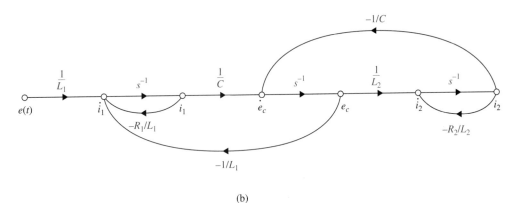

(b)

Figure 4-2 (a) Network of Example 4-2. (b) State diagram.

currents in the capacitor in terms of the three state variables. The state equations are

$$L_1\frac{di_1(t)}{dt} = -R_1 i_1(t) - e_c(t) + e(t) \tag{4-6}$$

$$L_2\frac{di_2(t)}{dt} = -R_2 i_2(t) + e_c(t) \tag{4-7}$$

$$C\frac{de_c(t)}{dt} = i_1(t) - i_2(t) \tag{4-8}$$

In vector-matrix form, the state equations are written as

$$\begin{bmatrix} \dfrac{di_1(t)}{dt} \\[2mm] \dfrac{di_2(t)}{dt} \\[2mm] \dfrac{de_c(t)}{dt} \end{bmatrix} = \begin{bmatrix} -\dfrac{R_1}{L_1} & 0 & -\dfrac{1}{L_1} \\[2mm] 0 & -\dfrac{R_2}{L_2} & \dfrac{1}{L_2} \\[2mm] \dfrac{1}{C} & -\dfrac{1}{C} & 0 \end{bmatrix} \begin{bmatrix} i_1(t) \\[2mm] i_2(t) \\[2mm] e_c(t) \end{bmatrix} + \begin{bmatrix} \dfrac{1}{L_1} \\[2mm] 0 \\[2mm] 0 \end{bmatrix} e(t) \tag{4-9}$$

The state diagram of the network, without the initial states, is shown in Fig. 4-2(b). The transfer functions between $I_1(s)$ and $E(s)$, $I_2(s)$ and $E(s)$, and $E_c(s)$ and $E(s)$, respectively, are written from the state diagram.

$$\frac{I_1(s)}{E(s)} = \frac{L_2 C s^2 + R_2 C s + 1}{\Delta} \tag{4-10}$$

$$\frac{I_2(s)}{E(s)} = \frac{1}{\Delta} \tag{4-11}$$

$$\frac{E_c(s)}{E(s)} = \frac{L_2 s + R_2}{\Delta} \tag{4-12}$$

where

$$\Delta = L_1 L_2 C s^3 + (R_1 L_2 + R_2 L_1) C s^2 + (L_1 + L_2 + R_1 R_2 C)s + R_1 + R_2 \qquad (4\text{-}13)$$

◀

▶ 4-3 MODELING OF MECHANICAL SYSTEMS ELEMENTS

Most control systems contain mechanical as well as electrical components; some systems even have hydraulic and pneumatic elements. From a mathematical viewpoint, the descriptions of electrical and mechanical elements are analogous. In fact, we can show that given an electrical device, there is usually a mathematically analogous mechanical counterpart, and vice versa.

The motion of mechanical elements can be described in various dimensions as **translational**, **rotational**, or a combination of both. The equations governing the motion of mechanical systems are often directly or indirectly formulated from Newton's Law of Motion.

4-3-1 Translational Motion

The motion of translation is defined as a motion that takes place along a straight or curved path. The variables that are used to describe translational motion are **acceleration**, **velocity**, and **displacement**.

Newton's Law of Motion states that the *algebraic sum of forces acting on a rigid body in a given direction is equal to the product of the mass of the body and its acceleration in the same direction*. The law can be expressed as

$$\sum \text{forces} = Ma \qquad (4\text{-}14)$$

where M denotes the mass, and a is the acceleration in the direction considered. For linear translational motion, the following system elements are usually involved.

1. **Mass.** *Mass is considered as a property of an element that stores the kinetic energy of translational motion.* Mass is analogous to the inductance of electric networks. If W denotes the weight of a body, then M is given by

$$M = \frac{W}{g} \qquad (4\text{-}15)$$

where g is the acceleration of free fall of the body due to gravity ($g = 32.174$ ft/sec^2 in British units, and $g = 9.8066$ m/sec^2 in SI units.)

The consistent sets of basic units in the British and SI units are as follows:

Units	Mass M	Acceleration	Force
SI	kilogram (kg)	m/sec^2	Newton (N)
British	slug	ft/sec^2	pound (lb force)

Conversion factors between these and other secondary units are as follows:

Force:

$$1 \text{ N} = 0.2248 \text{ lb (force)} = 3.5969 \text{ oz (force)}$$

Mass:

$$1 \text{ kg} = 1000 \text{ g} = 2.2046 \text{ lb (mass)}$$
$$= 35.274 \text{ oz (mass)}$$
$$= 0.06852 \text{ slug}$$

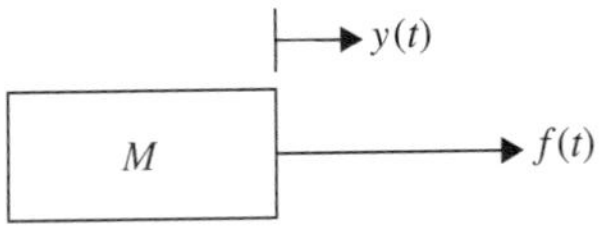

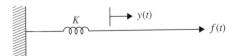

Figure 4-3 Force-mass system.

Distance:

$$1 \text{ m} = 3.2808 \text{ ft} = 39.37 \text{ in.}$$

$$1 \text{ in.} = 25.4 \text{ mm}$$

$$1 \text{ ft} = 0.3048 \text{ m}$$

Figure 4-3 illustrates the situation where a force is acting on a body with mass M. The force equation is written

$$f(t) = Ma(t) = M\frac{d^2y(t)}{dt^2} = M\frac{dv(t)}{dt} \tag{4-16}$$

where $v(t)$ denotes linear velocity.

2. **Linear spring.** In practice, a linear spring may be a model of an actual spring or a compliance of a cable or a belt. In general, *a spring is considered to be an element that stores potential energy*. It is analogous to a capacitor in electric networks. All springs in real life are nonlinear to some extent. However, if the deformation of the spring is small, its behavior can be approximated by a linear relationship:

$$f(t) = Ky(t) \tag{4-17}$$

where K is the **spring constant**, or simply **stiffness**.

The two basic unit systems for the spring constant are as follows:

Units	Spring Constant K
SI	N/m
British	lb/ft

Equation (4-17) implies that the force acting on the spring is directly proportional to the displacement (deformation) of the spring. The model representing a linear spring element is shown in Fig. 4-4. If the spring is preloaded with a preload tension of T, then Eq. (4-17) should be modified to

$$f(t) - T = Ky(t) \tag{4-18}$$

Friction for Translation Motion. Whenever there is motion or tendency of motion between two physical elements, frictional forces exist. The frictional forces encountered in physical systems are usually of a nonlinear nature. The characteristics of the frictional forces between two contacting surfaces often depend on such factors as the composition of the surfaces, the pressure between the surfaces, and their relative velocity among others, so an

Figure 4-4 Force-spring system.

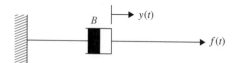

Figure 4-5 Dashpot for viscous friction.

exact mathematical description of the frictional force is difficult. Three different types of friction are commonly used in practical systems: **viscous friction**, **static friction**, and **Coulomb friction**. These are discussed separately in the following paragraphs.

1. **Viscous friction.** *Viscous friction represents a retarding force that is a linear relationship between the applied force and velocity.* The schematic diagram element for viscous friction is often represented by a dashpot, such as that shown in Fig. 4-5. The mathematical expression of viscous friction is

$$f(t) = B\frac{dy(t)}{dt} \tag{4-19}$$

where B is the **viscous frictional coefficient**.
The units of B are as follows:

Units	Viscous Frictional Coefficient B
SI	N/m/sec
British	lb/ft/sec

Figure 4-6(a) shows the functional relation between the viscous frictional force and velocity.

2. **Static friction.** *Static friction represents a retarding force that tends to prevent motion from beginning.* The static frictional force can be represented by the expression

$$f(t) = \pm (F_s)|_{\dot{y}=0} \tag{4-20}$$

which is defined as a frictional force that exists only when the body is stationary but has a tendency of moving. The sign of the friction depends on the direction of motion or the initial direction of velocity. The force-to-velocity relation of static friction is illustrated in Fig. 4-6(b). Notice that once motion begins, the static frictional force vanishes and other frictions take over.

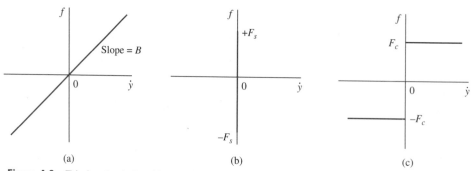

Figure 4-6 Frictional relationships of linear and nonlinear frictional forces. (a) Viscous friction. (b) Static friction. (c) Coulomb friction.

3. **Coulomb friction.** *Coulomb friction is a retarding force that has a constant amplitude with respect to the change of velocity, but the sign of the frictional force changes with the reversal of the direction of velocity.* The mathematical relation for the Coulomb friction is given by

$$f(t) = F_c \frac{\left(\dfrac{dy(t)}{dt} \right)}{\left| \left(\dfrac{dy(t)}{dt} \right) \right|} \tag{4-21}$$

where F_c is the **Coulomb friction coefficient**. The functional description of the friction-to-velocity relation is shown in Fig. 4-6(c).

It should be pointed out that the three types of frictions cited here are merely practical models that have been devised to portray frictional phenomena found in physical systems. They are by no means exhaustive or guaranteed to be accurate. In many unusual situations, we have to use other frictional models to represent the actual phenomenon accurately. One such example is rolling dry friction [3, 4], which is used to model friction in high-precision ball bearings used in spacecraft systems. It turns out that rolling dry friction has nonlinear hysteresis properties that make it impossible for use in linear system modeling.

4-3-2 Rotational Motion

The rotational motion of a body can be defined as motion about a fixed axis. The extension of Newton's Law of Motion for rotational motion states that the *algebraic sum of moments or torque about a fixed axis is equal to the product of the inertia and the angular acceleration about the axis.* Or,

$$\sum \text{torques} = J\alpha \tag{4-22}$$

where J denotes the inertia, and α is the angular acceleration. The other variables generally used to describe the motion of rotation are **torque** T, **angular velocity** ω, and **angular displacement** θ. The elements involved with the rotational motion are as follows:

1. **Inertia.** *Inertia, J, is considered as a property of an element that stores the kinetic energy of rotational motion.* The inertia of a given element depends on the geometric composition about the axis of rotation and its density. For instance, the inertia of a circular disk or shaft about its geometric axis is given by

$$J = \frac{1}{2}Mr^2 \tag{4-23}$$

When a torque is applied to a body with inertia J, as shown in Fig. 4-7, the torque equation is written

$$T(t) = J\alpha(t) = J\frac{d\omega(t)}{dt} = J\frac{d^2\theta(t)}{dt^2} \tag{4-24}$$

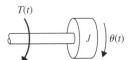

Figure 4-7 Torque-inertia system.

where $\theta(t)$ is the angular displacement; $\omega(t)$, the angular velocity; and $\alpha(t)$, the angular acceleration.

The SI and British units for inertia and the variables in Eq. (4-24) are tabulated as follows:

Units	Inertia	Torque	Angular Displacement
SI	kg-m^2	N-m	radian
		dyne-cm	radian
British	slug-ft^2	lb-ft	
	lb-ft-sec^2	oz-in.	radian
	oz-in.-sec^2		

The following conversion factors are often found to be useful:

Angular Displacement

$$1 \text{ rad} = \frac{180}{\pi} = 57.3 \text{ deg}$$

Angular Velocity

$$1 \text{ rpm} = \frac{2\pi}{60} = 0.1047 \text{ rad/sec}$$

$$1 \text{ rpm} = 6 \text{ deg/sec}$$

Torque

$$1 \text{ g-cm} = 0.0139 \text{ oz-in.}$$
$$1 \text{ lb-ft} = 192 \text{ oz-in.}$$
$$1 \text{ oz-in.} = 0.00521 \text{ lb-ft}$$

Inertia

$$1 \text{ g-cm} = 1.417 \times 10^{-5} \text{ oz-in.-sec}^2$$
$$1 \text{ lb-ft-sec}^2 = 192 \text{ oz-in.-sec}^2 = 32.2 \text{ lb-ft}^2$$
$$1 \text{ oz-in.-sec}^2 = 386 \text{ oz-in.}^2$$
$$1 \text{ g-cm-sec}^2 = 980 \text{ g-cm}^2$$

You must use the unit conversion software within **ACSYS** for a more comprehensive picture. See Appendix K.

2. **Torsional Spring.** As with the linear spring for translational motion, a **torsional spring constant** K, in torque per unit angular displacement, can be devised to represent the compliance of a rod or a shaft when it is subject to an applied torque. Figure 4-8 illustrates a simple torque-spring system that can be represented by the equation

$$T(t) = K\theta(t) \tag{4-25}$$

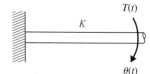

$\theta(t)$ **Figure 4-8**   Torque torsional spring system.

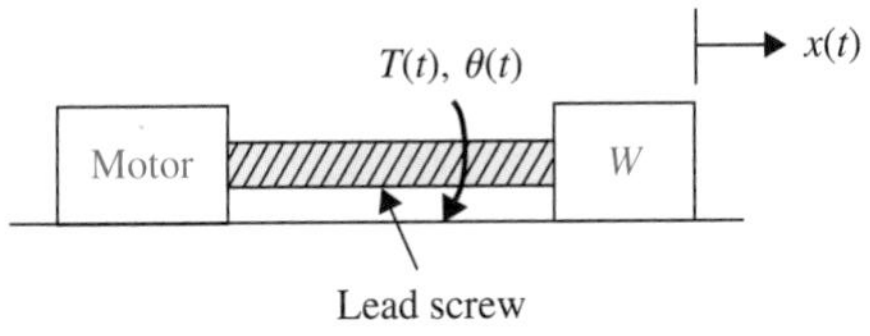

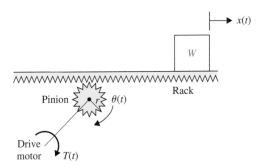

Figure 4-9 Rotary-to-linear motion control system (lead screw).

The units of the spring constant K in the SI and British systems are as follows:

Units	Spring Constant K
SI	N-m/rad
British	ft-lb/rad

If the torsional spring is preloaded by a preload torque of TP, Eq. (4-25) is modified to

$$T(t) - TP = K\theta(t) \tag{4-26}$$

3. **Friction for Rotational Motion.** The three types of friction described for translational motion can be carried over to the motion of rotation. Therefore, Eqs. (4-19), (4-20), and (4-21) can be replaced, respectively, by their counterparts:

Viscous friction:

$$T(t) = B\frac{d\theta(t)}{dt} \tag{4-27}$$

Static friction:

$$T(t) = \pm(F_s)\big|_{\dot{\theta}=0} \tag{4-28}$$

Coulomb friction:

$$T(t) = F_c\frac{\dfrac{d\theta(t)}{dt}}{\left|\dfrac{d\theta(t)}{dt}\right|} \tag{4-29}$$

4-3-3 Conversion Between Translational and Rotational Motions

In motion-control systems, it is often necessary to convert rotational motion into translation. For instance, a load may be controlled to move along a straight line through a rotary motor-and-screw assembly, such as that shown in Fig. 4-9. Figure 4-10 shows a similar situation in which a rack-and-pinion assembly is used as a mechanical linkage. Another familiar system in motion control is the control of a mass through a pulley by a rotary motor, such as the control of a printwheel in an electric typewriter with a belt-and-pulley

Figure 4-10 Rotary-to-linear motion control system (rack and pinion).

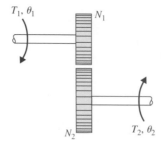

Figure 4-11 Rotary-to-linear motion-control system (belt and pulley).

system as shown in Fig. 4-11. The systems shown in Figs. 4-9, 4-10, and 4-11 can all be represented by a simple system with an equivalent inertia connected directly to the drive motor. For instance, the mass in Fig. 4-11 can be regarded as a point mass that moves about the pulley, which has a radius r. By disregarding the inertia of the pulley, the equivalent inertia that the motor sees is

$$J = Mr^2 = \frac{W}{g}r^2 \qquad (4\text{-}30)$$

If the radius of the pinion in Fig. 4-10 is r, the equivalent inertia that the motor sees is also given by Eq. (4-30).

Now consider the system of Fig. 4-9. The lead of the screw, L, is defined as the linear distance that the mass travels per revolution of the screw. In principle, the two systems in Figs. 4-10 and 4-11 are equivalent. In Fig. 4-10, the distance traveled by the mass per revolution of the pinion is $2\pi r$. By using Eq. (4-30) as the equivalent inertia for the system of Fig. 4-9, we have

$$J = \frac{W}{g}\left(\frac{L}{2\pi}\right)^2 \qquad (4\text{-}31)$$

where, in British units,

$$J = \text{inertia (oz-in.-sec}^2)$$
$$W = \text{weight (oz)}$$
$$L = \text{screw lead (in.)}$$
$$g = \text{gravitational force (386.4 in./sec}^2)$$

4-3-4 Gear Trains

Gear train, lever, or timing belt over a pulley is a mechanical device that transmits energy from one part of the system to another in such a way that force, torque, speed, and displacement may be altered. These devices can also be regarded as matching devices used to attain maximum power transfer. Two gears are shown coupled together in Fig. 4-12. The inertia and friction of the gears are neglected in the ideal case considered.

T_1, θ_1 N_1

N_2 T_2, θ_2 **Figure 4-12** Gear train.

The relationships between the torques T_1 and T_2, angular displacement θ_1 and θ_2, and the teeth numbers N_1 and N_2 of the gear train are derived from the following facts:

1. *The number of teeth on the surface of the gears is proportional to the radii r_1 and r_2 of the gears; that is,*

$$r_1 N_2 = r_2 N_1 \tag{4-32}$$

2. *The distance traveled along the surface of each gear is the same. Thus,*

$$\theta_1 r_1 = \theta_2 r_2 \tag{4-33}$$

3. *The work done by one gear is equal to that of the other since there are assumed to be no losses. Thus,*

$$T_1 \theta_1 = T_2 \theta_2 \tag{4-34}$$

If the angular velocities of the two gears ω_1 and ω_2 are brought into the picture, Eqs. (4-32) through (4-34) lead to

$$\frac{T_1}{T_2} = \frac{\theta_2}{\theta_1} = \frac{N_1}{N_2} = \frac{\omega_2}{\omega_1} = \frac{r_1}{r_2} \tag{4-35}$$

In practice, gears do have inertia and friction between the coupled gear teeth that often cannot be neglected. An equivalent representation of a gear train with viscous friction, Coulomb friction, and inertia considered as lumped parameters, is shown in Fig. 4-13, where T denotes the applied torque, T_1 and T_2 are the transmitted torque, F_{c1}, and F_{c2} are the Coulomb friction coefficients, and B_1 and B_2 are the viscous friction coefficients. The torque equation for gear 2 is

$$T_2(t) = J_2 \frac{d^2\theta_2(t)}{dt^2} + B_2 \frac{d\theta_2(t)}{dt} + F_{c2} \frac{\omega_2}{|\omega_2|} \tag{4-36}$$

The torque equation on the side of gear 1 is

$$T(t) = J_1 \frac{d^2\theta_1(t)}{dt^2} + B_1 \frac{d\theta_1(t)}{dt} + F_{c1} \frac{\omega_1}{|\omega_1|} + T_1(t) \tag{4-37}$$

Using Eq. (4-35), Eq. (4-36) is converted to

$$T_1(t) = \frac{N_1}{N_2} T_2(t) = \left(\frac{N_1}{N_2}\right)^2 J_2 \frac{d^2\theta_1(t)}{dt^2} + \left(\frac{N_1}{N_2}\right)^2 B_2 \frac{d\theta_1(t)}{dt} + \frac{N_1}{N_2} F_{c2} \frac{\omega_2}{|\omega_2|} \tag{4-38}$$

Equation (4-38) indicates that *it is possible to reflect inertia, friction, compliance, torque, speed, and displacement from one side of a gear train to the other.* The following quantities

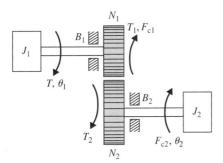

Figure 4-13 Gear train with friction and inertia.

are obtained when reflecting from gear 2 to gear 1:

Inertia: $\left(\dfrac{N_1}{N_2}\right)^2 J_2$

Viscous friction coefficient: $\left(\dfrac{N_1}{N_2}\right)^2 B_2$

Torque: $\dfrac{N_1}{N_2} T_2$

Angular displacement: $\dfrac{N_1}{N_2}\theta_2$

Angular velocity: $\dfrac{N_1}{N_2}\omega_2$

Coulomb friction torque: $\dfrac{N_1}{N_2}F_{c2}\dfrac{\omega_2}{|\omega_2|}$

Similarly, gear parameters and variables can be reflected from gear 1 to gear 2 by simply interchanging the subscripts in the preceding expressions.

If a torsional spring effect is present, the spring constant is also multiplied by $(N_1/N_2)^2$ in reflecting from gear 2 to gear 1. Now substituting Eq. (4-38) into Eq. (4-37), we get

$$T(t) = J_{1e}\frac{d^2\theta_1(t)}{dt^2} + B_{1e}\frac{d\theta_1(t)}{dt} + T_F \qquad (4\text{-}39)$$

where

$$J_{1e} = J_1 + \left(\frac{N_1}{N_2}\right)^2 J_2 \qquad (4\text{-}40)$$

$$B_{1e} = B_1 + \left(\frac{N_1}{N_2}\right)^2 B_2 \qquad (4\text{-}41)$$

$$T_F = F_{c1}\frac{\omega_1}{|\omega_1|} + \frac{N_1}{N_2}F_{c2}\frac{\omega_2}{|\omega_2|} \qquad (4\text{-}42)$$

▶ **EXAMPLE 4-3** Given a load that has inertia of 0.05 oz-in.-sec^2 and a Coulomb friction torque of 2 oz-in., find the inertia and frictional torque reflected through a 1:5 gear train ($N_1/N_2 = 1/5$, with N_2 on the load side.) The reflected inertia on the side of N_1 is $(1/5)^2 \times 0.05 = 0.002$ oz-in.-sec^2. The reflected Coulomb friction is $(1/5) \times 2 = 0.4$ oz-in. ◀

4-3-5 Backlash and Dead Zone (Nonlinear Characteristics)

Backlash and dead zone are commonly found in gear trains and similar mechanical linkages where the coupling is not perfect. In a majority of situations, backlash may give rise to undesirable inaccuracy, oscillations, and instability in control systems. In addition, it has a tendency to wear out the mechanical elements. Regardless of the actual mechanical elements, a physical model of backlash or dead zone between an input and an output member is shown in Fig. 4-14. The model can be used for a rotational system as

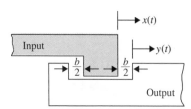

Figure 4-14 Physical model of backlash between two mechanical elements.

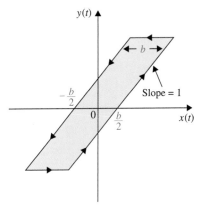

Figure 4-15 Input-output characteristic of backlash.

well as for a translational system. The amount of backlash is $b/2$ on either side of the reference position.

In general, the dynamics of the mechanical linkage with backlash depend upon the relative inertia-to-friction ratio of the output member. If the inertia of the output member is very small compared with that of the input member, the motion is controlled predominantly by friction. This means that the output member will not coast whenever there is no contact between the two members. When the output is driven by the input, the two members will travel together until the input member reverses its direction; then the output member will be at a standstill until the backlash is taken up on the other side, at which time it is assumed that the output member instantaneously takes on the velocity of the input member. The transfer characteristic between the input and output displacements of a system with backlash with negligible output inertia is shown in Fig. 4-15.

► 4-4 EQUATIONS OF MECHANICAL SYSTEMS

The equations of a linear mechanical system are written by first constructing a model of the system containing interconnected linear elements, and then by applying Newton's Law of Motion to the **free-body diagram**. For translational motion, the equation of motion is Eq. (4-14), and for rotational motion, Eq. (4-22) is used. The following examples illustrate how equations of mechanical systems are written.

► **EXAMPLE 4-4** Consider the mass-spring-friction system shown in Fig. 4-16(a). The linear motion concerned is in the horizontal direction. The free-body diagram of the system is shown in Fig. 4-16(b). The force equation of the system is

$$f(t) = M\frac{d^2y(t)}{dt^2} + B\frac{dy(t)}{dt} + Ky(t) \tag{4-43}$$

The last equation is rearranged by equating the highest-order derivative term to the rest of the terms:

$$\frac{d^2y(t)}{dt^2} = -\frac{B}{M}\frac{dy(t)}{dt} - \frac{K}{M}y(t) + \frac{1}{M}f(t) \tag{4-44}$$

The state diagram of the system is constructed as shown in Fig. 4-16(c). By defining the outputs of the integrators on the state diagram as state variables x_1 and x_2, the state

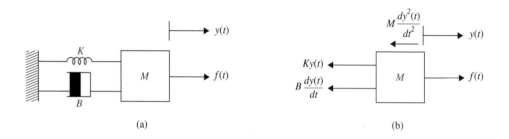

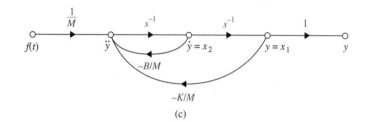

Figure 4-16 (a) Mass-
spring-friction system.
(b) Free-body diagram.
(c) State diagram.

equations are

$$\frac{dx_1(t)}{dt} = x_2(t) \tag{4-45}$$

$$\frac{dx_2(t)}{dt} = -\frac{K}{M}x_1(t) - \frac{B}{M}x_2(t) + \frac{1}{M}f(t) \tag{4-46}$$

It is not difficult to see that this mechanical system is analogous to a series *RLC* electric network. With this analogy, the state equations can be written directly using a different set of state variables. Consider that mass *M* is analogous to inductance *L*, the spring constant *K* is analogous to the inverse of capacitance $1/C$, and the viscous-friction coefficient *B* is analogous to resistance *R*. It is logical to assign $v(t)$, the velocity, and $f_k(t)$, the force acting on the spring, as state variables, since the former is analogous to the current in *L* and the latter is analogous to the voltage across *C*. Writing the force on *M* and the velocity of the spring as functions of the state variables and the input force $f(t)$, we have

• The state variables and
state equations of a
dynamic system are not
unique.

Force on mass:

$$M\frac{dv(t)}{dt} = -Bv(t) - f_k(t) + f(t) \tag{4-47}$$

Velocity of spring:

$$\frac{1}{K}\frac{df_k(t)}{dt} = v(t) \tag{4-48}$$

The state equations are obtained by dividing both sides of Eq. (4-47) by *M* and multiplying Eq. (4-48) by *K*.

This simple example illustrates that the state equation and state variables of a dynamic system are not unique. The transfer function between $Y(s)$ and $F(s)$ is obtained by taking the Laplace transform on both sides of Eq. (4-43) with zero initial conditions:

$$\frac{Y(s)}{F(s)} = \frac{1}{Ms^2 + Bs + K} \tag{4-49}$$

The same result is obtained by applying the gain formula to Fig. 4-16(c). ◀

▶ **EXAMPLE 4-5** As another example of writing the dynamic equations of a mechanical system with translational motion, consider the system shown in Fig. 4-17(a). Since the spring is deformed when it is subject to a force $f(t)$, two displacements, y_1 and y_2, must be assigned to the end points of the spring. The free-body diagrams of the system are shown in Fig. 4-17(b). The force equations are

$$f(t) = K[y_1(t) - y_2(t)] \tag{4-50}$$

$$K[y_1(t) - y_2(t)] = M\frac{d^2y_2(t)}{dt^2} + B\frac{dy_2(t)}{dt} \tag{4-51}$$

These equations are rearranged as

$$y_1(t) = y_2(t) + \frac{1}{K}f(t) \tag{4-52}$$

$$\frac{d^2y_2(t)}{dt^2} = -\frac{B}{M}\frac{dy_2(t)}{dt} + \frac{K}{M}[y_1(t) - y_2(t)] \tag{4-53}$$

By using the last two equations, the state diagram of the system is drawn in Fig. 4-17(c). The state variables are defined as $x_1(t) = y_2(t)$ and $x_2(t) = dy_2(t)/dt$. The state equations are written directly from the state diagram:

$$\frac{dx_1(t)}{dt} = x_2(t) \tag{4-54}$$

$$\frac{dx_2(t)}{dt} = -\frac{B}{M}x_2(t) + \frac{1}{M}f(t) \tag{4-55}$$

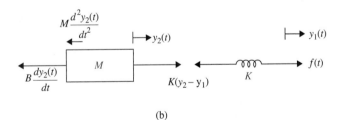

(a)

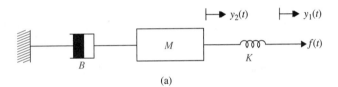

(b)

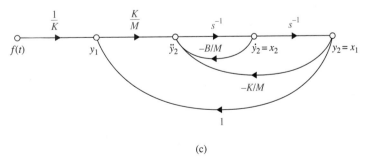

(c)

Figure 4-17 Mechanical system for Example 4-5. (a) Mass-spring-friction system. (b) Free-body diagram. (c) State diagram.

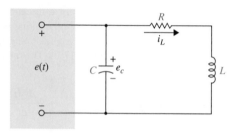

Figure 4-18 Electric network analogous to the mechanical system in Fig. 4-17.

As an alternative, we can assign the velocity $v(t)$ of the mass M as one state variable and the force $f_k(t)$ on the spring as the other state variable. We have

$$\frac{dv(t)}{dt} = -\frac{B}{M}v(t) + \frac{1}{M}f_k(t) \tag{4-56}$$

$$f_k(t) = f(t) \tag{4-57}$$

One may wonder why there is only one state equation in Eq. (4-47), whereas there are two state variables in $v(t)$ and $f_k(t)$. The two state equations of Eqs. (4-54) and (4-55) clearly show that the system is of the second order. The situation is better explained by referring to the analogous electric network of the system shown in Fig. 4-18. Although the network has two energy-storage elements in L and C, and thus there should be two state variables, the voltage across the capacitance $e_c(t)$ in this case is redundant, since it is equal to the applied voltage $e(t)$. Equations (4-56) and (4-55) can provide only the solutions to the velocity of M, $v(t)$, which is the same as $dy_2(t)/dt$, once $f(t)$ is specified. Then $y_2(t)$ is determined by integrating $v(t)$ with respect to t. The displacement $y_1(t)$ is then found using Eq. (4-50). On the other hand, Eqs. (4-54) and (4-55) give the solutions to $y_2(t)$ and $dy_2(t)/dt$ directly, and $y_1(t)$ is obtained from Eq. (4-50).

The transfer functions of the system are obtained by applying the gain formula to the state diagram in Fig. 4-17(c).

$$\frac{Y_2(s)}{F(s)} = \frac{1}{s(Ms + B)} \tag{4-58}$$

$$\frac{Y_1(s)}{F(s)} = \frac{Ms^2 + Bs + K}{Ks(Ms + B)} \tag{4-59}$$

◀

▶ **EXAMPLE 4-6** The rotational system shown in Fig 4-19(a) consists of a disk mounted on a shaft that is fixed at one end. The moment of inertia of the disk about the axis of rotation is J. The edge of the disk is riding on the surface, and the viscous friction coefficient between the two surfaces is B. The inertia of the shaft is negligible, but the torsional spring constant is K.

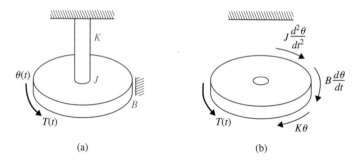

Figure 4-19 Rotational system for Example 4-6.

(a)   (b)

Assume that a torque is applied to the disk, as shown; then the torque or moment equation about the axis of the shaft is written from the free-body diagram of Fig. 4-19(b):

$$T(t) = J\frac{d^2\theta(t)}{dt^2} + B\frac{d\theta(t)}{dt} + K\theta(t) \tag{4-60}$$

Notice that this system is analogous to the translational system in Fig. 4-16. The state equations may be written by defining the state variables as $x_1(t) = \theta(t)$ and $x_2(t) = dx_1(t)/dt$. ◄

► **EXAMPLE 4-7**

Figure 4-20(a) shows the diagram of a motor coupled to an inertial load through a shaft with a spring constant K. A nonrigid coupling between two mechanical components in a control system often causes torsional resonances that can be transmitted to all parts of the system. The system variables and parameters are defined as follows:

$T_m(t)$ = motor torque J_m = motor inertia
B_m = motor viscous friction coefficient J_L = load inertia
K = spring constant of the shaft $\theta_m(t)$ = motor displacement
$\theta_L(t)$ = load displacement $\omega_m(t)$ = motor velocity
$\omega_L(t)$ = load velocity

The free-body diagrams of the system are shown in Fig. 4-20(b). The torque equations of the system are

$$\frac{d^2\theta_m(t)}{dt^2} = -\frac{B_m}{J_m}\frac{d\theta_m(t)}{dt} - \frac{K}{J_m}[\theta_m(t) - \theta_L(t)] + \frac{1}{J_m}T_m(t) \tag{4-61}$$

$$K[\theta_m(t) - \theta_L(t)] = J_L\frac{d^2\theta_L(t)}{dt^2} \tag{4-62}$$

In this case, the system contains three energy-storage elements in J_m, J_L, and K. Thus, there should be three state variables. Care should be taken in constructing the state diagram and assigning the state variables so that a minimum number of the latter are incorporated. Equations (4-61) and (4-62) are rearranged as

$$\frac{d^2\theta_m(t)}{dt^2} = -\frac{B_m}{J_m}\frac{d\theta_m(t)}{dt} - \frac{K}{J_m}[\theta_m(t) - \theta_L(t)] + \frac{1}{J_m}T_m(t) \tag{4-63}$$

$$\frac{d^2\theta_L(t)}{dt^2} = \frac{K}{J_L}[\theta_m(t) - \theta_L(t)] \tag{4-64}$$

The state diagram with three integrators is shown in Fig. 4-21. The clue given by Eqs. (4-63) and (4-64) is that $\theta_m(t)$ and $\theta_L(t)$ appear only as the difference, $\theta_m(t) - \theta_L(t)$ in these equations. From

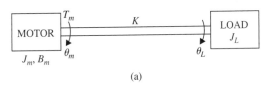

(a)

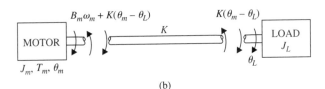

(b)

Figure 4-20 (a) Motor-load system. (b) Free-body diagram.

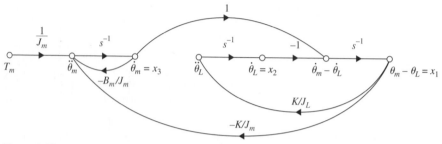

Figure 4-21 State diagram of the system of Fig. 4-20.

the state diagram in Fig. 4-21, the state variables are defined as $x_1(t) = \theta_m(t) - \theta_L(t)$, $x_2(t) = d\theta_L(t)/dt$, and $x_3(t) = d\theta_m(t)/dt$. The state equations are

$$\frac{dx_1(t)}{dt} = x_3(t) - x_2(t) \tag{4-65}$$

$$\frac{dx_2(t)}{dt} = \frac{K}{J_L}x_1(t) \tag{4-66}$$

$$\frac{dx_3(t)}{dt} = -\frac{K}{J_m}x_1(t) - \frac{B_m}{J_m}x_3(t) + \frac{1}{J_m}T_m(t) \tag{4-67}$$

The transfer functions between $\Theta_m(s)$ and $T_m(t)$, and $\Theta_L(s)$ and $T_m(s)$ are written by applying the gain formula to the state diagram in Fig. 4-21:

$$\frac{\Theta_m(s)}{T_m(s)} = \frac{X_3(s)}{sT_m(s)} = \frac{J_Ls^2 + K}{s[J_mJ_Ls^3 + B_mJ_Ls^2 + K(J_m + J_L)s + B_mK]} \tag{4-68}$$

$$\frac{\Theta_L(s)}{T_m(s)} = \frac{X_2(s)}{sT_m(s)} = \frac{K}{s[J_mJ_Ls^3 + B_mJ_Ls^2 + K(J_m + J_L)s + B_mK]} \tag{4-69}$$

◀

▶ 4-5 SENSORS AND ENCODERS IN CONTROL SYSTEMS

Sensors and encoders are important components used to monitor the performance and for feedback in control systems. In this section, the principle of operation and applications of some of the sensors and encoders that are commonly used in control systems are described.

4-5-1 Potentiometer

A potentiometer is an electromechanical transducer that converts mechanical energy into electrical energy. The input to the device is in the form of a mechanical displacement, either linear or rotational. When a voltage is applied across the fixed terminals of the potentiometer, the output voltage, which is measured across the variable terminal and ground, is proportional to the input displacement, either linearly or according to some nonlinear relation.

Rotary potentiometers are available commercially in single-revolution or multirevolution form, with limited or unlimited rotational motion. The potentiometers are commonly

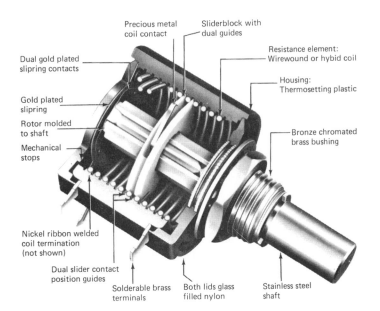

Figure 4-22 Ten-turn rotary potentiometer (courtesy of Helipot Division of Beckman Instruments, Inc.).

made with wirewound or conductive plastic resistance material. Figure 4-22 shows a cutaway view of a rotary potentiometer, and Fig. 4-23 shows a linear potentiometer that also contains a built-in operational amplifier. For precision control, the conductive plastic potentiometer is preferable, since it has infinite resolution, long rotational life, good output smoothness, and low static noise.

Figure 4-24 shows the equivalent circuit representation of a potentiometer, linear or rotary. Since the voltage across the variable terminal and reference is proportional to the shaft displacement of the potentiometer, when a voltage is applied across the fixed terminals, the device can be used to indicate the absolute position of a system or the relative position of two mechanical outputs. Figure 4-25(a) shows the arrangement when the housing of the potentiometer is fixed at reference; the output voltage $e(t)$ will be proportional to the shaft position $\theta_c(t)$ in the case of a rotary motion. Then

$$e(t) = K_s\theta_c(t) \tag{4-70}$$

Figure 4-23 Linear motion potentiometer with built-in operational amplifier (courtesy of Waters Manufacturing, Inc.).

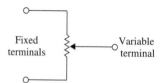

Figure 4-24 Electric circuit representation of a potentiometer.

where K_s is the proportional constant. For an N-turn potentiometer, the total displacement of the variable arm is $2\pi N$ radians. The proportional constant K_s is given by

$$K_s = \frac{E}{2\pi N} \quad \text{V/rad} \tag{4-71}$$

where E is the magnitude of the reference voltage applied to the fixed terminals. A more flexible arrangement is obtained by using two potentiometers connected in parallel, as shown in Fig. 4-25(b). This arrangement allows the comparison of two remotely located shaft positions. The output voltage is taken across the variable terminals of the two potentiometers and is given by

$$e(t) = K_s[\theta_1(t) - \theta_2(t)] \tag{4-72}$$

Figure 4-26 illustrates the block diagram representation of the setups in Fig. 4-25.

    In dc-motor control systems, potentiometers are often used for position feedback. Figure 4-27(a) shows the schematic diagram of a typical dc-motor, position-control system. The potentiometers are used in the feedback path to compare the actual load position with the desired reference position. If there is a discrepancy between the load position and the reference input, an error signal is generated by the potentiometers that will drive the motor in such a way that this error is minimized quickly. As shown in Fig. 4-27(a), the error signal is amplified by a dc amplifier whose output drives the armature of a permanent-magnet dc motor. Typical waveforms of the signals in the system when the input $\theta_r(t)$ is a step function are shown in Fig. 4-27(b). Note that the electric signals are all unmodulated. *In control-systems terminology, a dc signal usually refers to an unmodulated signal. On the other hand, an ac signal refers to signals that are modulated by a modulation process.* These definitions are different from those commonly used in electrical engineering, where dc simply refers to unidirectional signals and ac indicates alternating signals.

• In control-system terminology, a dc signal usually refers to an unmodulated signal.

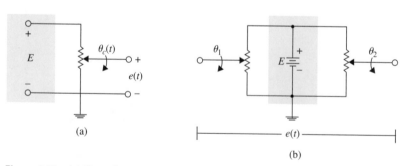

(a)

(b)

Figure 4-25 (a) Potentiometer used as a position indicator. (b) Two potentiometers used to sense the positions of two shafts.

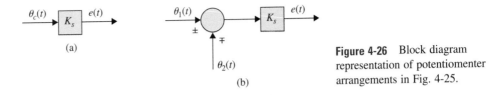

Figure 4-26 Block diagram representation of potentiomenter arrangements in Fig. 4-25.

Figure 4-28(a) illustrates a control system that serves essentially the same purpose as that of the system in Fig. 4-27(a), except that ac signals prevail. In this case, the voltage applied to the error detector is sinusoidal. The frequency of this signal is usually much higher than that of the signal that is being transmitted through the system.

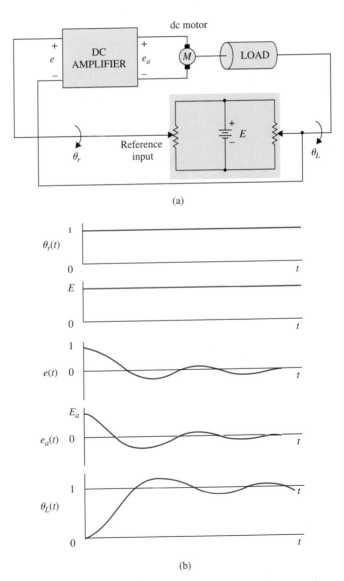

Figure 4-27 (a) A dc-motor, position-control system with potentiometers as error sensors. (b) Typical waveforms of signals in the control system of part (a).

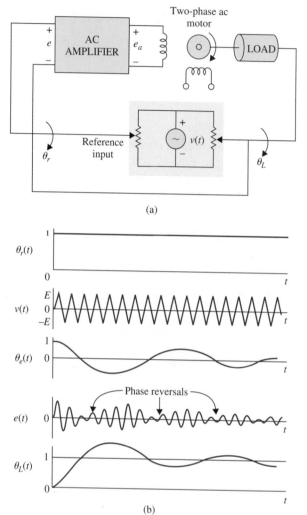

Figure 4-28 (a) An ac control system with potentiometers as error detectors. (b) Typical waveforms of signals in the control system of part (a).

Control systems with ac signals are usually found in aerospace systems that are more susceptible to noise.

Typical signals of an ac control system are shown in Fig. 4-28(b). The signal $v(t)$ is referred to as the carrier whose frequency is ω_c, or

$$v(t) = E \sin \omega_c t \tag{4-73}$$

Analytically, the output of the error signal is given by

$$e(t) = K_s \theta_e(t) v(t) \tag{4-74}$$

where $\theta_e(t)$ is the difference between the input displacement and the load displacement, or

$$\theta_e(t) = \theta_r(t) - \theta_L(t) \tag{4-75}$$

For the $\theta_e(t)$ shown in Fig. 4-28(b), $e(t)$ becomes a **suppressed-carrier-modulated** signal. A reversal in phase of $e(t)$ occurs whenever the signal crosses the zero-magnitude axis. This reversal in phase causes the ac motor to reverse in direction according to the desired sense of correction of the error signal $\theta_e(t)$. The term *suppressed-carrier modulation* stems from the fact that when a signal $\theta_e(t)$ is modulated by a carrier signal $v(t)$ according to Eq. (4-74), the resultant signal $e(t)$ no longer contains the original carrier frequency ω_c. To illustrate this, let us assume that $\theta_e(t)$ is also a sinusoid given by

$$\theta_e(t) = \sin \omega_s t \tag{4-76}$$

where, normally, $\omega_s \ll \omega_c$. Using familiar trigonometric relations and substituting Eqs. (4-73) and (4-76) into Eq. (4-74), we get

$$e(t) = \tfrac{1}{2} K_s E[\cos(\omega_c - \omega_s)t - \cos(\omega_c + \omega_s)t] \tag{4-77}$$

Therefore, $e(t)$ no longer contains the carrier frequency ω_c or the signal frequency ω_s, but has only the two sidebands $\omega_c + \omega_s$ and $\omega_c - \omega_s$.

When the modulated signal is transmitted through the system, the motor acts as a demodulator, so that the displacement of the load will be of the same form as the dc signal before modulation. This is clearly seen from the waveforms of Fig. 4-28(b). It should be pointed out that a control system need not contain all dc or all ac components. It is quite common to couple a dc component to an ac component through a modulator, or an ac device to a dc device through a demodulator. For instance, the dc amplifier of the system in Fig. 4-28(a) may be replaced by an ac amplifier that is preceded by a modulator and followed by a demodulator.

4-5-2   Tachometers

Tachometers are electromechanical devices that convert mechanical energy into electrical energy. The device works essentially as a voltage generator, with the output voltage proportional to the magnitude of the angular velocity of the input shaft. In control systems, most of the tachometers used are of the dc variety; that is, the output voltage is a dc signal. DC tachometers are used in control systems in many ways; they can be used as velocity indicators to provide shaft-speed readout, velocity feedback, speed control, or stabilization. Figure 4-29 is a block diagram of a typical velocity-control system in which the tachometer output is compared with the reference voltage, which represents the desired velocity to be achieved. The difference between the two signals, or the error, is amplified and used to drive the motor so that the velocity will eventually reach the desired value. In this type of application, the accuracy of the tachometer is highly critical, as the accuracy of the speed control depends on it.

• Note: Tachometers are often used in control systems to improve stability.

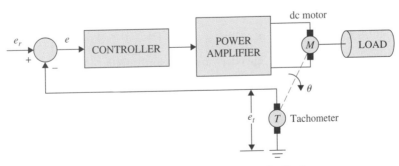

Figure 4-29   Velocity-control system with tachometer feedback.

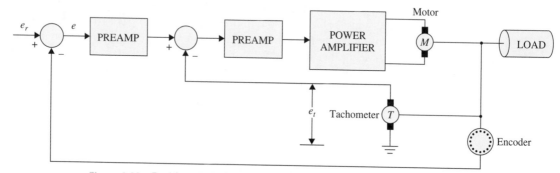

Figure 4-30   Position-control system with tachometer feedback.

In a position-control system, velocity feedback is often used to improve the stability or the damping of the closed-loop system. Figure 4-30 shows the block diagram of such an application. In this case, the tachometer feedback forms an inner loop to improve the damping characteristics of the system, and the accuracy of the tachometer is not so critical.

The third and most traditional use of a dc tachometer is in providing the visual speed readout of a rotating shaft. Tachometers used in this capacity are generally connected directly to a voltmeter calibrated in revolutions per minute (rpm).

Mathematical Modeling of Tachometers

The dynamics of the tachometer can be represented by the equation

$$e_t(t) = K_t \frac{d\theta(t)}{dt} = K_t \omega(t) \tag{4-78}$$

where $e_t(t)$ is the output voltage; $\theta(t)$, the rotor displacement in radians; $\omega(t)$, the rotor velocity in rad/sec; and K_t, the **tachometer constant in** V/rad/sec. The value of K_t is usually given as a catalog parameter in **volts per 1000 rpm** (V/krpm).

The transfer function of a tachometer is obtained by taking the Laplace transform on both sides of Eq. (4-78). The result is

$$\frac{E_t(s)}{\Theta(s)} = K_t s \tag{4-79}$$

where $E_t(s)$ and $\Theta(s)$ are the Laplace transforms of $e_t(t)$ and $\theta(t)$, respectively.

4-5-3 Incremental Encoder

Incremental encoders are frequently found in modern control systems for converting linear or rotary displacement into digitally coded or pulse signals. The encoders that output a digital signal are known as **absolute encoders**. In the simplest terms, absolute encoders provide as output a distinct digital code indicative of each particular least significant increment of resolution. **Incremental encoders**, on the other hand, provide a pulse for each increment of resolution, but do not make distinctions between the increments. In practice, the choice of which type of encoder to use depends on economics

Figure 4-31 Rotary incremental encoder (courtesy of DISC Instruments, Inc.).

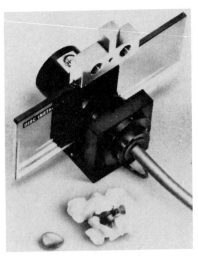

Figure 4-32 Linear incremental encoder (courtesy of DISC Instruments, Inc.).

and control objectives. For the most part, the need for absolute encoders has much to do with the concern for data loss during power failure or the applications involving periods of mechanical motion without the readout under power. However, the incremental encoder's simplicity in construction, low cost, ease of application, and versatility have made it by far one of the most popular encoders in control systems.

Incremental encoders are available in rotary and linear forms. Figures 4-31 and 4-32 show typical rotary and linear incremental encoders.

A typical rotary incremental encoder has four basic parts: a light source, a rotary disk, a stationary mask, and a sensor, as shown in Fig. 4-33. The disk has alternate opaque and transparent sectors. Any pair of these sectors represents an incremental period. The mask is used to pass or block a beam of light between the light source and the photosensor located behind the mask. For encoders with relatively low resolution, the mask is not necessary. For fine-resolution encoders (up to thousands of increments per evolution), a multiple-slit mask is often used to maximize reception of the shutter light. The waveforms of the sensor outputs are generally triangular or sinusoidal, depending on the resolution required. Square-wave signals compatible with digital logic are derived by using a linear

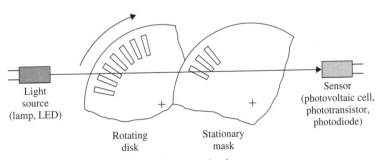

Figure 4-33 Typical incremental optomechanics.

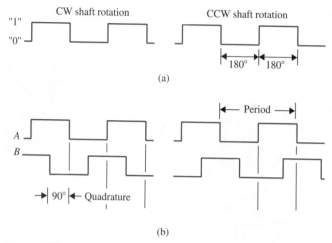

(a)

(b)

Figure 4-34 (a) Typical rectangular output waveform of a single-channel encoder device (bidirectional). (b) Typical dual-channel encoder signals in quadrature (bidirectional).

amplifier followed by a comparator. Figure 4-34(a) shows a typical rectangular output waveform of a single-channel incremental encoder. In this case, pulses are produced for both directions of shaft rotation. A dual-channel encoder with two sets of output pulses is necessary for direction sensing and other control functions. When the phase of the two-output pulse train is 90° apart electrically, the two signals are said to be in quadrature, as shown in Fig. 4-34(b); the signals uniquely define 0-to-1 and 1-to-0 logic transitions with respect to the direction of rotation of the encoder disk so that a direction-sending logic circuit can be constructed to decode the signals. Figure 4-35 shows the single-channel output and the quadrature outputs with sinusoidal waveforms. The sinusoidal signals from the incremental encoder can be used for fine position control in feedback control systems. The following example illustrates some applications of the incremental encoder in control systems.

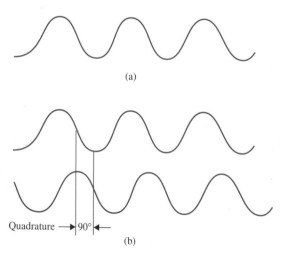

Figure 4-35 (a) Typical sinusoidal output waveform of a single-channel encoder device. (b) Typical dual-channel encoder signals in quadrature.

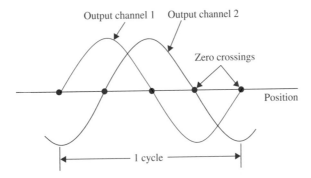

Figure 4-36 One cycle of the output signals of a dual-channel incremental encoder.

▶ **EXAMPLE 4-8** Consider an incremental encoder that generates two sinusoidal signals in quadrature as the encoder disk rotates. The output signals of the two channels are shown in Fig 4-36 over one cycle. Note that the two encoder signals generate 4 zero crossings per cycle. These zero crossings can be used for position indication, position control, or speed measurements in control systems. Let us assume that the encoder shaft is coupled directly to the rotor shaft of a motor that directly drives the printwheel of an electronic typewriter or word processor. The printwheel has 96 character positions on its periphery, and the encoder has 480 cycles. Thus, there are $480 \times 4 = 1920$ zero crossings per revolution. For the 96-character printwheel, this corresponds to $1920/96 = 20$ zero crossings per character; that is, there are 20 zero crossings between two adjacent characters.

One way of measuring the velocity of the printwheel is to count the number of pulses generated by an electronic clock that occur between consecutive zero crossings of the encoder outputs. Let us assume that a 500-kHz clock is used, that is, the clock generates 500,000 pulses/sec. If the counter records, say, 500 clock pulses while the encoder rotates from the zero crossing to the next, the shaft speed is

$$\frac{500{,}000 \text{ pulses/sec}}{500 \text{ pulses/zero crossing}} = 1000 \text{ zero crossings/sec}$$

$$= \frac{1000 \text{ zero crossings/sec}}{1920 \text{ zero crossings/rev}} = 0.52083 \text{ rev/sec}$$

$$= 31.25 \text{ rpm} \tag{4-80}$$

The encoder arrangement described can be used for fine position control of the printwheel. Let the zero crossing A of the waveforms in Fig. 4-36 correspond to a character position on the printwheel (the next character position is 20 zero crossings away) and the point correspond to a stable equilibrium point. The coarse position control of the system must first drive the printwheel position to within 1 zero crossing on either side of position A; then by using the slope of the sine wave at position, A, the control system should null the error quickly. ◄

▶▶ **4-6 DC MOTORS IN CONTROL SYSTEMS**

Direct-current (dc) motors are one of the most widely used prime movers in industry today. Years ago, the majority of the small servomotors used for control purposes were ac. In reality, ac motors are more difficult to control, especially for position control, and their characteristics are quite nonlinear, which makes the analytical task more difficult. DC motors, on the other hand, are more expensive, because of their brushes and commutators, and variable-flux dc motors are suitable only for certain types of control applications. Before permanent-magnet technology was fully developed, the torque per unit volume or weight of a dc motor with a permanent-magnet (PM) field was far from desirable. Today, with the development of the rare-earth magnet, it is possible to achieve very high torque-to-volume PM dc motors at reasonable cost. Furthermore, the advances made in brush-and-commutator technology have made these wearable parts practically maintenance-free. The advancements made in power electronics have made brushless dc motors quite popular in high-performance control systems. Advanced manufacturing techniques have also produced dc motors with ironless rotors that have very low inertia, thus achieving very high torque-to-inertia ratio. Low-time-constant properties have opened new applications for dc motors in computer peripheral equipment such as tape drives, printers, disk drives, and word processors, as well as in the automation and machine-tool industries.

4-6-1 Basic Operational Principles of DC Motors

The dc motor is basically a torque transducer that converts electric energy into mechanical energy. The torque developed on the motor shaft is directly proportional to the field flux and the armature current. As shown in Fig. 4-37, a current-carrying conductor is established in a magnetic field with flux ϕ, and the conductor is located at a distance r from the center of rotation. The relationship among the developed torque, flux ϕ, and current i_a is

$$T_m = K_m \phi i_a \qquad (4\text{-}81)$$

where T_m is the motor torque (in N-m, lb-ft, or oz-in.); ϕ, the magnetic flux (in webers); i_a, the armature current (in amperes); and K_m, a proportional constant.

In addition to the torque developed by the arrangement shown in Fig. 4-37, when the conductor moves in the magnetic field, a voltage is generated across its terminals. This voltage, the **back emf**, which is proportional to the shaft velocity, tends to oppose the current flow. The relationship between the back emf and the shaft velocity is

$$e_b = K_m \phi \omega_m \qquad (4\text{-}82)$$

where e_b denotes the back emf (volts), and ω_m is the shaft velocity (rad/sec) of the motor. Equations (4-81) and (4-82) form the basis of the dc-motor operation.

4-6-2 Basic Classifications of PM DC Motors

In general, the magnetic field of a dc motor can be produced by field windings or permanent magnets. Due to the popularity of PM dc motors in control system applications, we shall concentrate on this type of motor.

PM dc motors can be classified according to commutation scheme and armature design. Conventional dc motors have mechanical brushes and commutators. However, an

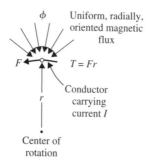

Figure 4-37 Torque production in a dc motor.

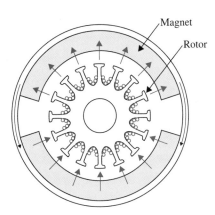

Figure 4-38 Cross-section view of a permanent-magnet (PM) iron-core dc motor.

important type of dc motors in which the commutation is done electronically is called **brushless dc**.

According to the armature construction, the PM dc motor can be broken down into three types of armature design: **iron-core**, **surface-wound**, and **moving-coil** motors.

Iron-Core PM DC Motors

The rotor and stator configuration of an iron-core PM dc motor is shown in Fig. 4-38. The permanent-magnet material can be barium ferrite, Alnico, or a rare-earth compound. The magnetic flux produced by the magnet passes through a laminated rotor structure that contains slots. The armature conductors are placed in the rotor slots. This type of dc motor is characterized by relatively high rotor inertia (since the rotating part consists of the armature windings), high inductance, low cost, and high reliability.

Surface-Wound DC Motors

Figure 4-39 shows the rotor construction of a surface-wound PM dc motor. The armature conductors are bonded to the surface of a cylindrical rotor structure, which is made of laminated disks fastened to the motor shaft. Since no slots are used on the rotor in this design, the armature has no "cogging" effect. The conductors are laid out in the air gap between the rotor and the (PM) field, so this type of motor has lower inductance than that of the iron-core structure.

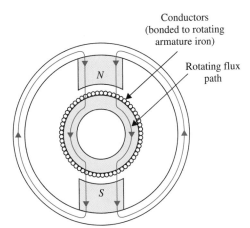

Figure 4-39 Cross-section view of a surface-wound permanent-magnet (PM) dc motor.

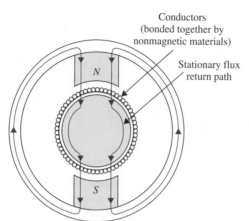

Figure 4-40 Cross-section view of a surface-wound permanent-magnet (PM) dc motor.

Moving-Coil DC Motors

Moving-coil motors are designed to have very low moments of inertia and very low armature inductance. This is achieved by placing the armature conductors in the air gap between a stationary flux return path and the (PM) structure, as shown in Fig. 4-40. In this case, the conductor structure is supported by nonmagnetic material—usually epoxy resins or fiberglass—to form a hollow cylinder. One end of the cylinder forms a hub, which is attached to the motor shaft. A cross-section view of such a motor is shown in Fig. 4-41. Since all unnecessary elements have been removed from the armature of the moving-coil motor, its moment of inertia is very low. Since the conductors in the moving-coil armature are not in direct contact with iron, the motor inductance is very low, and values of less than 100 μH are common in this type of motor. Its low-inertia and low-inductance properties make the moving coil motor one of the best actuator choices for high-performance control systems.

Brushless DC Motors

Brushless dc motors differ from the previously mentioned dc motors in that they employ electrical (rather than mechanical) commutation of the armature current. The most common configuration of brushless dc motors—especially for incremental-motion applications—is one in which the rotor consists of magnets and "back-iron" support, and whose commutated

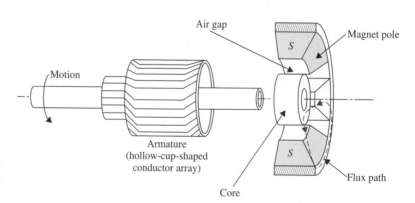

Figure 4-41 Cross-section side view of a moving-coil dc motor.

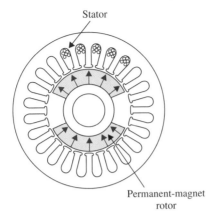

Figure 4-42   Cross-section view of a brushless,
permanent-magnet (PM), iron-core dc motor.

windings are located external to the rotating parts, as shown in Fig. 4-42. Compared to the
conventional dc motors, such as the one shown in Fig. 4-38, it is an inside-out configuration. Depending on the specific application, brushless dc motors can be used when a low
moment of inertia is needed, such as the spindle drive in high-performance disk drives used
in computers.

4-6-3   Mathematical Modeling of PM DC Motors

Since dc motors are used extensively in control systems, it is necessary to establish mathematical models for analytical purposes dc motors for controls applications. We use the
equivalent circuit diagram in Fig. 4-43 to represent a PM dc motor. The armature is modeled as a circuit with resistance R_a connected in series with an inductance L_a, and a voltage source e_b representing the back emf (electromotive force) in the armature when the
rotor rotates. The motor variables and parameters are defined as follows:

$$i_a(t) = \text{armature current} \qquad L_a = \text{armature inductance}$$
$$R_a = \text{armature resistance} \qquad e_a(t) = \text{applied voltage}$$
$$e_b(t) = \text{back emf} \qquad K_b = \text{back-emf constant}$$
$$T_L(t) = \text{load torque} \qquad \phi = \text{magnetic flux in the air gap}$$
$$T_m(t) = \text{motor torque} \qquad \omega_m(t) = \text{rotor angular velocity}$$
$$\theta_m(t) = \text{rotor displacement} \qquad J_m = \text{rotor inertia}$$
$$K_i = \text{torque constant} \qquad B_m = \text{viscous-friction coefficient}$$

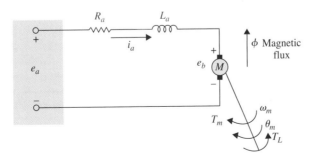

Figure 4-43   Model of a
separately excited dc motor.

With reference to the circuit diagram of Fig. 4-43, the control of the dc motor is applied at the armature terminals in the form of the applied voltage $e_a(t)$. For linear analysis, we assume that the torque developed by the motor is proportional to the air-gap flux and the armature current. Thus,

$$T_m(t) = K_m(t)\phi i_a(t) \tag{4-83}$$

Since ϕ is constant, Eq. (4-83) is written

$$T_m(t) = K_i i_a(t) \tag{4-84}$$

where K_i is the **torque constant** in N-m/A, lb-ft/A, or oz-in/A.

Starting with the control input voltage $e_a(t)$, the cause-and-effect equations for the motor circuit in Fig. 4-43 are

$$\frac{di_a(t)}{dt} = \frac{1}{L_a}e_a(t) - \frac{R_a}{L_a}i_a(t) - \frac{1}{L_a}e_b(t) \tag{4-85}$$

$$T_m(t) = K_i i_a(t) \tag{4-86}$$

$$e_b(t) = K_b\frac{d\theta_m(t)}{dt} = K_b\omega_m(t) \tag{4-87}$$

$$\frac{d^2\theta_m(t)}{dt^2} = \frac{1}{J_m}T_m(t) - \frac{1}{J_m}T_L(t) - \frac{B_m}{J_m}\frac{d\theta_m(t)}{dt} \tag{4-88}$$

where $T_L(t)$ represents a load frictional torque such as Coulomb friction.

Equations (4-85) through (4-88) consider that the applied voltage $e_a(t)$ is the cause; Eq. (4-85) considers that $di_a(t)/dt$ is the immediate effect due to $e_a(t)$; in Eq. (4-86), $i_a(t)$ causes the torque $T_m(t)$; Eq. (4-87) defines the back emf; and, finally, in Eq. (4-88), the torque $T_m(t)$ causes the angular velocity $\omega_m(t)$ and displacement $\theta_m(t)$.

The state variables of the system can be defined as $i_a(t)$, $\omega_m(t)$, and $\theta_m(t)$. By direct substitution and eliminating all the nonstate variables from Eqs. (4-85) through (4-88), the state equations of the dc-motor system are written in vector-matrix form:

$$
\begin{bmatrix} \dfrac{di_a(t)}{dt} \\[2mm] \dfrac{d\omega_m(t)}{dt} \\[2mm] \dfrac{d\theta_m(t)}{dt} \end{bmatrix} =
\begin{bmatrix} -\dfrac{R_a}{L_a} & -\dfrac{K_b}{L_a} & 0 \\[2mm] \dfrac{K_i}{J_m} & -\dfrac{B_m}{J_m} & 0 \\[2mm] 0 & 1 & 0 \end{bmatrix}
\begin{bmatrix} i_a(t) \\[2mm] \omega_m(t) \\[2mm] \theta_m(t) \end{bmatrix} +
\begin{bmatrix} \dfrac{1}{L_a} \\[2mm] 0 \\[2mm] 0 \end{bmatrix} e_a(t) +
\begin{bmatrix} 0 \\[2mm] -\dfrac{1}{J_m} \\[2mm] 0 \end{bmatrix} T_L(t) \tag{4-89}
$$

Notice that in this case, $T_L(t)$ is treated as a second input in the state equations.

The state diagram of the system is drawn as shown in Fig. 4-44, using Eq. (4-89). The transfer function between the motor displacement and the input voltage is obtained

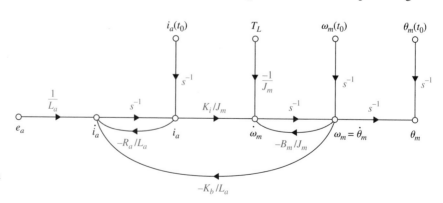

Figure 4-44 State diagram of a dc-motor system.

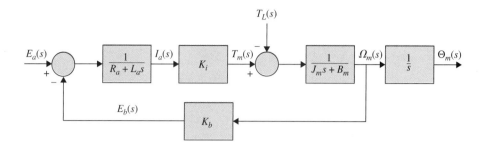

Figure 4-45 Block diagram of a dc-motor system.

from the state diagram as

$$\frac{\Theta_m(s)}{E_a(s)} = \frac{K_i}{L_a J_m s^3 + (R_a J_m + B_m L_a)s^2 + (K_b K_i + R_a B_m)s} \quad (4\text{-}90)$$

where $T_L(t)$ has been set to zero.

• A dc motor is essentially an integrating device.

Figure 4-45 shows a block-diagram representation of the dc-motor system. The advantage of using the block diagram is that it gives a clear picture of the transfer function relation between each block of the system. Since an s can be factored out of the denominator of Eq. (4-90), *the significance of the transfer function $\Theta_m(s)/E_a(s)$ is that the dc motor is essentially an integrating device between these two variables.* This is expected, since if $e_a(t)$ is a constant input, the output motor displacement will behave as the output of an integrator; that is, it will increase linearly with time.

• Back emf generally improves system stability.

Although a dc motor by itself is basically an open-loop system, the state diagram of Fig. 4-44 and the block diagram of Fig. 4-45 show that the motor has a "built-in" feedback loop caused by the back emf. Physically, the back emf represents the feedback of a signal that is proportional to the negative of the speed of the motor. As seen from Eq. (4-90), the back-emf constant K_b represents an added term to the resistance R_a and the viscous-friction coefficient B_m. Therefore, *the back-emf effect is equivalent to an "electric friction," which tends to improve the stability of the motor, and in general, the stability of the system.*

Relation between K_i and K_b

Although functionally the torque constant K_i and back-emf constant K_b are two separate parameters, for a given motor their values are closely related. To show the relationship, we write the mechanical power developed in the armature as

$$P = e_b(t)i_a(t) \quad (4\text{-}91)$$

The mechanical power is also expressed as

$$P = T_m(t)\omega_m(t) \quad (4\text{-}92)$$

where, in SI units, $T_m(t)$ is in N-m, and $\omega_m(t)$ is in rad/sec. Now substituting Eqs. (4-86) and (4-87) in Eq. (4-91), we get

$$P = T_m(t)\omega_m(t) = K_b \omega_m(t)\frac{T_m(t)}{K_i} \quad (4\text{-}93)$$

from which we get

$$K_b \,(\text{V/rad/sec}) = K_i (\text{N-m/A}) \quad (4\text{-}94)$$

Thus, we see that in SI units, the values of K_b and K_i are identical if K_b is represented in V/rad/sec and K_i is in N-m/A.

In the British unit system, we convert Eq. (4-91) into horsepower (hp); that is,

$$P = \frac{e_b(t)i_a(t)}{746} \text{ hp} \tag{4-95}$$

In terms of torque and angular velocity, P is

$$P = \frac{T_m(t)\omega_m(t)}{550} \text{ hp} \tag{4-96}$$

where $T_m(t)$ is in ft-1b, and $\omega_m(t)$ is in rad/sec. Using Eq. (4-86) and (4-87), and equating Eq. (4-95) to Eq. (4-96), we get

$$\frac{K_b\omega_m(t)T_m(t)}{746K_i} = \frac{T_m(t)\omega_m(t)}{550} \tag{4-97}$$

Thus,

$$K_b = \frac{746}{550}K_i = 1.356K_i \tag{4-98}$$

where K_b is in V/rad/sec, and K_i is in ft-lb/A.

▶ 4-7 LINEARIZATION OF NONLINEAR SYSTEMS

From the discussions given in the preceding sections on system modeling, we should realize that most components and actuators found in physical systems have nonlinear characteristics. In practice, we may find that some devices have moderate nonlinear characteristics, or nonlinear properties that would occur if they were driven into certain operating regions. For these devices, the modeling by linear-system models may give quite accurate analytical results over a relatively wide range of operating conditions. However, there are numerous physical devices that possess strong nonlinear characteristics. For these devices, a linearized model is valid only for limited range of operation, and often only at the operating point at which the linearization is carried out. More importantly, when a nonlinear system is linearized at an operating point, the linear model may contain time-varying elements.

Let us represent a nonlinear system by the following vector-matrix state equations:

$$\frac{d\mathbf{x}(t)}{dt} = \mathbf{f}[\mathbf{x}(t), \mathbf{r}(t)] \tag{4-99}$$

where $\mathbf{x}(t)$ represents the $n \times 1$ state vector; $\mathbf{r}(t)$, the $p \times 1$ input vector; and $\mathbf{f}[\mathbf{x}(t), \mathbf{r}(t)]$, an $n \times 1$ function vector. In general, $\mathbf{f}$ is a function of the state vector and the input vector.

Being able to represent a nonlinear and/or time-varying system by state equations is a distinct advantage of the state-variable approach over the transfer-function method, since the latter is strictly defined only for linear time-invariant systems.

As a simple example, the following nonlinear state equations are given:

$$\frac{dx_1(t)}{dt} = x_1(t) + x_2^2(t) \tag{4-100}$$

$$\frac{dx_2(t)}{dt} = x_1(t) + r(t) \tag{4-101}$$

Since nonlinear systems are usually difficult to analyze and design, it is desirable to perform a linearization whenever the situation justifies it.

A linearization process that depends on expanding the nonlinear state equations into a Taylor series about a nominal operating point or trajectory is now described. All the terms of the Taylor series of order higher than the first are discarded, and the linear approximation of the nonlinear state equations at the nominal point results.

Let the nominal operating trajectory be denoted by $\mathbf{x}_0(t)$, which corresponds to the nominal input $\mathbf{r}_0(t)$ and some fixed initial states. Expanding the nonlinear state equation of Eq. (4-99) into a Taylor series about $\mathbf{x}(t) = \mathbf{x}_0(t)$ and neglecting all the higher-order terms yields

$$\dot{x}_i(t) = f_i(\mathbf{x}_0, \mathbf{r}_0) + \sum_{j=1}^{n} \frac{\partial f_i(\mathbf{x}, \mathbf{r})}{\partial x_j}\bigg|_{\mathbf{x}_0, \mathbf{r}_0} (x_j - x_{0j}) + \sum_{j=1}^{p} \frac{\partial f_i(\mathbf{x}, \mathbf{r})}{\partial r_j}\bigg|_{\mathbf{x}_0, \mathbf{r}_0} (r_j - r_{0j}) \quad (4\text{-}102)$$

where $i = 1, 2, \ldots, n$. Let

$$\Delta x_i = x_i - x_{0i} \quad (4\text{-}103)$$

and

$$\Delta r_j = r_j - r_{0j} \quad (4\text{-}104)$$

Then

$$\Delta \dot{x}_i = \dot{x}_i - \dot{x}_{0i} \quad (4\text{-}105)$$

Since

$$\dot{x}_{0i} = f_i(\mathbf{x}_0, \mathbf{r}_0) \quad (4\text{-}106)$$

Equation (4-102) is written

$$\Delta \dot{x}_i = \sum_{j=1}^{n} \frac{\partial f_i(\mathbf{x}, \mathbf{r})}{\partial x_j}\bigg|_{\mathbf{x}_0, \mathbf{r}_0} \Delta x_j + \sum_{j=1}^{p} \frac{\partial f_i(\mathbf{x}, \mathbf{r})}{\partial r_j}\bigg|_{\mathbf{x}_0, \mathbf{r}_0} \Delta r_j \quad (4\text{-}107)$$

Equation (4-107) may be written in vector-matrix form:

$$\Delta \dot{\mathbf{x}} = \mathbf{A}^* \Delta \mathbf{x} + \mathbf{B}^* \Delta r \quad (4\text{-}108)$$

where

$$\mathbf{A}^* = \begin{bmatrix} \dfrac{\partial f_1}{\partial x_1} & \dfrac{\partial f_1}{\partial x_2} & \cdots & \dfrac{\partial f_1}{\partial x_n} \\[2mm] \dfrac{\partial f_2}{\partial x_1} & \dfrac{\partial f_2}{\partial x_2} & \cdots & \dfrac{\partial f_2}{\partial x_n} \\[2mm] \cdot & \cdot & \cdots & \cdot \\[2mm] \dfrac{\partial f_n}{\partial x_1} & \dfrac{\partial f_n}{\partial x_2} & \cdots & \dfrac{\partial f_n}{\partial x_n} \end{bmatrix} \quad (4\text{-}109)$$

$$\mathbf{B}^* = \begin{bmatrix} \dfrac{\partial f_1}{\partial r_1} & \dfrac{\partial f_1}{\partial r_2} & \cdots & \dfrac{\partial f_1}{\partial r_p} \\[2mm] \dfrac{\partial f_2}{\partial r_1} & \dfrac{\partial f_2}{\partial r_2} & \cdots & \dfrac{\partial f_2}{\partial r_p} \\[2mm] \cdot & \cdot & \cdots & \cdot \\[2mm] \dfrac{\partial f_n}{\partial r_1} & \dfrac{\partial f_n}{\partial r_2} & \cdots & \dfrac{\partial f_n}{\partial r_p} \end{bmatrix} \quad (4\text{-}110)$$

The following examples serve to illustrate the linearization procedure just described.

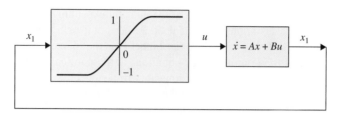

Figure 4-46 Nonlinear control system.

▶ **EXAMPLE 4-9** Figure 4-46 shows the block diagram of a control system with a saturation nonlinearity. The state equations of the system are

$$\dot{x}_1(t) = f_1(t) = x_2(t) \tag{4-111}$$

$$\dot{x}_2(t) = f_2(t) = u(t) \tag{4-112}$$

where the input-output relation of the saturation nonlinearity is represented by

$$u(t) = (1 - e^{-K|x_1(t)|})\text{SGN}\, x_1(t) \tag{4-113}$$

where

$$\text{SGN}\, x_1(t) = \begin{cases} +1 & x_1(t) > 0 \\ -1 & x_1(t) < 0 \end{cases} \tag{4-114}$$

Substituting Eq. (4-113) into Eq. (4-112) and using Eq. (4-107), we have the linearized state equations:

$$\Delta\dot{x}_1(t) = \frac{\partial f_1(t)}{\partial x_2}\Delta x_2(t) = \Delta x_2(t) \tag{4-115}$$

$$\Delta\dot{x}_2(t) = \frac{\partial f_2(t)}{\partial x_1(t)}\Delta x_1(t) = Ke^{-K|x_{01}|}\Delta x_1(t) \tag{4-116}$$

where x_{01} denotes a nominal value of $x_1(t)$. Notice that the last two equations are linear and are valid only for small signals. In vector-matrix form, these linearized state equations are written as

$$\begin{bmatrix} \Delta\dot{x}_1(t) \\ \Delta\dot{x}_2(t) \end{bmatrix} = \begin{bmatrix} 0 & 1 \\ a & 0 \end{bmatrix}\begin{bmatrix} \Delta x_1(t) \\ \Delta x_2(t) \end{bmatrix} \tag{4-117}$$

where

$$a = Ke^{-K|x_{01}|} = \text{constant} \tag{4-118}$$

It is of interest to check the significance of the linearization. If x_{01} is chosen to be at the origin of the nonlinearity, $x_{01} = 0$, then $a = K$; Eq. (4-116) becomes

$$\Delta\dot{x}_2(t) = K\Delta x_1(t) \tag{4-119}$$

Thus, the linearized model is equivalent to having a linear amplifier with a constant gain K. On the other hand, if x_{01} is a large number, the nominal operating point will lie on the saturated portion of the nonlinearity, and $a = 0$. This means that any small variation in $x_1(t)$ (that is, small $\Delta x_1(t)$ will give rise to practically no change in $\Delta\dot{x}_2(t)$. ◀

▶ **EXAMPLE 4-10** In Example 4-9, the linearized system turns out to be time-invariant. As mentioned earlier, linearization of a nonlinear system often results in a linear time-varying system. Consider the following nonlinear system:

$$\dot{x}_1(t) = \frac{-1}{x_2^2(t)} \tag{4-120}$$

$$\dot{x}_2(t) = u(t)x_1(t) \tag{4-121}$$

These equations are to be linearized about the norminal trajectory $[x_{01}(t), x_{02}(t)]$, which is the solution to the equations with initial conditions $x_1(0) = x_2(0) = 1$ and input $u(t) = 0$.

Integrating both sides of Eq. (4-121) with respect to t, we have

$$x_2(t) = x_2(0) = 1 \tag{4-122}$$

Then Eq. (4-120) gives

$$x_1(t) = -t + 1 \tag{4-123}$$

Therefore, the nominal trajectory about which Eqs. (4-120) and (4-121) are to be linearized is described by

$$x_{01}(t) = -t + 1 \tag{4-124}$$

$$x_{02}(t) = 1 \tag{4-125}$$

Now evaluating the coefficients of Eq. (4-107), we get

$$\frac{\partial f_1(t)}{\partial x_1(t)} = 0 \quad \frac{\partial f_1(t)}{\partial x_2(t)} = \frac{2}{x_2^3(t)} \quad \frac{\partial f_2(t)}{\partial x_1(t)} = u(t) \quad \frac{\partial f_2(t)}{\partial u(t)} = x_1(t) \tag{4-126}$$

Equation (4-107) gives

$$\Delta \dot{x}_1(t) = \frac{2}{x_{02}^3(t)} \Delta x_2(t) \tag{4-127}$$

$$\Delta \dot{x}_2(t) = u_0(t) \Delta x_1(t) + x_{01}(t) \Delta u(t) \tag{4-128}$$

By substituting Eqs. (4-124) and (4-125) into Eqs. (4-127) and (4-128), the linearized equations are

$$\begin{bmatrix} \Delta \dot{x}_1(t) \\ \Delta \dot{x}_2(t) \end{bmatrix} = \begin{bmatrix} 0 & 2 \\ 0 & 0 \end{bmatrix} \begin{bmatrix} \Delta x_1(t) \\ \Delta x_2(t) \end{bmatrix} + \begin{bmatrix} 0 \\ 1 - t \end{bmatrix} \Delta u(t) \tag{4-129}$$

which is a set of linear state equations with time-varying coefficients.   ◀

▶ **EXAMPLE 4-11**   Figure 4-47 shows the diagram of a magnetic-ball suspension system. The objective of the system is to control the position of the steel ball by adjusting the current in the electromagnet through the input voltage $e(t)$. The differential equations of the system are

$$M \frac{d^2 y(t)}{dt^2} = Mg - \frac{i^2(t)}{y(t)} \tag{4-130}$$

$$e(t) = Ri(t) + L \frac{di(t)}{dt} \tag{4-131}$$

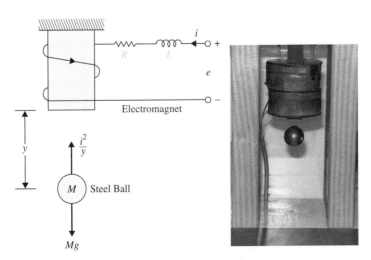

Figure 4-47 Magnetic-ball-suspension system.

where

$e(t)$ = input voltage $\qquad$ $y(t)$ = ball position

$i(t)$ = winding current $\qquad$ R = winding resistance

L = winding inductance $\qquad$ M = mass of ball

g = gravitational acceleration

Let us define the state variables as $x_1(t) = y(t)$, $x_2(t) = dy(t)/dt$, and $x_3(t) = i(t)$. The state equations of the system are

$$\frac{dx_1(t)}{dt} = x_2(t) \tag{4-132}$$

$$\frac{dx_2(t)}{dt} = g - \frac{1}{M}\frac{x_3^2(t)}{x_1(t)} \tag{4-133}$$

$$\frac{dx_3(t)}{dt} = -\frac{R}{L}x_3(t) + \frac{1}{L}e(t) \tag{4-134}$$

Let us linearize the system about the equilibrium point $y_0(t) = x_{01}$ = constant. Then,

$$x_{02}(t) = \frac{dx_{01}(t)}{dt} = 0 \tag{4-135}$$

$$\frac{d^2 y_0(t)}{dt^2} = 0 \tag{4-136}$$

The nominal value of $i(t)$ is determined by substituting Eq. (4-136) into Eq. (4-130). Thus,

$$i_0(t) = x_{03}(t) = \sqrt{Mgx_{01}} \tag{4-137}$$

The linearized state equation is expressed in the form of Eq. (4-108), with the coefficient matrices **A*** and **B*** evaluated as

$$\mathbf{A}^* = \begin{bmatrix} 0 & 1 & 0 \\ \dfrac{x_{03}^2}{Mx_{01}^2} & 0 & \dfrac{-2x_{03}}{Mx_{01}} \\ 0 & 0 & -\dfrac{R}{L} \end{bmatrix} = \begin{bmatrix} 0 & 1 & 0 \\ \dfrac{g}{x_{01}} & 0 & -2\left(\dfrac{g}{Mx_{01}}\right)^{1/2} \\ 0 & 0 & -\dfrac{R}{L} \end{bmatrix} \tag{4-138}$$

$$\mathbf{B}^* = \begin{bmatrix} 0 \\ 0 \\ \dfrac{1}{L} \end{bmatrix} \tag{4-139}$$

◀

▶ 4-8 SYSTEMS WITH TRANSPORTATION LAGS (TIME DELAYS)

Thus far the systems considered all have transfer functions that are quotients of polynomials. In practice, pure time delays may be encountered in various types of systems, especially systems with hydraulic, pneumatic, or mechanical transmissions. Systems with computer control also have time delays, since it takes time for the computer to execute numerical operations. In these systems, the output will not begin to respond to an input until after a given time interval. Figure 4-48 illustrates systems in which transportation lags or pure time delays are observed. Figure 4-48(a) outlines an arrangement in which two different fluids are to be mixed in appropriate proportions. To assure that a homogeneous solution is measured, the monitoring point is located some distance from the mixing point. A time delay therefore exists between the mixing point and the place where the change in concentration is detected. If the rate of flow of the mixed solution is v inches

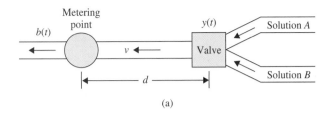

(a)

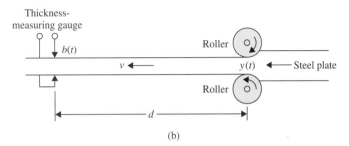

Figure 4-48 Systems with transportation lag.

(b)

per second and d is the distance between the mixing and the metering points, the time lag is given by

$$T_d = \frac{d}{v} \text{ seconds} \tag{4-140}$$

If it is assumed that the concentration of the mixing point is $y(t)$ and that it is reproduced without change T_d seconds later at the monitoring point, the measured quantity is

$$b(t) = y(t - T_d) \tag{4-141}$$

The Laplace transform of Eq. (4-141) is

$$B(s) = e^{-T_d s} Y(s) \tag{4-142}$$

where $Y(s)$ is the Laplace transform of $y(t)$. The transfer function between $b(t)$ and $y(t)$ is

$$\frac{B(s)}{Y(s)} = e^{-T_d s} \tag{4-143}$$

Figure 4-48(b) illustrates the control of thickness of rolled steel plates. The transfer function between the thickness at the rollers and the measuring point is again given by Eq. (4-143).

4-8-1 Approximation of the Time-Delay Function by Rational Functions

Systems that are described inherently by transcendental transfer functions are more difficult to handle. Many analytical tools such as the Routh-Hurwitz criterion (Chapter 6) are restricted to rational transfer functions. The root-locus technique (Chapter 8) is also more easily applied only to systems with rational transfer functions.

There are many ways of approximating $e^{-T_d s}$ by a rational function. One way is to approximate the exponential function by a Maclaurin series; that is,

$$e^{-T_d s} \cong 1 - T_d s + \frac{T_d^2 s^2}{2} \tag{4-144}$$

or

$$e^{-T_d s} \cong \frac{1}{1 + T_d s + T_d^2 s^2/2} \tag{4-145}$$

where only three terms of the series are used. Apparently, the approximations are not valid when the magnitude of $T_d s$ is large.

A better approximation is to use the Pade approximation [5, 6], which is given in the following for a two-term approximation:

$$e^{-T_d s} \cong \frac{1 - T_d s/2}{1 + T_d s/2} \tag{4-146}$$

The approximation of the transfer function in Eq. (4-146) contains a zero in the right-half s-plane so that the step response of the approximating system may exhibit a small negative undershoot near $t = 0$.

▶ 4-9 A SUN-SEEKER SYSTEM

In this section we shall model a sun-seeker control system whose purpose is to control the attitude of a space vehicle so that it will track the sun with high accuracy. In the system described here, tracking the sun in only one plane is accomplished. A schematic diagram of the system is shown in Fig. 4-49. The principal elements of the error discriminator are two small rectangular silicon photovoltaic cells mounted behind a rectangular slit in an enclosure. The cells are mounted in such a way that when the sensor is pointed at the sun, a beam of light from the slit overlaps both cells. Silicon cells are used as current sources and connected in opposite polarity to the input of an op-amp. Any difference in the short-circuit current of the two cells is sensed and amplified by the op-amp. Since the current of each cell is proportional to the illumination on the cell, an error signal will be present at the output

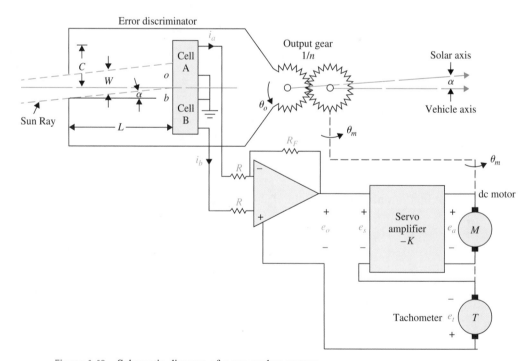

Figure 4-49 Schematic diagram of a sun-seeker system.

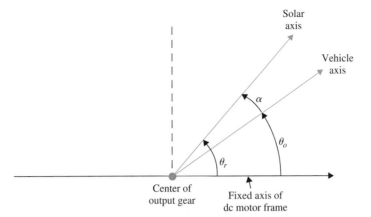

Figure 4-50 Coordinate system of the sun-seeker system.

of the amplifier when the light from the slit is not precisely centered on the cells. This error voltage, when fed to the servoamplifier, will cause the motor to drive the system back into alignment. The description of each part of the system is given in following sections.

4-9-1 Coordinate System

The center of the coordinate system is considered to be at the output gear of the system. The reference axis is taken to be the fixed frame of the dc motor, and all rotations are measured with respect to this axis. The solar axis or the line from the output gear to the sun makes an angle $\theta_r(t)$ with respect to the reference axis, and $\theta_o(t)$ denotes the vehicle axis with respect to the reference axis. The objective of the control system is to maintain the error between $\theta_r(t)$ and $\theta_o(t)$, $\alpha(t)$, near zero:

$$\alpha(t) = \theta_r(t) - \theta_o(t) \tag{4-147}$$

The coordinate system described is illustrated in Fig. 4-50.

4-9-2 Error Discriminator

When the vehicle is aligned perfectly with the sun, $\alpha(t) = 0$, and $i_a(t) = i_b(t) = I$, or $i_a(t) = i_b(t) = 0$. From the geometry of the sun ray and the photovoltaic cells shown in Fig. 4-49, we have

$$oa = \frac{W}{2} + L \tan \alpha(t) \tag{4-148}$$

$$ob = \frac{W}{2} - L \tan \alpha(t) \tag{4-149}$$

where oa denotes the width of the sun ray that shines on cell A, and ob is the same on cell B, for a given $\alpha(t)$. Since the current $i_a(t)$ is proportional to oa, and $i_b(t)$ is proportional to ob, we have

$$i_a(t) = I + \frac{2LI}{W} \tan \alpha(t) \tag{4-150}$$

$$i_b(t) = I - \frac{2LI}{W} \tan \alpha(t) \tag{4-151}$$

for $0 \leq \tan \alpha(t) \leq W/2L$. For $W/2L \leq \tan \alpha(t) \leq (C - W/2)/L$, the sun ray is completely on cell A, and $i_a(t) = 2I$, $i_b(t) = 0$. For $(C - W/2)L \leq \tan \alpha(t) \leq (C + W/2)L$, $i_a(t)$ decreases

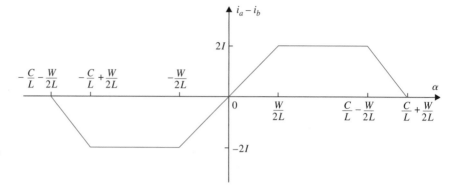

Figure 4-51 Nonlinear characteristic of the error discriminator. The abscissa is tan α, but it is approximated by α for small values of α.

linearly from $2I$ to zero. $i_a(t) = i_b(t) = 0$ for tan $\alpha(t) \geq (C + W/2)/L$. Therefore, the error discriminator may be represented by the nonlinear characteristic of Fig. 4-51, where for small angle $\alpha(t)$, tan $\alpha(t)$ has been approximated by $\alpha(t)$ on the abscissa.

4-9-3 Op-Amp

The relationship between the output of the op-amp and the currents $i_a(t)$ and $i_b(t)$ is

$$e_o(t) = -R_F[i_a(t) - i_b(t)] \tag{4-152}$$

4-9-4 Servoamplifier

The gain of the servoamplifier is $-K$. With reference to Fig. 4-49, the output of the servoamplifier is expressed as

$$e_a(t) = -K[e_o(t) + e_t(t)] = -Ke_s(t) \tag{4-153}$$

4-9-5 Tachometer

The output voltage of the tachometer e_t is related to the angular velocity of the motor through the tachometer constant K_t:

$$e_t(t) = K_t \omega_m(t) \tag{4-154}$$

The angular position of the output gear is related to the motor position through the gear ratio $1/n$. Thus,

$$\theta_o = \frac{1}{n}\theta_m \tag{4-155}$$

4-9-6 DC Motor

The dc motor has been modeled in Section 4-6. The equations are

$$
\begin{aligned}
e_a(t) &= R_a i_a(t) + e_b(t) \\
e_b(t) &= K_b \omega_m(t) \\
T_m(t) &= K_i i_a(t) \\
T_m(t) &= J\frac{d\omega_m(t)}{dt} + B\omega_m(t)
\end{aligned} \tag{4-156}
$$

where J and B are the inertia and viscous-friction coefficient seen at the motor shaft. The inductance of the motor is neglected in Eq. (4-156). A block diagram that characterizes all the functional relations of the system is shown in Fig. 4-52.

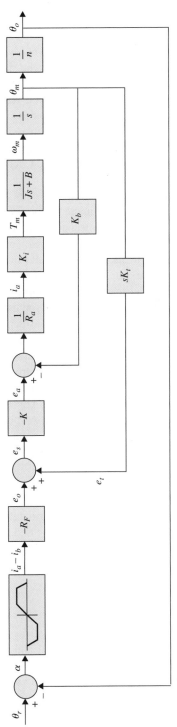

Figure 4-52 Block diagram of the sun-seeker system.

▶ 4-10 MATLAB TOOLS AND CASE STUDIES

Since the focus in this chapter is on the modeling aspects, no specific software description will appear in this section. However, it is very important to reiterate that a good model is the backbone of any realistic computer simulation. In practice, it is best to design and test a controller off-line, by evaluating the system performance in the "safety" of the simulation environment. The simulation model can be based on system parameters, which may be available or may be identified experimentally. The response of the actual system to the same test input will then verify the validity of the model. If the actual response to the test input were significantly different from the predicted response, certain model parameter values would have to be revised, or the model structure would have to be refined to better reflect the observed system behavior. Once satisfactory model performance has been achieved, various control schemes can be implemented.

The key components of a control system include actuators, sensors, feedback, amplifiers, and analog circuits, to name a few. For example, in a realistic system including an actuator (e.g., a dc motor), mechanical parts (gears), and electrical components (amplifiers), accurate modeling of these components must be completed before designing a controller. A decision to include all aspects such as amplifier saturation, friction in the motor, backlash in gears, or circuit nonlinearities may improve the model. However, the complexity of the model may result in a more complicated controller design, which will ultimately increase the cost and degree of sophistication of the system. So, as in any other field, controller-design engineers must consider all the issues at hand before designing a control system. Computer simulation techniques provide designers with a powerful means to test the degree of sophistication and accuracy required.

The **A**utomatic **C**ontrol **Sys**tems software (**ACSYS**) consists of a number of m-files and GUIs (graphical user Interfaces) for the analysis of simple control transfer functions. It can be invoked from the MATLAB command line by simply typing **ACSYS**, and then clicking the appropriate button. A specific MATLAB tool has been developed for most chapters of this textbook. Throughout this chapter we have identified subjects that may be solved using **ACSYS** with a box in the left margin of the text, called MATLAB Toolbox.

The components of **ACSYS** most relevant to the problems in this chapter are Virtual Lab and SIMLab, which are discussed in detail in Chapter 11. These simulation tools provide the user with virtual experiments and design projects using systems involving dc motors, sensors, and electronic and mechanical components. Virtual Lab and SIMLab also address the practical issues discussed in this chapter (see, for example, section 4-10).

▶ 4-11 SUMMARY

This chapter was devoted to the mathematical modeling of physical systems. The basic mathematical relations of linear electrical and mechanical systems were described, because linear systems, differential equations, state equations, and transfer functions are the fundamental tools of modeling. The operations and mathematical descriptions of some of the commonly used components in control systems, such as error detectors, tachometers, and dc motors, were presented in this chapter.

Due to space limitations and the intended scope of this text, only some of the physical devices used in practice were described. The main purpose of this chapter was to illustrate the methods of system modeling, and the coverage was not intended to be exhaustive.

Since nonlinear systems cannot be ignored in the real world, and this book is not devoted to the subject, Section 4-7 introduced the linearization of nonlinear systems at a nominal operating point. Once the linearized model is determined, the performance of the nonlinear system can be investigated under small-signal conditions at the designated operating point.

Systems with pure time delays were modeled, and methods of approximating the transfer functions by rational ones were described.

Finally, The components of ACSYS most relevant to the problems in this chapter, Virtual Lab and SIMLab, were described. These simulation tools provide the user with virtual experiments and design projects using systems involving dc motors, sensors, and electronic and mechanical components.

► REVIEW
QUESTIONS

1. Among the three types of friction described, which type is governed by a linear mathematical relation?

2. Given a two-gear system with angular displacement θ_1 and θ_2, numbers of teeth N_1 and N_2, and torques T_1 and T_2, write the mathematical relations between these variables and parameters.

3. How are potentiometers used in control systems?

4. Digital encoders are used in control systems for position and speed detection. Consider that an encoder is set up to output 3600 zero crossings per revolution. What is the angular rotation of the encoder shaft in degrees if 16 zero crossings are detected?

5. The same encoder described in Question 4 and an electronic clock with a frequency of 1 MHz are used for speed measurement. What is the average speed of the encoder shaft in rpm if 500 clock pulses are detected between two consecutive zero crossings of the encoder?

6. Give the advantages of dc motors for control-system applications.

7. What are the sources of nonlinearities in a dc motor?

8. What are the effects of inductance and inertia in a dc motor?

9. What is back emf in a dc motor, and how does it affect the performance of a control system?

10. What are the electrical and mechanical time constants of an electric motor?

11. Under what condition is the torque constant K_i of a dc motor valid, and how is it related to the back-emf constant K_b?

12. An inertial and frictional load is driven by a dc motor with torque T_m. The dynamic equation of the system is

$$T_m(t) = J_m \frac{d\omega_m(t)}{dt} + B_m \omega_m$$

If the inertia is doubled, how will it affect the steady-state speed of the motor? How will the steady-state speed be affected if, instead, the frictional coefficient B_m is doubled? What is the mechanical constant of the system?

13. What is a tachometer, and how is it used in control systems?

14. Give the transfer function of a pure time delay T_d.

15. Does the linearization technique described in this chapter always result in a linear time-invariant system?

► **REFERENCES**

Mechanical Systems

1. R. Cannon, *Dynamics of Physical Systems*, McGraw-Hill, New York, 1967.

DC Motors

2. B. C. Kuo and J. Tal, eds., *Incremental Motion Control*, Vol. 1, *DC Motors and Control Systems*, SRL Publishing, Champaign, IL, 1979.

Rolling Dry Friction

3. P. B. Dahl, *A Solid Friction Model*, Report No. TOR-0158 (3107-18)-1, Aerospace Corporation, El Segundo, CA, May 1968.
4. N. A. Osborn and D. L Rittenhouse, "The Modeling of Friction and Its Effects on Fine Pointing Control," *AIAA Mechanics and Control of Flight Conference*, Paper No. 74-875, August, 1974.

Pade Approximation

5. J. G. Truxal, *Automatic Feedback Control System Synthesis*, McGraw-Hill, New York, 1955.
6. H. S. Wall, *Continued Fractions*, Chapter 20, D. Van Nostrand, New York, 1948.

► **PROBLEMS**

• State equations, state diagrams, transfer functions

4-1. Write the force equations of the linear translational systems shown in Fig. 4P-1.

(a) Draw state diagrams using a minimum number of integrators. Write the state equations from the state diagrams.

(b) Define the state variables as follows:

   (i) $x_1 = y_2$, $x_2 = dy_2/dt$, $x_3 = y_1$, and $x_4 = dy_1/dt$

   (ii) $x_1 = y_2$, $x_2 = y_1$, and $x_3 = dy_1/dt$

   (iii) $x_1 = y_1$, $x_2 = y_2$, and $x_3 = dy_2/dt$

Write the state equations and draw the state diagram with these state variables. Find the transfer functions $Y_1(s)/F(s)$ and $Y_2(s)/F(s)$.

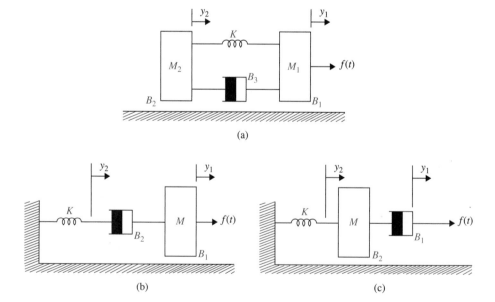

Figure 4P-1

(a)

(b)   (c)

• Force equations, state diagram, state equations, transfer functions

4-2. Write the force equations of the linear translational system shown in Fig. 4P-2. Draw the state diagram using a minimum number of integrators. Write the state equations from the state diagram. Find the transfer functions $Y_1(s)/F(s)$ and $Y_2(s)/F(s)$. Set $Mg = 0$ for the transfer functions.

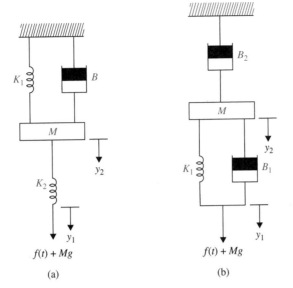

Figure 4P-2

(a)　　　　(b)

• Rotational system, torque equations, state equations, transfer functions

4-3. Write the torque equations of the rotational systems shown in Fig. 4P-3. Draw state diagrams using a minimum number of integrators. Write the state equations from the state diagrams. Find the transfer function $\Theta(s)/T(s)$ for the system in (a). Find the transfer functions $\Theta_1(s)/T(s)$ and $\Theta_2(s)/T(s)$ for the systems in parts (b), (c), (d), and (e).

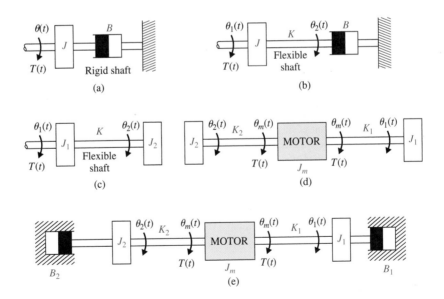

Figure 4P-3

• Motor-control system

4-4. An open-loop motor control system is shown in Fig. 4P-4. The potentiometer has a maximum range of 10 turns (20π rad). Find the transfer functions $E_o(s)/T_m(s)$. The following parameters and variables are defined: $\theta_m(t)$ is the motor displacement; $\theta_L(t)$, the load displacement; $T_m(t)$, the motor torque; J_m, the motor inertia; B_m, the motor viscous-friction coefficient; B_p, the potentiometer viscous-friction coefficient; $e_o(t)$, the output voltage; and K, the torsional spring constant.

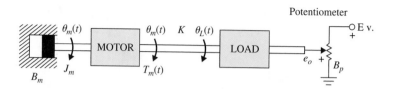

Figure 4P-4

• Gear-train system

4-5. Write the torque equations of the gear-train system shown in Fig. 4P-5. The moments of inertia of gears are lumped as J_1, J_2, and J_3. $T_m(t)$ is the applied torque; N_1, N_2, N_3, and N_4 are the number of gear teeth. Assume rigid shafts.

(a) Assume that J_1, J_2, and J_3 are negligible. Write the torque equations of the system. Find the total inertia of the motor.

(b) Repeat part (a) with the moments of inertia J_1, J_2, and J_3.

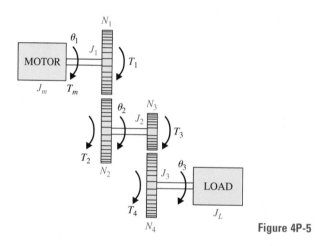

Figure 4P-5

• Trailer-towing system

4-6. A vehicle towing a trailer through a spring-damper coupling hitch is shown in Fig. 4P-6. The following parameters and variables are defined: M is the mass of the trailer; K_h, the spring constant of the hitch; B_h, the viscous damping coefficient of the hitch; B_t, the viscous-friction coefficient of the trailer; $y_1(t)$, the displacement of the towing vehicle; $y_2(t)$, the displacement of the trailer; and $f(t)$, the force of the towing vehicle.

(a) Write the differential equation of the system.

(b) Write the state equations by defining the following state variables: $x_1(t) = y_1(t) - y_2(t)$ and $x_2(t) = dy_2(t)dt$.

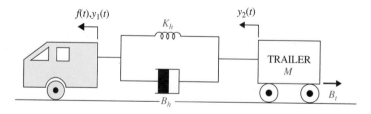

Figure 4P-6

• Motor-load system

4-7. Figure 4P-7 shows a motor-load system coupled through a gear train with gear ratio $n = N_1/N_2$. The motor torque is $T_m(t)$, and $T_L(t)$ represents a load torque.

(a) Find the optimum gear ratio n^* such that the load acceleration $\alpha_L = d^2\theta_L/dt^2$ is maximized.

(b) Repeat part (a) when the load torque is zero.

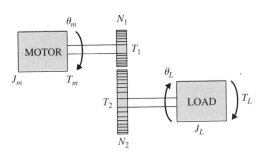

Figure 4P-7

Printwheel control system

4-8. Figure 4P-8 shows the simplified diagram of the printwheel control system of a word processor. The printwheel is controlled by a dc motor through belts and pulleys. Assume that the belts are rigid. The following parameters and variables are defined: $T_m(t)$ is the motor torque; $\theta_m(t)$, the motor displacement; $y(t)$, the linear displacement of the printwheel; J_m, the motor inertia; B_m, the motor viscous-friction coefficient; r, the pulley radius; and M, the mass of the printwheel.

(a) Write the differential equation of the system.

(b) Find the transfer function $Y(s)/T_m(s)$.

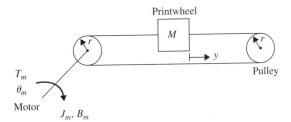

Figure 4P-8

• Printwheel control system

4-9. Figure 4P-9 shows the diagram of a printwheel system with belts and pulleys. The belts are modeled as linear springs with spring constants K_1 and K_2.

(a) Write the differential equations of the system using θ_m and y as the dependent variables.

(b) Write the state equations using $x_1 = r\theta_m - y$, $x_2 = dy/dt$, and $x_3 = \omega_m = d\theta_m/dt$ as the state variables.

(c) Draw a state diagram for the system.

(d) Find the transfer function $Y(s)/T_m(s)$.

(e) Find the characteristic equation of the system.

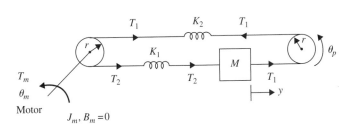

Figure 4P-9

• Motor-load system

4-10. The schematic diagram of a motor-load system is shown in Fig. 4P-10. The following parameters and variables are defined: $T_m(t)$ is the motor torque; $\omega_m(t)$, the motor velocity; $\theta_m(t)$, the motor displacement; $\omega_L(t)$, the load velocity; $\theta_L(t)$, the load displacement; K, the torsional spring constant; J_m, the motor inertia; B_m, the motor viscous-friction coefficient; and B_L, the load viscous-friction coefficient.

(a) Write the torque equations of the system.

(b) Find the transfer functions $\Theta_L(s)/T_m(s)$ and $\Theta_m(s)/T_m(s)$.

(c) Find the characteristic equation of the system.

(d) Let $T_m(t) = T_m$ be a constant applied torque; show that $\omega_m = \omega_L = $ constant in the steady state. Find the steady-state speeds ω_m and ω_L.

(e) Repeat part (d) when the value of J_L is doubled, but J_m stays the same.

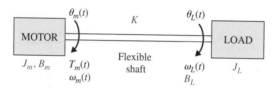

Figure 4P-10

• Motor-tachometer system

4-11. The schematic diagram of a control system containing a motor coupled to a tachometer and an inertial load is shown in Fig. 4P-11. The following parameters and variables are defined: T_m is the motor torque; J_m, the motor inertia; J_t, the tachometer inertia; J_L, the load inertia; K_1 and K_2, the spring constants of the shafts; θ_t, the tachometer displacement; θ_m, the motor velocity; θ_L, the load displacement; ω_t, the tachometer velocity; ω_L, the load velocity; and B_m, the motor viscous-friction coefficient.

(a) Write the state equations of the system using θ_L, ω_L, θ_t, ω_t, θ_m, and ω_m as the state variables (in the listed order). The motor torque T_m is the input.

(b) Draw a state diagram with T_m at the left and ending with θ_L on the far right. The state diagram should have a total of 10 nodes. Leave out the initial states.

(c) Find the following transfer functions:

$$\frac{\Theta_L(s)}{T_m(s)} \qquad \frac{\Theta_t(s)}{T_m(s)} \qquad \frac{\Theta_m(s)}{T_m(s)}$$

(d) Find the characteristic equation of the system.

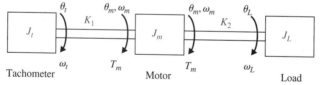

Figure 4P-11

• DC-motor system

4-12. The voltage equation of a dc motor is written as

$$e_a(t) = R_a i_a(t) + L_a \frac{di_a(t)}{dt} + K_b \omega_m(t)$$

where $e_a(t)$ is the applied voltage; $i_a(t)$, the armature current; R_a, the armature resistance; L_a, the armature inductance; K_b, the back-emf constant; $\omega_m(t)$, the motor velocity; and $\omega_r(t)$, the reference input voltage. Taking the Laplace transform on both sides of the voltage equation, with zero initial conditions, and solving for $\Omega_m(s)$, we get

$$\Omega_m(s) = \frac{E_a(s) - (R_a + L_a s)I_a(s)}{K_b}$$

which shows that the velocity information can be generated by feeding back the armature voltage and current. The block diagram in Fig. 4P-12 shows a dc-motor system, with voltage and current feedbacks, for speed control.

(a) Let K_1 be a very high gain amplifier. Show that when $H_i(s)/H_e(s) = -(R_a + L_a s)$, the motor velocity $\omega_m(t)$ is totally independent of the load-disturbance torque T_L.

(b) Find the transfer function between $\Omega_m(s)$ and $\Omega_r(s)$ ($T_L = 0$) when $H_i(s)$ and $H_e(s)$ are selected as in part (a).

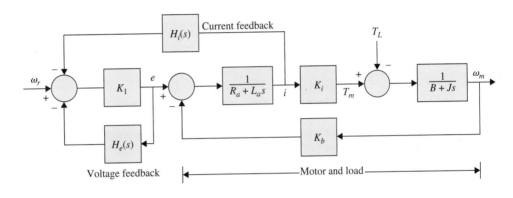

Figure 4P-12

• Attitude control of guided missile

4-13. This problem deals with the attitude control of a guided missile. When traveling through the atmosphere, a missile encounters aerodynamic forces that tend to cause instability in the attitude of the missile. The basic concern from the flight-control standpoint is the lateral force of the air, which tends to rotate the missile about its center of gravity. If the missile centerline is not aligned with the direction in which the center of gravity C is traveling, as shown in Fig. 4P-13, with angle θ, which is also called the angle of attack, a side force is produced by the drag of the air through which the missile travels. The total force F_α may be considered to be applied at the center of pressure P. As shown in Fig. 4P-13, this side force has a tendency to cause the missile to tumble end over end, especially if the point P is in front of the center of gravity C. Let the angular acceleration of the missile about the point C, due to the side force, be denoted by α_F. Normally, α_F is directly proportional to the angle of attack θ and is given by

$$\alpha_F = \frac{K_F d_1}{J}\theta$$

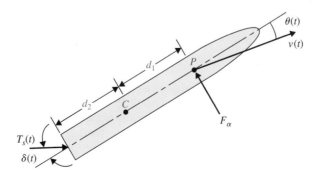

Figure 4P-13

where K_F is a constant that depends on such parameters as dynamic pressure, velocity of the missile, air density, and so on, and

$$J = \text{missile moment of inertia about } C$$
$$d_1 = \text{distance between } C \text{ and } P$$

The main objective of the flight-control system is to provide the stabilizing action to counter the effect of the side force. One of the standard control means is to use gas injection at the tail of the missile to deflect the direction of the rocket engine thrust T_s, as shown in the figure.

(a) Write a torque differential equation to relate T_s, δ, θ, and the system parameters given. Assume that δ is very small, so $\sin \delta(t)$ is approximated by $\delta(t)$.

(b) Assume that T_s is a constant torque. Find the transfer function $\Theta(s)/\Delta(s)$, where $\Theta(s)$ and $\Delta(s)$ are the Laplace transforms of $\theta(t)$ and $\delta(t)$, respectively. Assume that $\delta(t)$ is very small.

(c) Repeat parts (a) and (b) with points C and P interchanged. The d_1 in the expression of α_F should be changed to d_2.

Printwheel control system

4-14. Figure 4P-14(a) shows the schematic diagram of a dc-motor control system for the control of a printwheel of a word processor. The load in this case is the printwheel, which is directly coupled to the motor shaft. The following parameters and variables are defined: K_s is the error-detector gain (V/rad); K_i, the torque constant (oz-in./A); K, the amplifier gain (V/V); K_b, the back-emf constant (V/rad/sec); n, the gear train ratio $= \theta_2/\theta_m = T_m/T_2$; B_m, the motor viscous-friction coefficient (oz-in.-sec); J_m, the motor inertia (oz-in.-sec²); K_L the torsional spring constant of the motor shaft (oz-in./rad); and J_L, the load inertia (oz-in.-sec²).

(a) Write the cause-and-effect equations of the system. Rearrange these equations into the form of state equations with $x_1 = \theta_o$, $x_2 = \omega_o$, $x_3 = \theta_m$, $x_4 = \omega_m$, and $x_5 = i_a$.

(b) Draw a state diagram using the nodes shown in Fig. 3P-38(b).

(c) Derive the forward-path transfer function (with the outer feedback path open): $G(s) = \Theta_o(s)/\Theta_e(s)$. Find the closed-loop transfer function $M(s) = \Theta_o(s)/\Theta_r(s)$.

(e) Repeat part (c) when the motor shaft is rigid, that is, $K_L = \infty$. Show that you can obtain the solutions by taking the limit as K_L approaches infinity in the results in part (c).

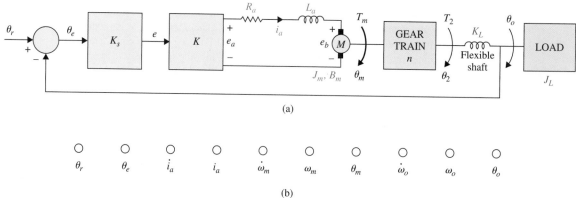

(a)

(b)

Figure 4P-14

• Voice-coil motor

4-15. The schematic diagram of a voice-coil motor (VCM), used as a linear actuator in a disk memory-storage system, is shown in Fig. 4P-15(a). The VCM consists of a cylindrical permanent magnet (PM) and a voice coil. When current is sent through the coil, the magnetic field of the PM interacts with the current-carrying conductor, causing the coil to move linearly. The voice coil of the VCM in Fig. 4P-15(a) consists of a primary coil and a shorted-turn coil. The latter is

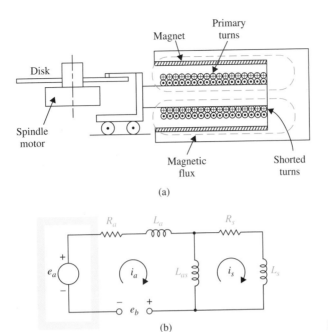

Figure 4P-15

installed for the purpose of effectively reducing the electric constant of the device. Figure 4P-15(b) shows the equivalent circuit of the coils. The following parameters and variables are defined: $e_a(t)$ is the applied coil voltage; $i_a(t)$, the primary-coil current; $i_s(t)$, the shorted-turn coil current; R_a, the primary-coil resistance; L_a, the primary-coil inductance; L_{as}, the mutual inductance between the primary and shorted-turn coils; $v(t)$, the velocity of the voice coil; $y(t)$, the displacement of the voice coil; $f(t) = K_i v(t)$, the force of the voice coil; K_i, the force constant; K_b, the back-emf constant; $e_b(t), = K_b v(t)$ the back emf; M_T, the total mass of the voice coil and load; and B_T, the total viscous-friction coefficient of the voice coil and load.

(a) Write the differential equations of the system.

(b) Draw a block diagram of the system with $E_a(s)$, $I_a(s)$, $I_s(s)$, $V(s)$, and $Y(s)$ as variables.

(c) Derive the transfer function $Y(s)/E_a(s)$.

• DC-motor control system **4-16.** A dc-motor position-control system is shown in Fig. 4P-16(a). The following parameters and variables are defined: e is the error voltage; e_r, the reference input; θ_L, the load position; K_A, the amplifier gain; e_a, the motor input voltage; e_b, the back emf; i_a, the motor current; T_m, the motor torque; J_m, the motor inertia = 0.03 oz-in.-sec²; B_m, the motor viscous-friction coefficient = 10 oz-in.-sec; K_L, the torsional spring constant = 50,000 oz-in./rad; J_L, the load inertia = 0.05 oz-in.-sec²; K_i, the motor torque constant = 21 oz-in./A; K_b, the back-emf constant = 15.5 V/1000 rpm; K_s, the error-detector gain = $E/2\pi$; E, the error-detector applied voltage = 2π V; R_a, the motor resistance = 1.15 Ω; and $\theta_e = \theta_r - \theta_L$.

(a) Write the state equations of the system using the following state variables:

$$x_1 = \theta_L, x_2 = d\theta_L/dt = \omega_L, x_3 = \theta_3, \text{ and } x_4 = d\theta_m/dt = \omega_m$$

(b) Draw a state diagram using the nodes shown in Fig. 4P-16(b).

(c) Derive the forward-path transfer function $G(s) = \Theta_L(s)/\Theta_e(s)$ when the outer feedback path from θ_L is opened. Find the poles of $G(s)$.

(d) Derive the closed-loop transfer function $M(s) = \Theta_L(s)/\Theta_e(s)$. Find the poles of $M(s)$ when $K_A = 1$, 2738, and 5476. Locate these poles in the s-plane, and comment on the significance of these values of K_A.

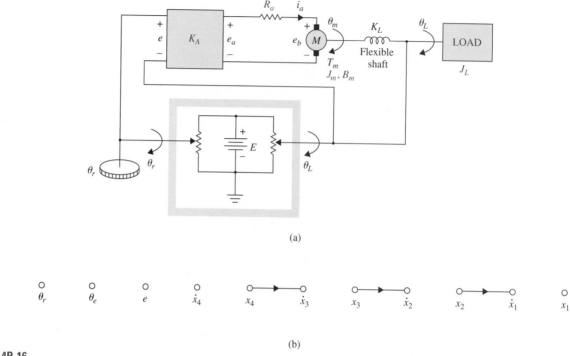

(a)

(b)

Figure 4P-16

• Temperature control of air-flow system

4-17. Figure 4P-17(a) shows the setup of the temperature control of an air-flow system. The hot-water reservoir supplies the water that flows into the heat exchanger for heating the air. The temperature sensor senses the air temperature T_{AO} and sends it to be compared with the reference temperature T_r. The temperature error T_e is sent to the controller, which has the transfer function $G_c(s)$. The output of the controller, $u(t)$, which is an electric signal, is converted to a pneumatic signal by a transducer. The output of the actuator controls the water-flow rate through the three-way valve. Figure 4P-17(b) shows the block diagram of the system.

The following parameters and variables are defined: dM_w is the flow rate of the heating fluid $= K_M u$, $K_M = 0.054$ kg/sec/V; T_W, the water temperature $= K_R dM_W$; $K_R = 65°C$/kg/sec; and T_{AO} the output air temperature.

Heat-transfer equation between water and air:

$$\tau_c \frac{dT_{AO}}{dt} = T_w - T_{AO} \qquad \tau_c = 10 \text{ seconds}$$

Temperature sensor equation:

$$\tau_s \frac{dT_s}{dt} = T_{AO} - T_s \qquad \tau_s = 2 \text{ seconds}$$

(a) Draw a functional block diagram that includes all the transfer functions of the system.

(b) Derive the transfer function $T_{AO}(s)/T_r(s)$ when $G_c(s) = 1$.

Idle-speed control system

4-18. The objective of this problem is to develop a linear analytical model of the automobile engine for idle-speed control system shown in Fig. 1-2. The input of the system is the throttle position that controls the rate of air flow into the manifold (see Fig. 4P-18). Engine torque is developed from the buildup of manifold pressure due to air intake and the intake of the air/gas mixture into the cylinder. The engine variations are as follows:

$q_i(t)$ = amount of air flow across throttle into manifold

$dq_i(t)/dt$ = rate of air flow across throttle into manifold

(a)

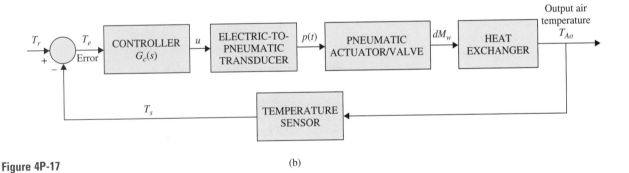

(b)

Figure 4P-17

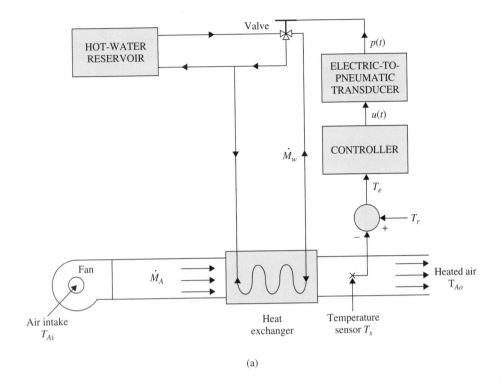

Figure 4P-18

$q_m(t)$ = average air mass in manifold
$q_o(t)$ = amount of air leaving intake manifold through intake valves
$dq_o(t)/dt$ = rate of air leaving intake manifold through intake valves
$T(t)$ = engine torque
T_d = disturbance torque due to application of auto accessories, a constant
$\omega(t)$ = engine speed
$\alpha(t)$ = throttle position
τ_D = time delay in engine
J_e = inertia of engine

The following assumptions and mathematical relations between the engine variables are given:

1. The rate of air flow into the manifold is linearly dependent on the throttle position:

$$\frac{dq_i(t)}{dt} = K_1\alpha(t) \qquad K_1 = \text{proportional constant}$$

2. The rate of air flow leaving the manifold depends linearly on the air mass in the manifold and the engine speed:

$$\frac{dq_o(t)}{dt} = K_2 q_m(t) + K_3\omega(t) \qquad K_2, K_3 = \text{constants}$$

3. A pure time delay of τ_D seconds exists between the change in the manifold air mass and the engine torque:

$$T(t) = K_4 q_m(t - \tau_D) \qquad K_4 = \text{constant}$$

4. The engine drag is modeled by a viscous-friction torque $B\omega(t)$, where B is the viscous-friction coefficient.

5. The average air mass $q_m(t)$ is determined from

$$q_m(t) = \int \left(\frac{dq_i(t)}{dt} - \frac{dq_o(t)}{dt} \right) dt$$

6. The equation describing the mechanical components is

$$T(t) = J\frac{d\omega(t)}{dt} + B\omega(t) + T_d$$

(a) Draw a functional block diagram of the system with $\alpha(t)$ as input, $\omega(t)$ as output, and T_d as the disturbance input. Show the transfer function of each block.

(b) Find the transfer function $\Omega(s)/\alpha(s)$ of the system.

(c) Find the characteristic equation and show that it is not rational with constant coefficients.

(d) Approximate the engine time delay by

$$e^{-\tau_D s} \cong \frac{1 - \tau_D s/2}{1 + \tau_D s/2}$$

and repeat parts (b) and (c).

• Phase-locked loop control

4-19. Phase-locked loops are control systems used for precision motor-speed control. The basic elements of a phase-locked loop system incorporating a dc motor is shown in Fig. 4P-19(a). An input pulse train represents the reference frequency or desired output speed. The digital encoder produces digital pulses that represent motor speed. The phase detector compares the motor speed and the reference frequency and sends an error voltage to the filter (controller) that governs the dynamic response of the system. Phase detector gain = K_p, encoder gain = K_e, counter gain = $1/N$, and dc-motor torque constant = K_i. Assume zero inductance and zero friction for the motor.

(a) Derive the transfer function $E_c(s)/E(s)$ of the filter shown in Fig. 4P-19(b). Assume that the filter sees infinite impedance at the output and zero impedance at the input.

(b) Draw a functional block diagram of the system with gains or transfer functions in the blocks.

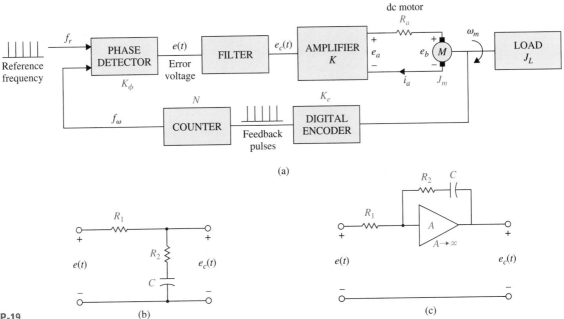

Figure 4P-19

(c) Derive the forward-path transfer function $\Omega_m(s)/E(s)$ when the feedback path is open.

(d) Find the closed-loop transfer function $\Omega_m(s)/F_r(s)$.

(e) Repeat parts (a), (c), and (d) for the filter shown in Fig. 4P-19(c).

(f) The digital encoder has an output of 36 pulses per revolution. The reference frequency f_r is fixed at 120 pulse/sec. Find K_e in pulse/rad. The idea of using the counter N is that with f_r fixed, various desired output speeds can be attained by changing the value of N. Find N if the desired output speed is 200 rpm. Find N if the desired output speed is 1800 rpm.

• Robot arm control

4-20.   The linearized model of a robot arm system driven by a dc motor is shown in Fig. 4P-20. The system parameters and variables are given as follows:

DC Motor:	Robot Arm:
T_m = motor torque = $K_i i_a$	J_L = inertia of arm
K_i = torque constant	T_L = disturbance torque on arm
i_a = armature current of motor	θ_L = arm displacement
J_m = motor inertia	K = torsional spring constant
B_m = motor viscous-friction coefficient	B = viscous-friction coefficient of shaft between the motor and arm
θ_m = motor-shaft displacement	B_L = viscous-friction coefficient of the robot arm shaft

(a) Write the differential equations of the system with $i_a(t)$ and $T_L(t)$ as input and $\theta_m(t)$ and $\theta_L(t)$ as outputs.

(b) Draw a SFG using $I_a(s)$, $T_L(s)$, $\Theta_m(s)$, and $\Theta_L(s)$ as node variables.

(c) Express the transfer-function relations as

$$\begin{bmatrix} \Theta_m(s) \\ \Theta_L(s) \end{bmatrix} = \mathbf{G}(s) \begin{bmatrix} I_a(s) \\ -T_L(s) \end{bmatrix}$$

Find $\mathbf{G}(s)$.

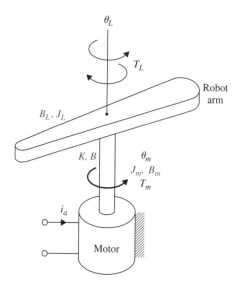

Figure 4P-20

4-21. The following differential equations describe the motion of an electric train in a traction system:

$$\frac{dx(t)}{dt} = v(t)$$

$$\frac{dv(t)}{dt} = -k(v) - g(x) + f(t)$$

where

$x(t)$ = linear displacement of train

$v(t)$ = linear velocity of train

$k(v)$ = resistance force on train [odd function of v, with the properties: $k(0) = 0$ and $dk(v)/dv = 0$].

$g(x)$ = gravitational force for a nonlevel track or due to curvature of track

$f(t)$ = tractive force

The electric motor that provides the tractive force is described by the following equations:

$$e(t) = K_b \phi(t) v(t) + R_a i_a(t)$$
$$f(t) = K_i \phi(t) i_a(t)$$

where $e(t)$ is the applied voltage; $i_a(t)$, the armature current; $i_f(t)$, the field current; R_a, the armature resistance; $\phi(t)$, the magnetic flux from a separately excited field $= K_f i_f(t)$; and K_i, the force constant.

(a) Consider that the motor is a dc series motor with the armature and field windings connected in series, so that $i_a(t) = i_f(t)$, $g(x) = 0$, $k(v) = Bv(t)$, and $R_a = 0$. Show that the system is described by the following nonlinear state equations:

$$\frac{dx(t)}{dt} = v(t)$$

$$\frac{dv(t)}{dt} = -Bv(t) + \frac{K_i}{K_b^2 K_f v^2(t)} e^2(t)$$

(b) Consider that for the conditions stated in part (a), $i_a(t)$ is the input of the system [instead of $e(t)$]. Derive the state equations of the system.

(c) Consider the same conditions as in part (a), but with $\phi(t)$ as the input. Derive the state equations.

• Broom-balancing system

4-22. Figure 4P-22(a) shows a well-known "broom-balancing" system. The objective of the control system is to maintain the broom in the upright position by means of the force $u(t)$ applied to the car as shown. In practical applications, the system is analogous to a one-dimensional control problem of balancing a unicycle or a missile immediately after launching. The free-body diagram of the system is shown in Fig. 4P-22(b), where

f_x = force at broom base in horizontal direction

f_y = force at broom base in vertical direction

M_b = mass of broom

g = gravitational acceleration

M_c = mass of car

J_b = moment of inertia of broom about center of gravity $CG = M_b L^2/3$

(a) Write the force equations in the x and the y directions at the pivot point of the broom. Write the torque equation about the center of gravity CG of the broom. Write the force equation of the car in the horizontal direction.

(b) Express the equations obtained in part (a) as state equations by assigning the state variables as $x_1 = \theta$, $x_2 = d\theta/dt$, $x_3 = x$, and $x_4 = dx/dt$. Simplify these equations for small θ by making the approximations: $\sin \theta \cong \theta$ and $\cos \theta \cong 1$.

(c) Obtain a small-signal linearized state-equation model for the system in the form of

$$\frac{d\Delta \mathbf{x}(t)}{dt} = \mathbf{A} * \Delta \mathbf{x}(t) + \mathbf{B} * \Delta \mathbf{r}(t)$$

at the equilibrium point $x_{01}(t) = 1$, $x_{02}(t) = 0$, $x_{03}(t) = 0$, and $x_{04}(t) = 0$.

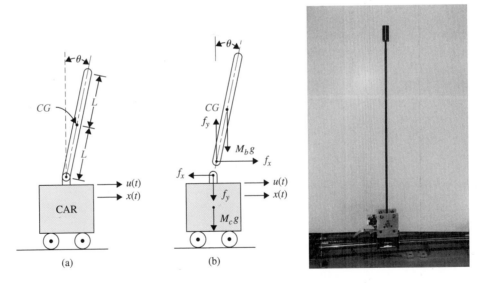

Figure 4P-22

(a)   (b)

• Ball-suspension control system

4-23. Figure 4P-23 shows the schematic diagram of a ball-suspension control system. The steel ball is suspended in the air by the electromagnetic force generated by the electromagnet. The objective of the control is to keep the metal ball suspended at the nominal equilibrium position by controlling the current in the magnet with the voltage $e(t)$. The practical application of this system is the magnetic levitation of trains or magnetic bearings in high-precision control systems.

The resistance of the coil is R, and the inductance is $L(y) = L/y(t)$, where L is a constant. The applied voltage $e(t)$ is a constant with amplitude E.

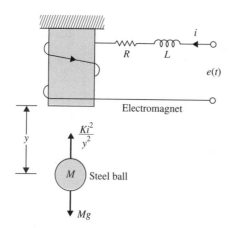

Figure 4P-23

(a) Let E_{eq} be a nominal value of E. Find the nominal values of $y(t)$ and $dy(t)/dt$ at equilibrium.

(b) Define the state variables at $x_1(t) = i(t)$, $x_2(t) = y(t)$, and $x_3(t) = dy(t)/dt$. Find the nonlinear state equations in the form of

$$\frac{d\mathbf{x}(t)}{dt} = \mathbf{f}(\mathbf{x}, e)$$

(c) Linearize the state equations about the equilibrium point and express the linearized state equations as

$$\frac{d\Delta\mathbf{x}(t)}{dt} = \mathbf{A}*\Delta\mathbf{x}(t) + \mathbf{B}*\Delta e(t)$$

The force generated by the electromagnet is $Ki^2(t)/y(t)$, where K is a proportional constant, and the gravitational force on the steel ball is Mg.

• Ball-suspension control
system

4-24. Figure 4P-24(a) shows the schematic diagram of a ball-suspension system. The steel ball is suspended in the air by the electromagnetic force generated by the electromagnet. The objective of the control is to keep the metal ball suspended at the nominal position by controlling the current in the electromagnet. When the system is at the stable equilibrium point, any small perturbation of the ball position from its floating equilibrium position will cause the control to return the ball to the equilibrium position. The free-body diagram of the system is shown in Fig. 4P-24(b), where

M_1 = mass of electromagnet = 2.0

M_2 = mass of steel ball = 1.0

B = viscous-friction coefficient of air = 0.1

K = proportional constant of electromagnet = 1.0

g = gravitational acceleration = 32.2

Assume all units are consistent. Let the stable equilibrium values of the variable, $i(t)$, $y_1(t)$, and $y_2(t)$ be I, Y_1, and Y_2, respectively. The state variables are defined as $x_1(t) = y_1(t)$, $x_2(t) = dy_1(t)/dt$, $x_3(t) = y_2(t)$, and $x_4(t) = dy_2(t)/dt$.

(a) Given $Y_1 = 1$, find I and Y_2.

(b) Write the nonlinear state equations of the system in the form of $dx(t)/dt = f(x, i)$.

(c) Find the state equations of the linearized system about the equilibrium state I, Y_1, and Y_2 in the form:

$$\frac{d\mathbf{x}(t)}{dt} = \mathbf{A}*\Delta\mathbf{x}(t) + \mathbf{B}*\Delta i(t)$$

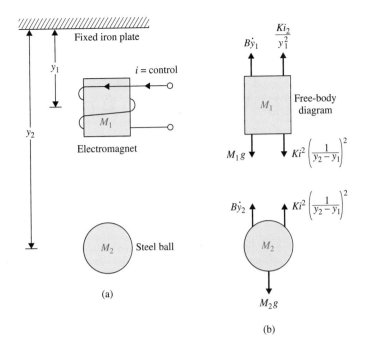

Figure 4P-24

• Steel-rolling process control

4-25. The schematic diagram of a steel-rolling process is shown in Fig. 4P-25. The steel plate is fed through the rollers at a constant speed of V ft/sec. The distance between the rollers and the point where the thickness is measured is d ft. The rotary displacement of the motor, $\theta_m(t)$, is converted to the linear displacement $y(t)$ by the gear box and linear-actuator combination; $y(t) = n\theta_m(t)$, where n is a positive constant in ft/rad. The equivalent inertia of the load that is reflected to the motor shaft is J_L.

(a) Draw a functional block diagram for the system.

(b) Derive the forward-path transfer function $Y(s)/E(s)$ and the closed-loop transfer function $Y(s)/R(s)$.

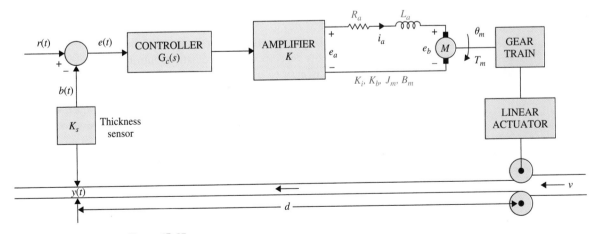

Figure 4P-25

CHAPTER 5

State Variable Analysis

▶ 5-1 INTRODUCTION

In Appendix B the concept and definition of state variables and state equations are presented for linear continuous-data and discrete-data dynamic systems. The signal-flow-graph (SFG) method is extended to the modeling of the state equations, and the result is the **state diagram**. In contrast to the transfer-function approach to the analysis and design of linear control systems, the state-variable method is regarded as modern, since it uses underlying force for optimal control. The basic characteristic of the state-variable formulation is that linear and nonlinear systems, time-invariant and time-varying systems, and single-variable and multivariable systems can all be modeled in a unified manner. Transfer functions, on the other hand, are defined only for linear time-invariant systems.

The objective of this chapter is to introduce the basic methods of state variables and state equations so that the reader can gain a working knowledge of the subject for further studies when the state-space approach is used for modern and optimal control design. Specifically, the closed-form solutions of linear time-invariant state equations are presented. Various transformations that may be used to facilitate the analysis and design of linear control systems in the state-variable domain are introduced. The relationship between the conventional transfer-function approach and the state-variable approach is established so that the analyst will be able to investigate a system problem with various alternative methods. Finally, the controllability and observability of linear systems are defined and their applications investigated. At the end of the chapter we present two MATLAB tools (statetool and tfsym) to solve most state-space problems. These tools apply to problems identified by a MATLAB Toolbox in the left margin of the text throughout this chapter. As with most tools developed for **A**utomatic **C**ontrol **Sys**tems software package (**ACSYS**), the **statetool** is easy to use and fully graphics-based to eliminate the users' need to write code. The **tfsym** tool is based on MATLAB Symbolic Toolbox, and it too does not require any programming by the user. These programs have been developed using MATLAB R12 and prerelease R13, and should also work with the student version of MATLAB (optimized for MATLAB 6.11).[1]

▶ 5-2 VECTOR-MATRIX REPRESENTATION OF STATE EQUATIONS

Let the n state equations of an nth-order dynamic system be represented as

$$\frac{dx_1(t)}{dt} = f_i[x_1(t), x_2(t), \ldots, x_n(t), u_2(t), \ldots, u_p(t), w_1(t), w_2(t), \ldots, w_v(t)] \qquad (5\text{-}1)$$

---

[1]Visit www.mathworks.com.

where $i = 1, 2, \ldots, n$. The ith state variable is represented by $x_i(t)$; $u_j(t)$ denotes the jth input for $j = 1, 2, \ldots, p$; and $w_k(t)$ denotes the kth disturbance input, with $k = 1, 2, \ldots, v$.

Let the variables $y_1(t), y_2(t), \ldots, y_q(t)$ be the q output variables of the system. In general, the output variables are functions of the state variables and the input variables. The **output equations** can be expressed as

$$y_j(t) = g_j[x_1(t), x_2(t), \ldots, x_n(t), u_1(t), u_2(t), \ldots, u_p(t), w_1(t), w_2(t), \ldots, w_v(t)] \quad (5\text{-}2)$$

where $j = 1, 2, \ldots, q$.

The set of n state equations in Eq. (5-1) and q output equations in Eq. (5-2) together form the **dynamic equations.**

For ease of expression and manipulation, it is convenient to represent the dynamic equations in vector-matrix form. Let us define the following vectors:

State vector:
$$\mathbf{x}(t) = \begin{bmatrix} x_1(t) \\ x_2(t) \\ \vdots \\ x_n(t) \end{bmatrix} \quad (n \times 1) \quad (5\text{-}3)$$

Input vector:
$$\mathbf{u}(t) = \begin{bmatrix} u_1(t) \\ u_2(t) \\ \vdots \\ u_p(t) \end{bmatrix} \quad (p \times 1) \quad (5\text{-}4)$$

Output vector:
$$\mathbf{y}(t) = \begin{bmatrix} y_1(t) \\ y_2(t) \\ \vdots \\ y_q(t) \end{bmatrix} \quad (q \times 1) \quad (5\text{-}5)$$

Disturbance vector
$$\mathbf{w}(t) = \begin{bmatrix} w_1(t) \\ w_2(t) \\ \vdots \\ w_v(t) \end{bmatrix} \quad (v \times 1) \quad (5\text{-}6)$$

By using these vectors, the n state equations of Eq. (5-1) can be written:

$$\frac{d\mathbf{x}(t)}{dt} = \mathbf{f}[\mathbf{x}(t), \mathbf{u}(t), \mathbf{w}(t)] \quad (5\text{-}7)$$

where $\mathbf{f}$ denotes an $n \times 1$ column matrix that contains the functions $f_1, f_2, \ldots, f_n$ as elements. Similarly, the q output equations in Eq. (5-2) become

$$\mathbf{y}(t) = \mathbf{g}[\mathbf{x}(t), \mathbf{u}(t), \mathbf{w}(t)] \quad (5\text{-}8)$$

where $\mathbf{g}$ denotes a $q \times 1$ column matrix that contains the functions $g_1, g_2, \ldots, g_q$ as elements.

For a linear time-invariant system, the dynamic equations are written as

State equations:
$$\frac{d\mathbf{x}(t)}{dt} = \mathbf{A}\mathbf{x}(t) + \mathbf{B}\mathbf{u}(t) + \mathbf{E}\mathbf{w}(t) \quad (5\text{-}9)$$

Output equations:
$$\mathbf{y}(t) = \mathbf{C}\mathbf{x}(t) + \mathbf{D}\mathbf{u}(t) + \mathbf{H}\mathbf{w}(t) \quad (5\text{-}10)$$

where

$$\mathbf{A} = \begin{bmatrix} a_{11} & a_{12} & \cdots & a_{1n} \\ a_{21} & a_{22} & \cdots & a_{2n} \\ \vdots & \vdots & \ddots & \vdots \\ a_{n1} & a_{n2} & \cdots & a_{nn} \end{bmatrix} \qquad (n \times n) \tag{5-11}$$

$$\mathbf{B} = \begin{bmatrix} b_{11} & b_{12} & \cdots & b_{1p} \\ b_{21} & b_{22} & \cdots & b_{2p} \\ \vdots & \vdots & \ddots & \vdots \\ b_{n1} & b_{n2} & \cdots & b_{np} \end{bmatrix} \qquad (n \times p) \tag{5-12}$$

$$\mathbf{C} = \begin{bmatrix} c_{11} & c_{12} & \cdots & c_{1n} \\ c_{21} & c_{22} & \cdots & c_{2n} \\ \vdots & \vdots & \ddots & \vdots \\ c_{q1} & c_{q2} & \cdots & c_{qn} \end{bmatrix} \qquad (q \times n) \tag{5-13}$$

$$\mathbf{D} = \begin{bmatrix} d_{11} & d_{12} & \cdots & d_{1p} \\ d_{21} & d_{22} & \cdots & d_{2p} \\ \vdots & \vdots & \ddots & \vdots \\ d_{q1} & d_{q2} & \cdots & d_{qp} \end{bmatrix} \qquad (q \times p) \tag{5-14}$$

$$\mathbf{E} = \begin{bmatrix} e_{11} & e_{12} & \cdots & e_{1v} \\ e_{21} & e_{22} & \cdots & e_{2v} \\ \vdots & \vdots & \ddots & \vdots \\ e_{n1} & e_{n2} & \cdots & e_{nv} \end{bmatrix} \qquad (n \times v) \tag{5-15}$$

$$\mathbf{H} = \begin{bmatrix} h_{11} & h_{12} & \cdots & h_{1v} \\ h_{12} & h_{22} & \cdots & h_{2v} \\ \vdots & \vdots & \ddots & \vdots \\ h_{q1} & h_{q2} & \cdots & h_{qv} \end{bmatrix} \qquad (q \times v) \tag{5-16}$$

▶ 5-3 STATE-TRANSITION MATRIX

Once the state equations of a linear time-invariant system are expressed in the form of Eq. (5-9), the next step often involves the solutions of these equations given the initial state vector $\mathbf{x}(t_0)$, the input vector $\mathbf{u}(t)$, and the disturbance vector $\mathbf{w}(t)$, for $t \geq t_0$. The first term on the right-hand side of Eq. (5-9) is known as the homogeneous part of the state equation, and the last two terms represent the forcing functions $\mathbf{u}(t)$ and $\mathbf{w}(t)$.

The **state-transition matrix** is defined as a matrix that satisfies the linear homogeneous state equation:

$$\frac{d\mathbf{x}(t)}{dt} = \mathbf{A}\mathbf{x}(t) \tag{5-17}$$

Let $\boldsymbol{\phi}(t)$ be the $n \times n$ matrix that represents the state-transition matrix; then it must satisfy the equation

$$\frac{d\boldsymbol{\phi}(t)}{dt} = \mathbf{A}\boldsymbol{\phi}(t) \tag{5-18}$$

Furthermore, let $\mathbf{x}(0)$ denote the initial state at $t = 0$; then $\boldsymbol{\phi}(t)$ is also defined by the matrix equation

$$\mathbf{x}(t) = \boldsymbol{\phi}(t)\mathbf{x}(0) \tag{5-19}$$

which is the solution of the homogeneous state equation for $t \geq 0$.

One way of determining $\boldsymbol{\phi}(t)$ is by taking the Laplace transform on both sides of Eq. (5-17); we have

$$s\mathbf{X}(s) - \mathbf{x}(0) = \mathbf{A}\mathbf{X}(s) \tag{5-20}$$

Solving for $\mathbf{X}(s)$ from Eq. (5-20), we get

$$\mathbf{X}(s) = (s\mathbf{I} - \mathbf{A})^{-1}\mathbf{x}(0) \tag{5-21}$$

where it is assumed that the matrix $(s\mathbf{I} - \mathbf{A})$ is nonsingular. Taking the inverse Laplace transform on both sides of Eq. (5-21) yields

$$\mathbf{x}(t) = \mathcal{L}^{-1}[(s\mathbf{I} - \mathbf{A})^{-1}]\mathbf{x}(0) \qquad t \geq 0 \tag{5-22}$$

By comparing Eq. (5-19) with Eq. (5-22), the state-transition matrix is identified to be

$$\boldsymbol{\phi}(t) = \mathcal{L}^{-1}[(s\mathbf{I} - \mathbf{A})^{-1}] \tag{5-23}$$

An alternative way of solving the homogeneous state equation is to assume a solution, as in the classical method of solving linear differential equations. We let the solution to Eq. (5-17) be

$$\mathbf{x}(t) = e^{\mathbf{A}t}\mathbf{x}(0) \tag{5-24}$$

for $t \geq 0$, where $e^{\mathbf{A}t}$ represents the following power series of the matrix $\mathbf{A}t$, and

$$e^{\mathbf{A}t} = \mathbf{I} + \mathbf{A}t + \frac{1}{2!}\mathbf{A}^2t^2 + \frac{1}{3!}\mathbf{A}^3t^3 + \cdots \tag{5-25}$$

It is easy to show that Eq. (5-24) is a solution of the homogeneous state equation, since, from Eq. (5-25),

$$\frac{de^{\mathbf{A}t}}{dt} = \mathbf{A}e^{\mathbf{A}t} \tag{5-26}$$

Therefore, in addition to Eq. (5-23), we have obtained another expression for the state-transition matrix:

$$\boldsymbol{\phi}(t) = e^{\mathbf{A}t} = \mathbf{I} + \mathbf{A}t + \frac{1}{2!}\mathbf{A}^2t^2 + \frac{1}{3!}\mathbf{A}^3t^3 + \cdots \tag{5-27}$$

Equation (5-27) can also be obtained directly from Eq. (5-23). This is left as an exercise for the reader (Problem 5-2).

5-3-1 Significance of the State-Transition Matrix

Since the state-transition matrix satisfies the homogeneous state equation, it represents the **free response** of the system. In other words, it governs the response that is excited by the initial conditions only. In view of Eqs. (5-23) and (5-27), the state-transition matrix is dependent only upon the matrix $\mathbf{A}$, and, therefore, is sometimes referred to as the **state-transition matrix of A**. As the name implies, the state-transition matrix $\boldsymbol{\phi}(t)$

completely defines the transition of the states from the initial time $t = 0$ to any time t when the inputs are zero.

5-3-2 Properties of the State-Transition Matrix

The state-transition matrix $\boldsymbol{\phi}(t)$ possesses the following properties:

1. $\boldsymbol{\phi}(0) = \mathbf{I}$ (the identity matrix)   (5-28)

    Proof: Equation (5-28) follows directly from Eq. (5-27) by setting $t = 0$.

2. $\boldsymbol{\phi}^{-1}(t) = \boldsymbol{\phi}(-t)$ (5-29)

    Proof: Postmultiplying both sides of Eq. (5-27) by $e^{-\mathbf{A}t}$, we get

$$\boldsymbol{\phi}(t)e^{-\mathbf{A}t} = e^{\mathbf{A}t}e^{-\mathbf{A}t} = \mathbf{I} \tag{5-30}$$

Then, premultiplying both sides of Eq. (5-30) by $\boldsymbol{\phi}^{-1}(t)$, we get

$$e^{-\mathbf{A}t} = \boldsymbol{\phi}^{-1}(t) \tag{5-31}$$

Thus,

$$\boldsymbol{\phi}(-t) = \boldsymbol{\phi}^{-1}(t) = e^{-\mathbf{A}t} \tag{5-32}$$

An interesting result from this property of $\boldsymbol{\phi}(t)$ is that Eq. (5-24) can be rearranged to read

$$\mathbf{x}(0) = \boldsymbol{\phi}(-t)\mathbf{x}(t) \tag{5-33}$$

which means that the state-transition process can be considered as bilateral in time. That is, the transition in time can take place in either direction.

3. $\boldsymbol{\phi}(t_2 - t_1)\boldsymbol{\phi}(t_1 - t_0) = \boldsymbol{\phi}(t_2 - t_0)$ for any t_0, t_1, t_2 (5-34)

    Proof:

$$\boldsymbol{\phi}(t_2 - t_1)\boldsymbol{\phi}(t_1 - t_0) = e^{\mathbf{A}(t_2 - t_1)}e^{\mathbf{A}(t_1 - t_0)} \tag{5-35}$$
$$= e^{\mathbf{A}(t_2 - t_0)} = \boldsymbol{\phi}(t_2 - t_0)$$

This property of the state-transition matrix is important since it implies that a state-transition process can be divided into a number of sequential transitions. Figure 5-1 illustrates that the transition from $t = t_0$ to $t = t_2$ is equal to the transition from t_0 to t_1, and then from t_1 to t_2. In general, of course, the state-transition process can be divided into any number of parts.

4. $[\boldsymbol{\phi}(t)]^k = \boldsymbol{\phi}(kt)$ for $k = $ positive integer   (5-36)

    Proof:

$$[\boldsymbol{\phi}(t)]^k = e^{\mathbf{A}t}e^{\mathbf{A}t} \ldots e^{\mathbf{A}t} \quad (k \text{ terms}) \tag{5-37}$$
$$= e^{k\mathbf{A}t} = \boldsymbol{\phi}(kt)$$

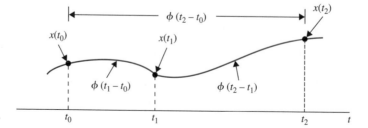

Figure 5-1 Property of the state-transition matrix.

▶ 5-4 STATE-TRANSITION EQUATION

The **state-transition equation** *is defined as the solution of the linear homogeneous state equation.* The linear time-invariant state equation

$$\frac{d\mathbf{x}(t)}{dt} = \mathbf{A}\mathbf{x}(t) + \mathbf{B}\mathbf{u}(t) + \mathbf{E}\mathbf{w}(t) \tag{5-38}$$

can be solved using either the classical method of solving linear differential equations or the Laplace transform method. The Laplace transform solution is presented in the following equations.

Taking the Laplace transform on both sides of Eq. (5-38), we have

$$s\mathbf{X}(s) - \mathbf{x}(0) = \mathbf{A}\mathbf{X}(s) + \mathbf{B}\mathbf{U}(s) + \mathbf{E}\mathbf{W}(s) \tag{5-39}$$

where $\mathbf{x}(0)$ denotes the initial-state vector evaluated at $t = 0$. Solving for $\mathbf{X}(s)$ in Eq. (5-39) yields

$$\mathbf{X}(s) = (s\mathbf{I} - \mathbf{A})^{-1}\mathbf{x}(0) + (s\mathbf{I} - \mathbf{A})^{-1}[\mathbf{B}\mathbf{U}(s) + \mathbf{E}\mathbf{W}(s)] \tag{5-40}$$

The state-transition equation of Eq. (5-38) is obtained by taking the inverse Laplace transform on both sides of Eq. (5-40):

$$\mathbf{x}(t) = \mathcal{L}^{-1}[(s\mathbf{I} - \mathbf{A})^{-1}]\mathbf{x}(0) + \mathcal{L}^{-1}\{(s\mathbf{I} - \mathbf{A})^{-1}[\mathbf{B}\mathbf{U}(s) + \mathbf{E}\mathbf{W}(s)]\}$$
$$= \boldsymbol{\phi}(t)\mathbf{x}(0) + \int_0^t \boldsymbol{\phi}(t - \tau)[\mathbf{B}\mathbf{u}(\tau) + \mathbf{E}\mathbf{w}(\tau)]d\tau \qquad t \geq 0 \tag{5-41}$$

The state-transition equation in Eq. (5-41) is useful only when the initial time is defined to be at $t = 0$. In the study of control systems, especially discrete-data control systems, it is often desirable to break up a state-transition process into a sequence of transitions, so a more flexible initial time must be chosen. Let the initial time be represented by t_0 and the corresponding initial state by $\mathbf{x}(t_0)$, and assume that the input $\mathbf{u}(t)$ and the disturbance $\mathbf{w}(t)$ are applied at $t \geq 0$. We start with Eq. (5-41) by setting $t = t_0$, and solving for $\mathbf{x}(0)$, we get

$$\mathbf{x}(0) = \boldsymbol{\phi}(-t_0)\mathbf{x}(t_0) - \boldsymbol{\phi}(-t_0)\int_0^{t_0} \boldsymbol{\phi}(t_0 - \tau)[\mathbf{B}\mathbf{u}(\tau) + \mathbf{E}\mathbf{w}(\tau)]d\tau \tag{5-42}$$

where the property on $\boldsymbol{\phi}(t)$ of Eq. (5-29) has been applied.

Substituting Eq. (5-42) into Eq. (5-41) yields

$$\mathbf{x}(t) = \boldsymbol{\phi}(t)\boldsymbol{\phi}(-t_0)\mathbf{x}(t_0) - \boldsymbol{\phi}(t)\boldsymbol{\phi}(-t_0)\int_0^{t_0} \boldsymbol{\phi}(t_0 - \tau)[\mathbf{B}\mathbf{u}(\tau) + \mathbf{E}\mathbf{w}(\tau)]d\tau$$

$$\tag{5-43}$$

$$+ \int_0^t \boldsymbol{\phi}(t - \tau)[\mathbf{B}\mathbf{u}(\tau) + \mathbf{E}\mathbf{w}(\tau)]d\tau$$

Now by using the property of Eq. (5-34), and combining the last two integrals, Eq. (5-43) becomes

$$\mathbf{x}(t) = \boldsymbol{\phi}(t - t_0)\mathbf{x}(t_0) + \int_{t_0}^t \boldsymbol{\phi}(t - \tau)[\mathbf{B}\mathbf{u}(\tau) + \mathbf{E}\mathbf{w}(\tau)]d\tau \qquad t \geq t_0 \tag{5-44}$$

It is apparent that Eq. (5-44) reverts to Eq. (5-41) when $t_0 = 0$.

Once the state-transition equation is determined, the output vector can be expressed as a function of the initial state and the input vector simply by substituting $\mathbf{x}(t)$ from

Eq. (5-44) into Eq. (5-10). Thus, the output vector is

$$\mathbf{y}(t) = \mathbf{C}\boldsymbol{\phi}(t - t_0)\mathbf{x}(t_0) + \int_{t_0}^{t} \mathbf{C}\boldsymbol{\phi}(t - \tau)[\mathbf{B}\mathbf{u}(\tau) + \mathbf{E}\mathbf{w}(\tau)]d\tau$$
$$+ \mathbf{D}\mathbf{u}(t) + \mathbf{H}\mathbf{w}(t) \qquad t \geq t_0$$
(5-45)

The following example illustrates the determination of the state-transition matrix and equation.

▶ **EXAMPLE 5-1** Consider the state equation

$$\begin{bmatrix} \dfrac{dx_1(t)}{dt} \\ \dfrac{dx_2(t)}{dt} \end{bmatrix} = \begin{bmatrix} 0 & 1 \\ -2 & -3 \end{bmatrix} \begin{bmatrix} x_1(t) \\ x_2(t) \end{bmatrix} + \begin{bmatrix} 0 \\ 1 \end{bmatrix} u(t)$$
(5-46)

The problem is to determine the state-transition matrix $\boldsymbol{\phi}(t)$ and the state vector $\mathbf{x}(t)$ for $t \geq 0$ when the input is $u(t) = 1$ for $t \geq 0$. The coefficient matrices are identified to be

$$\mathbf{A} = \begin{bmatrix} 0 & 1 \\ -2 & -3 \end{bmatrix} \quad \mathbf{B} = \begin{bmatrix} 0 \\ 1 \end{bmatrix} \quad \mathbf{E} = 0$$
(5-47)

Therefore,

$$s\mathbf{I} - \mathbf{A} = \begin{bmatrix} s & 0 \\ 0 & s \end{bmatrix} - \begin{bmatrix} 0 & 1 \\ -2 & -3 \end{bmatrix} = \begin{bmatrix} s & -1 \\ 2 & s+3 \end{bmatrix}$$
(5-48)

The inverse matrix of $(s\mathbf{I} - \mathbf{A})$ is

$$(s\mathbf{I} - \mathbf{A})^{-1} = \frac{1}{s^2 + 3s + 2}\begin{bmatrix} s+3 & 1 \\ -2 & s \end{bmatrix}$$
(5-49)

The state-transition matrix of $\mathbf{A}$ is found by taking the inverse Laplace transform of Eq. (5-49). Thus,

$$\boldsymbol{\phi}(t) = \mathcal{L}^{-1}[(s\mathbf{I} - \mathbf{A})^{-1}] = \begin{bmatrix} 2e^{-t} - e^{-2t} & e^{-t} - e^{-2t} \\ -2e^{-t} + 2e^{-2t} & -e^{-t} + 2e^{-2t} \end{bmatrix}$$
(5-50)

The state-transition equation for $t \geq 0$ is obtained by substituting Eq. (5-50), **B**, and $u(t)$ into Eq. (5-41). We have

$$\mathbf{x}(t) = \begin{bmatrix} 2e^{-t} - e^{-2t} & e^{-t} - e^{-2t} \\ -2e^{-t} + 2e^{-2t} & -e^{-t} + 2e^{-2t} \end{bmatrix}\mathbf{x}(0)$$
$$+ \int_{0}^{t}\begin{bmatrix} 2e^{-(t-\tau)} - e^{-2(t-\tau)} & e^{-(t-\tau)} - e^{-2(t-\tau)} \\ -2e^{-(t-\tau)} + e^{-2(t-\tau)} & -e^{-(t-\tau)} + 2e^{-2(t-\tau)} \end{bmatrix}\begin{bmatrix} 0 \\ 1 \end{bmatrix}d\tau$$
(5-51)

or

$$\mathbf{x}(t) = \begin{bmatrix} 2e^{-t} - e^{-2t} & e^{-t} - e^{-2t} \\ -2e^{-t} + 2e^{-2t} & -e^{-t} + 2e^{-2t} \end{bmatrix}\mathbf{x}(0) + \begin{bmatrix} 0.5 - e^{-t} + 0.5e^{-2t} \\ e^{-t} - e^{-2t} \end{bmatrix} \quad t \geq 0$$
(5-52)

As an alternative, the second term of the state-transition equation can be obtained by taking the inverse Laplace transform of $(s\mathbf{I} - \mathbf{A})^{-1}\mathbf{B}U(s)$. Thus, we have

$$\mathcal{L}^{-1}[(s\mathbf{I} - \mathbf{A})^{-1}]\mathbf{B}U(s) = \mathcal{L}^{-1}\left(\frac{1}{s^2 + 3s + 2}\begin{bmatrix} s+3 & 1 \\ -2 & s \end{bmatrix}\begin{bmatrix} 0 \\ 1 \end{bmatrix}\frac{1}{s}\right)$$
$$= \mathcal{L}^{-1}\left(\frac{1}{s^2 + 3s + 2}\begin{bmatrix} \dfrac{1}{s} \\ 1 \end{bmatrix}\right) = \begin{bmatrix} 0.5 - e^{-t} + 0.5e^{-2t} \\ e^{-t} - e^{-2t} \end{bmatrix} \quad t \geq 0$$
(5-53)

◀

5-4-1 State-Transition Equation Determined from the State Diagram

Equations (5-40) and (5-41) show that the Laplace transform method of solving the state equations requires the carrying out of the matrix inverse of $(s\mathbf{I} - \mathbf{A})$. We shall now show that the state diagram described in Chapter 3 and the SFG gain formula can be used to solve for the state-transition equation in the Laplace domain of Eq. (5-40). Let the initial time be t_0; then Eq. (5-40) is written

$$\mathbf{X}(s) = (s\mathbf{I} - \mathbf{A})^{-1}\mathbf{x}(t_0) + (s\mathbf{I} - \mathbf{A})^{-1}[\mathbf{B}U(s) + \mathbf{E}W(s)] \qquad t \ge t_0 \qquad (5\text{-}54)$$

The last equation can be written directly from the state diagram using the gain formula, with $X_i(s)$, $i = 1, 2, \ldots, n$, as the output nodes. The following example illustrates the state-diagram method of finding the state-transition equations for the system described in Example 5-1.

▶ **EXAMPLE 5-2** The state diagram for the system described by Eq. (5-46) is shown in Fig. 5-2 with t_0 as the initial time. The outputs of the integrators are assigned as state variables. Applying the gain formula to the state diagram in Fig. 5-2, with $X_1(s)$ and $X_2(s)$ as output nodes, and $x_1(t_0)$, $x_2(t_0)$, and $U(s)$ as input nodes, we have

$$X_1(s) = \frac{s^{-1}(1 + 3s^{-1})}{\Delta}x_1(t_0) + \frac{s^{-2}}{\Delta}x_2(t_0) + \frac{s^{-2}}{\Delta}U(s) \qquad (5\text{-}55)$$

$$X_2(s) = \frac{-2s^{-2}}{\Delta}x_1(t_0) + \frac{s^{-1}}{\Delta}x_2(t_0) + \frac{s^{-1}}{\Delta}U(s) \qquad (5\text{-}56)$$

where

$$\Delta = 1 + 3s^{-1} + 2s^{-2} \qquad (5\text{-}57)$$

After simplification, Eqs. (5-55) and (5-56) are presented in vector-matrix form:

$$\begin{bmatrix} X_1(s) \\ X_2(s) \end{bmatrix} = \frac{1}{(s+1)(s+2)}\begin{bmatrix} s+3 & 1 \\ -2 & s \end{bmatrix}\begin{bmatrix} x_1(t_0) \\ x_2(t_0) \end{bmatrix} + \frac{1}{(s+1)(s+2)}\begin{bmatrix} 1 \\ s \end{bmatrix}U(s) \qquad (5\text{-}58)$$

The state-transition equation for $t \ge t_0$ is obtained by taking the inverse Laplace transform on both sides of Eq. (5-58).

Consider that the input $u(t)$ is a unit-step function applied at $t = t_0$. Then the following inverse Laplace-transform relationships are identified:

$$\mathcal{L}^{-1}\left(\frac{1}{s}\right) = u_s(t - t_0) \qquad t \ge t_0 \qquad (5\text{-}59)$$

$$\mathcal{L}^{-1}\left(\frac{1}{s+a}\right) = e^{-a(t-t_0)}u_s(t - t_0) \qquad t \ge t_0 \qquad (5\text{-}60)$$

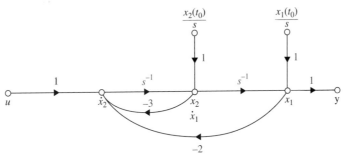

Figure 5-2 State diagram for Eq. (5-46).

Since the initial time is defined to be t_0, the Laplace-transform expressions here do not have the delay factor $e^{-t_0 s}$. The inverse Laplace transform of Eq. (5-58) is

$$
\begin{bmatrix} x_1(t) \\ x_2(t) \end{bmatrix} = \begin{bmatrix} 2e^{-(t-t_0)} - e^{-2(t-t_0)} & e^{-(t-t_0)} - e^{-2(t-t_0)} \\ -2e^{-(t-t_0)} + 2e^{-2(t-t_0)} & -e^{(t-t_0)} + 2e^{-2(t-t_0)} \end{bmatrix} \begin{bmatrix} x_1(t_0) \\ x_2(t_0) \end{bmatrix}
$$
$$
+ \begin{bmatrix} 0.5u_s(t - t_0) - e^{-(t-t_0)} + 0.5e^{-2(t-t_0)} \\ e^{-(t-t_0)} - e^{-2(t-t_0)} \end{bmatrix} \qquad t \ge t_0
$$

(5-61)

The reader should compare this result with that in Eq. (5-52), which is obtained for $t \ge 0$. ◄

► **EXAMPLE 5-3** In this example, we illustrate the utilization of the state-transition method to a system with input discontinuity. An RL network is shown in Fig. 5-3. The history of the network is completely specified by the initial current of the inductance, $i(0)$ at $t = 0$. Consider that at $t = 0$, the voltage shown in Fig. 5-4 is applied to the network. The state equation of the network for $t \ge 0$ is

$$
\frac{di(t)}{dt} = -\frac{R}{L}i(t) + \frac{1}{L}e_{in}(t)
$$

(5-62)

Comparing the last equation with Eq. (5-9), the scalar coefficients of the state equation are identified to be

$$
A = -\frac{R}{L} \qquad B = \frac{1}{L} \qquad E = 0
$$

(5-63)

The state-transition matrix is

$$
\phi(t) = e^{-At} = e^{-Rt/L}
$$

(5-64)

The conventional approach of solving for $i(t)$ for $t \ge 0$ is to express the input voltage as

$$
e(t) = E_{in}u_s(t) + E_{in}u_s(t - t_1)
$$

(5-65)

where $u_s(t)$ is the unit-step function. The Laplace transform of $e(t)$ is

$$
E_{in}(s) = \frac{E_{in}}{s}(1 + e^{-t_1 s})
$$

(5-66)

Then

$$
(s\mathbf{I} - \mathbf{A})^{-1}\mathbf{B}U(s) = \frac{E_{in}}{Ls(s + R/L)}(1 + e^{-t_1 s})
$$

(5-67)

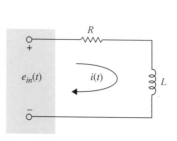

Figure 5-3 *RL* network.

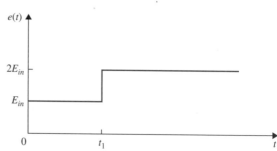

Figure 5-4 Input voltage waveform for the network in Figure 5-3.

By substituting Eq. (5-67) into Eq. (5-41), the state-transition equation, the current for $t \geq 0$ is obtained:

$$i(t) = e^{-Rt/L} i(0)u_s(t) + \frac{E_{in}}{R}(1 - e^{-Rt/L})u_s(t) + \frac{E_{in}}{R}(1 - e^{-R(t - t_1)/L})u_s(t - t_1) \qquad (5\text{-}68)$$

Using the state-transition approach, we can divide the transition period into two parts: $t = 0$ to $t = t_1$, and $t = t_1$ to $t = \infty$. First, for the time interval $0 \leq t \leq t_1$, the input is

$$e(t) = E_{in}u_s(t) \qquad 0 \leq t < t_1 \qquad (5\text{-}69)$$

Then

$$(s\mathbf{I} - \mathbf{A})^{-1}\mathbf{B}U(s) = \frac{E_{in}}{Ls(s + R/L)} = \frac{E_{in}}{Rs[1 + (L/R)s]} \qquad (5\text{-}70)$$

Thus, the state-transition equation for the time interval $0 \leq t \leq t_1$ is

$$i(t) = \left[e^{-Rt/L} i(0) + \frac{E_{in}}{R}(1 - e^{-Rt/L}) \right] u_s(t) \qquad (5\text{-}71)$$

Substituting $t = t_1$ into Eq. (5-71), we get

$$i(t_1) = e^{-Rt_1/L} i(0) + \frac{E_{in}}{R}(1 - e^{-Rt_1/L}) \qquad (5\text{-}72)$$

The value of $i(t)$ at $t = t_1$ is now used as the initial state for the next transition period of $t_1 \leq t < \infty$. The amplitude of the input for the interval is $2E_{in}$. The state-transition equation for the second transition period is

$$i(t) = e^{-R(t - t_1)/L} i(t_1) + \frac{2E_{in}}{R}(1 - e^{-R(t - t_1)/L}) \qquad t \geq t_1 \qquad (5\text{-}73)$$

where $i(t_1)$ is given by Eq. (5-72).

This example illustrates two possible ways of solving a state-transition problem. In the first approach, the transition is treated as one continuous process, whereas in the second, the transition is period is divided into parts over which the input can be more easily presented. Although the first approach requires only one operation, the second method yields relatively simple results to the state-transition equation, and it often presents computational advantages. Notice that in the second method, the state at $t = t_1$ is used as the initial state for the next transition period, which begins at t_1. ◀

▶ 5-5 RELATIONSHIP BETWEEN STATE EQUATIONS AND HIGH-ORDER DIFFERENTIAL EQUATIONS

In the preceding sections, we defined the state equations and their solutions for linear time-invariant systems. Although it is usually possible to write the state equations directly from the schematic diagram of a system, in practice the system may have been described by a high-order differential equation or transfer function. It becomes necessary to investigate how state equations can be written directly from the high-order differential equation or the transfer function. In Appendix B we illustrated how the state variables of an nth-order differential equation in Eq. (B-2) are intuitively defined in Eq. (B-8). The results are the n state equations in Eq. (B-9).

The state equations are written in vector-matrix form:

$$\frac{d\mathbf{x}(t)}{dt} = \mathbf{A}\mathbf{x}(t) + \mathbf{B}u(t) \qquad (5\text{-}74)$$

where

$$\mathbf{A} = \begin{bmatrix} 0 & 1 & 0 & \cdots & 0 \\ 0 & 0 & 1 & \cdots & 0 \\ \vdots & \vdots & \vdots & \ddots & \vdots \\ 0 & 0 & 0 & \cdots & 1 \\ -a_0 & -a_1 & -a_2 & \cdots & -a_{n-1} \end{bmatrix} \quad (n \times n) \tag{5-75}$$

$$\mathbf{B} = \begin{bmatrix} 0 \\ 0 \\ \vdots \\ 0 \\ 1 \end{bmatrix} \quad (n \times 1) \tag{5-76}$$

Notice that the last row of **A** contains the negative values of the coefficients of the homogeneous part of the differential equation in ascending order, except for the coefficient of the highest-order term, which is unity. **B** is a column matrix with the last row equal to one, and the rest of the elements are all zeros. The state equations in Eq. (5-74) with **A** and **B** given in Eqs. (5-75) and (5-76) are known as the **phase-variable canonical form (PVCF)**, or the **controllability canonical form (CCF)**.

The output equation of the system is written

$$y(t) = \mathbf{C}\mathbf{x}(t) = x_1(t) \tag{5-77}$$

where

$$\mathbf{C} = \begin{bmatrix} 1 & 0 & 0 & \cdots & 0 \end{bmatrix} \tag{5-78}$$

We have shown earlier that the state variables of a given system are not unique. In general, we seek the most convenient way of assigning the state variables as long as the definition of state variables is satisfied. Unfortunately, there is no intuitive way of defining the state variables for the general nth-order differential equation given in Eq. (2-86), which differs from Eq. (B-2) by having derivatives of the input on the right-hand side of the equation. However, in Section 5-9 we shall show that by first writing the transfer function and then drawing the state diagram of the system by decomposition of the transfer function, the state variables and state equations of any system can be found very easily. The following example serves the purpose of reviewing the method described in Eqs. (B-8) and (B-9) and above.

▶ **EXAMPLE 5-4** Consider the differential equation

$$\frac{d^3y(t)}{dt^3} + 5\frac{d^2y(t)}{dt^2} + \frac{dy(t)}{dt} + 2y(t) = u(t) \tag{5-79}$$

Rearranging the last equation so that the highest-order derivative term is set equal to the rest of the terms, we have

$$\frac{d^3y(t)}{dt^3} = -5\frac{d^2y(t)}{dt^2} - \frac{dy(t)}{dt} - 2y(t) + u(t) \tag{5-80}$$

The state variables are defined as

$$\begin{aligned} x_1(t) &= y(t) \\ x_2(t) &= \frac{dy(t)}{dt} \\ x_3(t) &= \frac{d^2y(t)}{dt^2} \end{aligned} \tag{5-81}$$

Then the state equations are represented by the vector-matrix equation

$$\frac{d\mathbf{x}(t)}{dt} = \mathbf{A}\mathbf{x}(t) + \mathbf{B}u(t) \tag{5-82}$$

where $\mathbf{x}(t)$ is the 2×1 state vector, $u(t)$ the scalar input, and

$$\mathbf{A} = \begin{bmatrix} 0 & 1 & 0 \\ 0 & 0 & 1 \\ -2 & -1 & -5 \end{bmatrix} \qquad \mathbf{B} = \begin{bmatrix} 0 \\ 0 \\ 1 \end{bmatrix} \tag{5-83}$$

The output equation is

$$y(t) = x_1(t) = \begin{bmatrix} 1 & 0 \end{bmatrix}\mathbf{x}(t) \tag{5-84}$$

◀

▶ 5-6 RELATIONSHIP BETWEEN STATE EQUATIONS AND TRANSFER FUNCTIONS

We have presented the methods of modeling a linear time-invariant system by transfer functions and dynamic equations. We now investigate the relationship between these two representations.

In Eq. (3-5) the transfer function of a linear single-input, single-output system is defined in terms of the coefficients of the system's differential equation. Similarly, Eq. (3-11) gives the transfer-function matrix relation for a multivariable system that has p inputs and q outputs. Now we investigate the transfer-function matrix relation using the dynamic equation notation.

Consider a linear time-invariant system described by the following dynamic equations:

$$\frac{d\mathbf{x}(t)}{dt} = \mathbf{A}\mathbf{x}(t) + \mathbf{B}u(t) + \mathbf{E}w(t) \tag{5-85}$$

$$y(t) = \mathbf{C}\mathbf{x}(t) + \mathbf{D}u(t) + \mathbf{H}w(t) \tag{5-86}$$

where
$\mathbf{x}(t) = n \times 1$ state vector
$\mathbf{u}(t) = p \times 1$ input vector
$\mathbf{y}(t) = q \times 1$ output vector
$\mathbf{w}(t) = v \times 1$ disturbance vector

and $\mathbf{A}$, $\mathbf{B}$, $\mathbf{C}$, $\mathbf{D}$, $\mathbf{E}$, and $\mathbf{H}$ are coefficient matrices of appropriate dimensions.

Taking the Laplace transform on both sides of Eq. (5-85) and solving for $\mathbf{X}(s)$, we have

$$\mathbf{X}(s) = (s\mathbf{I} - \mathbf{A})^{-1}\mathbf{x}(0) + (s\mathbf{I} - \mathbf{A})^{-1}[\mathbf{B}U(s) + \mathbf{E}W(s)] \tag{5-87}$$

The Laplace transform of Eq. (5-86) is

$$\mathbf{Y}(s) = \mathbf{C}\mathbf{X}(s) + \mathbf{D}U(s) + \mathbf{H}W(s) \tag{5-88}$$

Substituting Eq. (5-87) into Eq. (5-88), we have

$$\mathbf{Y}(s) = \mathbf{C}(s\mathbf{I} - \mathbf{A})^{-1}\mathbf{x}(0) + \mathbf{C}(s\mathbf{I} - \mathbf{A})[\mathbf{B}U(s) + \mathbf{E}W(s)] + \mathbf{D}U(s) + \mathbf{H}W(s) \tag{5-89}$$

Since the definition of a transfer function requires that the initial conditions be set to zero, $\mathbf{x}(0) = \mathbf{0}$; thus, Eq. (5-89) becomes

$$\mathbf{Y}(s) = [\mathbf{C}(s\mathbf{I} - \mathbf{A})^{-1}\mathbf{B} + \mathbf{D}]U(s) + [\mathbf{C}(s\mathbf{I} - \mathbf{A})\mathbf{E} + \mathbf{H}]W(s) \tag{5-90}$$

Let us define

$$\mathbf{G}_u(s) = \mathbf{C}(s\mathbf{I} - \mathbf{A})^{-1}\mathbf{B} + \mathbf{D} \tag{5-91}$$

$$\mathbf{G}_w(s) = \mathbf{C}(s\mathbf{I} - \mathbf{A})^{-1}\mathbf{E} + \mathbf{H} \tag{5-92}$$

where $\mathbf{G}_u(s)$ is a $q \times p$ transfer-function matrix between $\mathbf{u}(t)$ and $\mathbf{y}(t)$ when $\mathbf{w}(t) = \mathbf{0}$, and $\mathbf{G}_w(s)$ is a $q \times v$ transfer-function matrix between $\mathbf{w}(t)$ and $\mathbf{y}(t)$ when $\mathbf{u}(t) = \mathbf{0}$.
   Then, Eq. (5-90) becomes

$$\mathbf{Y}(s) = \mathbf{G}_u(s)\mathbf{U}(s) + \mathbf{G}_w(s)\mathbf{W}(s) \tag{5-93}$$

▶ **EXAMPLE 5-5** Consider that a multivariable system is described by the differential equations:

$$\frac{d^2 y_1(t)}{dt^2} + 4\frac{dy_1(t)}{dt} - 3y_2(t) = u_1(t) + 2w(t) \tag{5-94}$$

$$\frac{dy_1(t)}{dt} + \frac{dy_2(t)}{dt} + y_1(t) + 2y_2(t) = u_2(t) \tag{5-95}$$

The state variables of the system are assigned as:

$$x_1(t) = y_1(t)$$
$$x_2(t) = \frac{dy_1(t)}{dt} \tag{5-96}$$
$$x_3(t) = y_2(t)$$

These state variables are defined by mere inspection of the two differential equations, since no particular reasons for the definitions are given other than that these are the most convenient. Now equating the first term of each of the equations of Eqs. (5-94) and (5-95) to the rest of the terms and using the state-variable relations of Eq. (5-96), we arrive at the following state equations and output equations in vector-matrix form:

$$\begin{bmatrix} \dfrac{dx_1(t)}{dt} \\ \dfrac{dx_2(t)}{dt} \\ \dfrac{dx_3(t)}{dt} \end{bmatrix} = \begin{bmatrix} 0 & 1 & 0 \\ 0 & -4 & 3 \\ -1 & -1 & -2 \end{bmatrix} \begin{bmatrix} x_1(t) \\ x_2(t) \\ x_3(t) \end{bmatrix} + \begin{bmatrix} 0 & 0 \\ 1 & 0 \\ 0 & 1 \end{bmatrix} \begin{bmatrix} u_1(t) \\ u_2(t) \end{bmatrix} + \begin{bmatrix} 0 \\ 2 \\ 0 \end{bmatrix} w(t) \tag{5-97}$$

$$\begin{bmatrix} y_1(t) \\ y_2(t) \end{bmatrix} = \begin{bmatrix} 1 & 0 & 0 \\ 0 & 0 & 1 \end{bmatrix} \begin{bmatrix} x_1(t) \\ x_2(t) \\ x_3(t) \end{bmatrix} = \mathbf{Cx}(t) \tag{5-98}$$

   To determine the transfer-function matrix of the system using the state-variable formulation, we substitute the $\mathbf{A}$, $\mathbf{B}$, $\mathbf{C}$, $\mathbf{D}$, and $\mathbf{E}$ matrices into Eq. (5-90). First, we form the matrix $(s\mathbf{I} - \mathbf{A})$:

$$(s\mathbf{I} - \mathbf{A}) = \begin{bmatrix} s & -1 & 0 \\ 0 & s+4 & -3 \\ 1 & 1 & s+2 \end{bmatrix} \tag{5-99}$$

The determinant of $(s\mathbf{I} - \mathbf{A})$ is

$$|s\mathbf{I} - \mathbf{A}| = s^3 + 6s^2 + 11s + 3 \tag{5-100}$$

Thus,

$$(s\mathbf{I} - \mathbf{A})^{-1} = \frac{1}{|s\mathbf{I} - \mathbf{A}|} \begin{bmatrix} s^2 + 6s + 11 & s+2 & 3 \\ -3 & s(s+2) & 3s \\ -(s+4) & -(s+1) & s(s+4) \end{bmatrix} \tag{5-101}$$

The transfer-function matrix between $\mathbf{u}(t)$ and $\mathbf{y}(t)$ is

$$\mathbf{G}_u(s) = \mathbf{C}(s\mathbf{I} - \mathbf{A})^{-1}\mathbf{B} = \frac{1}{s^3 + 6s^2 + 11s + 3}\begin{bmatrix} s + 2 & 3 \\ -(s + 1) & s(s + 4) \end{bmatrix} \quad (5\text{-}102)$$

and that between $\mathbf{w}(t)$ and $\mathbf{y}(t)$ is

$$\mathbf{G}_w(s) = \mathbf{C}(s\mathbf{I} - \mathbf{A})^{-1}\mathbf{E} = \frac{1}{s^3 + 6s + 11s + 3}\begin{bmatrix} 2(s + 2) \\ -2(s + 1) \end{bmatrix} \quad (5\text{-}103)$$

Using the conventional approach, we take the Laplace transform on both sides of Eqs. (5-94) and (5-95) and assume zero initial conditions. The resulting transformed equations are written in vector-matrix form as

$$\begin{bmatrix} s(s + 4) & -3 \\ s + 1 & s + 2 \end{bmatrix}\begin{bmatrix} Y_1(s) \\ Y_2(s) \end{bmatrix} = \begin{bmatrix} U_1(s) \\ U_2(s) \end{bmatrix} + \begin{bmatrix} 2 \\ 0 \end{bmatrix}W(s) \quad (5\text{-}104)$$

Solving for $\mathbf{Y}(s)$ from Eq. (5-104), we obtain

$$\mathbf{Y}(s) = \mathbf{G}_u(s)\mathbf{U}(s) + \mathbf{G}_w(s)W(s) \quad (5\text{-}105)$$

where

$$\mathbf{G}_u(s) = \begin{bmatrix} s(s + 4) & -3 \\ s + 1 & s + 2 \end{bmatrix}^{-1} \quad (5\text{-}106)$$

$$\mathbf{G}_w(s) = \begin{bmatrix} s(s + 4) & -3 \\ s + 1 & s + 2 \end{bmatrix}^{-1}\begin{bmatrix} 2 \\ 0 \end{bmatrix} \quad (5\text{-}107)$$

which will give the same results as in Eqs. (5-102) and (5-103), respectively, when the matrix inverses are carried out.   ◀

▶ 5-7 CHARACTERISTIC EQUATIONS, EIGENVALUES, AND EIGENVECTORS

Characteristic equations play an important role in the study of linear systems. They can be defined with respect to differential equations, transfer functions, or state equations.

Characteristic Equation from a Differential Equation

Consider that a linear time-invariant system is described by the differential equation

$$\frac{d^n y(t)}{dt^n} + a_{n-1}\frac{d^{n-1}y(t)}{dt^{n-1}} + \cdots + a_1\frac{dy(t)}{dt} + a_0 y(t)$$
$$= b_m\frac{d^m u(t)}{dt^m} + b_{m-1}\frac{d^{m-1}u(t)}{dt^{m-1}} + \cdots + b_1\frac{du(t)}{dt} + b_0 u(t) \quad (5\text{-}108)$$

where $n > m$. By defining the operator s as

$$s^k = \frac{d^k}{dt^k} \quad k = 1, 2, \ldots, n \quad (5\text{-}109)$$

Eq. (5-108) is written

$$(s^n + a_{n-1}s^{n-1} + \cdots + a_1 s + a_0)y(t)$$
$$= (b_m s^m + b_{m-1}s^{m-1} + \cdots + b_1 s + b_0)u(t) \quad (5\text{-}110)$$

The **characteristic equation** of the system is defined as

$$s^n + a_{n-1}s^{n-1} + \cdots + a_1 s + a_0 = 0 \quad (5\text{-}111)$$

which is obtained by setting the homogeneous part of Eq. (5-110) to zero.

▶ **EXAMPLE 5-6** Consider the differential equation in Eq. (5-79). The characteristic equation is obtained by inspection,

$$s^3 + 5s^2 + s + 2 = 0 \tag{5-112}$$

◀

Characteristic Equation from a Transfer Function

The transfer function of the system described by Eq. (5-108) is

• The characteristic equation is obtained by setting the denominator polynomial of the transfer function to zero.

$$G(s) = \frac{b_m s^m + b_{m-1} s^{m-1} + \cdots + b_1 s + b_0}{s^n + a_{n-1} s^{n-1} + \cdots + a_1 s + a_0} \tag{5-113}$$

The characteristic equation is obtained by equating the denominator polynomial of the transfer function to zero.

▶ **EXAMPLE 5-7** The transfer function of the system described by the differential equation in Eq. (5-79) is

$$\frac{Y(s)}{U(s)} = \frac{1}{s^3 + 5s^2 + s + 2} \tag{5-114}$$

The same characteristic equation as in Eq. (5-112) is obtained by setting the denominator polynomial of Eq. (5-114) to zero. ◀

Characteristic Equation from State Equations

From the state-variable approach, we can write Eq. (5-91) as

$$\begin{aligned}
\mathbf{G}_r(s) &= \mathbf{C} \frac{\text{adj}(s\mathbf{I} - \mathbf{A})}{|s\mathbf{I} - \mathbf{A}|} \mathbf{B} + \mathbf{D} \\
&= \frac{\mathbf{C}[\text{adj}(s\mathbf{I} - \mathbf{A})]\mathbf{B} + |s\mathbf{I} - \mathbf{A}|\mathbf{D}}{|s\mathbf{I} - \mathbf{A}|}
\end{aligned} \tag{5-115}$$

Setting the denominator of the transfer-function matrix $\mathbf{G}_r(s)$ to zero, we get the characteristic equation

$$|s\mathbf{I} - \mathbf{A}| = 0 \tag{5-116}$$

which is an alternative form of the characteristic equation, but should lead to the same equation as in Eq. (5-111). *An important property of the characteristic equation is that if the coefficients of $\mathbf{A}$ are real, then the coefficients of $|s\mathbf{I} - \mathbf{A}|$ are also real.*

▶ **EXAMPLE 5-8** The matrix $\mathbf{A}$ for the state equations of the differential equation in Eq. (5-79) is given in Eq. (5-83). The characteristic equation of $\mathbf{A}$ is

$$|s\mathbf{I} - \mathbf{A}| = \begin{vmatrix} s & -1 & 0 \\ 0 & s & -1 \\ 2 & 1 & s+5 \end{vmatrix} = s^3 + 5s^2 + s + 2 = 0 \tag{5-117}$$

◀

5-7-1 Eigenvalues

The roots of the characteristic equation are often referred to as the eigenvalues of the matrix $\mathbf{A}$.

Some of the important properties of eigenvalues are given as follows.

• The eigenvalues of $\mathbf{A}$ are also the roots of the characteristic equation.

1. If the coefficients of $\mathbf{A}$ are all real, then its eigenvalues are either real or in complex conjugate pairs.

2. If $\lambda_1, \lambda_2, ..., \lambda_n$ are the eigenvalues of **A**, then

$$\text{tr}(\mathbf{A}) = \sum_{i=1}^{n} \lambda_i \tag{5-118}$$

That is, the trace of **A** is the sum of all the eigenvalues of **A**.

3. If λ_i, $i = 1, 2, ..., n$, is an eigenvalue of **A**, then it is an eigenvalue of **A′**.

4. If **A** is nonsingular, with eigenvalues λ_i, $i = 1, 2, ..., n$, then $1/\lambda_i$, $i = 1, 2, ..., n$, are the eigenvalues of $\mathbf{A}^{-1}$.

▶ **EXAMPLE 5-9** The eigenvalues or the roots of the characteristic equation of the matrix **A** in Eq. (5-83) are obtained by solving for the roots of Eq. (5-117). The results are

$$s = -0.06047 + j0.63738 \qquad s = -0.06047 - j0.63738 \qquad s = -4.87906 \tag{5-119}$$

◀

5-7-2 Eigenvectors

Eigenvectors are useful in modern control methods, one of which is the similarity transformation, which will be discussed in a later section.

Any nonzero vector $\mathbf{p}_i$ which satisfies the matrix equation

$$(\lambda_i \mathbf{I} - \mathbf{A})\mathbf{p}_i = \mathbf{0} \tag{5-120}$$

*where λ_i, $i = 1, 2, ..., n$, denotes the ith eigenvalue of **A**, called the **eigenvector** of **A** associated with the eigenvalue λ_i. If **A** has distinct eigenvalues, the eigenvectors can be solved directly from Eq. (5-120).*

▶ **EXAMPLE 5-10** Consider that the state equation of Eq. (5-38) has the coefficient matrices:

$$\mathbf{A} = \begin{bmatrix} 1 & -1 \\ 0 & -1 \end{bmatrix} \qquad \mathbf{B} = \begin{bmatrix} 1 \\ 1 \end{bmatrix} \qquad \mathbf{E} = \mathbf{0} \tag{5-121}$$

The characteristic equation of **A** is

$$|s\mathbf{I} - \mathbf{A}| = s^2 - 1 \tag{5-122}$$

The eigenvalues are $\lambda_1 = 1$ and $\lambda_2 = -1$. Let the eigenvectors be written as

$$\mathbf{p}_1 = \begin{bmatrix} p_{11} \\ p_{21} \end{bmatrix} \qquad \mathbf{p}_2 = \begin{bmatrix} p_{12} \\ p_{22} \end{bmatrix} \tag{5-123}$$

Substituting $\lambda_1 = 1$ and $\mathbf{p}_1$ into Eq. (5-120), we get

$$\begin{bmatrix} 0 & 1 \\ 0 & 2 \end{bmatrix}\begin{bmatrix} p_{11} \\ p_{21} \end{bmatrix} = \begin{bmatrix} 0 \\ 0 \end{bmatrix} \tag{5-124}$$

Thus, $p_{21} = 0$, and p_{11} is arbitrary which in this case can be set equal to 1.

Similarly, for $\lambda_2 = -1$, Eq. (5-120) becomes

$$\begin{bmatrix} -2 & 1 \\ 0 & 0 \end{bmatrix}\begin{bmatrix} p_{12} \\ p_{22} \end{bmatrix} = \begin{bmatrix} 0 \\ 0 \end{bmatrix} \tag{5-125}$$

which leads to

$$-2p_{12} + p_{22} = 0 \tag{5-126}$$

The last equation has two unknowns, which means that one can be set arbitrarily. Let $p_{12} = 1$, then $p_{22} = 2$. The eigenvectors are

$$\mathbf{p}_1 = \begin{bmatrix} 1 \\ 0 \end{bmatrix} \qquad \mathbf{p}_2 = \begin{bmatrix} 1 \\ 2 \end{bmatrix} \tag{5-127}$$

◀

Generalized Eigenvectors

It should be pointed out that if **A** has multiple-order eigenvalues and is nonsymmetric, not all the **eigenvectors** can be found using Eq. (5-120). Let us assume that there are $q\ (< n)$ distinct eigenvalues among the n eigenvalues of **A**. The eigenvectors that correspond to the q distinct eigenvalues can be determined in the usual manner from

$$(\lambda_i \mathbf{I} - \mathbf{A})\mathbf{p}_i = \mathbf{0} \tag{5-128}$$

where λ_i denotes the ith distinct eigenvalue, $i = 1, 2, \ldots, q$. Among the remaining high-order eigenvalues, let λ_j be of the mth-order ($m \le n - q$). The corresponding eigenvectors are called the **generalized eigenvectors** and can be determined from the following m vector equations:

$$
\begin{aligned}
(\lambda_j \mathbf{I} - \mathbf{A})\mathbf{p}_{n-q+1} &= \mathbf{0} \\
(\lambda_j \mathbf{I} - \mathbf{A})\mathbf{p}_{n-q+2} &= -\mathbf{p}_{n-q+1} \\
(\lambda_j \mathbf{I} - \mathbf{A})\mathbf{p}_{n-q+3} &= -\mathbf{p}_{n-q+2} \\
&\vdots \\
(\lambda_j \mathbf{I} - \mathbf{A})\mathbf{p}_{n-q+m} &= -\mathbf{p}_{n-q+m-1}
\end{aligned}
\tag{5-129}
$$

▶ **EXAMPLE 5-11** Given the matrix

$$\mathbf{A} = \begin{bmatrix} 0 & 6 & -5 \\ 1 & 0 & 2 \\ 3 & 2 & 4 \end{bmatrix} \tag{5-130}$$

The eigenvalues of **A** are $\lambda_1 = 2$, $\lambda_2 = \lambda_3 = 1$. Thus, **A** is a second-order eigenvalue at 1. The eigenvector that is associated with $\lambda_1 = 2$ is determined using Eq. (5-128). Thus,

$$(\lambda_1 \mathbf{I} - \mathbf{A})\mathbf{p}_1 = \begin{bmatrix} 2 & -6 & 5 \\ -1 & 2 & -2 \\ -3 & -2 & -2 \end{bmatrix} \begin{bmatrix} p_{11} \\ p_{21} \\ p_{31} \end{bmatrix} = \mathbf{0} \tag{5-131}$$

Since there are only two independent equations in Eq. (5-131), we arbitrarily set $p_{11} = 2$, and we have $p_{21} = -1$ and $p_{31} = -2$. Thus,

$$\mathbf{p}_1 = \begin{bmatrix} 2 \\ -1 \\ -2 \end{bmatrix} \tag{5-132}$$

For the generalized eigenvectors that are associated with the second-order eigenvalues, we substitute $\lambda_2 = 1$ into the first equation of Eq. (5-129). We have

$$(\lambda_2 \mathbf{I} - \mathbf{A})\mathbf{p}_2 = \begin{bmatrix} 1 & -6 & 5 \\ -1 & 1 & -2 \\ -3 & -2 & -3 \end{bmatrix} \begin{bmatrix} p_{12} \\ p_{22} \\ p_{32} \end{bmatrix} = \mathbf{0} \tag{5-133}$$

Setting $p_{12} = 1$ arbitrarily, we have $p_{22} = -\frac{3}{7}$ and $p_{32} = -\frac{5}{7}$. Thus,

$$\mathbf{p}_2 = \begin{bmatrix} 1 \\ -\frac{3}{7} \\ -\frac{5}{7} \end{bmatrix} \tag{5-134}$$

Substituting $\lambda_3 = 1$ into the second equation of Eq. (5-129), we have

$$(\lambda_3 \mathbf{I} - \mathbf{A})\mathbf{p}_3 = \begin{bmatrix} 1 & -6 & -5 \\ -1 & 1 & -2 \\ -3 & -2 & -3 \end{bmatrix} \begin{bmatrix} p_{13} \\ p_{23} \\ p_{33} \end{bmatrix} = -\mathbf{p}_2 = \begin{bmatrix} -1 \\ \frac{3}{7} \\ \frac{5}{7} \end{bmatrix} \tag{5-135}$$

Setting p_{13} arbitrarily to 1, we have the generalized eigenvector

$$\mathbf{p}_3 = \begin{bmatrix} 1 \\ -\frac{22}{49} \\ -\frac{46}{49} \end{bmatrix} \tag{5-136}$$

◀

▶ 5-8 SIMILARITY TRANSFORMATION

The dynamic equations of a single-input-single-output (SISO) system are

$$\frac{d\mathbf{x}(t)}{dt} = \mathbf{A}\mathbf{x}(t) + \mathbf{B}u(t) \tag{5-137}$$

$$y(t) = \mathbf{C}\mathbf{x}(t) + \mathbf{D}u(t) \tag{5-138}$$

where $\mathbf{x}(t)$ is the $n \times 1$ state vector, $u(t)$ and $y(t)$ are the scalar input and output, respectively. When carrying out analysis and design in the state domain, it is often advantageous to transform these equations into particular forms. For example, as we will show later, the controllability canonical form (CCF) has many interesting properties that make it convenient for controllability tests and state-feedback design.

Let us consider that the dynamic equations of Eq. (5-137) and (5-138) are transformed into another set of equations of the same dimension by the following transformation:

$$\mathbf{x}(t) = \mathbf{P}\bar{\mathbf{x}}(t) \tag{5-139}$$

where $\mathbf{P}$ is an $n \times n$ nonsingular matrix, so

$$\bar{\mathbf{x}}(t) = \mathbf{P}^{-1}\mathbf{x}(t) \tag{5-140}$$

The transformed dynamic equations are written

$$\frac{d\bar{\mathbf{x}}(t)}{dt} = \bar{\mathbf{A}}\bar{\mathbf{x}}(t) + \bar{\mathbf{B}}u(t) \tag{5-141}$$

$$\bar{y}(t) = \bar{\mathbf{C}}\bar{\mathbf{x}}(t) + \bar{\mathbf{D}}u(t) \tag{5-142}$$

Taking the derivative on both sides of Eq. (140) with respect to t, we have

$$\frac{d\bar{\mathbf{x}}(t)}{dt} = \mathbf{P}^{-1}\frac{d\mathbf{x}(t)}{dt} = \mathbf{P}^{-1}\mathbf{A}\mathbf{x}(t) + \mathbf{P}^{-1}\mathbf{B}u(t)$$

$$= \mathbf{P}^{-1}\mathbf{A}\mathbf{P}\bar{\mathbf{x}}(t) + \mathbf{P}^{-1}\mathbf{B}u(t) \tag{5-143}$$

Comparing Eq. (5-143) with Eq. (5-141), we get

$$\bar{\mathbf{A}} = \mathbf{P}^{-1}\mathbf{A}\mathbf{P} \tag{5-144}$$

and

$$\bar{\mathbf{B}} = \mathbf{P}^{-1}\mathbf{B} \tag{5-145}$$

Using Eq. (5-139), Eq. (5-142) is written

$$\bar{y}(t) = \mathbf{C}\mathbf{P}\mathbf{x}(t) + \bar{\mathbf{D}}u(t) \tag{5-146}$$

Comparing Eq. (5-146) with (5-138), we see that

$$\overline{\mathbf{C}} = \mathbf{CP} \qquad \overline{\mathbf{D}} = \mathbf{D} \tag{5-147}$$

The transformation just described is called a **similarity transformation**, since in the transformed system such properties as the characteristic equation, eigenvectors, eigenvalues, and transfer function are all preserved by the transformation. We shall describe the controllability canonical form (CCF), the observability canonical form (OCF), and the diagonal canonical form (DCF) transformations in the following sections. The transformation equations are given without proofs.

5-8-1 Invariance Properties of the Similarity Transformations

One of the important properties of the similarity transformations is that the characteristic equation, eigenvalues, eigenvectors, and transfer functions are invariant under the transformations.

Characteristic Equations, Eigenvalues, Eigenvectors

The characteristic equation of the system described by Eq. (5-141) is $|s\mathbf{I} - \overline{\mathbf{A}}| = 0$ and is written

$$|s\mathbf{I} - \overline{\mathbf{A}}| = |s\mathbf{I} - \mathbf{P}^{-1}\mathbf{AP}| = |s\mathbf{P}^{-1}\mathbf{P} - \mathbf{P}^{-1}\mathbf{AP}| \tag{5-148}$$

Since the determinant of a product matrix is equal to the product of the determinants of the matrices, the last equation becomes

$$|s\mathbf{I} - \overline{\mathbf{A}}| = |\mathbf{P}^{-1}||s\mathbf{I} - \mathbf{A}||\mathbf{P}| = |s\mathbf{I} - \mathbf{A}| \tag{5-149}$$

Thus, the characteristic equation is preserved, which naturally leads to the same eigenvalues and eigenvectors.

Transfer-Function Matrix

From Eq. (5-91), the transfer-function matrix of the system of Eqs. (5-141) and (5-142) is

$$\overline{\mathbf{G}}(s) = \overline{\mathbf{C}}(s\mathbf{I} - \overline{\mathbf{A}})\overline{\mathbf{B}} + \overline{\mathbf{D}}$$
$$= \mathbf{CP}(s\mathbf{I} - \mathbf{P}^{-1}\mathbf{AP})\mathbf{P}^{-1}\mathbf{B} + \mathbf{D} \tag{5-150}$$

which is simplified to

$$\overline{\mathbf{G}}(s) = \mathbf{C}(s\mathbf{I} - \mathbf{A})\mathbf{B} + \mathbf{D} = \mathbf{G}(s) \tag{5-151}$$

5-8-2 Controllability Canonical Form (CCF)

Consider the dynamic equations given in Eqs. (5-137) and (5-138). The characteristic equation of $\mathbf{A}$ is

$$|s\mathbf{I} - \mathbf{A}| = s^n + a_{n-1}s^{n-1} + \cdots + a_1 s + a_0 = 0 \tag{5-152}$$

The dynamic equations in Eqs. (5-137) and (5-138) are transformed into CCF of the form of Eqs. (5-141) and (5-142) by the transformation of Eq. (5-139), with

$$\mathbf{P} = \mathbf{SM} \tag{5-153}$$

• To transform **A** and **B** into **CCF**, the matrix **S** must be nonsingular.

where

$$\mathbf{S} = [\mathbf{B} \quad \mathbf{AB} \quad \mathbf{A}^2\mathbf{B} \ldots \mathbf{A}^{n-1}\mathbf{B}] \tag{5-154}$$

and

$$\mathbf{M} = \begin{bmatrix} a_1 & a_2 & \cdots & a_{n-1} & 1 \\ a_2 & a_3 & \cdots & 1 & 0 \\ \vdots & \vdots & \ddots & \vdots & \vdots \\ a_{n-1} & 1 & \cdots & 0 & 0 \\ 1 & 0 & \cdots & 0 & 0 \end{bmatrix} \tag{5-155}$$

Then,

$$\overline{\mathbf{A}} = \mathbf{P}^{-1}\mathbf{A}\mathbf{P} = \begin{bmatrix} 0 & 1 & 0 & \cdots & 0 \\ 0 & 0 & 1 & \cdots & 0 \\ \vdots & \vdots & \vdots & \ddots & \vdots \\ 0 & 0 & 0 & \cdots & 1 \\ -a_0 & -a_1 & -a_2 & \cdots & -a_{n-1} \end{bmatrix} \tag{5-156}$$

$$\overline{\mathbf{B}} = \mathbf{P}^{-1}\mathbf{B} = \begin{bmatrix} 0 \\ 0 \\ \vdots \\ 0 \\ 1 \end{bmatrix} \tag{5-157}$$

The matrices $\overline{\mathbf{C}}$ and $\overline{\mathbf{D}}$ are given by Eq. (5-147) and do not follow any particular pattern. The CCF transformation requires that $\mathbf{P}^{-1}$ exists, which implies that the matrix $\mathbf{S}$ must have an inverse, since the inverse of $\mathbf{M}$ always exists because its determinant is $(-1)^{n-1}$, which is nonzero. The $n \times n$ matrix $\mathbf{S}$ in Eq. (5-154) is later defined as the **controllability matrix**.

▶ **EXAMPLE 5-12** Consider the coefficient matrices of the state equations in Eq. (5-137):

$$\mathbf{A} = \begin{bmatrix} 1 & 2 & 1 \\ 0 & 1 & 3 \\ 1 & 1 & 1 \end{bmatrix} \qquad \mathbf{B} = \begin{bmatrix} 1 \\ 0 \\ 1 \end{bmatrix} \tag{5-158}$$

The state equations are to be transformed to CCF.

The characteristic equation of $\mathbf{A}$ is

$$|s\mathbf{I} - \mathbf{A}| = \begin{vmatrix} s-1 & -2 & -1 \\ 0 & s-1 & -3 \\ -1 & -1 & s-1 \end{vmatrix} = s^3 - 3s^2 - s - 3 = 0 \tag{5-159}$$

Thus, the coefficients of the characteristic equation are identified as $a_0 = -3$, $a_1 = -1$, $a_2 = -3$. From Eq. (5-155),

$$\mathbf{M} = \begin{bmatrix} a_1 & a_2 & 1 \\ a_2 & 1 & 0 \\ 1 & 0 & 0 \end{bmatrix} = \begin{bmatrix} -1 & -3 & 1 \\ -3 & 1 & 0 \\ 1 & 0 & 0 \end{bmatrix} \tag{5-160}$$

The controllability matrix is

$$\mathbf{S} = \begin{bmatrix} \mathbf{B} & \mathbf{AB} & \mathbf{A}^2\mathbf{B} \end{bmatrix} = \begin{bmatrix} 1 & 2 & 10 \\ 0 & 3 & 9 \\ 1 & 2 & 7 \end{bmatrix} \tag{5-161}$$

We can show that S is nonsingular, so the system can be transformed into the CCF. Substituting S and M into Eq. (5-153), we get

$$P = SM = \begin{bmatrix} 3 & -1 & 1 \\ 0 & 3 & 0 \\ 0 & -1 & 1 \end{bmatrix} \tag{5-162}$$

Thus, from Eqs. (5-156) and (5-157), the CCF model is given by

$$\overline{A} = P^{-1}AP = \begin{bmatrix} 0 & 1 & 0 \\ 0 & 0 & 1 \\ 3 & 1 & 3 \end{bmatrix} \qquad \overline{B} = P^{-1}B = \begin{bmatrix} 0 \\ 0 \\ 1 \end{bmatrix} \tag{5-163}$$

which could have been determined once the coefficients of the characteristic equation are known; however, the exercise is to show how the CCF transformation matrix P is obtained. ◄

5-8-3 Observability Canonical Form (OCF)

A dual form of transformation of the CCF is the **observability canonical form** (OCF). The system described by Eqs. (5-137) and (5-138) is transformed to the OCF by the transformation

$$x(t) = Q\overline{x}(t) \tag{5-164}$$

The transformed equations are as given in Eqs. (5-141) and (5-142). Thus,

$$\overline{A} = Q^{-1}AQ \qquad \overline{B} = Q^{-1}B \qquad \overline{C} = CQ \qquad \overline{D} = D \tag{5-165}$$

where

$$\overline{A} = Q^{-1}AQ = \begin{bmatrix} 0 & 0 & \cdots & 0 & -a_0 \\ 1 & 0 & \cdots & 0 & -a_1 \\ 0 & 1 & \cdots & 0 & -a_2 \\ \vdots & \vdots & \ddots & \vdots & \vdots \\ 0 & 0 & \cdots & 1 & -a_{n-1} \end{bmatrix} \tag{5-166}$$

$$\overline{C} = CQ = \begin{bmatrix} 0 & 0 & \cdots & 0 & 1 \end{bmatrix} \tag{5-167}$$

The elements of the matrices $\overline{B}$ and $\overline{D}$ are not restricted to any form. Notice that $\overline{A}$ and $\overline{C}$ are the transpose of the $\overline{A}$ and $\overline{B}$ in Eqs. (5-156) and (5-157), respectively.

The OCF transformation matrix Q is given by

$$Q = (MV)^{-1} \tag{5-168}$$

where M is as given in Eq. (5-155), and

$$V = \begin{bmatrix} C \\ CA \\ CA^2 \\ \vdots \\ CA^{n-1} \end{bmatrix} \qquad (n \times n) \tag{5-169}$$

The matrix V is often defined as the **observability matrix**, and V^{-1} must exist in order for the OCF transformation to be possible.

► **EXAMPLE 5-13** Consider that the coefficient matrices of the system described by Eqs. (5-137) and (5-138) are

MATLAB

$$
\mathbf{A} = \begin{bmatrix} 1 & 2 & 1 \\ 0 & 1 & 3 \\ 1 & 1 & 1 \end{bmatrix} \quad \mathbf{B} = \begin{bmatrix} 1 \\ 0 \\ 1 \end{bmatrix} \quad \mathbf{C} = \begin{bmatrix} 1 & 1 & 0 \end{bmatrix} \quad \mathbf{D} = \mathbf{0} \tag{5-170}
$$

Since the matrix **A** is identical to that of the system in Example 5-12, the matrix **M** the same as that in Eq. (5-160). The observability matrix is

$$
\mathbf{V} = \begin{bmatrix} \mathbf{C} \\ \mathbf{CA} \\ \mathbf{CA}^2 \end{bmatrix} = \begin{bmatrix} 1 & 1 & 0 \\ 1 & 3 & 4 \\ 5 & 9 & 14 \end{bmatrix} \tag{5-171}
$$

We can show that **V** is nonsingular, so the system can be transformed into the OCF. Substituting **V** and **M** into Eq. (5-168), we have the OCF transformation matrix,

$$
\mathbf{Q} = (\mathbf{MV})^{-1} = \begin{bmatrix} 0.3333 & -0.1667 & 0.3333 \\ -0.3333 & 0.1667 & 0.6667 \\ 0.1667 & 0.1667 & 0.1667 \end{bmatrix} \tag{5-172}
$$

From Eq. (5-156), the OCF model of the system is described by

$$
\overline{\mathbf{A}} = \mathbf{Q}^{-1}\mathbf{AQ} = \begin{bmatrix} 0 & 0 & 3 \\ 1 & 0 & 1 \\ 0 & 1 & 3 \end{bmatrix} \quad \overline{\mathbf{C}} = \mathbf{CQ} = \begin{bmatrix} 0 & 0 & 1 \end{bmatrix} \quad \overline{\mathbf{B}} = \mathbf{Q}^{-1}\mathbf{B} = \begin{bmatrix} 3 \\ 2 \\ 1 \end{bmatrix} \tag{5-173}
$$

Thus, $\overline{\mathbf{A}}$ and $\overline{\mathbf{C}}$ are of the OCF form given in Eqs. (5-166) and (5-167), respectively, and $\overline{\mathbf{B}}$ does not conform to any particular form. ◄

5-8-4 Diagonal Canonical Form (DCF)

Given the dynamic equations in Eqs. (5-137) and (5-138), if **A** has distinct eigenvalues, there is a nonsingular transformation

$$
\mathbf{x}(t) = \mathbf{T}\overline{\mathbf{x}}(t) \tag{5-174}
$$

which transforms these equations to the dynamic equations of Eqs. (5-141) and (5-142), where

$$
\overline{\mathbf{A}} = \mathbf{T}^{-1}\mathbf{AT} \quad \overline{\mathbf{B}} = \mathbf{T}^{-1}\mathbf{B} \quad \overline{\mathbf{C}} = \mathbf{CT} \quad \overline{\mathbf{D}} = \mathbf{D} \tag{5-175}
$$

The matrix $\overline{\mathbf{A}}$ is a diagonal matrix,

$$
\overline{\mathbf{A}} = \begin{bmatrix} \lambda_1 & 0 & 0 & \cdots & 0 \\ 0 & \lambda_2 & 0 & \cdots & 0 \\ 0 & 0 & \lambda_3 & \cdots & 0 \\ \vdots & \vdots & \vdots & \ddots & \vdots \\ 0 & 0 & 0 & \cdots & \lambda_n \end{bmatrix} \quad (n \times n) \tag{5-176}
$$

where $\lambda_1, \lambda_2, \ldots, \lambda_n$ are the n distinct eigenvalues of **A**. The coefficient matrices $\overline{\mathbf{B}}, \overline{\mathbf{C}}$, and $\overline{\mathbf{D}}$ are given in Eq. (5-175), and do not follow any particular form.

It is apparent that one of the advantages of the DCF is that the transformed state equations are *decoupled* from each other, and, therefore, can be solved individually.

We show in the following that the DCF transformation matrix **T** can be formed by use of the the eigenvectors of **A** as its columns; that is,

$$
\mathbf{T} = \begin{bmatrix} \mathbf{p}_1 & \mathbf{p}_2 & \mathbf{p}_3 & \cdots & \mathbf{p}_n \end{bmatrix} \tag{5-177}
$$

where $\mathbf{p}_i$, $i = 1, 2, \ldots, n$, denotes the eigenvector associated with the eigenvalue λ_i. This is proved by use of Eq. (5-120), which is written as

$$\lambda_i \mathbf{p}_i = \mathbf{A}\mathbf{p}_i \qquad i = 1, 2, \ldots, n \qquad (5\text{-}178)$$

Now forming the $n \times n$ matrix,

$$\begin{aligned}
\begin{bmatrix} \lambda_1 \mathbf{p}_1 & \lambda_2 \mathbf{p}_2 & \cdots & \lambda_n \mathbf{p}_n \end{bmatrix} &= \begin{bmatrix} \mathbf{A}\mathbf{p}_1 & \mathbf{A}\mathbf{p}_2 & \cdots & \mathbf{A}\mathbf{p}_n \end{bmatrix} \\
&= \mathbf{A}\begin{bmatrix} \mathbf{p}_1 & \mathbf{p}_2 & \cdots & \mathbf{p}_n \end{bmatrix}
\end{aligned} \qquad (5\text{-}179)$$

The last equation is written

$$\begin{bmatrix} \mathbf{p}_1 & \mathbf{p}_2 & \cdots & \mathbf{p}_n \end{bmatrix}\overline{\mathbf{A}} = \mathbf{A}\begin{bmatrix} \mathbf{p}_1 & \mathbf{p}_2 & \cdots & \mathbf{p}_n \end{bmatrix} \qquad (5\text{-}180)$$

where $\overline{\mathbf{A}}$ is as given in Eq. (5-176). Thus, if we let

$$\mathbf{T} = \begin{bmatrix} \mathbf{p}_1 & \mathbf{p}_2 & \cdots & \mathbf{p}_n \end{bmatrix} \qquad (5\text{-}181)$$

Eq. (5-180) is written

$$\overline{\mathbf{A}} = \mathbf{T}^{-1}\mathbf{A}\mathbf{T} \qquad (5\text{-}182)$$

If the matrix $\mathbf{A}$ is of the CCF, and $\mathbf{A}$ has distinct eigenvalues, then the DCF transformation matrix is the Vandermonde matrix,

$$\mathbf{T} = \begin{bmatrix}
1 & 1 & 1 & \cdots & 1 \\
\lambda_1 & \lambda_2 & \lambda_3 & \cdots & \lambda_n \\
\lambda_1^2 & \lambda_2^2 & \lambda_3^2 & \cdots & \lambda_n^2 \\
\vdots & \vdots & \vdots & \ddots & \vdots \\
\lambda_1^{n-1} & \lambda_2^{n-1} & \lambda_3^{n-1} & \cdots & \lambda_n^{n-1}
\end{bmatrix} \qquad (5\text{-}183)$$

where $\lambda_1, \lambda_2, \ldots, \lambda_n$ are the eigenvalues of $\mathbf{A}$. This can be proven by substituting the CCF of $\mathbf{A}$ in Eq. (5-75) into Eq. (5-120). The result is that the ith eigenvector $\mathbf{p}_i$ is equal to the ith column of $\mathbf{T}$ in Eq. (5-183).

▶ **EXAMPLE 5-14** Consider the matrix

$$\mathbf{A} = \begin{bmatrix} 0 & 1 & 0 \\ 0 & 0 & 1 \\ -6 & -11 & -6 \end{bmatrix} \qquad (5\text{-}184)$$

which has eigenvalues $\lambda_1 = -1$, $\lambda_2 = -2$, and $\lambda_3 = -3$. Since $\mathbf{A}$ is CCF, to transform it into DCF the transformation matrix can be the Vandermonde matrix in Eq. (5-183). Thus,

$$\mathbf{T} = \begin{bmatrix} 1 & 1 & 1 \\ \lambda_1 & \lambda_2 & \lambda_3 \\ \lambda_1^2 & \lambda_2^2 & \lambda_3^2 \end{bmatrix} = \begin{bmatrix} 1 & 1 & 1 \\ -1 & -2 & -3 \\ 1 & 4 & 9 \end{bmatrix} \qquad (5\text{-}185)$$

Thus, the DCF of $\mathbf{A}$ is written

$$\overline{\mathbf{A}} = \mathbf{T}^{-1}\mathbf{A}\mathbf{T} = \begin{bmatrix} -1 & 0 & 0 \\ 0 & -2 & 0 \\ 0 & 0 & -3 \end{bmatrix} \qquad (5\text{-}186)$$

◀

5-8-5 Jordan Canonical Form (JCF)

In general, when the matrix $\mathbf{A}$ has multiple-order eigenvalues, unless the matrix is symmetric with real elements, it cannot be transformed into a diagonal matrix. However, there exists a similarity transformation in the form of Eq. (5-182) such that the matrix $\overline{\mathbf{A}}$ is almost diagonal. The matrix $\overline{\mathbf{A}}$ is called the **Jordan canonical form** (**JCF**). A typical JCF is shown below.

$$\overline{\mathbf{A}} = \begin{bmatrix} \lambda_1 & 1 & 0 & 0 & 0 \\ 0 & \lambda_1 & 1 & 0 & 0 \\ 0 & 0 & \lambda_1 & 0 & 0 \\ \hdashline 0 & 0 & 0 & \lambda_2 & 0 \\ 0 & 0 & 0 & 0 & \lambda_3 \end{bmatrix} \tag{5-187}$$

where it is assumed that $\mathbf{A}$ has a third-order eigenvalue λ_1 and distinct eigenvalues λ_2 and λ_3. The JCF generally has the following properties:

1. The elements on the main diagonal are the eigenvalues.

2. All the elements below the main diagonal are zero.

3. Some of the elements immediately above the multiple-order eigenvalues on the main diagonal are 1s, as shown in Eq. (5-187).

4. The 1s together with the eigenvalues form typical blocks called the **Jordan blocks**. As shown in Eq. (5-187) the Jordan blocks are enclosed by dashed lines.

5. When the nonsymmetrical matrix $\mathbf{A}$ has multiple-order eigenvalues, its eigenvectors are not linearly independent. For an $\mathbf{A}$ that is $n \times n$, there are only r (where r is an integer that is less than n and is dependent on the number of multiple-order eigenvalues) linearly independent eigenvectors.

6. The number of Jordan blocks is equal to the number of independent eigenvectors r. There is one and only one linearly independent eigenvector associated with each Jordan block.

7. The number of 1s above the main diagonal is equal to $n - r$.

To perform the JCF transformation, the transformation matrix $\mathbf{T}$ is again formed by using the eigenvectors and generalized eigenvectors as its columns.

▶ **EXAMPLE 5-15** Consider the matrix given in Eq. (5-130). We have shown that the matrix has eigenvalues 2, 1, and 1. Thus, the DCF transformation matrix can be formed by using the eigenvector and generalized eigenvector given in Eqs. (5-132), (5-134), and (5-136), respectively. That is,

$$\mathbf{T} = \begin{bmatrix} \mathbf{p}_1 & \mathbf{p}_2 & \mathbf{p}_3 \end{bmatrix} = \begin{bmatrix} 2 & 1 & 1 \\ -1 & -\frac{3}{7} & -\frac{22}{49} \\ -2 & -\frac{5}{7} & -\frac{46}{49} \end{bmatrix} \tag{5-188}$$

Thus, the DCF is

$$\overline{\mathbf{A}} = \mathbf{T}^{-1}\mathbf{A}\mathbf{T} = \begin{bmatrix} 2 & 0 & 0 \\ 0 & 1 & 1 \\ 0 & 0 & 1 \end{bmatrix} \tag{5-189}$$

Note that in this case there are two Jordan blocks and there is one element of 1 above the main diagonal. ◀

▶ 5-9 DECOMPOSITIONS OF TRANSFER FUNCTIONS

Up to this point, various methods of characterizing linear systems have been presented. To summarize, it has been shown that the starting point of modeling of a linear system may be the system's differential equation, transfer function, or dynamic equations; all

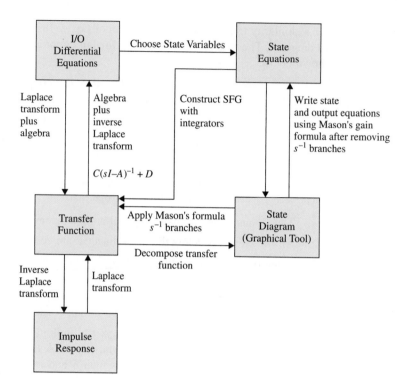

Figure 5-5 Block diagram showing the relationships among various methods of describing linear systems.

these methods are closely related. Furthermore, the state diagram defined in Chapter 3 is also a useful tool that can not only lead to the solutions of state equations, but also serve as a vehicle of transformation from one form of description to the others. The block diagram of Fig. 5-5 shows the relationships among the various ways of describing a linear system. For example, the block diagram shows that starting with the differential equation of a system, one can find the solution by the transfer-function or state-equation method. The block diagram also shows that the majority of the relationships are bilateral, so a great deal of flexibility exists between the methods.

One subject remains to be discussed, which involves the construction of the state diagram from the transfer function between the input and the output. The process of going from the transfer function to the state diagram is called **decomposition**. In general, there are three basic ways to decompose of transfer functions. These are **direct decomposition**, **cascade decomposition**, and **parallel decomposition**. Each of these three schemes of decomposition has its own merits and is best suited for a particular purpose.

5-9-1 Direct Decomposition

Direct decomposition is applied to an input-output transfer function that is not in factored form. Consider the transfer function of an nth-order SISO system between the input $U(s)$ and output $Y(s)$:

$$\frac{Y(s)}{U(s)} = \frac{b_{n-1}s^{n-1} + b_{n-2}s^{n-2} + \cdots + b_1 s + b_0}{s^n + a_{n-1}s^{n-1} + \cdots + a_1 s + a_0} \qquad (5\text{-}190)$$

where we have assumed that the order of the denominator is at least one degree higher than that of the numerator.

We shall show that the direct decomposition can be conducted in at least two ways, one leading to a state diagram that corresponds to the CCF, and the other, to the OCF.

Direct Decomposition to CCF

The objective is to construct a state diagram from the transfer function of Eq. (5-190). The following steps are outlined:

1. Express the transfer function in negative powers of s. This is done by multiplying the numerator and the denominator of the transfer function by s^{-n}.

2. Multiply the numerator and the denominator of the transfer function by a dummy variable $X(s)$. By implementing the last two steps, Eq. (5-190) becomes

$$\frac{Y(s)}{U(s)} = \frac{b_{n-1}s^{-1} + b_{n-2}s^{-2} + \cdots + b_1 s^{-n+1} + b_0 s^{-n}}{1 + a_{n-1}s^{-1} + \cdots + a_1 s^{-n+1} + a_0 s^{-n}} \frac{X(s)}{X(s)} \qquad (5\text{-}191)$$

3. The numerators and the denominators on both sides of Eq. (5-191) are equated to each other, respectively. The results are:

$$Y(s) = (b_{n-1}s^{-1} + b_{n-2}s^{-2} + \cdots + b_1 s^{-n+1} + b_0 s^{-n}) X(s) \qquad (5\text{-}192)$$
$$U(s) = (1 + a_{n-1}s^{-1} + \cdots + a_1 s^{-n+1} + a_0 s^{-n}) X(s) \qquad (5\text{-}193)$$

4. To construct a state diagram using the two equations in Eqs. (5-192) and (5-193), they must first be in the proper cause-and-effect relation. It is apparent that Eq. (5-192) already satisfies this prerequisite. However, Eq. (5-193) has the input on the left-hand side of the equation and must be rearranged. Equation (5-193) is rearranged as

$$X(s) = U(s) - (a_{n-1}s^{-1} + a_{n-2}s^{-2} + \cdots + a_1 s^{-n+1} + a_0 s^{-n})X(s) \qquad (5\text{-}194)$$

The state diagram is drawn as shown in Fig. 5-6 using Eqs. (5-192) and (5-194). For simplicity, the initial states are not drawn on the diagram. The state variables $x_1(t)$, $x_2(t)$, ..., $x_n(t)$ are defined as the outputs of the integrators, and are arranged in order from the right to the left on the state diagram. The state equations are obtained by applying the SFG gain formula to Fig. 5-6 with the derivatives of the state variables as the outputs and the state variables and $u(t)$ as the inputs, and overlooking the integrator branches. The

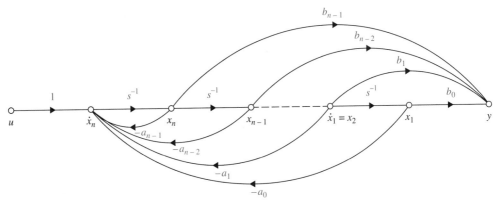

Figure 5-6 CCF state diagram of the transfer function in Eq. (5-190) by direct decomposition.

output equation is determined by applying the gain formula among the state variables, the input, and the output $y(t)$. The dynamic equations are written

$$\frac{d\mathbf{x}(t)}{dt} = \mathbf{A}\mathbf{x}(t) + \mathbf{B}u(t) \tag{5-195}$$

$$y(t) = \mathbf{C}\mathbf{x}(t) + Du(t) \tag{5-196}$$

where

$$\mathbf{A} = \begin{bmatrix} 0 & 1 & 0 & \cdots & 0 \\ 0 & 0 & 1 & \cdots & 0 \\ \vdots & \vdots & \vdots & \ddots & \vdots \\ 0 & 0 & 0 & 0 & 1 \\ -a_0 & -a_1 & -a_2 & \cdots & -a_{n-1} \end{bmatrix} \qquad \mathbf{B} = \begin{bmatrix} 0 \\ 0 \\ \vdots \\ 0 \\ 1 \end{bmatrix} \tag{5-197}$$

$$\mathbf{C} = \begin{bmatrix} b_0 & b_1 & \cdots & b_{n-2} & b_{n-1} \end{bmatrix} \qquad D = 0 \tag{5-198}$$

Apparently, $\mathbf{A}$ and $\mathbf{B}$ in Eq. (5-197) are of the CCF.

Direct Decomposition to OCF

Multiplying the numerator and the denominator of Eq. (5-190) by s^{-n}, the equation is expanded as

$$(1 + a_{n-1}s^{-1} + \cdots + a_1 s^{-n+1} + a_0 s^{-n})Y(s)$$
$$= (b_{n-1}s^{-1} + b_{n-2}s^{-2} + \cdots + b_1 s^{-n+1} + b_0 s^{-n})U(s) \tag{5-199}$$

or

$$Y(s) = -(a_{n-1}s^{-1} + \cdots + a_1 s^{-n+1} + a_0 s^{-n})Y(s)$$
$$+ (b_{n-1}s^{-1} + b_{n-2}s^{-2} + \cdots + b_1 s^{-n+1} + b_0 s^{-n})U(s) \tag{5-200}$$

Figure 5-7 shows the state diagram that results from using Eq. (5-200). The outputs of the integrators are designated as the state variables. However, unlike the usual convention,

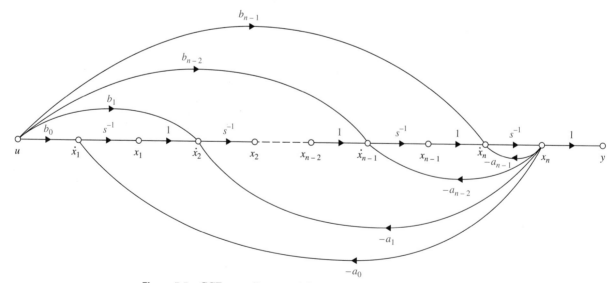

Figure 5-7 CCF state diagram of the transfer function in Eq. (5-190) by direct decomposition.

the state variables are assigned in *descending order* from right to left. Applying the SFG gain formula to the state diagram, the dynamic equations are written as in Eqs. (5-195) and (5-196), with

$$\mathbf{A} = \begin{bmatrix} 0 & 0 & \cdots & 0 & -a_0 \\ 1 & 0 & \cdots & 0 & -a_1 \\ 0 & 1 & \cdots & 0 & -a_2 \\ \vdots & \vdots & \ddots & \vdots & \vdots \\ 0 & 0 & \cdots & 1 & -a_{n-1} \end{bmatrix} \qquad \mathbf{B} = \begin{bmatrix} b_0 \\ b_1 \\ b_2 \\ \vdots \\ b_{n-1} \end{bmatrix} \qquad (5\text{-}201)$$

and

$$\mathbf{C} = \begin{bmatrix} 0 & 0 & \cdots & 0 & 1 \end{bmatrix} \qquad\qquad D = 0 \qquad (5\text{-}202)$$

The matrices **A** and **C** are in OCF.

It should be pointed out that given the dynamic equations of a system, the input-output transfer function is unique. However, given the transfer function, the state model is not unique, as shown by the CCF, OCF, and DCF, and many other possibilities. In fact, even for any one of these canonical forms (for example, CCF), while matrices **A** and **B** are defined, the elements of **C** and **D** could still be different depending on how the state diagram is drawn, that is, how the transfer function is decomposed. In other words, referring to Fig. 5-6, whereas the feedback branches are fixed, the feedforward branches that contain the coefficients of the numerator of the transfer function can still be manipulated to change the contents of **C**.

► **EXAMPLE 5-16** Consider the following input-output transfer function:

$$\frac{Y(s)}{U(s)} = \frac{2s^2 + s + 5}{s^3 + 6s^2 + 11s + 4} \qquad (5\text{-}203)$$

The CCF state diagram of the system is shown in Fig. 5-8, which is drawn from the following equations:

$$Y(s) = (2s^{-1} + s^{-2} + 5s^{-3})X(s) \qquad (5\text{-}204)$$
$$X(s) = U(s) - (6s^{-1} + 11s^{-2} + 4s^{-3})X(s) \qquad (5\text{-}205)$$

The dynamic equations of the system in CCF are

$$\begin{bmatrix} \dfrac{dx_1(t)}{dt} \\[2mm] \dfrac{dx_2(t)}{dt} \\[2mm] \dfrac{dx_3(t)}{dt} \end{bmatrix} = \begin{bmatrix} 0 & 1 & 0 \\ 0 & 0 & 1 \\ -4 & -11 & -6 \end{bmatrix} \begin{bmatrix} x_1(t) \\ x_2(t) \\ x_3(t) \end{bmatrix} + \begin{bmatrix} 0 \\ 0 \\ 1 \end{bmatrix} u(t) \qquad (5\text{-}206)$$

$$y(t) = \begin{bmatrix} 5 & 1 & 2 \end{bmatrix} \mathbf{x}(t) \qquad (5\text{-}207)$$

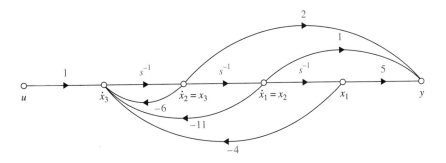

Figure 5-8 CCF state diagram of the transfer function in Eq. (5-203).

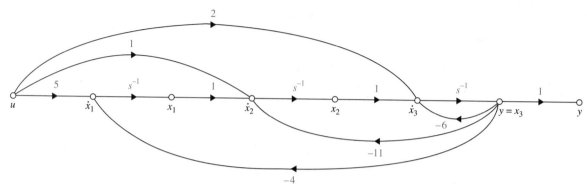

Figure 5-9 OCF state diagram of the transfer function in Eq. (5-203).

For the OCF, Eq. (5-203) is expanded to

$$Y(s) = (2s^{-1} + s^{-2} + 5s^{-3})U(s) - (6s^{-1} + 11s^{-2} + 4s^{-3})Y(s) \tag{5-208}$$

which leads to the OCF state diagram shown in Fig. 5-9. The OCF dynamic equations are written

$$\begin{bmatrix} \dfrac{dx_1(t)}{dt} \\[2mm] \dfrac{dx_2(t)}{dt} \\[2mm] \dfrac{dx_3(t)}{dt} \end{bmatrix} = \begin{bmatrix} 0 & 0 & -4 \\ 1 & 0 & -11 \\ 0 & 1 & -6 \end{bmatrix} \begin{bmatrix} x_1(t) \\ x_2(t) \\ x_3(t) \end{bmatrix} + \begin{bmatrix} 5 \\ 1 \\ 2 \end{bmatrix} u(t) \tag{5-209}$$

$$y(t) = \begin{bmatrix} 0 & 0 & 1 \end{bmatrix} \mathbf{x}(t) \tag{5-210}$$

◀

5-9-2 Cascade Decomposition

Cascade compensation refers to transfer functions that are written as products of simple first-order or second-order components. Consider the following transfer function, which is the product of two first-order transfer functions.

$$\frac{Y(s)}{U(s)} = K\left(\frac{s + b_1}{s + a_1}\right)\left(\frac{s + b_2}{s + a_2}\right) \tag{5-211}$$

where a_1, a_2, b_1, and b_2 are real constants. Each of the first-order transfer functions is decomposed by the direct decomposition, and the two state diagrams are connected in cascade, as shown in Fig. 5-10. The state equations are obtained by regarding the

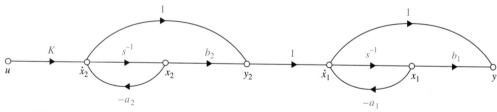

Figure 5-10 State diagram of the transfer function in Eq. (5-211) by cascade decomposition.

derivatives of the state variables as outputs and the state variables and $u(t)$ as inputs, and then applying the SFG gain formula to the state diagram in Fig. 5-10. The integrator branches are neglected when applying the gain formula. The results are

$$\begin{bmatrix} \dfrac{dx_1(t)}{dt} \\ \dfrac{dx_2(t)}{dt} \end{bmatrix} = \begin{bmatrix} -a_1 & b_2 - a_2 \\ 0 & -a_2 \end{bmatrix} \begin{bmatrix} x_1(t) \\ x_2(t) \end{bmatrix} + \begin{bmatrix} K \\ K \end{bmatrix} u(t) \tag{5-212}$$

The output equation is obtained by regarding the state variables and $u(t)$ as inputs and $y(t)$ as the output, and applying the gain formula to Fig. 5-10. Thus,

$$y(t) = [b_1 - a_1 \quad b_2 - a_2]\mathbf{x}(t) + Ku(t) \tag{5-213}$$

When the overall transfer function has complex poles or zeros, the individual factors related to these poles or zeros should be in second-order form. As an example, consider the following transfer function:

$$\frac{Y(s)}{U(s)} = \left(\frac{s + 5}{s + 2}\right)\left(\frac{s + 1.5}{s^2 + 3s + 4}\right) \tag{5-214}$$

where the poles of the second term are complex. The state diagram of the system with the two subsystems connected in cascade is shown in Fig. 5-11. The dynamic equations of the system are

$$\begin{bmatrix} \dfrac{dx_1(t)}{dt} \\ \dfrac{dx_2(t)}{dt} \\ \dfrac{dx_3(t)}{dt} \end{bmatrix} = \begin{bmatrix} 0 & 1 & 0 \\ -4 & -3 & 3 \\ 0 & 0 & -2 \end{bmatrix} \begin{bmatrix} x_1(t) \\ x_2(t) \\ x_3(t) \end{bmatrix} + \begin{bmatrix} 0 \\ 1 \\ 1 \end{bmatrix} u(t) \tag{5-215}$$

$$y(t) = [1.5 \quad 1 \quad 0]\mathbf{x}(t) \tag{5-216}$$

5-9-3 Parallel Decomposition

When the denominator of the transfer function is in factored form, the transfer function may be expanded by partial-fraction expansion. The resulting state diagram will consist of simple first- or second-order systems connected in parallel, which leads to the state equations in DCF or JCF, the latter in the case of multiple-order eigenvalues.

Consider that a second-order system is represented by the transfer function

$$\frac{Y(s)}{U(s)} = \frac{Q(s)}{(s + a_1)(s + a_2)} \tag{5-217}$$

Figure 5-11 State
diagram of the transfer
function in Eq. (5-214)
by cascade
decomposition.

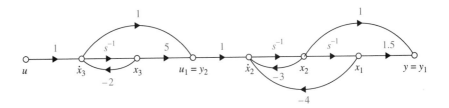

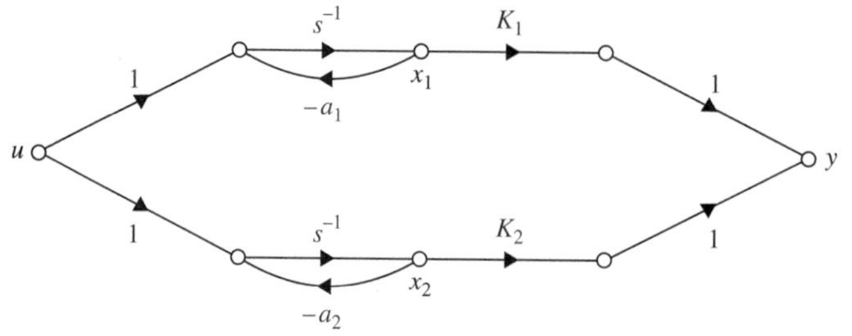

Figure 5-12 State diagram of the transfer function of Eq. (5-217) by parallel decomposition.

where $Q(s)$ is a polynomial of order less than 2, and a_1 and a_2 are real and distinct. Although, analytically, a_1 and a_2 may be complex, in practice, complex numbers are difficult to implement on a computer. Equation (5-219) is expansion by partial fractions:

$$\frac{Y(s)}{U(s)} = \frac{K_1}{s + a_1} + \frac{K_2}{s + a_2} \tag{5-218}$$

where K_1 and K_2 are real constants.

The state diagram of the system is drawn by the parallel combination of the state diagrams of each of the first-order terms in Eq. (5-218), as shown in Fig. 5-12. The dynamic equations of the system are

$$\begin{bmatrix} \dfrac{dx_1(t)}{dt} \\ \dfrac{dx_2(t)}{dt} \end{bmatrix} = \begin{bmatrix} -a_1 & 0 \\ 0 & -a_2 \end{bmatrix} \begin{bmatrix} x_1(t) \\ x_2(t) \end{bmatrix} + \begin{bmatrix} 1 \\ 1 \end{bmatrix} u(t) \tag{5-219}$$

$$y(t) = \begin{bmatrix} K_1 & K_2 \end{bmatrix} \mathbf{x}(t) \tag{5-220}$$

Thus, the state equations are of the DCF.

The conclusion is that for transfer functions with distinct poles, parallel decomposition will lead to the DCF for the state equations. For transfer functions with multiple-order eigenvalues, parallel decomposition to a state diagram with a minimum number of integrators will lead to the JCF state equations. The following example will clarify this point.

▶ **EXAMPLE 5-17** Consider the following transfer function and its partial-fraction expansion:

$$\frac{Y(s)}{U(s)} = \frac{2s^2 + 6s + 5}{(s + 1)^2(s + 2)} = \frac{1}{(s + 1)^2} + \frac{1}{s + 1} + \frac{1}{s + 2} \tag{5-221}$$

Note that the transfer function is of the third order, and although the total order of the terms on the right-hand side of Eq. (5-221) is four, only three integrators should be used in the state diagram, which is drawn as shown in Fig. 5-13. The minimum number of three integrators is used, with one integrator being shared by two channels. The state equations of the system are written directly from Fig. 5-13.

$$\begin{bmatrix} \dfrac{dx_1(t)}{dt} \\ \dfrac{dx_2(t)}{dt} \\ \dfrac{dx_3(t)}{dt} \end{bmatrix} = \begin{bmatrix} -1 & 1 & 0 \\ 0 & -1 & 0 \\ 0 & 0 & -2 \end{bmatrix} \begin{bmatrix} x_1(t) \\ x_2(t) \\ x_3(t) \end{bmatrix} + \begin{bmatrix} 0 \\ 1 \\ 1 \end{bmatrix} u(t) \tag{5-222}$$

which is recognized to be the JCF.

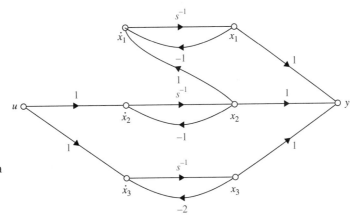

Figure 5-13 State diagram of the transfer function of Eq. (5-221) by parallel decomposition.

◀

▶ 5-10 CONTROLLABILITY OF CONTROL SYSTEMS

The concepts of **controllability** and **observability** introduced first by Kalman [3] play an important role in both theoretical and practical aspects of modern control. The conditions on controllability and observability essentially govern the existence of a solution to an optimal control problem. This seems to be the basic difference between optimal control theory and classical control theory. In the classical control theory, the design techniques are dominated by trial-and-error methods so that given a set of design specifications the designer at the outset does not know if any solution exists. Optimal control theory, on the other hand, has criteria for determining at the outset if the design solution exists for the system parameters and design objectives.

We shall show that the condition of controllability of a system is closely related to the existence of solutions of state feedback for assigning the values of the eigenvalues of the system arbitrarily. The concept of *observability* relates to the condition of observing or estimating the state variables from the output variables, which are generally measurable.

The block diagram shown in Fig. 5-14 illustrates the motivation behind investigating controllability and observability. Figure 5-14(a) shows a system with the process dynamics described by

$$\frac{d\mathbf{x}(t)}{dt} = \mathbf{A}\mathbf{x}(t) + \mathbf{B}\mathbf{u}(\mathbf{t}) \tag{5-223}$$

The closed-loop system is formed by feeding back the state variables through a constant feedback gain matrix **K**. Thus, from Fig. 5-14,

$$\mathbf{u}(t) = -\mathbf{K}\mathbf{x}(t) + \mathbf{r}(t) \tag{5-224}$$

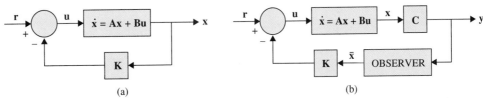

(a)   (b)

Figure 5-14 (a) Control system with state feedback. (b) Control system with observer and state feedback.

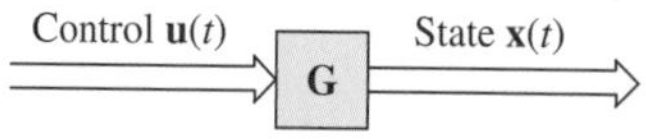

Control $\mathbf{u}(t)$ State $\mathbf{x}(t)$

G

Figure 5-15   Linear time-invariant system.

where $\mathbf{K}$ is a $p \times n$ feedback matrix with constant elements. The closed-loop system is thus described by

$$\frac{d\mathbf{x}(t)}{dt} = (\mathbf{A} - \mathbf{BK})\mathbf{x}(t) + \mathbf{B}r(t) \qquad (5\text{-}225)$$

This problem is also known as the **pole-placement design** through state feedback. The design objective in this case is to find the feedback matrix $\mathbf{K}$ such that the eigenvalues of $(\mathbf{A} - \mathbf{BG})$, or of the closed-loop system, are of certain prescribed values. The word *pole* refers here to the poles of the closed-loop transfer function, which are the same as the eigenvalues of $(\mathbf{A} - \mathbf{BG})$.

We shall show later that the existence of the solution to the pole-placement design with arbitrarily assigned pole values through state feedback is directly based on the controllability of the states of the system. The result is that *if the system of Eq. (5-225) is controllable, then there exists a constant feedback matrix $\mathbf{K}$ that allows the eigenvalues of $(\mathbf{A} - \mathbf{BG})$ to be arbitrarily assigned.*

Once the closed-loop system is designed, the practical problems of implementing the feeding back of the state variables must be considered. There are two problems with implementing state feedback control: First, the number of state variables may be excessive, which will make the cost of sensing each of these state variables for feedback prohibitive. Second, not all the state variables are physically accessible, and so it may be necessary to design and construct an **observer** that will estimate the state vector from the output vector $\mathbf{y}(t)$. Figure 5-14(b) shows the block diagram of a closed-loop system with an observer. The observed state vector $\bar{\mathbf{x}}(t)$ is used to generate the control $\mathbf{u}(t)$ through the feedback matrix $\mathbf{K}$. *The condition that such an observer can be designed for the system is called the observability of the system.*

5-10-1 General Concept of Controllability

The concept of controllability can be stated with reference to the block diagram of Fig. 5-15. *The process is said to be* ***completely controllable*** *if every state variable of the process can be controlled to reach a certain objective in finite time by some unconstrained control* $\mathbf{u}(t)$. Intuitively, it is simple to understand that if any one of the state variables is independent of the control $\mathbf{u}(t)$, there would be no way of driving this particular state variable to a desired state in finite time by means of a control effort. Therefore, this particular state is said to be uncontrollable, and as long as there is at least one uncontrollable state, the system is said to be not completely controllable or, simply, uncontrollable.

As a simple example of an uncontrollable system, Fig. 5-16 illustrates the state diagram of a linear system with two state variables. Since the control $u(t)$ affects only the

Figure 5-16   State diagram of the system that is not state controllable.

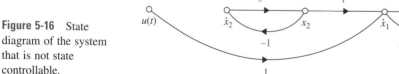

state $x_1(t)$, the state $x_2(t)$ is uncontrollable. In other words, it would be impossible to drive $x_2(t)$ from an initial state $x_2(t_0)$ to a desired state $x_2(t_f)$ in finite time interval $t_f - t_0$ by the control $u(t)$. Therefore, the entire system is said to be uncontrollable.

The concept of controllability given here refers to the states and is sometimes referred to as **state controllability**. Controllability can also be defined for the outputs of the system, so there is a difference between state controllability and output controllability.

5-10-2 Definition of State Controllability

Consider that a linear time-invariant system is described by the following dynamic equations:

$$\frac{d\mathbf{x}(t)}{dt} = \mathbf{A}\mathbf{x}(t) + \mathbf{B}\mathbf{u}(t) \tag{5-226}$$

$$\mathbf{y}(t) = \mathbf{C}\mathbf{x}(t) + \mathbf{D}\mathbf{u}(t) \tag{5-227}$$

where $\mathbf{x}(t)$ is the $n \times 1$ state vector, $\mathbf{u}(t)$ is the $r \times 1$ input vector, and $\mathbf{y}(t)$ is the $p \times 1$ output vector. $\mathbf{A}$, $\mathbf{B}$, $\mathbf{C}$, and $\mathbf{D}$ are coefficients of appropriate dimensions.

The state $\mathbf{x}(t)$ is said to be controllable at $t = t_0$ if there exists a piecewise continuous input $\mathbf{u}(t)$ that will drive the state to any final state $\mathbf{x}(t_f)$ for a finite time $(t_f - t_0) \geq 0$. If every state $\mathbf{x}(t_0)$ of the system is controllable in a finite time interval, the system is said to be completely state controllable or, simply, controllable.

The following theorem shows that the condition of controllability depends on the coefficient matrices $\mathbf{A}$ and $\mathbf{B}$ of the system. The theorem also gives one method of testing for state controllability.

◼ **Theorem 5-1.** *For the system described by the state equation of Eq. (5-226) to be* **completely state controllable**, *it is necessary and sufficient that the following $n \times nr$ **controllability matrix** has a rank of n:*

$$\mathbf{S} = \begin{bmatrix} \mathbf{B} & \mathbf{A}\mathbf{B} & \mathbf{A}^2\mathbf{B} & \cdots & \mathbf{A}^{n-1}\mathbf{B} \end{bmatrix} \tag{5-228}$$

Since the matrices $\mathbf{A}$ and $\mathbf{B}$ are involved, sometimes we say that the pair $[\mathbf{A}, \mathbf{B}]$ is controllable, which implies that $\mathbf{S}$ is of rank n.

The proof of this theorem is given in any standard textbook on optimal control systems [7]. The idea is to start with the state-transition equation of Eq. (5-44) and then proceed to show that Eq. (5-230) must be satisfied in order that all the states are accessible by the input.

Although the criterion of state controllability given in Theorem 5-1 is quite straightforward, manually, it is not very easy to test for high-order systems and/or systems with many inputs. If $\mathbf{S}$ is nonsquare, we can form the matrix $\mathbf{S}\mathbf{S}'$, which is $n \times n$; then if $\mathbf{S}\mathbf{S}'$ is nonsingular, $\mathbf{S}$ has rank n.

5-10-3 Alternate Tests on Controllability

There are several alternate methods of testing controllability, and some of these may be more convenient to apply than the condition in Eq. (5-228).

◼ **Theorem 5-2.** *For a single-input-single-output (SISO) system described by the state equation of Eq. (5-226) with $r = 1$, the pair $[\mathbf{A}, \mathbf{B}]$ is completely controllable if $\mathbf{A}$ and $\mathbf{B}$ are in CCF or transformable into CCF by a similarity transformation.*

The proof of this theorem is straightforward, since it was established in Sec. 5-8 that the CCF transformation requires that the controllability matrix $\mathbf{S}$ be nonsingular. Since the CCF transformation in Sec. 5-8 was defined only for SISO systems, the theorem applies only to this type of systems.

■ **Theorem 5-3.** *For a system described by the state equation of Eq. (5-226), if* **A** *is in DCF or JCF, the pair* [**A**, **B**] *is completely controllable if all the elements in the rows of* **B** *that correspond to the last row of each Jordan block are nonzero.*

The proof of this theorem comes directly from the definition of controllability. Let us assume that **A** is diagonal and that it has distinct eigenvalues. Then, the pair [**A**, **B**] is controllable if **B** does not have any row with all zeros. The reason is that if **A** is diagonal, all the states are decoupled from each other, and if any row of **B** contains all zero elements, the corresponding state would not be accessed from any of the inputs, and that state would be uncontrollable.

For a system in JCF, such as the **A** and **B** matrices illustrated in Eq. (5-229), for controllability only the elements in the row of **B** that corresponds to the last row of the Jordan block are not all zeros. The elements in the other rows of **B** need not all be nonzero, since the corresponding states are still coupled through the 1s in the Jordan blocks of **A**.

$$\mathbf{A} = \begin{bmatrix} \lambda_1 & 1 & 0 & 0 \\ 0 & \lambda_1 & 1 & 0 \\ 0 & 0 & \lambda_1 & 0 \\ \hline 0 & 0 & 0 & \lambda_2 \end{bmatrix} \qquad \mathbf{B} = \begin{bmatrix} b_{11} & b_{12} \\ b_{21} & b_{22} \\ b_{31} & b_{32} \\ b_{41} & b_{42} \end{bmatrix} \qquad (5\text{-}229)$$

Thus, the condition of controllability for the **A** and **B** in Eq. (5-229) is $b_{31} \neq 0$, $b_{32} \neq 0$, $b_{41} \neq 0$, and $b_{42} \neq 0$.

▶ **EXAMPLE 5-18** The following matrices are for a system with two identical eigenvalues, but the matrix **A** is diagonal.

$$\mathbf{A} = \begin{bmatrix} \lambda_1 & 0 \\ 0 & \lambda_1 \end{bmatrix} \qquad \mathbf{B} = \begin{bmatrix} b_{11} \\ b_{21} \end{bmatrix} \qquad (5\text{-}230)$$

The system is uncontrollable, since the two state equations are dependent; that is, it would not be possible to control the states independently by the input. We can easily show that in this case **S** = [**B** **AB**] is singular. ◀

▶ **EXAMPLE 5-19** Consider the system shown in Fig. 5-16, which was reasoned earlier to be uncontrollable. Let us investigate the same system using the condition of Eq. (5-230). The state equations of the system are written in the form of Eq. (5-228) with

$$\mathbf{A} = \begin{bmatrix} -2 & 1 \\ 0 & -1 \end{bmatrix} \qquad \mathbf{B} = \begin{bmatrix} 1 \\ 0 \end{bmatrix} \qquad (5\text{-}231)$$

Thus, from Eq. (5-228) the controllability matrix is

$$\mathbf{S} = \begin{bmatrix} \mathbf{B} & \mathbf{AB} \end{bmatrix} = \begin{bmatrix} 1 & -2 \\ 0 & 0 \end{bmatrix} \qquad (5\text{-}232)$$

which is singular, and the system is uncontrollable. ◀

▶ **EXAMPLE 5-20** Consider that a third-order system has the coefficient matrices

$$\mathbf{A} = \begin{bmatrix} 1 & 2 & -1 \\ 0 & 1 & 0 \\ 1 & -4 & 3 \end{bmatrix} \qquad \mathbf{B} = \begin{bmatrix} 0 \\ 0 \\ 1 \end{bmatrix} \qquad (5\text{-}233)$$

The controllability matrix is

$$S = [\mathbf{B} \quad \mathbf{AB} \quad \mathbf{A}^2\mathbf{B}] = \begin{bmatrix} 0 & -1 & -4 \\ 0 & 0 & 0 \\ 1 & 3 & 8 \end{bmatrix} \tag{5-234}$$

which is singular. Thus, the system is not controllable.

The eigenvalues of $\mathbf{A}$ are $\lambda_1 = 2$, $\lambda_2 = 2$, and $\lambda_3 = 1$. The JCF of $\mathbf{A}$ and $\mathbf{B}$ are obtained with the transformation $\mathbf{x}(t) = \mathbf{T}\bar{\mathbf{x}}(t)$, where

$$\mathbf{T} = \begin{bmatrix} 1 & 0 & 0 \\ 0 & 0 & 1 \\ -1 & 1 & 2 \end{bmatrix} \tag{5-235}$$

Then,

$$\bar{\mathbf{A}} = \mathbf{T}^{-1}\mathbf{AT} = \begin{bmatrix} 2 & -1 & 0 \\ 0 & 2 & 0 \\ 0 & 0 & 1 \end{bmatrix} \qquad \bar{\mathbf{B}} = \mathbf{T}^{-1}\mathbf{B} = \begin{bmatrix} 0 \\ -1 \\ 0 \end{bmatrix} \tag{5-236}$$

Since the last row of $\bar{\mathbf{B}}$, which corresponds to the Jordan block for the eigenvalue λ_3, is zero, the transformed state variable $\bar{x}_3(t)$ is uncontrollable. From the transformation matrix $\mathbf{T}$ in Eq. (5-235), $x_2 = \bar{x}_3$, which means that x_2 is uncontrollable in the original system. It should be noted that the minus sign in front of the 1 in the Jordan block does not alter the basis definition of the block. ◀

▶ 5-11 OBSERVABILITY OF LINEAR SYSTEMS

The concept of observability has been covered earlier in Sec. 5-10 on controllability and observability. Essentially, *a system is completely observable if every state variable of the system affects some of the outputs.* In other words, it is often desirable to obtain information on the state variables from the measurements of the outputs and the inputs. If any one of the states cannot be observed from the measurements of the outputs, the state is said to be unobservable, and the system is not completely observable, or, simply, unobservable. Figure 5-17 shows the state diagram of a linear system in which the state x_2 is not connected to the output $y(t)$ in any way. Once we have measured $y(t)$, we can observe the state $x_1(t)$, since $x_1(t) = y(t)$. However, the state x_2 cannot be observed from the information on $y(t)$. Thus, the system is unobservable.

5-11-1 Definition of Observability

Given a linear time-invariant system that is described by the dynamic equations of Eqs. (5-226) and (5-227), the state $\mathbf{x}(t_0)$ is said to be observable if given any input $u(t)$,

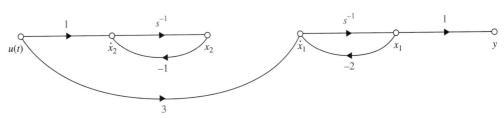

Figure 5-17 State diagram of a system that is not observable.

there exists a finite time $t_f \geq t_0$ such that the knowledge of $\mathbf{u}(t)$ for $t_0 \leq t < t_f$, matrices $\mathbf{A}$, $\mathbf{B}$, $\mathbf{C}$, and $\mathbf{D}$; and the output $\mathbf{y}(t)$ for $t_0 \leq t < t_f$ are sufficient to determine $\mathbf{x}(t_0)$. If every state of the system is observable for a finite t_f, we say that the system is completely observable, or, simply, observable.

The following theorem shows that the condition of observability depends on the matrices $\mathbf{A}$ and $\mathbf{C}$ of the system. The theorem also gives one method of testing observability.

■ **Theorem 5-4.** *For the system described by Eqs. (5-226) and (5-227) to be completely observable, it is necessary and sufficient that the following $n \times np$ **observability matrix** has a rank of n:*

$$\mathbf{V} = \begin{bmatrix} \mathbf{C} \\ \mathbf{CA} \\ \mathbf{CA}^2 \\ \vdots \\ \mathbf{CA}^{n-1} \end{bmatrix} \tag{5-237}$$

The condition is also referred to as the pair $[\mathbf{A}, \mathbf{C}]$ being observable. In particular, if the system has only one output, $\mathbf{C}$ is a $1 \times n$ row matrix; $\mathbf{V}$ is an $n \times n$ square matrix. Then the system is completely observable if $\mathbf{V}$ is nonsingular.

The proof of this theorem is not given here. It is based on the principle that Eq. (5-237) must be satisfied so that $\mathbf{x}(t_0)$ can be uniquely determined from the output $\mathbf{y}(t)$.

5-11-2 Alternate Tests on Observability

Just as with controllability, there are several alternate methods of testing observability. These are described in the following theorems.

■ **Theorem 5-5.** *For a SISO system, described by the dynamic equations of Eqs. (5-226) and (5-227) with $r = 1$ and $p = 1$, the pair $[\mathbf{A}, \mathbf{C}]$ is completely observable if $\mathbf{A}$ and $\mathbf{C}$ are in OCF or transformable into OCF by a similarity transformation.*

The proof of this theorem is straightforward, since it was established in Sec. 5-8 that the OCF transformation requires that the observability matrix $\mathbf{V}$ be nonsingular.

■ **Theorem 5-6.** *For a system described by the dynamic equations of Eqs. (5-226) and (5-227), if $\mathbf{A}$ is in DCF or JCF, the pair $[\mathbf{A}, \mathbf{C}]$ is completely observable if all the elements in the columns of $\mathbf{C}$ that correspond to the first row of each Jordan block are nonzero.*

Note that this theorem is a dual of the test of controllability given in Theorem 5-3. If the system has distinct eigenvalues, $\mathbf{A}$ is diagonal, then the condition on observability is that none of the columns of $\mathbf{C}$ can contain all zeros.

▶ **EXAMPLE 5-21** Consider the system shown in Fig. 5-17, which was earlier defined to be unobservable. The dynamic equations of the system are expressed in the form of Eqs. (5-226) and (5-227) with

$$\mathbf{A} = \begin{bmatrix} -2 & 0 \\ 0 & -1 \end{bmatrix} \qquad \mathbf{B} = \begin{bmatrix} 3 \\ 1 \end{bmatrix} \qquad \mathbf{C} = \begin{bmatrix} 1 & 0 \end{bmatrix} \tag{5-238}$$

Thus, the observability matrix is

$$\mathbf{V} = \begin{bmatrix} \mathbf{C} \\ \mathbf{CA} \end{bmatrix} = \begin{bmatrix} 1 & 0 \\ -2 & 0 \end{bmatrix} \tag{5-239}$$

which is singular. Thus, the pair [**A**, **C**] is unobservable. In fact, since **A** is of DCF, and the second column of **C** is zero, this means that the state $x_2(t)$ is unobservable, as conjectured from Fig. 5-16. ◀

▶ 5-12 RELATIONSHIP AMONG CONTROLLABILITY, OBSERVABILITY, AND TRANSFER FUNCTIONS

In the classical analysis of control systems, transfer functions are used for modeling of linear time-invariant systems. Although controllability and observability are concepts and tools of modern control theory, we shall show that they are closely related to the properties of transfer functions.

■ **Theorem 5-7.** *If the input-output transfer function of a linear system has polezero cancellation, the system will be either uncontrollable or unobservable, or both, depending on how the state variables are defined. On the other hand, if the input-output transfer function does not have pole-zero cancellation, the system can always be represented by dynamic equations as a completely controllable and observable system.*

The proof of this theorem is not given here. The importance of this theorem is that if a linear system is modeled by a transfer function with no pole-zero cancellation, then we are assured that it is a controllable and observable system, no matter how the state-variable model is derived. Let us amplify this point further by referring to the following SISO system.

$$\mathbf{A} = \begin{bmatrix} -1 & 0 & 0 & 0 \\ 0 & -2 & 0 & 0 \\ 0 & 0 & -3 & 0 \\ 0 & 0 & 0 & -4 \end{bmatrix} \quad \mathbf{B} = \begin{bmatrix} 1 \\ 1 \\ 0 \\ 0 \end{bmatrix} \quad \mathbf{C} = \begin{bmatrix} 1 & 0 & 1 & 0 \end{bmatrix} \quad D = 0$$

(5-240)

Since **A** is a diagonal matrix, the controllability and observability conditions of the four states are determined by inspection. They are

x_1:   Controllable and Observable (C and O)
x_2:   Controllable but Unobservable (C but UO)
x_3:   Uncontrollable but Observable (UC but O)
x_4:   Uncontrollable and Unobservable (UC and UO)

The block diagram of the system in Fig. 5-18 shows the DCF decomposition of the system. Clearly, the transfer function of the controllable and observable system should be

$$\frac{Y(s)}{U(s)} = \frac{1}{s+1}$$

(5-241)

whereas the transfer function that corresponds to the dynamics described in Eq. (5-240) is

$$\frac{Y(s)}{U(s)} = \mathbf{C}(s\mathbf{I} - \mathbf{A})^{-1}\mathbf{B} = \frac{(s+2)(s+3)(s+4)}{(s+1)(s+2)(s+3)(s+4)}$$

(5-242)

which has three pole-zero cancellations. This simple-minded example illustrates that "minimum-order" transfer function without pole-zero cancellation is the only component that corresponds to a system that is controllable and observable.

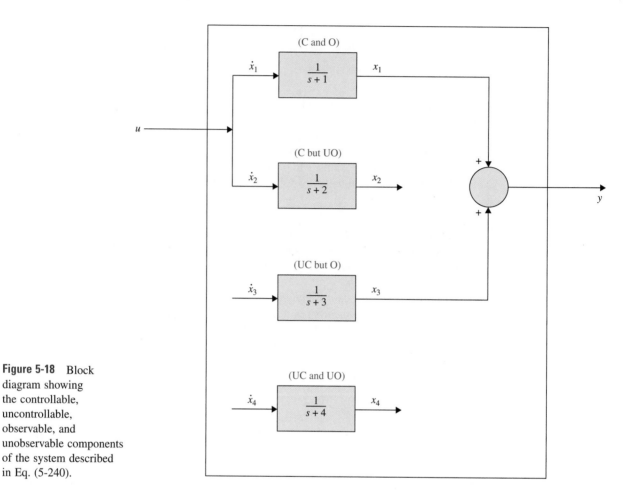

Figure 5-18 Block diagram showing the controllable, uncontrollable, observable, and unobservable components of the system described in Eq. (5-240).

▶ **EXAMPLE 5-22** Let us consider the transfer function

$$\frac{Y(s)}{U(s)} = \frac{s + 2}{(s + 1)(s + 2)} \tag{5-243}$$

which is a reduced form of Eq. (5-242). Equation (5-243) is decomposed into CCF and OCF as follows:

CCF:

$$\mathbf{A} = \begin{bmatrix} 0 & 1 \\ -2 & -3 \end{bmatrix} \qquad \mathbf{B} = \begin{bmatrix} 0 \\ 1 \end{bmatrix} \qquad \mathbf{C} = \begin{bmatrix} 1 & 1 \end{bmatrix} \tag{5-244}$$

Since the CCF transformation can be made, the pair [**A**, **B**] of the CCF is controllable. The observability matrix is

$$\mathbf{V} = \begin{bmatrix} \mathbf{C} \\ \mathbf{CA} \end{bmatrix} = \begin{bmatrix} 1 & 1 \\ -2 & -2 \end{bmatrix} \tag{5-245}$$

which is singular, and the pair [**A**, **C**] of the CCF is unobservable.

OCF:

$$\mathbf{A} = \begin{bmatrix} 0 & -2 \\ 1 & -3 \end{bmatrix} \qquad \mathbf{B} = \begin{bmatrix} 1 \\ 1 \end{bmatrix} \qquad \mathbf{C} = \begin{bmatrix} 0 & 1 \end{bmatrix} \tag{5-246}$$

Since the OCF transformation can be made, the pair [**A**, **C**] of the OCF is observable. However, the controllability matrix is

$$\mathbf{S} = [\mathbf{B} \quad \mathbf{AB}] = \begin{bmatrix} 1 & -2 \\ 1 & -2 \end{bmatrix} \tag{5-247}$$

which is singular, and the pair [**A**, **B**] of the OCF is uncontrollable.

The conclusion that can be drawn from this example is that given a system that is modeled by transfer function, the controllability and observability conditions of the system depend on how the state variables are defined. ◄

► 5-13 INVARIANT THEOREMS ON CONTROLLABILITY AND OBSERVABILITY

We now investigate the effects of the similarity transformations on controllability and observability. The effects of controllability and observability due to state feedback will be investigated.

■ **Theorem 5-8.** *Invariant theorem on similarity transformations: Consider that the system described by the dynamic equations of Eqs. (5-226) and (5-227). The similarity transformation* $\mathbf{x}(t) = \mathbf{P}\overline{\mathbf{x}}(t)$, *where* $\mathbf{P}$ *is nonsingular, transforms the dynamic equations to*

$$\frac{d\overline{\mathbf{x}}(t)}{dt} = \overline{\mathbf{A}}\overline{\mathbf{x}}(t) + \overline{\mathbf{B}}\mathbf{u}(t) \tag{5-248}$$

$$\overline{\mathbf{y}}(t) = \overline{\mathbf{C}}\mathbf{x}(t) + \overline{\mathbf{D}}\mathbf{u}(t) \tag{5-249}$$

where

$$\overline{\mathbf{A}} = \mathbf{P}^{-1}\mathbf{AP} \qquad \overline{\mathbf{B}} = \mathbf{P}^{-1}\mathbf{B} \tag{5-250}$$

The controllability of $[\overline{\mathbf{A}}, \overline{\mathbf{B}}]$ *and the observability of* $[\overline{\mathbf{A}}, \overline{\mathbf{C}}]$ *are not affected by the transformation.*

In other words, controllability and observability are preserved through similar transformations. The theorem is easily proven by showing that the ranks of $\overline{\mathbf{S}}$ and $\mathbf{S}$ and the ranks of $\overline{\mathbf{V}}$ and $\mathbf{V}$ are identical, where $\overline{\mathbf{S}}$ and $\overline{\mathbf{V}}$ are the controllability and observability matrices, respectively, of the transformed system.

■ **Theorem 5-9.** *Theorem on controllability of closed-loop systems with state feedback: If the open-loop system*

$$\frac{d\mathbf{x}(t)}{dt} = \mathbf{A}\mathbf{x}(t) + \mathbf{B}\mathbf{u}(t) \tag{5-251}$$

is completely controllable, then the closed-loop system obtained through state feedback,

$$\mathbf{u}(t) = \mathbf{r}(t) - \mathbf{K}\mathbf{x}(t) \tag{5-252}$$

so that the state equation becomes

$$\frac{d\mathbf{x}(t)}{dt} = (\mathbf{A} - \mathbf{BK})\mathbf{x}(t) + \mathbf{B}\mathbf{r}(t) \tag{5-253}$$

is also completely controllable. On the other hand, if [**A**, **B**] *is uncontrollable, then there is no* **K** *that will make the pair* [**A** − **BK**, **B**] *controllable. In other words, if an open-loop system is uncontrollable, it cannot be made controllable through state feedback.*

Proof. The controllability of [**A**, **B**] implies that there exists a control **u**(t) over the time interval [t_0, t_f] such that **the initial state x(t_0)** is driven to the final state **x**(t_f) over the finite time interval **x**(t_f) $t_f - t_0$. We can write Eq. (5-254) as

$$\mathbf{r}(t) = \mathbf{u}(t) + \mathbf{Kx}(t) \tag{5-254}$$

which is the control of the closed-loop system. Thus, if **u**(t) exists that can drive **x**(t_0) to any **x**(t_f) in finite time, then we cannot find an input **r**(t) that will do the same to **x**(t), because otherwise we can set **u**(t) as in Eq. (5-252) to control the open-loop system.

■ **Theorem 5-10.** *Theorem on observability of closed-loop systems with state feedback: If an open-loop system is controllable and observable, then state feedback of the form of Eq. (5-252) could destroy observability. In other words, the observability of open-loop and closed-loop systems due to state feedback is unrelated.*

The following example illustrates the relation between observability and state feedback.

▶ **EXAMPLE 5-23** Let the coefficient matrices of a linear system be

$$\mathbf{A} = \begin{bmatrix} 0 & 1 \\ -2 & -3 \end{bmatrix} \quad \mathbf{B} = \begin{bmatrix} 1 \\ 1 \end{bmatrix} \quad \mathbf{C} = [1 \quad 2] \tag{5-255}$$

We can show that the pair [**A**, **B**] is controllable and [**A**, **C**] is observable.

Let the state feedback be defined as

$$u(t) = r(t) - \mathbf{Kx}(t) \tag{5-256}$$

where

$$\mathbf{K} = [k_1 \quad k_2] \tag{5-257}$$

Then the closed-loop system is described by the state equation

$$\frac{d\mathbf{x}(t)}{dt} = (\mathbf{A} - \mathbf{BK})\mathbf{x}(t) + \mathbf{B}r(t) \tag{5-258}$$

$$\mathbf{A} - \mathbf{BK} = \begin{bmatrix} -k_1 & 1 - k_2 \\ -2 - k_1 & -3 - g_2 \end{bmatrix} \tag{5-259}$$

The observability matrix of the closed-loop system is

$$\mathbf{V} = \begin{bmatrix} \mathbf{C} \\ \mathbf{C}(\mathbf{A} - \mathbf{BK}) \end{bmatrix} = \begin{bmatrix} 1 & 2 \\ -k_1 - 4 & -3k_2 - 5 \end{bmatrix} \tag{5-260}$$

The determinant of **V** is

$$|\mathbf{V}| = 6k_1 - 3k_2 + 3 \tag{5-261}$$

Thus, if k_1 and k_2 are chosen so that $|\mathbf{V}| = 0$, the closed-loop system would be uncontrollable.

◀

▶ 5-14 A FINAL ILLUSTRATIVE EXAMPLE: MAGNETIC-BALL SUSPENSION SYSTEM

As a final example to illustrate some of the material presented in this chapter, let us consider the magnetic-ball suspension system shown in Fig. 5-19. The objective of the system is to regulate the current of the electromagnet so that the ball will be suspended at a fixed distance from the end of the magnet. The dynamic equations of

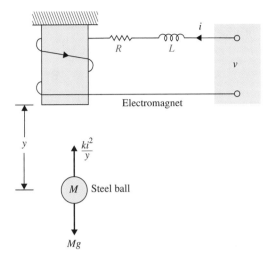

Figure 5-19 Ball-suspension system.

the system are

$$M\frac{d^2x(t)}{dt^2} = Mg - \frac{ki^2(t)}{x(t)} \tag{5-262}$$

$$v(t) = Ri(t) + L\frac{di(t)}{dt} \tag{5-263}$$

where Eq. (5-262) is nonlinear. The system variables and parameters are as follows:

$v(t)$ = input voltage (V) $x(t)$ = ball position (m)
$i(t)$ = winding current (A) k = proportional constant = 1.0
R = winding resistance = 1 Ω L = winding inductance = 0.01 H
M = ball mass = 1.0 Kg g = gravitational acceleration = 32.2 m/sec^2

The state variables are defined as

$$x_1(t) = x(t)$$
$$x_2(t) = \frac{dx(t)}{dt} \tag{5-264}$$
$$x_3(t) = i(t)$$

The state equations are

$$\frac{dx_1(t)}{dt} = x_2(t) \tag{5-265}$$

$$\frac{dx_2(t)}{dt} = g - \frac{k}{M}\frac{x_3^2(t)}{x_1(t)} \tag{5-266}$$

$$\frac{dx_3(t)}{dt} = -\frac{R}{L}x_3(t) + \frac{v(t)}{L} \tag{5-267}$$

These nonlinear state equations are linearized about the equilibrium point, $x_1(t) = x(t) = 0.5$ m, using the method described in Sec. 4-8. After substituting the parameter values, the linearized equations are written

$$\Delta\dot{\mathbf{x}}(t) = \mathbf{A}^*\Delta\mathbf{x}(t) + \mathbf{B}^*\Delta v(t) \tag{5-268}$$

where $\Delta x(t)$ denotes the state vector, and $\Delta v(t)$ is the input voltage of the linearized system. The coefficient matrices are

$$\mathbf{A}^* = \begin{bmatrix} 0 & 1 & 0 \\ 64.4 & 0 & -16 \\ 0 & 0 & -100 \end{bmatrix} \quad \mathbf{B}^* = \begin{bmatrix} 0 \\ 0 \\ 100 \end{bmatrix} \quad (5\text{-}269)$$

All the computations done in the following section can be carried out with a computer program, such as the MATLAB Toolbox. To show the analytical method, we carry out the steps of the derivations as follows.

The Characteristic Equation

$$|s\mathbf{I} - \mathbf{A}^*| = \begin{vmatrix} s & -1 & 0 \\ -64.4 & s & 16 \\ 0 & 0 & s + 100 \end{vmatrix} = s^3 + 100s^2 - 64.4s - 6440 = 0 \quad (5\text{-}270)$$

Eigenvalues: The eigenvalues of $\mathbf{A}^*$, or the roots of the characteristic equation, are

$$s = -100 \qquad s = -8.025 \qquad s = 8.025$$

The State-Transition Matrix: The state-transition matrix of $\mathbf{A}^*$ is

$$\phi(t) = \mathcal{L}^{-1}[(s\mathbf{I} - \mathbf{A}^*)^{-1}] = \mathcal{L}^{-1}\left(\begin{bmatrix} s & -1 & 0 \\ -64.4 & s & 16 \\ 0 & 0 & s + 100 \end{bmatrix}^{-1} \right) \quad (5\text{-}271)$$

or

$$\phi(t) = \mathcal{L}^{-1}\left(\frac{1}{(s + 100)(s + 8.025)(s - 8.025)} \begin{bmatrix} s(s + 100) & s + 100 & -16 \\ 64.4(s + 100) & s(s + 100) & -16s \\ 0 & 0 & s^2 - 64.4 \end{bmatrix} \right) \quad (5\text{-}272)$$

By performing the partial-fraction expansion and carrying out the inverse Laplace transform, the state-transition matrix is

$$\phi(t) = \begin{bmatrix} 0 & 0 & -0.0016 \\ 0 & 0 & 0.16 \\ 0 & 0 & 1 \end{bmatrix} e^{-100t} + \begin{bmatrix} 0.5 & -0.062 & 0.0108 \\ -4.012 & 0.5 & -0.087 \\ 0 & 0 & 0 \end{bmatrix} e^{-8.025t}$$

$$+ \begin{bmatrix} 0.5 & 0.062 & -0.0092 \\ 4.012 & 0.5 & -0.074 \\ 0 & 0 & 0 \end{bmatrix} e^{8.025t} \quad (5\text{-}273)$$

Since the last term in Eq. (5-273) has a positive exponent, the response of $\phi(t)$ increases with time, and the system is unstable. This is expected, since without control, the steel ball would be attracted by the magnet until it hits the bottom of the magnet.

Transfer Function: Let us define the ball position $x(t)$ as the output $y(t)$; then given the input, $v(t)$, the input-output transfer function of the system is

$$\frac{Y(s)}{V(s)} = \mathbf{C}^*(s\mathbf{I} - \mathbf{A}^*)^{-1}\mathbf{B}^* = \begin{bmatrix} 1 & 0 & 0 \end{bmatrix}(s\mathbf{I} - \mathbf{A}^*)^{-1}\mathbf{B}^*$$

$$= \frac{-1600}{(s + 100)(s + 8.025)(s - 8.025)} \quad (5\text{-}274)$$

Controllability: The controllability matrix is

$$\mathbf{S} = \begin{bmatrix} \mathbf{B^*} & \mathbf{A^*B^*} & \mathbf{A^{*2}B^*} \end{bmatrix} = \begin{bmatrix} 0 & 0 & -1{,}600 \\ 0 & -1{,}600 & 160{,}000 \\ 100 & -10{,}000 & 1{,}000{,}000 \end{bmatrix} \quad (5\text{-}275)$$

Since the rank of **S** is 3, the system is completely controllable.

Observability: The observability of the system depends on which variable is defined at the output. For state-feedback control, which will be discussed in Chapter 10, the full controller requires feeding back all three state variables, x_1, x_2, x_3. However, for reasons of economy, we may want to feed back only one of the three state variables. To make the problem more general, we may want to investigate which state, if chosen as the output, would render the system unobservable.

1. $y(t) =$ ball position $= x(t)$: $\mathbf{C^*} = \begin{bmatrix} 1 & 0 & 0 \end{bmatrix}$
   The observability matrix is

   $$\mathbf{V} = \begin{bmatrix} \mathbf{C^*} \\ \mathbf{C^*A^*} \\ \mathbf{C^*A^{*2}} \end{bmatrix} = \begin{bmatrix} 1 & 0 & 0 \\ 0 & 1 & 0 \\ 64.4 & 0 & -16 \end{bmatrix} \quad (5\text{-}276)$$

   which has a rank of 3. Thus, the system is completely observable.

2. $y(t) =$ ball velocity $= dx(t)/dt$: $\mathbf{C^*} = \begin{bmatrix} 0 & 1 & 0 \end{bmatrix}$
   The observability matrix is

   $$\mathbf{V} = \begin{bmatrix} \mathbf{C^*} \\ \mathbf{C^*A^*} \\ \mathbf{C^*A^{*2}} \end{bmatrix} = \begin{bmatrix} 0 & 1 & 0 \\ 64.4 & 0 & -16 \\ 0 & 64.4 & 1600 \end{bmatrix} \quad (5\text{-}277)$$

   which has a rank of 3. Thus, the system is completely observable.

3. $y(t) =$ winding current $= i(t)$: $\mathbf{C^*} = \begin{bmatrix} 0 & 0 & 1 \end{bmatrix}$
   The observability matrix is

   $$\mathbf{V} = \begin{bmatrix} \mathbf{C^*} \\ \mathbf{C^*A^*} \\ \mathbf{C^*A^{*2}} \end{bmatrix} = \begin{bmatrix} 0 & 0 & 1 \\ 0 & 0 & -100 \\ 0 & 0 & -10{,}000 \end{bmatrix} \quad (5\text{-}278)$$

   which has a rank of 1. Thus, the system is unobservable. The physical interpretation of this result is that if we choose the current $i(t)$ as the measurable output, we would not be able to reconstruct the state variables from the measured information.

The interested reader can enter the data of this system into any available computer program and verify the results obtained.

► 5-15 MATLAB TOOLS AND CASE STUDIES

In this section we present a MATLAB tool to solve most problems addressed in this chapter. The reader is encouraged to apply this tool to all the problems identified by a MATLAB Toolbox in the left margin of the text throughout this chapter. As in Chapter 2, we use MATLAB's Symbolic Tool to solve some of the initial problems in this chapter involving inverse Laplace transformations. Finally, using the **tfcal** tool, discussed in Chapter 3, we can convert from transfer functions to state-space representation. These programs allow the user to conduct the following tasks:

- Enter the state matrices.
- Find the system's characteristic polynomial, eigenvalues, and eigenvectors.
- Find the similarity transformation matrices.
- Examine the system controllability and observability properties.
- Obtain the step, impulse, and natural (response to initial conditions) responses, as well as the time response to any function of time.
- Use MATLAB Symbolic Tool to find the state-transition matrix using the inverse Laplace command.
- Convert a transfer function to state-space form or vice versa.

To better illustrate how to use the software, let us go through some of the steps involved in solving earlier examples in this chapter.

5-15-1 Description and Use of the State-Space Analysis Tool

The State-Space Analysis Tool (statetool) consists of a number of m-files and GUIs for the analysis of state-space systems. The statetool can be invoked either from the MATLAB

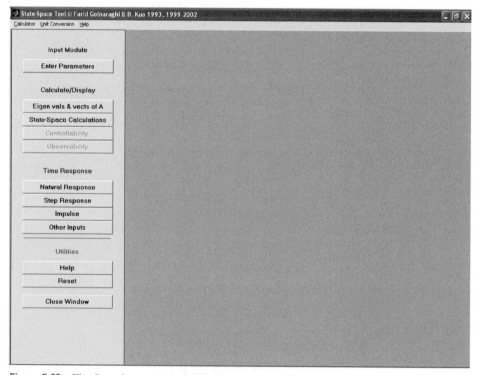

Figure 5-20 The State-Space Analysis Window.

command line by simply typing **statetool** or from the **A**utomatic **C**ontrol **Sys**tems launch applet (**ACSYS**) by clicking on the appropriate button. We use the example in Sec. 5-14 and Examples 1 and 2 to describe how to use the statetool.

First consider the example in Sec. 5-14. To enter the following coefficient matrices

$$\mathbf{A}^* = \begin{bmatrix} 0 & 1 & 0 \\ 64.4 & 0 & -16 \\ 0 & 0 & -100 \end{bmatrix} \quad \mathbf{B}^* = \begin{bmatrix} 0 \\ 0 \\ 100 \end{bmatrix} \quad \mathbf{C}^* = \begin{bmatrix} 1 & 0 & 0 \end{bmatrix} \quad (5\text{-}279)$$

click the "Enter Parameters" button as shown in Fig. 5-20.

Next, the State-Space Input Window will appear, as in Fig. 5-21. Enter the coefficient matrices by clicking the appropriate buttons. Note that the default value of initial conditions is set to zero and you do not have to adjust it for this example. Follow the instructions on screen very carefully. The elements in the row of a matrix may be separated by a space or a comma, while the rows themselves must be separated by a semicolon. For example, to enter matrix **A**, enter

$$0 \quad 1 \quad 0; 64.4 \quad 0 \quad -16; 0 \quad 0 \quad -100$$

and to enter matrix **B**, enter

$$0; \quad 0; \quad 100$$

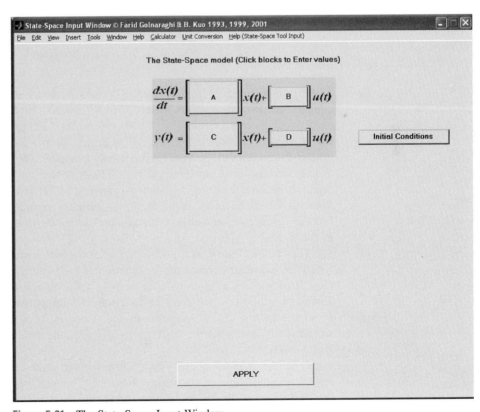

Figure 5-21 The State-Space Input Window.

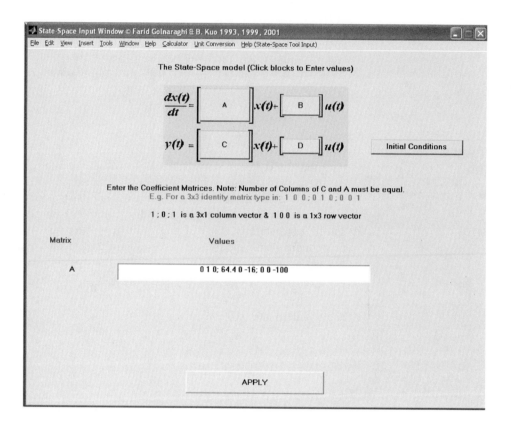

Figure 5-22 Entering matrix **A** in State-Space Input Window.

as shown in Figure 5-22. In this case the **D** matrix is set to zero (default value). Once you complete entering all matrices, press "APPLY" to return to the main window.

To find the characteristic Eq. (5-270), eigenvalues, and eigenvectors, click the "Eigenvals & vects of A" button. You should get Fig. 5-23. As shown, to get the detailed solution, you must refer to the MATLAB command window. The **A** matrix, eigenvalues of **A**, and eigenvectors of **A** are shown in Fig. 5-24. Note that the matrix representation of the eigenvalues corresponds to the diagonal canonical form (DCF) of **A**, while matrix **T**, representing the eigenvectors, is the DCF transformation matrix discussed in Sec. 5-8-4. To find the state-transition matrix $\phi(t)$ you must use the tfsym tool, which will be discussed in Sec. 5-15-2.

The choice of the **C** in Eq. (5-279) makes the ball position the output $y(t)$ for input $v(t)$. Then the input-output transfer function of the system can be obtained by clicking the "State-Space Calculations" button. The final output appearing in the MATLAB command window is the transfer function in both polynomial and factored forms, as shown in Fig. 5-25. As you can see, there is a small error due to numerical simulation. You may set the small terms to zero in the resulting transfer function to get Eq. (5-274).

Press the "Controllability" and "Observability" buttons to determine whether the system is controllable or observable. Note these buttons are only enabled after pressing the "State-Space Calculations" button. After clicking the "Controllability" button you get the MATLAB command window display, shown in Fig. 5-26. The **S** matrix in this case is the same as Eq. (5-275) with the rank of 3. As a result, the system is completely

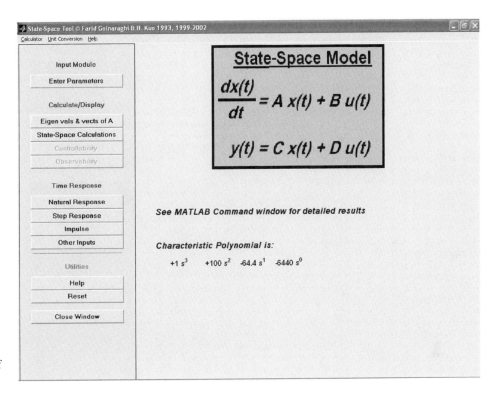

Figure 5-23 The State-Space Tool after clicking the "Eigenvals & vects of A" button.

The A matrix is:

Amat =

0	1.0000	0
64.4000	0	-16.0000
0	0	-100.0000

Characteristic polynomial:

ans =

s^3+100*s^2-2265873562520787/35184372088832*s-6440

Eigenvalues of A = diagonal canonical form of A is:

Abar =

8.0250	0	0
0	−8.0250	0
0	0	−100.0000

Eigenvectors are

T =

0.1237	-0.1237	-0.0016
0.9923	0.9923	0.1590
0	0	0.9873

Figure 5-24 The MAT-LAB command window display, after clicking the "Eigenvals & vects of A" button.

State-space model is:

a =

```
       x1   x2    x3
x1      0    1     0
x2    64.4   0   -16
x3      0    0  -100
```

b =

```
       ul
x1      0
x2      0
x3    100
```

c =

```
      x1   x2   x3
y1     1    0    0
```

d =

```
      u1
y1     0
```

Continuous-time model.
Characteristic polynomial:

ans =

s^3+100*s^2-2265873562520787/35184372088832*s − 6440

Equivalent transfer-function model is:

Transfer function:
$$\frac{4.263e − 014\ s^2 + 8.527e{-}014\ s − 1600}{s^3 + 100s^2 − 64.4s − 6440}$$

Pole, zero form:

Zero/pole/gain:
$$\frac{4.2633e − 014\ (s+1.937e008)(s − 1.937e008)}{(s+100)(s+8.025)(s − 8.025)}$$

Figure 5-25 The MATLAB command window after clicking the "State-Space Calculations" button.

The Controllibility matrix [B AB A^2B ...] is =

Smat =

```
     0        0   −1600
     0    −1600   160000
   100   −10000   1000000
```

The system is therefore controllable, rank of S matrix is =

rankS =

   3

Mmat =

```
  −64.4000    100.0000    1.0000
  100.0000      1.0000         0
    1.0000           0         0
```

The controllability canonical form (CCF) transformation matrix is:

Ptran =

```
  −1600        0     0
      0    −1600     0
  −6440        0   100
```

The transformed matrices using CCF are:

Abar =

1.0e+003 *

```
       0   0.0010        0
       0        0   0.0010
  6.4400   0.0644  −0.1000
```

Bbar =

```
0
0
1
```

Cbar =

```
  −1600     0     0
```

Dbar =

   0

Figure 5-26 The MATLAB command window after clicking the "Controllability" button.

controllable. The program also provides the **M** and **P** matrices and the system controlla-bility canonical form (CCF) representation as defined in Sec. 5-8.

Once you press the "Observability" button, the system observability is assessed in the MATLAB command window, as shown in Fig. 5-27. The system is completely observable,

---

The observability matrix (transpose:[C CA CA^2 ...]) is =

Vmat =

```
    1.0000        0         0
         0   1.0000         0
   64.4000        0  -16.0000
```

The system is therefore observable, rank of V matrix is =

rankV =

   3

Mmat =

```
  -64.4000   100.0000    1.0000
  100.0000     1.0000         0
    1.0000          0         0
```

The observability canonical form (OCF) transformation matrix is:

Qtran =

```
        0         0      1.0000
        0    1.0000   -100.0000
  -0.0625    6.2500   -625.0000
```

The transformed matrices using OCF are:

Abar =

1.0e+003 *

```
   0.0000   -0.0000    6.4400
   0.0010   -0.0000    0.0644
        0    0.0010   -0.1000
```

Bbar =

```
  -1600
      0
      0
```

Cbar =

```
0   0   1
```

Dbar =

0

Figure 5-27 The MATLAB Command window after clicking the "Observability" button.

since the **V** matrix has a rank of 3. Note the **V** matrix in Fig. 5-27 is the same as in Eq. (5-276). The program also provides the **M** and **Q** matrices and the system observability canonical form (OCF) representation as defined in Sec. 5-8. As an exercise, the user is urged to reproduce Eqs. (5-277) and (5-278) using this software.

You may obtain the output $y(t)$ natural time response (response to initial conditions only), the step response, the impulse response, or the time response to any other input function by clicking the appropriate buttons. In Chapter 10, we use these time response options to better assess the system performance.

The statetool program may be used on all the examples identified by a MATLAB Toolbox in the left margin of the text throughout this chapter, except problems involving inverse Laplace transformations and closed-form solutions. To address the analytical solutions, we need to use the tfsym tool, which requires the Symbolic Tool of MATLAB 6, R12.

5-15-2   Description and Use of tfsym for State-Space Applications

You may run the Transfer Function Symbolic Tool directly by typing **tfsym** in the MATLAB command window, or by clicking the "Transfer Function Symbolic" button in the **ACSYS** window. In either case, you should get the window in Fig. 5-28. The program is designed to solve two classes of state-space problems, one without specifying initial conditions and the other with initial conditions.

Let us continue our solution to the example in Sec. 5-14. Since this program is based on the MATLAB Symbolic Tool, it was decided to provide the user interface only through the MATLAB command window. As a result, you need to move to the MATLAB command window after clicking the "State-Space" button for entering the coefficient matrices in Eq. (5-279). The input and output displays in the MATLAB command window are selectively shown in Fig. 5-29. Note that at the first glance the $(s\mathbf{I}\text{-}\mathbf{A})^{-1}$ and $\phi(t)$ matrices may appear different from Eqs. (5-271) and (5-272). However, after minor manipulations, you may be able to verify that they are the same. This difference in representation is because of MATLAB symbolic approach. You may further simplify these matrices by using the "simple" command in the MATLAB command window. For example, to simplify $\phi(t)$ type in "simple(phi)" in the MATLAB command window. If the desired format has not been achieved, you may have reached the software limit.

5-15-3   Another Example

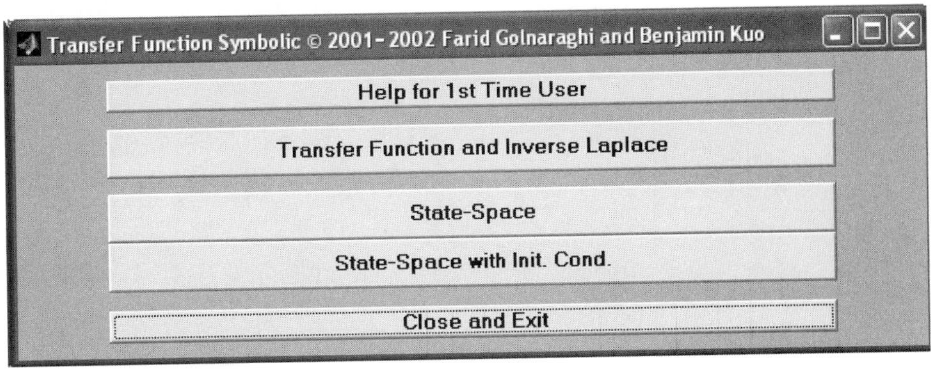

Figure 5-28   The Transfer Function Symbolic window.

Enter A = [0 1 0;64.4 0-16;0 0 -100]

Asym =

```
        0   1.0000          0
  64.4000        0    -16.0000
        0        0   -100.0000
```

Determinant of (s*I-A) is:

detSIA =

s^3+100*s^2-322/5*s-6440

the eigenvalues of A are:

eigA =

```
 -100.0000
    8.0250
   -8.0250
```

Inverse of (s*I-A) is:

```
[        s                5                      80                  ]
[5 ----------------- , ----------------- , -----------------------------------]
[        2                 2                  3       2              ]
[    5 s  -  322        5 s  - 322        5 s  + 500 s  - 322 s  - 32200]
[                                                                     ]
[      322                 s                         s                ]
[----------------- , 5 ----------------- , -80 -----------------------------------]
[      2                   2                   3       2              ]
[  5 s  -  322         5 s  - 322          5 s  + 500 s  - 322 s  - 32200]
[                                                                     ]
[                                                                1    ]
[0 ,                           0 ,                          -----------]
[                                                            s + 100]
```

State transition matrix of A:

```
[                       40                 2000              40       ]
[%2 ,  1/322 %1, - ---------- exp (-100 t) - --------------- %1 + ---------- %2]
[                     24839                 3999079           24839    ]
[                                                                     ]
[                       4000                4000                      ]
[1/5 %1 ,    %2 ,   ---------- exp (-100 t) - --------- %2 + 8/24839 %1]
[                     24839                 24839                     ]
[                                                                     ]
[0 ,                           0 ,                         exp (-100 t)]
```

$$\%1 : = 1610^{1/2} \ \sinh (1/5 \ 1610^{1/2} \ t)$$

$$\%2 : = \cosh (1/5 \ 1610^{1/2} \ t)$$

Transfer function between u(t) and y(t) is:

```
                8000
-  ----------------------------------------
      3       2
   5 s + 500 s - 322 s - 32200
```

Figure 5-29 Selective display of the MATLAB command window for the tfsym tool.

▶ **EXAMPLE 5-24** For the system in Examples 5-1 and 5-2,

$$\begin{bmatrix} \dfrac{dx_1(t)}{dt} \\ \dfrac{dx_2(t)}{dt} \end{bmatrix} = \begin{bmatrix} 0 & 1 \\ -2 & -3 \end{bmatrix} \begin{bmatrix} x_1(t) \\ x_2(t) \end{bmatrix} + \begin{bmatrix} 0 \\ 1 \end{bmatrix} u(t) \tag{5-280}$$

let us define the following system with four different input scenarios

$$\mathbf{A} = \begin{bmatrix} 0 & 1 \\ -2 & -3 \end{bmatrix} \quad \mathbf{B} = \begin{bmatrix} 0 \\ 1 \end{bmatrix} \quad \mathbf{C} = [1 \quad 0] \quad x(0) = \begin{bmatrix} 0 \\ 1 \end{bmatrix} \tag{5-281}$$

$$u(t) = 0; u(t) = u_s; u(t) = \delta; u(t) = e^{-t}$$

Use both tfsym and statetool to solve the problem.

First start up the statetool and enter all matrices and the initial condition vector. Press the "AP-PLY" button to return to the state-space main window. Next press the "Eigenvals & vects of A" button to find the characteristic equation, eigenvalues, and eigenvectors of **A**. The results appear in the MATLAB command window and are as follows:

Characteristic polynomial is the determinant of (s**I-A**): $s^2 + 3s + 2$

The eigenvalues of **A** are: -1 and -2

Diagonal Canonical Form of **A** is: $\begin{bmatrix} -1 & 0 \\ 0 & -2 \end{bmatrix}$

Eigenvectors are: $\mathbf{T} = \begin{bmatrix} 0.707 & -0.4472 \\ -0.707 & 0.8944 \end{bmatrix}$

To assess the controllability and observability of the system, press the corresponding buttons in the state-space main window. The results appear in the MATLAB command window and are as follows:

Equivalent transfer-function model is: $\dfrac{1}{s^2 + 3s + 2}$

Transfer function in pole, zero form: $\dfrac{1}{(s + 1)(s + 2)}$

The controllability matrix $[\mathbf{A} \quad \mathbf{AB} \quad \mathbf{A^2B} \quad \cdots]$ is: $\mathbf{S} = \begin{bmatrix} 0 & 1 \\ 1 & -3 \end{bmatrix}$

The system is therefore controllable, since rank of **S** is 2.

$\mathbf{M} = \begin{bmatrix} 3 & 1 \\ 1 & 0 \end{bmatrix} \quad \mathbf{P} = \mathbf{SM}$

The controllability canonical form (CCF) transformation matrix is $\mathbf{P} = \begin{bmatrix} 1 & 0 \\ 0 & 1 \end{bmatrix}$

The transformed matrices using CCF are $\overline{\mathbf{A}} = \begin{bmatrix} 0 & 1 \\ -2 & -3 \end{bmatrix} \quad \overline{\mathbf{B}} = \begin{bmatrix} 0 \\ 1 \end{bmatrix} \quad \overline{\mathbf{C}} = [1 \quad 0]$

The observability matrix $\begin{bmatrix} \mathbf{C} \\ \mathbf{CA} \\ \mathbf{CA^2} \\ \cdots \end{bmatrix}$ is: $\mathbf{V} = \begin{bmatrix} 1 & 0 \\ 0 & 1 \end{bmatrix}$

The system is therefore observable, since rank of **V** is 2.

$\mathbf{M} = \begin{bmatrix} 3 & 1 \\ 1 & 0 \end{bmatrix} \quad \mathbf{Q} = (\mathbf{MV})^{-1}$

The observability canonical form (OCF) transformation matrix is $\mathbf{Q} = \begin{bmatrix} 0 & 1 \\ 1 & -3 \end{bmatrix}$

The transformed matrices using OCF are $\overline{\mathbf{A}} = \begin{bmatrix} 0 & -2 \\ 1 & -3 \end{bmatrix} \quad \overline{\mathbf{B}} = \begin{bmatrix} 1 \\ 0 \end{bmatrix} \quad \overline{\mathbf{C}} = [0 \quad 1]$

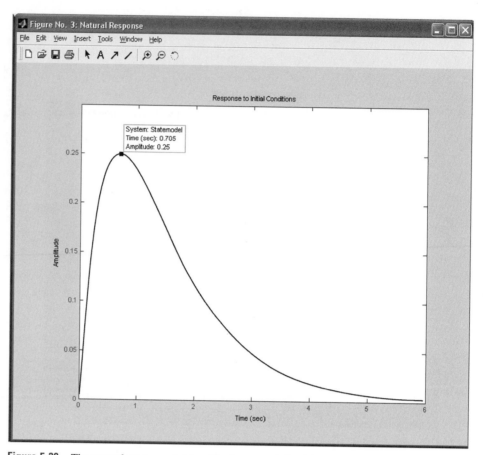

Figure 5-30 The natural response ($u(t) = 0$) of system defined by Eq. (5-281).

To plot various responses you can use the plot options. To plot the natural response (i.e., response due to the initial conditions alone while no input $u(t)$ is applied), click the "Natural Response" button. The plot in Fig. 5-30 shows the output $y(t) = x_1(t)$ response subject to initial conditions $x_1(0) = 0$ and $x_2(0) = 1$. If you wish to plot $x_2(t)$, you should change **C** to [0   1] in the input module, re-calculate the system model, and plot the response again.

The $y(t)$ response to a unit-step input is shown in Fig. 5-31, and the $y(t)$ response to an impulse input is shown in Fig. 5-32. Note that in this case the impulse response is identical to the natural response shown in Figure 5-30 (verify).

Finally, to get the response to $u(t) = e^{-t}$, you need to click the "Other Inputs" button in the State-Space Tool main window. This action causes the Time Response window to appear, allowing you to enter the numerical simulation start time, step size, and final time. In addition, you can enter the input type in symbolic form. Note that the input in this case **must** be a function of time, such as $sin(t)$ and $exp(-t)$. In this example we set the start time to zero, the step size to 0.01, the simulation time to 10 sec, and the input to $u(t) = e^{-t}$, as shown in Fig. 5-33. The time response of the system is shown in Fig. 5-34. In order to get an accurate reading of $y(t)$ at any point on the graph, right-click the mouse. A box containing the amplitude and time values will appear on the graph for the designated point. Further, you may use the edit window to edit the graph properties for all the plots discussed here.

Next, start the tfsym tool, and click the "State-Space with init. Cond." button. Move to the MATLAB command window and enter the system matrices as directed to get the following results.

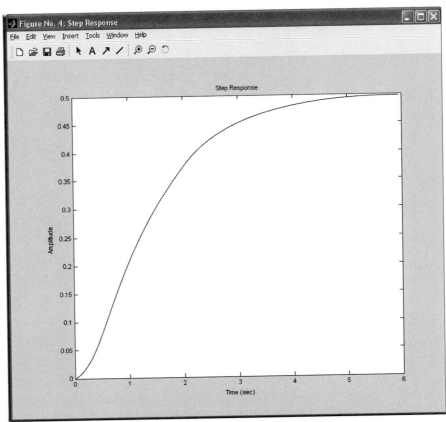

Figure 5-31 The unit-step response ($u(t) = u_s$) of system defined by Eq. (5-281).

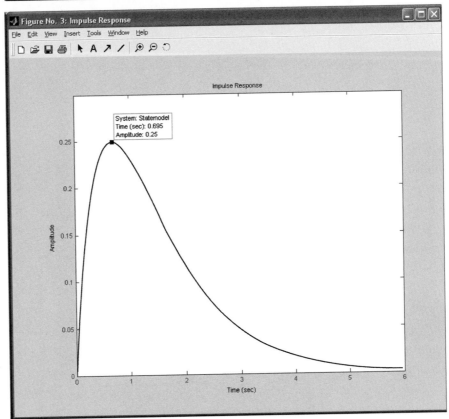

Figure 5-32 The impulse response ($u(t) = \delta$) of system defined by Eq. (5-281).

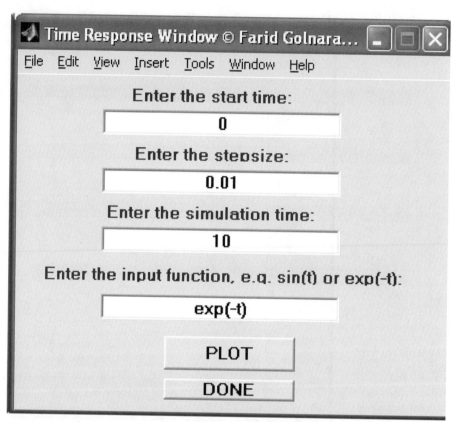

Figure 5-33 Entering the simulation parameters for the system defined by Eq. (5-281), $u(t) = e^{-t}$.

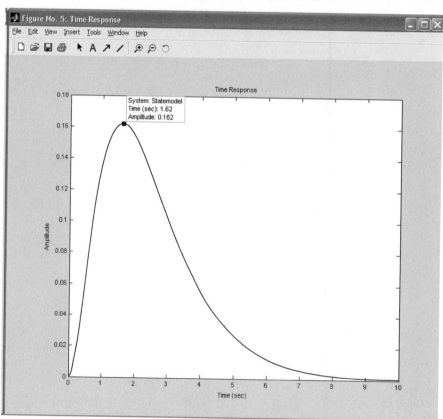

Figure 5-34 Time response of the system Eq. (5-281), $u(t) = e^{-t}$.

Characteristic polynomial is the determinant of (sI-A): $s^2 + 3s + 2$

The eigenvalues of **A** are: -1 and -2

$$(sI\text{-}A)^{-1} = \frac{1}{s^2 + 3s + 2} \begin{bmatrix} s+3 & 1 \\ 2 & s \end{bmatrix}$$

State transition matrix (phi) of A: $\begin{bmatrix} -e^{-2t} + 2e^{-t} & -e^{-2t} + e^{-t} \\ 2e^{-2t} - 2e^{-t} & 2e^{-2t} - e^{-t} \end{bmatrix}$

Equivalent transfer function model is: $\dfrac{1}{s^2 + 3s + 2}$

For $u(t) = e^{-t}$, $U(s) = \dfrac{1}{s+1}$, state vectors in Laplace domain are: $X(s) = \begin{bmatrix} \dfrac{1}{(s+1)^2} \\ \dfrac{s}{(s+1)^2} \end{bmatrix}$

Inverse Laplace: $x(t) = \begin{bmatrix} te^{-t} \\ (1-t)e^{-t} \end{bmatrix}$

Output $Y(s)$ is the same as $X_1(s)$: $Y(s) = \dfrac{1}{(s+1)^2}$

Inverse Laplace: $y(t) = te^{-t}$

Comparing both cases, it is clear that tfsym and statetool have the same results. ◀

This section demonstrated the functionality of the tfsym and statetool programs. The user is urged to solve the examples and homework problems in this chapter using these tools. These programs will be revisited in Chapter 10, where we study controller design. It is also recommended that the reader use the Transfer Function Calculator tool (tfcal) in the **ACSYS** software to solve problems involving conversion from transfer-function representation to state-space form (e.g., Examples 5-16 and 5-17). Comparing the results, you should notice that the state-space representation of a transfer function may not be unique.

▶ 5-16 SUMMARY

This chapter was devoted to the state-variable analysis of linear systems. The fundamentals on state variables and state equations were introduced in Chapters 2 and 3, and formal discussions on these subjects were covered in this chapter. Specifically, the state-transition matrix and state-transition equations were introduced and the relationship between the state equations and transfer functions was established. Given the transfer function of a linear system, the state equations of the system can be obtained by decomposition of the transfer function. Given the state equations and the output equations, the transfer function can be determined either analytically or directly from the state diagram.

Characteristic equations and eigenvalues were defined in terms of the state equations and the transfer function. Eigenvectors of **A** were also defined for distinct and multiple-order eigenvalues. Similarity transformations to controllability canonical form (CCF), observability canonical form (OCF), diagonal canonical form (DCF), and Jordan canonical form (JCF) were discussed. State controllability and observability of linear time-invariant systems were defined and illustrated, and a final example, on the magnetic-ball-suspension system, summarized the important elements of the state-variable analysis of linear systems.

The MATLAB software tools statetool, tfsym, and tfcal were described in the last section (see also Chapter 3). The program functionality was discussed with two examples.

Together these tools can solve most of the homework problems and examples in this chapter.

► REVIEW
QUESTIONS

1. What are the components of the dynamic equations of a linear system?

2. Given the state equations of a linear system as

$$\frac{d\mathbf{x}(t)}{dt} = \mathbf{A}\mathbf{x}(t) + \mathbf{B}\mathbf{u}(t)$$

give two expressions of the state-transition matrix $\phi(t)$ in terms of $\mathbf{A}$.

3. List the properties of the state-transition matrix $\phi(t)$.

4. Given the state equations as in Review Question 2, write the state-transition equation.

5. List the advantages of expressing a linear system in the controllability canonical form (CCF). Give an example of $\mathbf{A}$ and $\mathbf{B}$ in CCF.

6. Given the state equations in the form as in Review Question 2, give the conditions for $\mathbf{A}$ and $\mathbf{B}$ to be transformable into CCF.

7. Express the characteristic equation in terms of the matrix $\mathbf{A}$.

8. List the three methods of decomposition of a transfer function.

9. What special forms will the state equations be in if the transfer function is decomposed by direct decomposition?

10. What special form will the state equations be in if the transfer function is decomposed by parallel decomposition?

11. What is the advantage of using cascade decomposition?

12. State the relationship between the CCF and controllability.

13. For controllability, does the magnitude of the inputs have to be finite?

14. Give the condition of controllability in terms of the matrices $\mathbf{A}$ and $\mathbf{B}$.

15. What is the motivation behind the concept of observability?

16. Give the condition of observability in terms of the matrices $\mathbf{A}$ and $\mathbf{C}$.

17. What can be said about the controllability and observability conditions if the transfer function has pole-zero cancellation?

18. State the relationship between OCF and observability.

► REFERENCES

State Variables and State Equations
1. B. C. Kuo, *Linear Networks and Systems*, McGraw-Hill, New York, 1967.
2. R. A. Gabel and R. A. Roberts, *Signals and Linear Systems*, 3rd ed., John Wiley & Sons, New York, 1987.

Controllability and Observability
3. R. E. Kalman, "On the General Theory of Control Systems," *Proc. IFAC*, Vol. 1, pp. 481–492, Butterworths, London, 1961.
4. W. L. Brogan, *Modern Control Theory*, Second Edition, Prentice Hall, Englewood Cliffs, NJ, 1985.

► PROBLEMS

• Dynamic equations

5-1. The following differential equations represent linear time-invariant systems. Write the dynamic equations (state equations and output equations) in vector-matrix form.

(a) $\dfrac{d^2y(t)}{dt^2} + 4\dfrac{dy(t)}{dt} + y(t) = 5r(t)$

(b) $2\dfrac{d^3y(t)}{dt^3} + 3\dfrac{d^2y(t)}{dt^2} + 5\dfrac{dy(t)}{dt} + 2y(t) = r(t)$

(c) $\dfrac{d^3y(t)}{dt^3} + 5\dfrac{d^2y(t)}{dt^2} + 3\dfrac{dy(t)}{dt} + y(t) + \displaystyle\int_0^t y(\tau)d\tau = r(t)$

(d) $\dfrac{d^4y(t)}{dt^4} + 1.5\dfrac{d^3y(t)}{dt^3} + 2.5\dfrac{dy(t)}{dt} + y(t) = 2r(t)$

• State-transition matrix   **5-2.** By use of Eq. (5-23), show that

$$\phi(t) = \mathbf{I} + \mathbf{A}t + \frac{1}{2!}\mathbf{A}^2t^2 + \frac{1}{3!}\mathbf{A}^3t^3 + \cdots$$

• State-transition matrix   **5-3.** The state equations of a linear time-invariant system are represented by

$$\frac{d\mathbf{x}(t)}{dt} = \mathbf{A}\mathbf{x}(t) + \mathbf{B}\mathbf{u}(t)$$

Find the state-transition matrix $\phi(t)$, the characteristic equation and the eigenvalues of $\mathbf{A}$ for the following cases:

(a) $\mathbf{A} = \begin{bmatrix} 0 & 1 \\ -2 & -1 \end{bmatrix}$ $\mathbf{B} = \begin{bmatrix} 0 & 1 \\ 1 & 0 \end{bmatrix}$

(b) $\mathbf{A} = \begin{bmatrix} 0 & 1 \\ -4 & -5 \end{bmatrix}$ $\mathbf{B} = \begin{bmatrix} 1 \\ 1 \end{bmatrix}$

(c) $\mathbf{A} = \begin{bmatrix} -3 & 0 \\ 0 & -3 \end{bmatrix}$ $\mathbf{B} = \begin{bmatrix} 0 \\ 1 \end{bmatrix}$

(d) $\mathbf{A} = \begin{bmatrix} 3 & 0 \\ 0 & -3 \end{bmatrix}$ $\mathbf{B} = \begin{bmatrix} 0 \\ 1 \end{bmatrix}$

(e) $\mathbf{A} = \begin{bmatrix} 0 & 2 \\ -2 & 0 \end{bmatrix}$ $\mathbf{B} = \begin{bmatrix} 0 \\ 1 \end{bmatrix}$

(f) $\mathbf{A} = \begin{bmatrix} -1 & 0 & 0 \\ 0 & -2 & 1 \\ 0 & 0 & -2 \end{bmatrix}$ $\mathbf{B} = \begin{bmatrix} 0 \\ 1 \\ 0 \end{bmatrix}$

(g) $\mathbf{A} = \begin{bmatrix} -5 & 1 & 0 \\ 0 & -5 & 1 \\ 0 & 0 & -5 \end{bmatrix}$ $\mathbf{B} = \begin{bmatrix} 0 \\ 0 \\ 1 \end{bmatrix}$

Find $\phi(t)$ and the characteristic equation using a computer program if available.

• State-transition equation   **5-4.** Find the state-transition equation of each of the systems described in Problem 5-3 for $t \geq 0$. Assume that $\mathbf{x}(0)$ is the initial state vector, and the components of the input vector $\mathbf{u}(t)$ are all unit-step functions.

• State-transition matrix   **5-5.** Find out if the matrices given in the following can be state-transition matrices. [Hint: check the properties of $\phi(t)$.]

(a) $\begin{bmatrix} -e^{-t} & 0 \\ 0 & 1 - e^{-t} \end{bmatrix}$ **(b)** $\begin{bmatrix} 1 - e^{-t} & 0 \\ 1 & e^{-t} \end{bmatrix}$

(c) $\begin{bmatrix} 1 & 0 \\ 1 - e^{-t} & e^{-t} \end{bmatrix}$ **(d)** $\begin{bmatrix} e^{-2t} & te^{-2t} & t^2e^{-2t}/2 \\ 0 & e^{-2t} & te^{-2t} \\ 0 & 0 & e^{-2t} \end{bmatrix}$

5-6. Given a system described by the dynamic equations:

$$\frac{d\mathbf{x}(t)}{dt} = \mathbf{A}\mathbf{x}(t) + \mathbf{B}u(t) \qquad y(t) = \mathbf{C}\mathbf{x}(t)$$

(a) $\mathbf{A} = \begin{bmatrix} 0 & 1 & 0 \\ 0 & 0 & 1 \\ -1 & -2 & -3 \end{bmatrix}$ $\quad \mathbf{B} = \begin{bmatrix} 0 \\ 0 \\ 1 \end{bmatrix}$ $\quad \mathbf{C} = \begin{bmatrix} 1 & 0 & 0 \end{bmatrix}$

(b) $\mathbf{A} = \begin{bmatrix} -1 & 1 \\ 0 & -1 \end{bmatrix}$ $\quad \mathbf{B} = \begin{bmatrix} 0 \\ 1 \end{bmatrix}$ $\quad \mathbf{C} = \begin{bmatrix} 1 & 1 \end{bmatrix}$

(c) $\mathbf{A} = \begin{bmatrix} 0 & 1 & 0 \\ 0 & 0 & 1 \\ 0 & -1 & -2 \end{bmatrix}$ $\quad \mathbf{B} = \begin{bmatrix} 0 \\ 0 \\ 1 \end{bmatrix}$ $\quad \mathbf{C} = \begin{bmatrix} 1 & 1 & 0 \end{bmatrix}$

(1) Find the eigenvalues of **A**. Use the **ACSYS** computer program to check the answers. You may get the characteristic equation and solve for the roots using tfsim or tcal components of **ACSYS**.

(2) Find the transfer-function relation between $\mathbf{X}(s)$ and $U(s)$.

(3) Find the transfer function $Y(s)/U(s)$.

5-7. Given the dynamic equations of a time-invariant system:

$$\frac{d\mathbf{x}(t)}{dt} = \mathbf{A}\mathbf{x}(t) + \mathbf{B}u(t) \qquad y(t) = \mathbf{C}\mathbf{x}(t)$$

where

$$\mathbf{A} = \begin{bmatrix} 0 & 1 & 0 \\ 0 & 0 & 1 \\ -1 & -2 & -3 \end{bmatrix} \quad \mathbf{B} = \begin{bmatrix} 0 \\ 0 \\ 1 \end{bmatrix} \quad \mathbf{C} = \begin{bmatrix} 1 & 1 & 0 \end{bmatrix}$$

Find the matrices $\mathbf{A}_1$ and $\mathbf{B}_1$ so that the state equations are written as

$$\frac{d\overline{\mathbf{x}}(t)}{dt} = \mathbf{A}_1\overline{\mathbf{x}}(t) + \mathbf{B}_1 u(t)$$

where

$$\overline{\mathbf{x}}(t) = \begin{bmatrix} x_1(t) \\ y(t) \\ \dfrac{dy(t)}{dt} \end{bmatrix}$$

5-8. Given the dynamic equations

$$\frac{d\mathbf{x}(t)}{dt} = \mathbf{A}\mathbf{x}(t) + \mathbf{B}u(t) \qquad y(t) = \mathbf{C}\mathbf{x}(t)$$

(a) $\mathbf{A} = \begin{bmatrix} 0 & 2 & 0 \\ 1 & 2 & 0 \\ -1 & 0 & 1 \end{bmatrix}$ $\quad \mathbf{B} = \begin{bmatrix} 0 \\ 1 \\ 1 \end{bmatrix}$ $\quad \mathbf{C} = \begin{bmatrix} 1 & 0 & 1 \end{bmatrix}$

(b) $\mathbf{A} = \begin{bmatrix} 0 & 2 & 0 \\ 1 & 2 & 0 \\ -1 & 1 & 1 \end{bmatrix}$ $\quad \mathbf{B} = \begin{bmatrix} 1 \\ 1 \\ 0 \end{bmatrix}$ $\quad \mathbf{C} = \begin{bmatrix} 1 & 0 & 1 \end{bmatrix}$

(c) $\mathbf{A} = \begin{bmatrix} -2 & 1 & 0 \\ 0 & -2 & 0 \\ -1 & -2 & -3 \end{bmatrix}$ $\quad \mathbf{B} = \begin{bmatrix} 1 \\ 1 \\ 1 \end{bmatrix}$ $\quad \mathbf{C} = \begin{bmatrix} 1 & 0 & 0 \end{bmatrix}$

(d) $\mathbf{A} = \begin{bmatrix} -1 & 1 & 0 \\ 0 & -1 & 1 \\ 0 & 0 & -1 \end{bmatrix}$ $\quad \mathbf{B} = \begin{bmatrix} 0 \\ 1 \\ 1 \end{bmatrix}$ $\quad \mathbf{C} = \begin{bmatrix} 1 & 0 & 1 \end{bmatrix}$

(e) $\mathbf{A} = \begin{bmatrix} 1 & 1 \\ -2 & -3 \end{bmatrix}$ $\quad \mathbf{B} = \begin{bmatrix} 0 \\ 1 \end{bmatrix}$ $\quad \mathbf{C} = \begin{bmatrix} 1 & 0 \end{bmatrix}$

Find the transformation $\mathbf{x}(t) = \mathbf{P\bar{x}}(t)$ that transforms the state equations into the controllability canonical form (CCF).

- OCF

5-9. For the systems described in Problem 5-8, find the transformation $\mathbf{x}(t) = \mathbf{Q\bar{x}}(t)$ so that the state equations are transformed into the observability canonical form (OCF).

- DCF, JCF

5-10. For the systems described in Problem 5-8, find the transformation $\mathbf{x}(t) = \mathbf{T\bar{x}}(t)$ so that the state equations are transformed into the diagonal canonical form (DCF) if $\mathbf{A}$ has distinct eigenvalues, and Jordan canonical form (JCF) if $\mathbf{A}$ has at least one multiple-order eigenvalue.

- CCF

5-11. The state equation of a linear system is described by

$$\frac{d\mathbf{x}(t)}{dt} = \mathbf{Ax}(t) + \mathbf{Bu}(t)$$

The coefficient matrices are given as follows. Explain why the state equations cannot be transformed into the controllability canonical form (CCF).

(a) $\mathbf{A} = \begin{bmatrix} -2 & 0 \\ 0 & -1 \end{bmatrix}$ $\quad \mathbf{B} = \begin{bmatrix} 0 \\ 1 \end{bmatrix}$

(b) $\mathbf{A} = \begin{bmatrix} -1 & 0 & 0 \\ 0 & -1 & 0 \\ 0 & 0 & -1 \end{bmatrix}$ $\quad \mathbf{B} = \begin{bmatrix} 1 \\ 2 \\ 3 \end{bmatrix}$

(c) $\mathbf{A} = \begin{bmatrix} 1 & 2 \\ 1 & 1 \end{bmatrix}$ $\quad \mathbf{B} = \begin{bmatrix} 2 \\ \sqrt{2} \end{bmatrix}$

(d) $\mathbf{A} = \begin{bmatrix} -2 & 1 & 0 \\ 0 & -2 & 0 \\ -1 & -2 & -3 \end{bmatrix}$ $\quad \mathbf{B} = \begin{bmatrix} 1 \\ 0 \\ 1 \end{bmatrix}$

- State equations, transfer functions

5-12. The equations that describe the dynamics of a motor control system are

$$e_a(t) = R_a i_a(t) + L_a \frac{di_a(t)}{dt} + K_b \frac{d\theta_m(t)}{dt}$$

$$T_m(t) = K_i i_a(t)$$

$$T_m(t) = J \frac{d^2\theta_m(t)}{dt^2} + B \frac{d\theta_m(t)}{dt} + K\theta_m(t)$$

$$e_a(t) = K_a e(t)$$

$$e(t) = K_s[\theta_r(t) - \theta_m(t)]$$

(a) Assign the state variables as $x_1(t) = \theta_m(t)$, $x_2(t) = d\theta_m(t)/dt$, and $x_3(t) = i_a(t)$. Write the state equations in the form of

$$\frac{d\mathbf{x}(t)}{dt} = \mathbf{Ax}(t) + \mathbf{B}\theta_r(t)$$

Write the output equation in the form $y(t) = \mathbf{Cx}(t)$, where $y(t) = \theta_m(t)$.

(b) Find the transfer function $G(s) = \Theta_m(s)/E(s)$ when the feedback path from $\Theta_m(s)$ to $E(s)$ is broken. Find the closed-loop transfer function $M(s) = \Theta_m(s)/\Theta_r(s)$.

• State-transition matrix

5-13. Given the matrix **A** of a linear system described by the state equation

$$\frac{d\mathbf{x}(t)}{dt} = \mathbf{A}\mathbf{x}(t) + \mathbf{B}u(t)$$

(a) $\mathbf{A} = \begin{bmatrix} 0 & 1 \\ -1 & 0 \end{bmatrix}$ (b) $\mathbf{A} = \begin{bmatrix} -1 & 0 \\ 0 & -2 \end{bmatrix}$ (c) $\mathbf{A} = \begin{bmatrix} 0 & 1 \\ 1 & 0 \end{bmatrix}$

Find the state-transition matrix $\phi(t)$ using the following methods:
(1) Infinite-series expansion of $e^{\mathbf{A}t}$, expressed in closed form
(2) The inverse Laplace transform of $(s\mathbf{I} - \mathbf{A})^{-1}$

• State equations, inverse Laplace transform, characteristic equation

5-14. The schematic diagram of a feedback control system using a dc motor is shown in Fig. 5P-14. The torque developed by the motor is $T_m(t) = K_i i_a(t)$, where K_i is the torque constant. The constants of the system are

$$K_s = 2 \qquad\qquad R = 2\,\Omega \qquad\qquad R_s = 0.1\,\Omega$$
$$K_b = 5\ \text{V/rad/sec} \qquad K_i = 5\ \text{N-m/A} \qquad L_a \cong 0\ \text{H}$$
$$J_m + J_L = 0.1\ \text{N-m-sec}^2 \qquad B_m \cong 0\ \text{N-m-sec}$$

Assume that all the units are consistent so that no conversion is necessary.
(a) Let the state variables be assigned as $x_1 = \theta_y$ and $x_2 = d\theta_y/dt$. Let the output be $y = \theta_y$. Write the state equations in vector-matrix form. Show that the matrices **A** and **B** are in CCF.
(b) Let $\theta_r(t)$ be a unit-step function. Find $\mathbf{x}(t)$ in terms of $\mathbf{x}(0)$, the initial state. Use the Laplace-transform table.
(c) Find the characteristic equation and the eigenvalues of **A**.
(d) Comment on the purpose of the feedback resistor R_s.

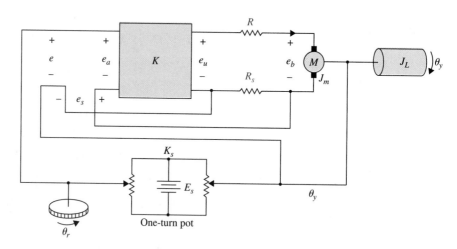

Figure 5P-14

• State equations, inverse Laplace transform, characteristic equation

5-15. Repeat Problem 5-14 with the following system parameters:

$$K_s = 1 \qquad K = 9 \qquad\qquad R_a = 0.1\,\Omega$$
$$R_s = 0.1\,\Omega \qquad K_b = 1\ \text{V/rad/s} \qquad K_i = 1\ \text{N-m/A}$$
$$L_a \cong 0\ \text{H} \qquad J_m + J_L = 0.01\ \text{N-m-sec}^2 \qquad B_m \cong 0\ \text{N-m-sec}$$

• Transfer functions, dynamic equations, final value

5-16. The block diagram of a feedback control system is shown in Fig. 5P-16.
(a) Find the forward-path transfer function $Y(s)/E(s)$ and the closed-loop transfer function $Y(s)/R(s)$.

(b) Write the dynamic equations in the form of

$$\frac{dx(t)}{dt} = Ax(t) + Br(t) \qquad y(t) = Cx(t) + Dr(t)$$

Find **A**, **B**, **C**, and **D** in terms of the system parameters.

(c) Apply the final-value theorem to find the steady-state value of the output $y(t)$ when the input $r(t)$ is a unit-step function. Assume that the closed-loop system is stable.

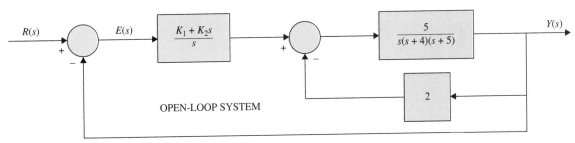

Figure 5P-16

• Adjoint matrix, characteristic equation

5-17. For the linear time-invariant system whose state equations have the coefficient matrices given by Eqs. (5-156) and (5-157) (CCF), show that

$$\mathrm{adj}(sI - A)B = \begin{bmatrix} 1 \\ s \\ s^2 \\ \vdots \\ s^{n-1} \end{bmatrix}$$

and the characteristic equation of **A** is

$$s^n + a_{n-1}s^{n-1} + \cdots + a_1 s + a_0 = 0$$

• State equations, state-transition equation, state-transition matrix, characteristic equation

5-18. A linear time-invariant system is described by the differential equation

$$\frac{d^3 y(t)}{dt^3} + 3\frac{d^2 y(t)}{dt^2} + 3\frac{dy(t)}{dt} + y(t) = r(t)$$

(a) Let the state variables be defined as $x_1 = y$, $x_2 = dy/dt$, and $x_3 = d^2y/dt^2$. Write the state equations of the system in vector-matrix form.

(b) Find the state-transition matrix $\phi(t)$ of **A**.

(c) Let $y(0) = 1$, $dy(0)/dt = 0$, $d^2y(0)/dt^2 = 0$, and $r(t) = u_s(t)$. Find the state-transition equation of the system.

(d) Find the characteristic equation and the eigenvalues of **A**.

• State equations, state-transition matrix, characteristic equation

5-19. A spring-mass-friction system is described by the following differential equation.

$$\frac{d^2 y(t)}{dt^2} + 2\frac{dy(t)}{dt} + y(t) = r(t)$$

(a) Define the state variables as $x_1(t) = y(t)$ and $x_2(t) = dy(t)/dt$. Write the state equations in vector-matrix form. Find the state-transition matrix $\phi(t)$ of **A**.

(b) Define the state variables as $x_1(t) = y(t)$ and $x_2(t) = y(t) + dy(t)/dt$. Write the state equations in vector-matrix form. Find the state-transition matrix $\phi(t)$ of **A**.

(c) Show that the characteristic equations, $|sI - A| = 0$, for parts (a) and (b) are identical.

• State transition matrix, eigenvalues

5-20. Given the state equations $dx(t)/dt = Ax(t)$, where σ and ω are real numbers:

(a) Find the state transition matrix of **A**.

(b) Find the eigenvalues of **A**.

• Dynamic equations

5-21. (a) Show that the input-output transfer functions of the two systems shown in Fig. 5P-21 are the same.

(b) Write the dynamic equations of the system in Fig. 5P-21 as

$$\frac{d\mathbf{x}(t)}{dt} = \mathbf{A}_1\mathbf{x}(t) + \mathbf{B}_1 u_1(t) \qquad y_1(t) = \mathbf{C}_1\mathbf{x}(t)$$

and those of the system in Fig. 5-21(b) as

$$\frac{d\mathbf{x}(t)}{dt} = \mathbf{A}_2\mathbf{x}(t) + \mathbf{B}_2 u_2(t) \qquad y_2(t) = \mathbf{C}_2\mathbf{x}(t)$$

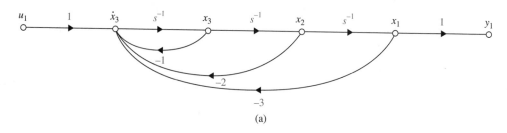

(a)

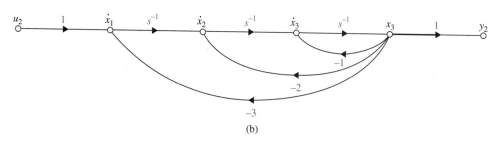

(b)

Figure 5P-21

• State diagrams

5-22. Draw the state diagrams for the following systems.

$$\frac{d\mathbf{x}(t)}{dt} = \mathbf{A}\mathbf{x}(t) + \mathbf{B}\mathbf{u}(t)$$

(a) $\mathbf{A} = \begin{bmatrix} -3 & 2 & 0 \\ -1 & 0 & 1 \\ -2 & -3 & -4 \end{bmatrix}$ $\mathbf{B} = \begin{bmatrix} 0 \\ 0 \\ 1 \end{bmatrix}$

(b) Same $\mathbf{A}$ as in part (a), but with

$$\mathbf{B} = \begin{bmatrix} 0 & 1 \\ 1 & 0 \\ 1 & 0 \end{bmatrix}$$

• State diagrams, state equations

5-23. Draw state diagrams for the following transfer functions by direct decomposition. Assign the state variables from right to left for $x_1, x_2, \ldots$. Write the state equations from the state diagram and show that the equations are in CCF.

(a) $G(s) = \dfrac{10}{s^3 + 8.5s^2 + 20.5s + 15}$ \qquad (b) $G(s) = \dfrac{10(s + 2)}{s^2(s + 1)(s + 3.5)}$

(c) $G(s) = \dfrac{5(s + 1)}{s(s + 2)(s + 10)}$ \qquad (d) $G(s) = \dfrac{1}{s(s + 5)(s^2 + 2s + 2)}$

• State diagrams, state equations

5-24. Draw state diagrams for the systems described in Problem 5-23 by parallel decomposition. Make certain that the state diagrams contain a minimum number of integrators. The constant branch gains must be real. Write the state equations from the state diagram.

• State diagrams, state equations

5-25. Draw the state diagrams for the systems described in Problem 5-23 by using cascade decomposition. Assign the state variables in ascending order from right to left. Write the state equations from the state diagram.

• State diagrams, dynamic equations, state-transition equations

5-26. The block diagram of a feedback control system is shown in Fig. 5P-26.

(a) Draw a state diagram for the system by first decomposing $G(s)$ by direct decomposition. Assign the state variables in ascending order, $x_1, x_2, \ldots$, from right to left. In addition to the state-variable-related nodes, the state diagram should contain nodes for $R(s)$, $E(s)$, and $C(s)$.

(b) Write the dynamic equations of the system in vector-matrix form.

(c) Find the state-transition equations of the system using the state equations found in part (b). The initial state vector is $\mathbf{x}(0)$, and $r(t) = u_s(t)$.

(d) Find the output $y(t)$ for $t \geq 0$ with the initial state $\mathbf{x}(0)$, and $r(t) = u_s(t)$.

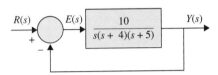

Figure 5P-26

• Transfer function, state diagram, state-transition equations

5-27. **(a)** Find the closed-loop transfer function $Y(s)/R(s)$, and draw the state diagram.

(b) Perform a direct decomposition to $Y(s)/R(s)$, and draw the state diagram.

(c) Assign the state variables from right to left in ascending order, and write the state equations in vector-matrix form.

(d) Find the state-transition equations of the system using the state equations found in part (c). The initial state vector is $\mathbf{x}(0)$, and $r(t) = u_s(t)$.

(e) Find the output $y(t)$ for $t \geq 0$ with the initial state $\mathbf{x}(0)$, and $r(t) = u_s(t)$.

State diagram, state equations, transfer function

5-28. The block diagram of a linearized idle-speed engine-control system of an automobile is shown in Fig. 5P-28. (For a discussion on linearization of nonlinear systems, refer to Sec. 4-8.) The system is linearized about a nominal operating point, so all the variables represent linear-perturbed quantities. The following variables are defined: $T_m(t)$ is the engine torque; T_D, the constant load-disturbance torque; $\omega(t)$, the engine speed; $u(t)$, the input-voltage to the throttle actuator; and α, the throttle angle. The time delay in the engine model can be approximated by

$$e^{-0.2s} \cong \frac{1 - 0.1s}{1 + 0.1s}$$

(a) Draw a state diagram for the system by decomposing each block individually. Assign the state variables from right to left in ascending order.

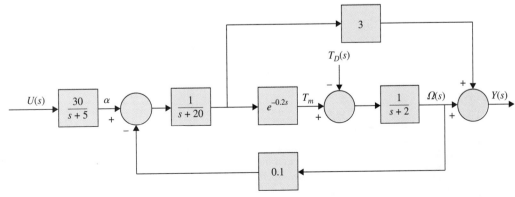

Figure 5P-28

(b) Write the state equations from the state diagram obtained in part (a), in the form of

$$\frac{d\mathbf{x}(t)}{dt} = \mathbf{A}\mathbf{x}(t) + \mathbf{B}\begin{bmatrix} u(t) \\ T_D(t) \end{bmatrix}$$

(c) Write $Y(s)$ as a function of $U(s)$ and $T_D(s)$. Write $\Omega(s)$ as a function of $U(s)$ and $T_D(s)$.

• Dynamic equations

5-29. The state diagram of a linear system is shown in Fig. 5P-29.

(a) Assign state variables on the state diagram from right to left in ascending order. Create additional artificial nodes if necessary so that the state-variable nodes satisfy as "input nodes" after the integrator branches are deleted.

(b) Write the dynamic equations of the system from the state diagram in part (a).

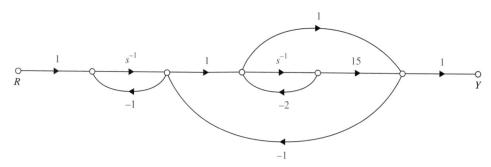

Figure 5P-29

Transfer function,
characteristic equation,
state diagram,
controllability,
observability

5-30. The block diagram of a linear spacecraft-control system is shown in Fig. 5P-30.

(a) Determine the transfer function $Y(s)/R(s)$.

(b) Find the characteristic equation and its roots of the system. Show that the roots of the characteristic equation are not dependent on K.

(c) When $K = 1$, draw a state diagram for the system by decomposing $Y(s)/R(s)$, using a minimum number of integrators.

(d) Repeat part (c) when $K = 4$.

(e) Determine the values of K that must be avoided if the system is to be both state controllable and observable.

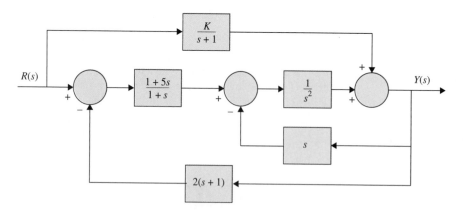

Figure 5P-30

State diagram,
characteristic equation

5-31. A considerable amount of effort is being spent by automobile manufacturers to meet the exhaust-emission-performance standards set by the government. Modern automobile-power-plant systems consist of an internal combustion engine that has an internal cleanup device called a catalytic converter. Such a system requires control of such variables as the engine air-fuel (A/F) ratio, ignition-spark timing, exhaust-gas recirculation, and injection air. The control-system problem considered in this problem deals with the control of the A/F ratio. In general, depending on fuel composition and other factors, a typical stoichiometric A/F is 14.7:1, that is, 14.7 grams of air to each gram of fuel. An A/F greater or less than stoichiometry will cause high hydrocarbons, carbon monoxide, and nitrous oxides in the tailpipe emission. The control system shown in Fig. 5P-31 is devised to control the air-fuel ratio so that a desired output is achieved for a given input command.

The sensor senses the composition of the exhaust-gas mixture entering the catalytic converter. The electronic controller detects the difference or the error between the command and the error and computes the control signal necessary to achieve the desired exhaust-gas composition. The output $y(t)$ denotes the effective air-fuel ratio. The transfer function of the engine is given by

$$G_p(s) = \frac{Y(s)}{U(s)} = \frac{e^{-T_d s}}{1 + 0.5s}$$

where $T_d = 0.2$ sec is the time delay and is approximated by

$$e^{-T_d s} = \frac{1}{e^{T_d s}} = \frac{1}{1 + T_d s + T_d^2 s^2/2! + \cdots} \cong \frac{1}{1 + T_d s + T_d^2 s^2/2!}$$

The gain of the sensor is 1.0.

(a) Using the approximation for $e^{-T_d s}$ given, find the expression for $G_p(s)$. Decompose $G_p(s)$ by direct decomposition, and draw the state diagram with $u(t)$ as the input and $y(t)$ as the output. Assign state variables from right to left in ascending order, and write the state equations in vector-matrix form.

(b) Assuming that the controller is a simple amplifier with a gain of 1, i.e., $u(t) = e(t)$, find the characteristic equation and its roots of the closed-loop system.

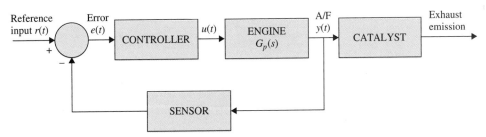

Figure 5P-31

• State diagram, characteristic equation

5-32. Repeat Problem 5-31 when the time delay of the automobile engine is approximated as

$$e^{-T_d s} \cong \frac{1 - T_d s/3}{1 + \frac{2}{3}T_d s + \frac{1}{6}T_d^2 s^2} \qquad T_d = 0.2 \text{ sec}$$

• State equation, state diagram, transfer functions

5-33. The schematic diagram in Fig. 5P-33 shows a permanent-magnet dc-motor-control system with a viscous-inertia damper. The system can be used for the control of the printwheel of an electronic word processor. A mechanical damper such as the viscous-inertia type is sometimes used in practice as a simple and economical way of stabilizing a control system. The damping effect is achieved by a rotor suspended in a viscous fluid. The differential and algebraic equations that describe the dynamics of the system are as follows:

$$e(t) = K_s[\omega_r(t) - \omega_m(t)] \qquad\qquad K_s = 1 \text{ V/rad/sec}$$
$$e_a(t) = Ke(t) = R_a i_a(t) + e_b(t) \qquad\qquad K = 10$$
$$e_b(t) = K_b \omega_m(t) \qquad\qquad K_b = 0.0706 \text{ V/rad/sec}$$
$$T_m(t) = J\frac{d\omega_m(t)}{dt} + K_D[\omega_m(t) - \omega_D(t)] \qquad J = J_h + J_m = 0.1 \text{ oz-in.-sec}^2$$
$$T_m(t) = K_i i_a(t) \qquad\qquad K_i = 10 \text{ oz-in./A}$$
$$K_D[\omega_m(t) - \omega_D(t)] = J_R \frac{d\omega_D(t)}{dt} \qquad J_R = 0.05 \text{ oz-in.-sec}^2$$
$$R_a = 1 \,\Omega \qquad\qquad K_D = 1 \text{ oz-in.-sec}$$

(a) Let the state variables be defined as $x_1(t) = \omega_m(t)$ and $x_2(t) = \omega_D(t)$. Write the state equations for the open-loop system with $e(t)$ as the input. (*Open-loop* refers to the feedback path from ω_m to e being open.)

(b) Draw the state diagram for the overall system using the state equations found in part (a) and $e(t) = K_s[\omega_r(t) - \omega_m(t)]$.

(c) Derive the open-loop transfer function $\Omega_m(s)/E(s)$ and the closed-loop transfer function $\Omega_m(s)/\Omega_r(s)$.

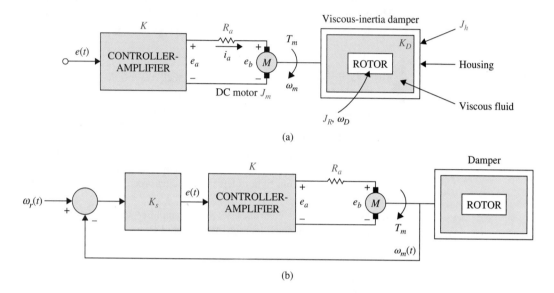

(a)

(b)

Figure 5P-33

• Controllability

5-34. Determine the state controllability of the system shown in Fig. 5P-34.

(a) $a = 1$, $b = 2$, $c = 2$, and $d = 1$.

(b) Are there any nonzero values for a, b, c, and d such that the system is uncontrollable?

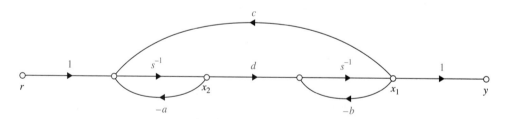

Figure 5P-34

• Controllability

5-35. Determine the controllability of the following systems:

(a) $\mathbf{A} = \begin{bmatrix} -1 & 0 & 0 \\ 0 & -1 & 0 \\ 0 & 0 & -1 \end{bmatrix}$ $\quad \mathbf{B} = \begin{bmatrix} 1 \\ 1 \\ 1 \end{bmatrix}$

(b) $\mathbf{A} = \begin{bmatrix} -1 & 0 & 0 \\ 0 & -2 & 0 \\ 0 & 0 & -3 \end{bmatrix}$ $\quad \mathbf{B} = \begin{bmatrix} 1 \\ 1 \\ 1 \end{bmatrix}$

• Controllability, observability

5-36. Determine the controllability and observability of the system shown in Fig. 5P-36 by the following methods:

(a) Conditions on the $\mathbf{A}$, $\mathbf{B}$, $\mathbf{C}$, and $\mathbf{D}$ matrices

(b) Conditions on the pole-zero cancellation of the transfer functions

Figure 5P-36

• Controllability, observability

5-37. The transfer function of a linear control system is

$$\frac{Y(s)}{R(s)} = \frac{s + \alpha}{s^3 + 7s^2 + 14s + 8}$$

(a) Determine the value(s) of α so that the system is either uncontrollable or unobservable.

(b) With the value(s) of α found in part (a), define the state variables so that one of them is uncontrollable.

(c) With the value(s) of α found in part (a), define the state variables so that one of them is unobservable.

• Controllability

5-38. Consider the system described by the state equation

$$\frac{d\mathbf{x}(t)}{dt} = \mathbf{A}\mathbf{x}(t) + \mathbf{B}u(t)$$

where

$$\mathbf{A} = \begin{bmatrix} 0 & 1 \\ -1 & a \end{bmatrix} \qquad \mathbf{B} = \begin{bmatrix} 1 \\ b \end{bmatrix}$$

Find the region in the a–b plane such that the system is completely controllable.

• Controllability, observability

5-39. Determine the condition on b_1, b_2, c_1, and c_2 so that the following system is completely controllable and observable.

$$\frac{d\mathbf{x}(t)}{dt} = \mathbf{A}\mathbf{x}(t) + \mathbf{B}u(t) \qquad y(t) = \mathbf{C}\mathbf{x}(t)$$

$$\mathbf{A} = \begin{bmatrix} 1 & 1 \\ 0 & 1 \end{bmatrix} \qquad \mathbf{B} = \begin{bmatrix} b_1 \\ b_2 \end{bmatrix} \qquad \mathbf{C} = [d_1 \quad d_2]$$

• State diagram, characteristic equation, eigenvalues, controllability, observability

5-40. The schematic diagram of Fig. 5P-40 represents a control system whose purpose is to hold the level of the liquid in the tank at a desired level. The liquid level is controlled by a float whose position $h(t)$ is monitored. The input signal of the open-loop system is $e(t)$. The system parameters and equations are as follows:

Motor resistance $R_a = 10 \ \Omega$	Motor inductance $L_a = 0$ H
Torque constant $K_i = 10$ oz-in./A	Rotor inertia $J_m = 0.005$ oz-in.-sec^2
Back-emf constant $K_b = 0.0706$ V/rad/sec	Gear ratio $n = N_1/N_2 = 1/100$
Load inertia $J_L = 10$ oz-in.-sec^2	Load and motor friction = negligible
Amplifier gain $K_a = 50$	Area of tank $A = 50$ ft^2

$$e_a(t) = R_a i_a(t) + K_b \omega_m(t) \qquad\qquad \omega_m(t) = \frac{d\theta_m(t)}{dt}$$

$$T_m(t) = K_i i_a(t) = (J_m + n^2 J_L)\frac{d\omega_m(t)}{dt} \qquad \theta_y(t) = n\theta_m(t)$$

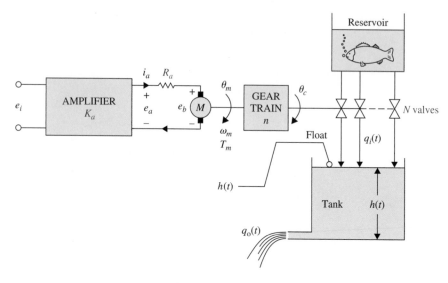

Figure 5P-40

The number of valves connected to the tank from the reservoir is $N = 10$. All the valves have the same characteristics and are controlled simultaneously by θ_y. The equations that govern the volume of flow are as follows:

$$q_i(t) = K_I N \theta_y(t) \qquad\qquad K_I = 10 \text{ ft}^3/\text{sec-rad}$$
$$q_o(t) = K_o h(t) \qquad\qquad K_o = 50 \text{ ft}^2/\text{sec}$$
$$h(t) = \frac{\text{volume of tank}}{\text{area of tank}} = \frac{1}{A}\int[q_i(t) - q_o(t)]dt$$

(a) Define the state variables as $x_1(t) = h(t)$, $x_2(t) = \theta_m(t)$, and $x_3(t) = d\theta_m(t)/dt$. Write the state equations of the system in the form of $d\mathbf{x}(t)/dt = \mathbf{A}\mathbf{x}(t) + \mathbf{B}e_i(t)$. Draw a state diagram for the system.

(b) Find the characteristic equation and the eigenvalues of the $\mathbf{A}$ matrix found in part (a).

(c) Show that the open-loop system is completely controllable; that is, the pair $[\mathbf{A}, \mathbf{B}]$ is controllable.

(d) For reasons of economy, only one of the three state variables is measured and fed back for control purposes. The output equation is $y = \mathbf{C}\mathbf{x}$, where $\mathbf{C}$ can be one of the following forms:
(1) $\mathbf{C} = \begin{bmatrix} 1 & 0 & 0 \end{bmatrix}$ **(2)** $\mathbf{C} = \begin{bmatrix} 0 & 1 & 0 \end{bmatrix}$ **(3)** $\mathbf{C} = \begin{bmatrix} 0 & 0 & 1 \end{bmatrix}$
Determine which case (or cases) corresponds to a completely observable system.

5-41. The "broom-balancing" control system described in Problem 3-47 has the following parameters:

$$M_b = 1 \text{ kg} \qquad M_c = 10 \text{ kg} \qquad L = 1 \text{ m} \qquad g = 32.2 \text{ ft/sec}^2$$

Characteristic equation,
controllability,
observability

The small-signal linearized state equation model of the system is

$$\Delta\dot{\mathbf{x}}(t) = \mathbf{A}^*\Delta\mathbf{x}(t) + \mathbf{B}^*\Delta r(t)$$

where

$$\mathbf{A}^* = \begin{bmatrix} 0 & 1 & 0 & 0 \\ 25.92 & 0 & 0 & 0 \\ 0 & 0 & 0 & 1 \\ -2.36 & 0 & 0 & 0 \end{bmatrix} \qquad \mathbf{B}^* = \begin{bmatrix} 0 \\ -0.0732 \\ 0 \\ 0.0976 \end{bmatrix}$$

(a) Find the characteristic equation of $\mathbf{A}^*$ and its roots.

(b) Determine the controllability of $[\mathbf{A}^*, \mathbf{B}^*]$.

(c) For reason of economy, only one of the state variables is to be measured for feedback. The output equation is written

$$\Delta y(t) = \mathbf{C}^* \Delta \mathbf{x}(t)$$

where

(1) $\mathbf{C}^* = \begin{bmatrix} 1 & 0 & 0 & 0 \end{bmatrix}$ (2) $\mathbf{C}^* = \begin{bmatrix} 0 & 1 & 0 & 0 \end{bmatrix}$

(3) $\mathbf{C}^* = \begin{bmatrix} 0 & 0 & 1 & 0 \end{bmatrix}$ (4) $\mathbf{C}^* = \begin{bmatrix} 0 & 0 & 0 & 1 \end{bmatrix}$

Determine which $\mathbf{C}^*$ corresponds to an observable system.

• Controllability

5-42. The double-inverted pendulum shown in Fig. 5P-42 is approximately modeled by the following linear state equation:

$$\frac{d\mathbf{x}(t)}{dt} = \mathbf{A}\mathbf{x}(t) + \mathbf{B}u(t)$$

where

$$\mathbf{x}(t) = \begin{bmatrix} \theta_1(t) \\ \dot{\theta}_1(t) \\ \theta_2(t) \\ \dot{\theta}_2(t) \\ x(t) \\ \dot{x}(t) \end{bmatrix}$$

$$\mathbf{A} = \begin{bmatrix} 0 & 1 & 0 & 0 & 0 & 0 \\ 16 & 0 & -8 & 0 & 0 & 0 \\ 0 & 0 & 0 & 1 & 0 & 0 \\ -16 & 0 & 16 & 0 & 0 & 0 \\ 0 & 0 & 0 & 0 & 0 & 1 \\ 0 & 0 & 0 & 0 & 0 & 0 \end{bmatrix} \qquad \mathbf{B} = \begin{bmatrix} 0 \\ -1 \\ 0 \\ 0 \\ 0 \\ 1 \end{bmatrix}$$

Determine the controllability of the states.

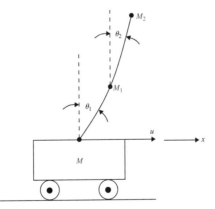

Figure 5P-42

State equations, state diagram, transfer function, characteristic equation

5-43. The block diagram of a simplified control system for the large space telescope (LST) is shown in Fig. 5P-43. For simulation and control purposes, model the system by state equations and by a state diagram.

(a) Draw a state diagram for the system and write the state equations in vector-matrix form. The state diagram should contain a minimum number of state variables, so it would be helpful if the transfer function of the system is written first.

(b) Find the characteristic equation of the system.

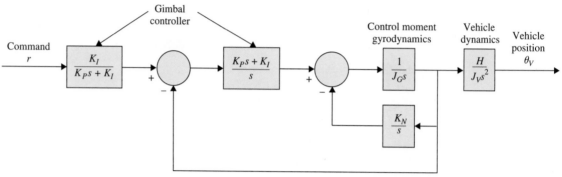

Figure 5P-43

• Controllability, observability, output feedback

5-44. The state diagram shown in Fig. 5P-44 represents two subsystems connected in cascade.

(a) Determine the controllability and observability of the system.

(b) Consider that output feedback is applied by feeding back y_2 to u_2; that is, $u_2 = -ky_2$, where k is a real constant. Determine how the value of k affects the controllability and observability of the system.

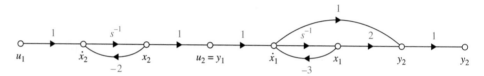

Figure 5P-44

• Controllability, state
feedback

5-45. Given the system

$$\frac{d\mathbf{x}(t)}{dt} = \mathbf{A}\mathbf{x}(t) + \mathbf{B}u(t) \qquad y(t) = \mathbf{C}\mathbf{x}(t)$$

where

$$\mathbf{A} = \begin{bmatrix} 0 & 1 \\ -1 & -3 \end{bmatrix} \qquad \mathbf{B} = \begin{bmatrix} 1 \\ 2 \end{bmatrix} \qquad \mathbf{C} = \begin{bmatrix} 1 & 1 \end{bmatrix}$$

(a) Determine the state controllability and observability of the system.

(b) Let $u(t) = -\mathbf{K}\mathbf{x}(t)$, where $\mathbf{K} = [k_1 \quad k_2]$, and k_1 and k_2 are real constants. Determine if and how controllability and observability of the closed-loop system are affected by the elements of $\mathbf{K}$.

▶ **ADDITIONAL**
COMPUTER
PROBLEMS

5-46. The torque equation for part (a) of Problem 5-13 is

$$J\frac{d^2\theta(t)}{dt^2} = K_Fd_1\theta(t) + T_sd_2\delta(t)$$

where $K_Fd_1 = 1$ and $J = 1$. Define the state variables as $x_1 = \theta$ and $x_2 = d\theta/dt$. Find the state-transition matrix $\phi(t)$ using any available computer program.

5-47. Starting with the state equation $d\mathbf{x}(t)/dt = \mathbf{A}\mathbf{x}(t) + \mathbf{B}\theta_r$ obtained in Problem 5-14, use **ACSYS/MATLAB**, or any other available computer program to do the following:

(a) Find the state-transition matrix of $\mathbf{A}$, $\phi(t)$.

(b) Find the characteristic equation of $\mathbf{A}$.

(c) Find the eigenvalues of $\mathbf{A}$.

(d) Compute and plot the unit-step response of $y(t) = \theta_y(t)$ for 3 seconds. Set all the initial conditions to zero.

CHAPTER 6

Stability of Linear Control Systems

▶ 6-1 INTRODUCTION

From the studies of linear differential equations with constant coefficients of SISO systems, we learned that the homogeneous solution that corresponds to the transient response of the system is governed by the roots of the characteristic equation. Basically, the design of linear control systems may be regarded as a problem of arranging the location of the poles and zeros of the system transfer function such that the system will perform according to the prescribed specifications.

Among the many forms of performance specifications used in design, the most important requirement is that the system must be stable. An unstable system is generally considered to be useless.

When all types of systems are considered—linear, nonlinear, time-invariant, and time-varying—the definition of stability can be given in many different forms. We shall deal only with the stability of linear SISO time-invariant systems in the following discussions.

For analysis and design purposes, we can classify stability as **absolute stability** and **relative stability**. Absolute stability refers to the condition whether the system is stable or unstable; it is a *yes* or *no* answer. Once the system is found to be stable, it is of interest to determine how stable it is, and this degree of stability is a measure of relative stability.

In preparation for the definition of stability, we define the two following types of responses for linear time-invariant systems:

1. **Zero-state response.** The zero-state response is due to the input only; all the initial conditions of the system are zero.

2. **Zero-input response.** The zero-input response is due to the initial conditions only; all the inputs are zero.

From the principle of superposition, when a system is subject to both inputs and initial conditions, the total response is written

$$\text{Total response} = \text{zero-state response} + \text{zero-input response}$$

The definitions just given apply to continuous-data as well as discrete-data systems.

▶ 6-2 BOUNDED-INPUT, BOUNDED-OUTPUT (BIBO) STABILITY—CONTINUOUS-DATA SYSTEMS

Let $u(t)$, $y(t)$, and $g(t)$ be the input, output, and the impulse response of a linear time-invariant system, respectively. *With zero initial conditions, the system is said to be BIBO (bounded-input bounded-output) stable, or simply stable, if its output $y(t)$ is bounded to a bounded input $u(t)$.*

The convolution integral relating $u(t)$, $y(t)$, and $g(t)$ is

$$y(t) = \int_0^\infty u(t - \tau)g(\tau)d\tau \tag{6-1}$$

Taking the absolute value of both sides of the equation, we get

$$|y(t)| = \left| \int_0^\infty u(t - \tau)g(\tau)d\tau \right| \tag{6-2}$$

or

$$|y(t)| \leq \int_0^\infty |u(t - \tau)||g(\tau)|d\tau \tag{6-3}$$

If $u(t)$ is bounded,

$$|u(t)| \leq M \tag{6-4}$$

where M is a finite positive number. Then,

$$|y(t)| \leq M \int_0^\infty |g(\tau)|d\tau \tag{6-5}$$

Thus, if $y(t)$ is to be bounded, or

$$|y(t)| \leq N < \infty \tag{6-6}$$

where N is a finite positive number, the following condition must hold:

$$M \int_0^\infty |g(\tau)|d\tau \leq N < \infty \tag{6-7}$$

Or, for any finite positive Q,

$$\int_0^\infty |g(\tau)|d\tau \leq Q < \infty \tag{6-8}$$

The condition given in Eq. (6-8) implies that the area under the $|g(\tau)|$–versus–τ-curve must be finite.

6-2-1 Relationship between Characteristic Equation Roots and Stability

To show the relation between the roots of the characteristic equation and the condition in Eq. (6-8), we write the transfer function $G(s)$, according to the Laplace transform definition, as

$$G(s) = \mathcal{L}[g(\tau)] = \int_0^\infty g(t)e^{-st}dt \tag{6-9}$$

Taking the absolute value on both sides of the last equation, we have

$$|G(s)| = \left| \int_0^\infty g(t)e^{-st}dt \right| \leq \int_0^\infty |g(t)||e^{-st}|dt \tag{6-10}$$

Since $|e^{-st}| = |e^{-\sigma t}|$, where σ is the real part of s. When s assumes a value of a pole of $G(s)$, $G(s) = \infty$, Eq. (6-10) becomes

$$\infty \leq \int_0^\infty |g(t)||e^{-\sigma t}|dt \tag{6-11}$$

- For BIBO stability, the roots of the characteristic equation must all lie in the left-half s-plane.

If one or more roots of the characteristic equation are in the right-half s-plane or on the $j\omega$-axis, $\sigma \geq 0$, then

$$|e^{-\sigma t}| \leq M = 1 \tag{6-12}$$

Equation (6-11) becomes

$$\infty \leq \int_0^\infty M|g(t)|dt = \int_0^\infty |g(t)|dt \tag{6-13}$$

- A system that is BIBO stable is simply called stable, otherwise it is unstable.

which violates the BIBO stability requirement. Thus, *for BIBO stability, the roots of the characteristic equation, or the poles of G(s), cannot be located in the right-half s-plane or on the jω-axis, in other words, they must all lie in the left-half s-plane. A system is said to be unstable if it is not BIBO stable.* When a system has roots on the $j\omega$-axis, say, at $s = j\omega_0$ and $s = -j\omega_0$, if the input is a sinusoid, $\sin \omega_0 t$, then the output will be of the form of $t \sin \omega_0 t$, which is unbounded, and the system is unstable.

▶ 6-3 ZERO-INPUT AND ASYMPTOTIC STABILITY OF CONTINUOUS-DATA SYSTEMS

In this section we shall define **zero-input stability** and **asymptotic stability**, and establish their relations with BIBO stability.

Zero-input stability refers to the stability condition when the input is zero, and the system is driven only by its initial conditions. We shall show that the zero-input stability also depends on the roots of the characteristic equation.

Let the input of an nth-order system be zero, and the output due to the initial conditions be $y(t)$. Then, $y(t)$ can be expressed as

$$y(t) = \sum_{k=0}^{n-1} g_k(t)y^{(k)}(t_0) \tag{6-14}$$

where

$$y^{(k)}(t_0) = \left. \frac{d^k y(t)}{dt^k} \right|_{t=t_0} \tag{6-15}$$

and $g_k(t)$ denotes the zero-input response due to $y^{(k)}(t_0)$. The zero-input stability is defined as follows: *If the zero-input response y(t), subject to the finite initial conditions, $y^{(k)}(t_0)$, reaches zero as t approaches infinity, the system is said to be zero-input stable, or stable; otherwise, the system is unstable.*

Mathematically, the foregoing definition can be stated: *A linear time-invariant system is zero-input stable if for any set of finite $y^{(k)}(t_0)$, there exists a positive number M, which depends on $y^{(k)}(t_0)$, such that*

$$\textbf{(1)} \qquad |y(t)| \leq M < \infty \qquad \textit{for all } t \geq t_0 \tag{6-16}$$

and

$$\textbf{(2)} \qquad \lim_{t\to\infty} |y(t)| = 0 \tag{6-17}$$

Because the condition in the last equation requires that the magnitude of $y(t)$ reaches zero as time approaches infinity, the zero-input stability is also known at the **asymptotic stability**.

Taking the absolute value on both sides of Eq. (6-14), we get

$$|y(t)| = \left| \sum_{k=0}^{n-1} g_k(t) y^{(k)}(t_0) \right| \le \sum_{k=0}^{n-1} |g_k(t)| |y^{(k)}(t_0)| \tag{6-18}$$

Since all the initial conditions are assumed to be finite, the condition in Eq. (6-16) requires that the following condition be true:

$$\sum_{k=0}^{n-1} |g_k(t)| < \infty \qquad \text{for all } t \ge 0 \tag{6-19}$$

Let the n characteristic equation roots be expressed as $s_i = \sigma_i + j\omega_i$, $i = 1, 2, ..., n$. Then, if m of the n roots are simple, and the rest are of multiple order, $y(t)$ will be of the form:

$$y(t) = \sum_{i=1}^{m} K_i e^{s_i t} + \sum_{i=0}^{n-m-1} L_i t^i e^{s_i t} \tag{6-20}$$

where K_i and L_i are constant coefficients. Since the exponential terms $e^{s_i t}$ in the last equation control the response $y(t)$ as $t \rightarrow \infty$, to satisfy the two conditions in Eqs. (6-16) and (6-17), the real parts of s_i must be negative. In other words, the roots of the characteristic equation must all be in the left-half s-plane.

From the preceding discussions, we see that *for linear time-invariant systems, BIBO, zero-input, and asymptotic stability all have the same requirement that the roots of the characteristic equation must all be located in the left-half s-plane. Thus, if a system is BIBO stable, it must also be zero-input or asymptotically stable.* For this reason, we shall simply refer to the stability condition of a linear system as **stable** or **unstable**. The latter condition refers to the condition that at least one of the characteristic equation roots is not in the left-half s-plane. For practical reasons, we often refer to the situation in which the characteristic equation has simple roots on the $j\omega$-axis and none in the right-half plane as **marginally stable** or **marginally unstable**. *An exception to this is if the system were intended to be an integrator (or in the case of control systems, a velocity control system), then the system would have root(s) at s = 0, and would be considered stable.* Similarly, if the system were designed to be an oscillator, the characteristic equation would have simple roots on the $j\omega$-axis, and the system would be regarded as stable.

• Systems with characteristic equation roots intentionally placed on the $j\omega$-axis are defined to be stable.

Since the roots of the characteristic equation are the same as the eigenvalues of **A** of the state equations, the stability condition places the same restrictions on the eigenvalues.

Let the characteristic equation roots or eigenvalues of **A** of a linear continuous-data time-invariant SISO system be $s_i = \sigma_i + j\omega_i$, $i = 1, 2, ..., n$. If any of the roots is complex, it is in complex conjugate pairs. The possible stability conditions of the system are summarized as follows with respect to the roots of the characteristic equation.

The following example illustrates the stability conditions of systems with reference to the poles of the system transfer functions that are also the roots of the characteristic equation.

TABLE 6-1   Stability Conditions of Linear Continuous-Data Time-Invariant SISO Systems

Stability Condition	Root Values
Asymptotically stable or simply stable	$\sigma_i < 0$ for all i, $i = 1, 2, \ldots, n$. (All the roots are in the left-half s-plane.)
Marginally stable or marginally unstable	$\sigma_i = 0$ for any i for simple roots, and no $\sigma_i > 0$
	For $i = 1, 2, \ldots, n$ (at least one simple root, no multiple-order roots on the $j\omega$-axis, and n roots in the right-half s-plane; note exceptions)
Unstable	$\sigma_i > 0$ for any i, or $\sigma_i = 0$ for any multiple-order root. $i = 1, 2, \ldots, n$ (at least one simple root in the right-half s-plane or at least one multiple-order root on the $j\omega$-axis)

▶ **EXAMPLE 6-1**   The following closed-loop transfer functions and their associated stability conditions are given.

$$M(s) = \frac{20}{(s + 1)(s + 2)(s + 3)}$$ BIBO or asymptotically stable (or, simply, stable.)

$$M(s) = \frac{20(s + 1)}{(s - 1)(s^2 + 2s + 2)}$$ Unstable due to the pole at $s = 1$

$$M(s) = \frac{20(s - 1)}{(s + 2)(s^2 + 4)}$$ Marginally stable or marginally unstable due to $s = \pm j2$

$$M(s) = \frac{10}{(s^2 + 4)^2(s + 10)}$$ Unstable due to the multiple-order pole at $s = \pm j2$

$$M(s) = \frac{10}{s^4 + 30s^3 + s^2 + 10s}$$ Stable if the pole at $s = 0$ is placed intentionally

◂

▶ 6-4 METHODS OF DETERMINING STABILITY

The discussions in the preceding sections lead to the conclusion that the stability of linear time-invariant SISO systems can be determined by checking on the location of the roots of the characteristic equation of the system. For all practical purposes, there is no need to compute the complete system response to determine stability. The regions of stability and instability in the s-plane are illustrated in Fig. 6-1. When the system parameters are all known, the roots of the characteristic equation can be found using various components within **ACSYS** such as the Transfer Function Tool (**tftool**), which was discussed in Chapter 2. A Routh-Hurwitz stability routine has also been developed for **ACSYS**, which is discussed at the end of this chapter. Most other tools developed for **ACSYS** may also be used to find the roots of the characteristic equation (the poles) of a transfer function. These include the Transfer Function Symbolic Tool (**tfsym**), the Transfer Function Calculator (**tfcal**), and the Controller Design Tool (**controls**). You may also conduct a more thorough stability study using **ACSYS** components such as the Time Response Analysis Tool (**timetool**) and the Frequency Response Tool (**freqtool**). These topics will be discussed in Chapters 7, 8, 9, and Appendix K.

For design purposes, there will be unknown or variable parameters imbedded in the characteristic equation, so it will not be feasible to use the root-finding programs. The methods outlined in the following list are well known for determining the stability of linear continuous-data systems without involving root solving.

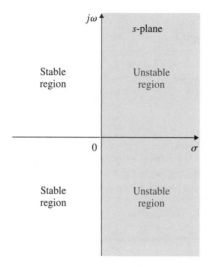

Figure 6-1   Stable and unstable regions in the s-plane.

1. **Route-Hurwitz criterion.**   This criterion is an algebraic method that provides information on the absolute stability of a linear time-invariant system that has a characteristic equation with constant coefficients. The criterion tests whether any of the roots of the characteristic equation lie in the right-half s-plane. The number of roots that lie on the $j\omega$-axis and in the right-half s-plane is also indicated.

2. **Nyquist criterion.**   This criterion is a semigraphical method that gives information on the difference between the number of poles and zeros of the closed-loop transfer function that are in the right-half s-plane by observing the behavior of the Nyquist plot of the loop transfer function.

3. **Bode diagram.**   This diagram is a plot of the magnitude of the loop transfer function $G(j\omega)H(j\omega)$ in dB and the phase of $G(j\omega)H(j\omega)$ in degrees, all versus frequency ω. The stability of the closed-loop system can be determined by observing the behavior of these plots.

Thus, as will be evident throughout the text, most of the analysis and design techniques on control systems represent alternate methods of solving the same problem. The designer simply has to choose the best analytical tool, depending on the particular situation.

Details of the Route-Hurwitz stability criterion are presented in the following section.

▶ 6-5 ROUTH-HURWITZ CRITERION

The Routh-Hurwitz criterion represents a method of determining the location of zeros of a polynomial with constant real coefficients with respect to the left half and right half of the s-plane, without actually solving for the zeros. *Since root-finding computer programs can solve for the zeros of a polynomial with ease, the value of the Routh-Hurwitz criterion is at best limited to equations with at least one unknown parameter.*

Consider that the characteristic equation of a linear time-variant SISO system is of the form

$$F(s) = a_n s^n + a_{n-1} s^{n-1} + \cdots + a_1 s + a_0 = 0 \qquad (6\text{-}21)$$

where all the coefficients are real. In order that the last equation does not have roots with positive real parts, it is *necessary (but not sufficient)* that the following conditions hold:

1. All the coefficients of the equation have the same sign.
2. None of the coefficients vanishes.

These conditions are based on the laws of algebra, which relate the coefficients of Eq. (6-21) as follows:

$$\frac{a_{n-1}}{a_n} = -\sum \text{all roots} \tag{6-22}$$

$$\frac{a_{n-2}}{a_n} = \sum \text{products of the roots taken two at a time} \tag{6-23}$$

$$\frac{a_{n-3}}{a_n} = -\sum \text{products of the roots taken three at a time} \tag{6-24}$$

$$\vdots$$

$$\frac{a_0}{a_n} = (-1)^n \text{ products of all the roots} \tag{6-25}$$

Thus, all these ratios must be positive and nonzero unless at least one of the roots has a positive real part.

The two necessary conditions for Eq. (6-21) to have no roots in the right-half *s*-plane can easily be checked by inspection of the equation. However, these conditions are not sufficient, for it is quite possible that an equation with all its coefficients nonzero and of the same sign still may not have all the roots in the left half of the *s*-plane.

6-5-1   Routh's Tabulation [1]

The Hurwitz criterion [1] gives the *necessary and sufficient condition for all roots of Eq. (6-21) to lie in the left half of the s-plane. The criterion requires that the equation's n **Hurwitz determinants** must all be positive.*

However, the evaluation of the *n* Hurwitz determinants is tedious to carry out. But Routh simplified the process by introducing a tabulation method in place of the Hurwitz determinants.

The first step in the simplification of the Hurwitz criterion, now called the Routh-Hurwitz criterion, is to arrange the coefficients of the equation in Eq. (6-21) into two rows. The first row consists of the first, third, fifth, ..., coefficients, and the second row consists of the second, fourth, sixth, ..., coefficients, all counting from the highest-order term, as shown in the following tabulation:

$$
\begin{array}{ccccc}
a_n & a_{n-2} & a_{n-4} & a_{n-6} & \cdots \\
a_{n-1} & a_{n-3} & a_{n-5} & a_{n-7} & \cdots
\end{array}
$$

The next step is to form the following array of numbers by the indicated operations, illustrated here for a sixth-order equation:

$$a_6 s^6 + a_5 s^5 + \cdots + a_1 s + a_0 = 0 \tag{6-26}$$

s^6	a_6	a_4	a_2 a_0
s^5	a_5	a_3	a_1 0
s^4	$\dfrac{a_5 a_4 - a_6 a_3}{a_5} = A$	$\dfrac{a_5 a_2 - a_6 a_1}{a_5} = B$	$\dfrac{a_5 a_0 - a_6 \times 0}{a_5} = a_0$ 0
s^3	$\dfrac{A a_3 - a_5 B}{A} = C$	$\dfrac{A a_1 - a_5 a_0}{A} = D$	$\dfrac{A \times 0 - a_5 \times 0}{A} = 0$ 0
s^2	$\dfrac{BC - AD}{C} = E$	$\dfrac{C a_0 - A \times 0}{C} = a_0$	$\dfrac{C \times 0 - A \times 0}{C} = 0$ 0
s^1	$\dfrac{ED - C a_0}{E} = F$	0	0 0
s^0	$\dfrac{F a_0 - E \times 0}{F} = a_0$	0	0 0

This array is called the **Routh's tabulation** or **Routh's array**. The column of s's on the left side is used for identification purposes. The reference column keeps track of the calculations, and the last row of the Routh's tabulation should always be the s^0 row.

Once the Routh's tabulation has been completed, the last step in the application of the criterion is to investigate the *signs* of the coefficients in the *first column* of the tabulation, which contains information on the roots of the equation. The following conclusions are made:

The roots of the equation are all in the left half of the s-plane if all the elements of the first column of the Routh's tabulation are of the same sign. The number of changes of signs in the elements of the first column equals the number of roots with positive real parts or in the right-half s-plane.

The following examples illustrate the applications of the Routh-Hurwitz criterion when the tabulation terminates without complications.

▶ **EXAMPLE 6-2** Consider the equation

$$2s^4 + s^3 + 3s^2 + 5s + 10 = 0 \tag{6-27}$$

Since the equation has no missing terms and the coefficients are all of the same sign, it satisfies the necessary condition for not having roots in the right-half or on the imaginary axis of the s-plane. However, the sufficient condition must still be checked. Routh's tabulation is made as follows:

s^4	2	3	10
s^3	1	5	0
Sign change			
s^2	$\dfrac{(1)(3) - (2)(5)}{1} = -7$	10	0
Sign change			
s^1	$\dfrac{(-7)(5) - (1)(10)}{-7} = 6.43$	0	0
s^0	10	0	0

Since there are two changes in sign in the first column of the tabulation, the equation has two roots in the right half of the s-plane. Solving for the roots of Eq. (6-27), we have the four roots at $s = -1.005 \pm j0.933$ and $s = 0.755 \pm j1.444$. Clearly, the last two roots are in the right-half s-plane, which cause the system to be unstable. ◀

6-5-2 Special Cases When Routh's Tabulation Terminates Prematurely

The equations considered in the two preceding examples are designed so that Routh's tabulation can be carried out without any complications. Depending on the coefficients of the equation, the following difficulties may occur that prevent Routh's tabulation from completing properly:

1. The first element in any one row of Routh's tabulation is zero, but the others are not.

2. The elements in one row of Routh's tabulation are all zero.

In the first case, if a zero appears in the first element of a row, the elements in the next row will all become infinite, and Routh's tabulation cannot continue. To remedy the situation, we *replace the zero element in the first column by an arbitrary small positive number ε, and then proceed with Routh's tabulation.* This is illustrated by the following example.

► **EXAMPLE 6-3**   Consider the characteristic equation of a linear system:

$$s^4 + s^3 + 2s^2 + 2s + 3 = 0 \tag{6-28}$$

Since all the coefficients are nonzero and of the same sign, we need to apply Routh-Hurwitz criterion. Routh's tabulation is carried as follows:

s^4	1	2	3
s^3	1	2	0
s^2	0	3	

Since the first element of the s^2 row is zero, the element in the s^1 row would all be infinite. To overcome this difficulty, we replace the zero in the s^2 row by a small positive number ε, and then proceed with the tabulation. Starting with the s^2 row, the results are as follows:

	s^2	ε	3
Sign change	s^1	$\dfrac{2\varepsilon - 3}{\varepsilon} \cong -\dfrac{3}{\varepsilon}$	0
Sign change	s^0	3	0

Since there are two sign changes in the first column of Routh's tabulation, the equation in Eq. (6-28) has two roots in the right-half s-plane. Solving for the roots of Eq. (6-28), we get $s = -0.091 \pm j0.902$ and $s = 0.406 \pm j1.293$; the last two roots are clearly in the right-half s-plane.   ◄

It should be noted that the ε-method described may not give correct results if the equation has pure imaginary roots [2, 3].

In the second special case, when all the elements in one row of Routh's tabulation are zeros before the tabulation is properly terminated, it indicates that one or more of the following conditions may exist:

• The coefficients of the auxiliary equation are those of the row just above the row of zeros in Routh's tabulation.

1. The equation has at least one pair of real roots with equal magnitude but opposite signs.

2. The equation has one or more pairs of imaginary roots.

3. The equation has pairs of complex-conjugate roots forming symmetry about the origin of the s-plane; for example, $s = -1 \pm j1$, $s = 1 \pm j1$.

• The roots of the auxiliary equation must also satisfy the original equation.

The situation with the entire row of zeros can be remedied by using the **auxiliary equation** $A(s) = 0$, which is formed from the coefficients of the row just above the row of zeros in Routh's tabulation. The auxiliary equation is always an even polynomial, that is, only even powers of s appear. *The roots of the auxiliary equation also satisfy the original equation.* Thus, by solving the auxiliary equation, we also get some of the roots of the original equation. To continue with Routh's tabulation when a row of zero appears, we conduct the following steps:

1. Form the auxiliary equation $A(s) = 0$ by using the coefficients from the row just preceding the row of zeros.

2. Take the derivative of the auxiliary equation with respect to s; this gives $dA(s)/ds = 0$.

3. Replace the row of zeros with the coefficients of $dA(s)/ds = 0$.

4. Continue with Routh's tabulation in the usual manner with the newly formed row of coefficients replacing the row of zeros.

5. Interpret the change of signs, if any, of the coefficients in the first column of the Routh's tabulation in the usual manner.

▶ **EXAMPLE 6-4** Consider the following equation, which may be the characteristic equation of a linear control system:

$$s^5 + 4s^4 + 8s^3 + 8s^2 + 7s + 4 = 0 \tag{6-29}$$

Routh's tabulation is

s^5	1	8	7
s^4	4	8	4
s^3	6	6	0
s^2	4	4	
s^1	0	0	

Since a row of zeros appears prematurely, we form the auxiliary equation using the coefficients of the s^2 row:

$$A(s) = 4s^2 + 4 = 0 \tag{6-30}$$

The derivative of $A(s)$ with respect to s is

$$\frac{dA(s)}{ds} = 8s = 0 \tag{6-31}$$

from which the coefficients 8 and 0 replace the zeros in the s^1 row of the original tabulation. The remaining portion of the Routh's tabulation is

s^1	8	0	coefficients of $dA(s)/ds$
s^0	4		

Since there are no sign changes in the first column of the entire Routh's tabulation, the equation in Eq. (6-31) does not have any root in the right-half s-plane. Solving the auxiliary equation in Eq. (6-30), we get the two roots at $s = j$ and $s = -j$, which are also two of the roots of Eq. (6-29). Thus, the equation has two roots on the $j\omega$-axis, and the system is marginally stable. These imaginary roots caused the initial Routh's tabulation to have the entire row of zeros in the s^1 row. ◀

Since all zeros occurring in a row that corresponds to an odd power of s creates an auxiliary equation that has only even powers of s, the roots of the auxiliary equation may all lie on the $j\omega$-axis. For design purposes, we can use the all-zero-row condition to solve for the marginal value of a system parameter for system stability. The following example illustrates the realistic value of the Routh-Hurwitz criterion in a simple design problem.

► **EXAMPLE 6-5** Consider that a third-order control system has the characteristic equation

$$s^3 + 3408.3s^2 + 1,204,000s + 1.5 \times 10^7 K = 0 \tag{6-32}$$

The Routh-Hurwitz criterion is best suited to determine the critical value of K for stability, that is, the value of K for which at least one root will lie on the $j\omega$-axis and none in the right-half s-plane. Routh's tabulation of Eq. (6-32) is made as follows:

s^3	1	1,204,000
s^2	3408.3	$1.5 \times 10^7 K$
s^1	$\dfrac{410.36 \times 10^7 - 1.5 \times 10^7 K}{3408.3}$	0
s^0	$1.5 \times 10^7 K$	

For the system to be stable, all the roots of Eq. (6-32) must be in the left-half s-plane, and, thus, all the coefficients in the first column of Routh's tabulation must have the same sign. This leads to the following conditions:

$$\frac{410.36 \times 10^7 - 1.5 \times 10^7 K}{3408.3} > 0 \tag{6-33}$$

and

$$1.5 \times 10^7 K > 0 \tag{6-34}$$

From the inequality of Eq. (6-33), we have $K < 273.57$, and the condition in Eq. (6-34) gives $K > 0$. Therefore, the condition of K for the system to be stable is

$$0 < K < 273.57 \tag{6-35}$$

If we let $K = 273.57$, the characteristic equation in Eq. (6-32) will have two roots on the $j\omega$-axis. To find these roots, we substitute $K = 273.57$ in the auxiliary equation, which is obtained from Routh's tabulation by using the coefficients of the s^2 row. Thus,

$$A(s) = 3408.3s^2 + 4.1036 \times 10^9 = 0 \tag{6-36}$$

which has roots at $s = j1097$ and $s = j1097$, and the corresponding value of K at these roots is 273.57. Also, if the system is operated with $K = 273.57$, the zero-input response of the system will be an undamped sinusoid with a frequency of 1097.27 rad/sec. ◄

► **EXAMPLE 6-6** As another example of using the Routh-Hurwitz criterion for simple design problems, consider that the characteristic equation of a closed-loop control system is

$$s^3 + 3Ks^2 + (K + 2)s + 4 = 0 \tag{6-37}$$

It is desired to find the range of K so that the system is stable. Routh's tabulation of Eq. (6-37) is

s^3	1	$K + 2$
s^2	$3K$	4
s^1	$\dfrac{3K(K + 2) - 4}{3K}$	0
s^0	4	

From the s^2 row, the condition of stability is $K > 0$, and from the s^1 row, the condition of stability is

$$3K^2 + 6K - 4 > 0 \tag{6-38}$$

or

$$K < -2.528 \quad \text{or} \quad K > 0.528$$

When the conditions of $K > 0$ and $K > 0.528$ are compared, it is apparent that the latter requirement is more stringent. Thus, for the closed-loop system to be stable, K must satisfy

$$K > 0.528$$

The requirement of $K < -2.528$ is disregarded since K cannot be negative. ◀

It should be reiterated that the Routh-Hurwitz criterion valid only if the characteristic equation is algebraic with real coefficients. If any one of the coefficients is complex, or if the equation is not algebraic, for example, containing exponential functions or sinusoidal functions of s, the Routh-Hurwitz criterion simply cannot be applied.

Another limitation of the Routh-Hurwitz criterion is that is valid only for the determination of roots of the characteristic equation with respect to the left-half or the right-half on the s-plane. The stability boundary is the $j\omega$-axis of the s-plane. The criterion *cannot* be applied to any other stability boundaries in a complex plane, such as the unit circle in the z-plane, which is the stability boundary of discrete-data systems (Appendix I).

▶ 6-6 MATLAB TOOLS AND CASE STUDIES

The Transfer Function Analysis Tool (tftool), which was introduced in Chapter 2, has been modified to **stabtool** for conducting Routh-Hurwitz stability tests. The **stabtool** can be invoked either from the MATLAB command line by simply typing stabtool or from the Automatic Control Systems launch applet (**ACSYS**) by clicking on the "Routh-Hurwitz" button.

Many of the other tools within ACSYS may also be used to find the poles of the closed-loop system transfer function, including the Transfer Function Symbolic Tool (tfsym), and the Transfer Function Calculator (tfcal). You may also conduct a more thorough stability study of your system using the root locus and phase-and-gain margin concepts utilizing the Time Response Tool (timetool) and the Frequency Response Tool (freqtool), respectively. These topics will be thoroughly discussed in Chapters 7, 8, and 9.

The steps involved in setting up and then solving a given stability problem using stabtool are as follows.

- Specify the transfer-function configuration; that is, polynomial or pole, zero, gain format.
- Enter the transfer-function values.
- Convert the transfer-function from polynomial form to poles and zeros or vice versa.
- Find the transfer-function poles and zeros.
- Form the Routh table and conduct the Routh-Hurwitz stability test.

To better illustrate how to use stabtool, let us solve some of the earlier examples in this chapter.

▶ **EXAMPLE 6-2 Revisited** Consider equation

$$2s^4 + s^3 + 3s^2 + 5s + 10 = 0 \qquad (6\text{-}39)$$

Start stabtool and use the "Input Module" to enter the characteristic equation (6-39) in "Polynomial" form as shown in Fig. 6-2. You must first click on the button designated by $G(s)$, and then enter the characteristic polynomial in the "Den $G(s)$" row. Once completed press "APPLY" to exit.

Next press "Calculate" in the stabtool main window to find the poles of the system. Refer to Chapter 2 for a more elaborated description of this step.

In order to obtain the Routh table and the Routh-Hurwitz stability test, press the appropriate button. This button can only be activated after you have pressed the "Calculate" button. The system stability numbers appear on screen, as shown in Fig. 6-3. The results match Example 6-2. The system is therefore unstable because of two positive poles. The first column of the Routh array also shows two sign changes, which confirm this result. In order to see the complete Routh table, refer to the MATLAB command window.

s^4:	2	3	10
s^3:	1	5	0
s^2:	−7	10	0
s^1:	6.4286	0	
s^0:	10		First column sign changes are: 2

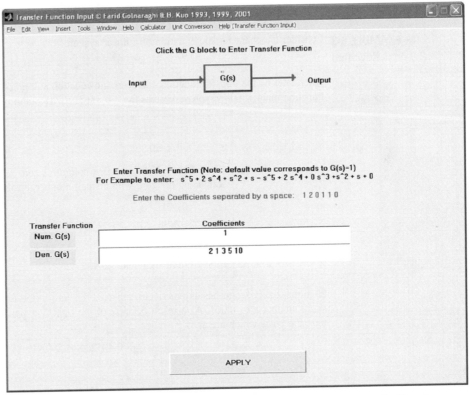

Figure 6-2 Entering characteristic polynomial for Example 6-2 using the Transfer Function Input module.

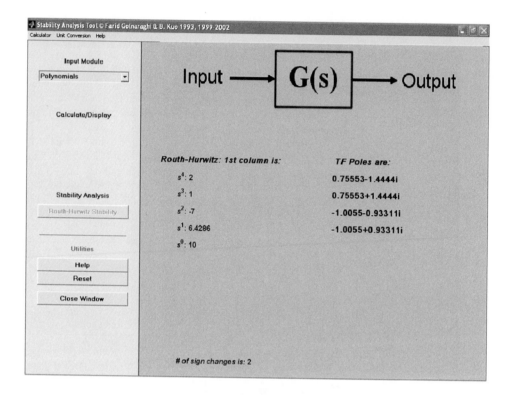

Figure 6-3 Stability results for Example 6-2, after using the Routh-Hurwitz test.

▶ **EXAMPLE 6-3 Revisited**

Consider the characteristic equation of a linear system:

$$s^4 + s^3 + 2s^2 + 2s + 3 = 0 \qquad (6\text{-}40)$$

After entering the transfer function characteristic equation and pressing "Calculate" in the Transfer Function main window, you may press the "Routh-Hurwitz Stability" button to get

s^4	1	2	3
s^3	1	2	0

Special case: Next row, first column is zero
Replacing the zero with a small number, epsilon = 0.001

s^2:	0.001	3	0
s^1:	−2998	0	
s^0:	3		

First column sign changes are: 2

Note that the program has automatically replaced the zero in the first column of the s^2 row with a small number $\varepsilon = 0.001$. As a result because of the final two sign changes we expect to see two unstable poles which are at

$$s_{12} = 0.4057 \pm j1.2928 \qquad (6\text{-}41)$$

◀

▶ **EXAMPLE 6-4 Revisited**

Use the stabtool to enter the following characteristic equation:

$$s^5 + 4s^4 + 8s^3 + 8s^2 + 7s + 4 = 0 \qquad (6\text{-}42)$$

After entering the transfer-function characteristic equation, and pressing "Calculate" in the Transfer Function main window, you may press the "Routh-Hurwitz Stability" button to get

$$
\begin{array}{c|cccc}
s^5 & 1 & 8 & 7 & 0 \\
s^4 & 4 & 8 & 4 & 0 \\
s^3 & 6 & 6 & 0 \\
s^2 & 4 & 4
\end{array}
$$

Special case: all zeros in the next row
Forming the auxiliary polynomial using the derivative of the above row

$$
\begin{array}{c|cc}
s^1: & 8 & 0 \\
s^0: & 4
\end{array}
$$

First column sign changes are: 0

In this case, the program has automatically replaced the whole row of zeros in s^1 with the coefficients of the polynomial formed from the derivative of an auxiliary polynomial formed from row s^2. As a result, the system is unstable. Further, because of the final zero sign changes we expect to see no additional unstable poles. The unstable poles of the system may be obtained directly by obtaining the roots of the auxiliary polynomial:

$$A(s) = 4s^2 + 4 = 0 \quad s_{12} = \pm j \tag{6-43}$$

◄

► **EXAMPLE 6-6**
Revisited

Considering the characteristic equation of a closed-loop control system

$$s^3 + 3Ks^2 + (K + 2)s + 4 = 0 \tag{6-44}$$

It is desired to find the range of K so that the system is stable.
Recall from the solution of Example 6-6 that for the closed-loop system to be stable, K must satisfy

$$K > 0.528$$

We substitute this value of K as shown in Fig. 6-4. After pressing the "Routh-Hurwitz Stability" button, the screen shown in Fig. 6-5 comes up. Recall that this button can only be activated after

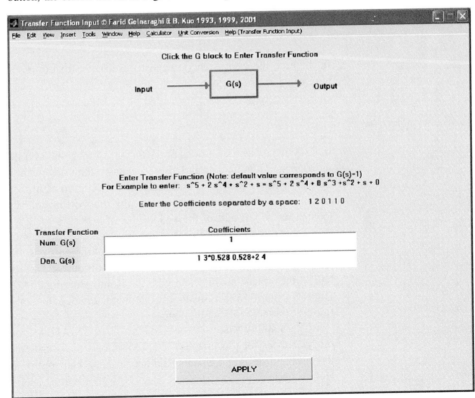

Figure 6-4 The Transfer Function Input window for entering $K = 0.528$.

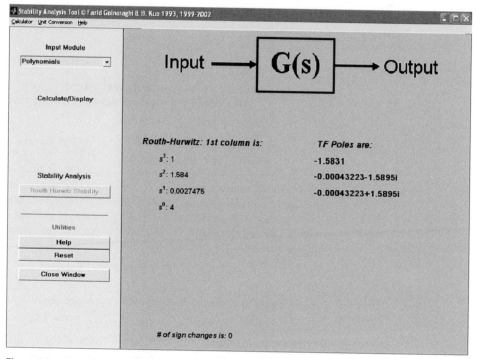

Figure 6-5 The Transfer Function main window, stability results for $K = 0.528$.

you have pressed the "Calculate" button. Clearly, the system is very close to becoming marginally unstable at $K = 0.528$. ◄

In the end, the user is encouraged to use the software to solve examples and problems appearing in this chapter.

►6-7 SUMMARY

In this chapter, the definitions of BIBO, zero-input, and asymptotic stability of linear time-invariant continuous-data and discrete-data systems are given. It is shown that the condition of these types of stability is related directly to the roots of the characteristic equation. For a continuous-data system to be stable, the roots of the characteristic equation must all be located in the left half of the s-plane.

The necessary condition for a polynomial $F(s)$ to have no zeros on the $j\omega$-axis and in the right-half of the s-plane is that all its coefficients must be of the same sign and none can vanish. The necessary and sufficient conditions of $F(s)$ to have zeros only in the left-half of the s-plane are checked with the Routh-Hurwitz criterion. The value of the Routh-Hurwitz criterion is diminished if the characteristic equation can be solved using a root-finding routine on the computer.

The Stability Analysis Tool (stabtool) is the MATLAB tool developed to conduct stability analysis for this chapter. This tool allows the user to enter a transfer function, find its poles and zeros and also its Routh array, and then confirm its stability using the Routh-Hurwitz criterion.

▶ REVIEW
QUESTIONS

1. Can the Routh-Hurwitz criterion be directly applied to the stability analysis of the following systems?

(a) Continuous-data system with the characteristic equation:

$$s^4 + 5s^3 + 2s^2 + 3s + 2e^{-2s} = 0$$

(b) Continuous-data system with the characteristic equation:

$$s^4 - 5s^3 + 3s^2 + Ks + K^2 = 0$$

2. The first two rows of Routh's tabulation of a third-order system are

$$\begin{array}{ccc} s^3 & 2 & 2 \\ s^2 & 4 & 4 \end{array}$$

Select the correct answer from the following choices:

(a) The equation has one root in the right-half s-plane.

(b) The equation has two roots on the $j\omega$-axis at $s = j$ and $-j$. The third root is in the left-half s-plane.

(c) The equation has two roots on the $j\omega$-axis at $s = 2j$ and $s = -2j$. The third root is in the left-half s-plane.

(d) The equation has two roots on the $j\omega$-axis at $s = 2j$ and $s = -2j$. The third root is in the right-half s-plane.

3. If the numbers in the first column of Routh's tabulation turn out to be all negative, then the equation for which the tabulation is made has at least one root not in the left-half of the s-plane.   **(T)**   **(F)**

4. The roots of the auxiliary equation, $A(s) = 0$, of Routh's tabulation of a characteristic equation must also be the roots of the latter.   **(T)**   **(F)**

5. The following characteristic equation of a continuous-data system represents an unstable system since it contains a negative coefficient.

$$s^3 - s^2 + 5s + 10 = 0$$ **(T)**   **(F)**

6. The following characteristic equation of a continuous-data system represents an unstable system since there is a zero coefficient.

$$s^3 + 5s^2 + 4 = 0$$ **(T)**   **(F)**

7. When a row of Routh's tabulation contains all zeros before the tabulation ends, this means that the equation has roots on the imaginary axis of the s-plane.   **(T)**   **(F)**

Answers to the true-false questions are found after the Problems section.

▶ REFERENCES

1. F. R. Gantmacher, *Matrix Theory*, vol. II, Chelsea Publishing Company, New York, 1964.

2. K. J. Khatwani, "On Routh-Hurwitz Criterion," *IEEE Trans. Automatic Control*, vol. AC-26, p. 583, April 1981.

3. S. K. Pillai, "The ε Method of the Routh-Hurwitz Criterion," *IEEE Trans. Automatic Control*, vol. AC-26, 584, April 1981.

▶ PROBLEMS

• System stability

6-1. Without using the Routh-Hurwitz criterion, determine if the following systems are asymptotically stable, marginally stable, or unstable. In each case, the closed-loop system transfer function is given.

(a) $M(s) = \dfrac{10(s + 2)}{s^3 + 3s^2 + 5s}$ (b) $M(s) = \dfrac{s - 1}{(s + 5)(s^2 + 2)}$

(c) $M(s) = \dfrac{K}{s^3 + 5s + 5}$ (d) $M(s) = \dfrac{100(s - 1)}{(s + 5)(s^2 + 2s + 2)}$

(e) $M(s) = \dfrac{100}{s^3 - 2s^2 + 3s + 10}$ (f) $M(s) = \dfrac{10(s + 12.5)}{s^4 + 3s^3 + 50s^2 + s + 10^6}$

• Routh-Hurwitz criterion

6-2. Using the Routh-Hurwitz criterion, determine the stability of the closed-loop system that has the following characteristic equations. Determine the number of roots of each equation that are in the right-half s-plane and on the $j\omega$-axis.

(a) $s^3 + 25s^2 + 10s + 450 = 0$ (b) $s^3 + 25s^2 + 10s + 50 = 0$

(c) $s^3 + 25s^2 + 250s + 10 = 0$ (d) $2s^4 + 10s^3 + 5.5s^2 + 5.5s + 10 = 0$

(e) $s^6 + 2s^5 + 8s^4 + 15s^3 + 20s^2 + 16s + 16 = 0$ (f) $s^4 + 2s^3 + 10s^2 + 20s + 5 = 0$

• Routh-Hurwitz criterion on margin of stability

6-3. For each of the characteristic equations of feedback control systems given, determine the range of K so that the system is asymptotically stable. Determine the value of K so that the system is marginally stable and the frequency of sustained oscillation if applicable.

(a) $s^4 + 25s^3 + 15s^2 + 20s + K = 0$ (b) $s^4 + Ks^3 + 2s^2 + (K + 1)s + 10 = 0$

(c) $s^3 + (K + 2)s^2 + 2Ks + 10 = 0$ (d) $s^3 + 20s^2 + 5s + 10K = 0$

(e) $s^4 + Ks^3 + 5s^2 + 10s + 10K = 0$ (f) $s^4 + 12.5s^3 + s^2 + 5s + K = 0$

• Region of stability in the parameter plane

6-4. The loop transfer function of a single-loop feedback control system is given as

$$G(s)H(s) = \dfrac{K(s + 5)}{s(s + 2)(1 + Ts)}$$

The parameters K and T may be represented in a plane with K as the horizontal axis and T as the vertical axis. Determine the regions in the T-versus-K parameter plane where the closed-loop system is asymptotically stable and where it is unstable. Indicate the boundary on which the system is marginally stable.

• Margin of stability, sustained oscillation

6-5. Given the forward-path transfer function of unity-feedback control systems,

(a) $G(s) = \dfrac{K(s + 4)(s + 20)}{s^3(s + 100)(s + 500)}$ (b) $G(s) = \dfrac{K(s + 10)(s + 20)}{s^2(s + 2)}$

(c) $G(s) = \dfrac{K}{s(s + 10)(s + 20)}$ (d) $G(s) = \dfrac{K(s + 1)}{s^3 + 2s^2 + 3s + 1}$

Apply the Routh-Hurwitz criterion to determine the stability of the closed-loop system as a function of K. Determine the value of K that will cause sustained constant-amplitude oscillations in the system. Determine the frequency of oscillation.

• Stability of system in state equation form, state feedback control

6-6. A controlled process is modeled by the following state equations.

$$\dfrac{dx_1(t)}{dt} = x_1(t) - 2x_2(t) \qquad \dfrac{dx_2(t)}{dt} = 10x_1(t) + u(t)$$

The control $u(t)$ is obtained from state feedback such that

$$u(t) = -k_1 x_1(t) - k_2 x_2(t)$$

where k_1 and k_2 are real constants. Determine the region in the k_1-versus-k_2 parameter plane in which the closed-loop system is asymptotically stable.

• Stability of system in state equation form, State feedback control

6-7. A linear time-invariant system is described by the following state equations.

$$\dfrac{d\mathbf{x}(t)}{dt} = \mathbf{A}\mathbf{x}(t) + \mathbf{B}u(t)$$

where

$$\mathbf{A} = \begin{bmatrix} 0 & 1 & 0 \\ 0 & 0 & 1 \\ 0 & -4 & -3 \end{bmatrix} \qquad \mathbf{B} = \begin{bmatrix} 0 \\ 0 \\ 1 \end{bmatrix}$$

The closed-loop system is implemented by state feedback, so that $u(t) = -\mathbf{K}\mathbf{x}(t)$, where $\mathbf{K} = [k_1 \quad k_2 \quad k_3]$; and k_1, k_2, and k_3 are real constants. Determine the constraints on the elements of $\mathbf{K}$ so that the closed-loop system is asymptotically stable.

• Stablizability with state feedback

6-8. Given the system in state equation form,

$$\frac{d\mathbf{x}(t)}{dt} = \mathbf{A}\mathbf{x}(t) + \mathbf{B}u(t)$$

(a) $\mathbf{A} = \begin{bmatrix} 1 & 0 & 0 \\ 0 & -3 & 0 \\ 0 & 0 & -2 \end{bmatrix}$ $\mathbf{B} = \begin{bmatrix} 1 \\ 0 \\ 1 \end{bmatrix}$

(b) $\mathbf{A} = \begin{bmatrix} 1 & 0 & 0 \\ 0 & -2 & 0 \\ 0 & 0 & 3 \end{bmatrix}$ $\mathbf{B} = \begin{bmatrix} 0 \\ 1 \\ 1 \end{bmatrix}$

Can the system be stabilized by state feedback $u(t) = -\mathbf{K}\mathbf{x}(t)$, where $\mathbf{K} = [k_1 \quad k_2 \quad k_3]$?

• System with tachometer feedback

6-9. The block diagram of a motor-control system with tachometer feedback is shown in Fig. 6P-9. Find the range of the tachometer constant K_t so that the system is asymptotically stable.

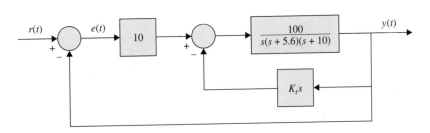

Figure 6P-9

• Stability in parameter plane

6-10. The block diagram of a control system is shown in Fig. 6P-10. Find the region in the K-versus-α plane for the system to be asymptotically stable. (Use K as the vertical and α as the horizontal axis.)

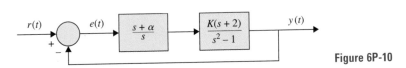

Figure 6P-10

• Stability of attitude control of a missile

6-11. The attitude control of a missile shown in Fig. 4P-13 is accomplished by thrust vectoring. The transfer function between the thrust angle $\Delta(s)$ and the angle of attack $\Theta(s)$ is represented by

$$G_p(s) = \frac{\Theta(s)}{\Delta(s)} = \frac{K}{s^2 - \alpha}$$

where K and α are positive real constants. The block diagram of the control system is shown in Fig. 6P-11.

(a) In Fig. 6P-11, consider that only the attitude-sensor loop is in operation, but $K_t = 0$. Determine the relationship between K, K_s, and α so that the missile will oscillate back and forth (marginally stable). Find the condition when the missile will tumble end over end (unstable).

(b) Consider that both loops are in operation. Determine relationships among K, K_s, K_t, and α so that the missile will oscillate back and forth. Repeat the problem for the missile tumbling end over end.

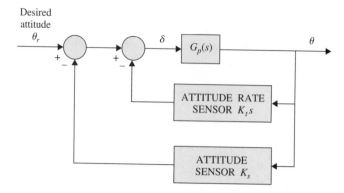

Figure 6P-11

• Application of Routh-Hurwitz criterion to relative stability

6-12. The conventional Routh-Hurwitz criterion gives information only on the location of the zeros of a polynomial $F(s)$ with respect to the left half and right half of the s-plane. Devise a linear transformation $s = f(p, \alpha)$, where p is a complex variable, so that the Routh-Hurwitz criterion can be applied to determine whether $F(s)$ has zeros to the right of the line $s = -\alpha$, where α is a positive real number. Apply the transformation to the following characteristic equations to determine how many roots are to the right of the line $s = -1$ in the s-plane.

(a) $F(s) = s^2 + 5s + 3 = 0$ **(b)** $s^3 + 3s^2 + 3s + 1 = 0$

(c) $F(s) = s^3 + 4s^2 + 3s + 10 = 0$ **(d)** $s^3 + 4s^2 + 4s + 4 = 0$

• Stability of liquid-level control system

6-13. Figure 6P-13 shows the closed-loop version of the liquid-level control system described in Problem 5-40. All the system parameters and equations are given in Problem 5-40. The error detector is modeled as

$$K_s = 1 \quad \text{V/ft} \qquad e(t) = K_s[r(t) - h(t)]$$

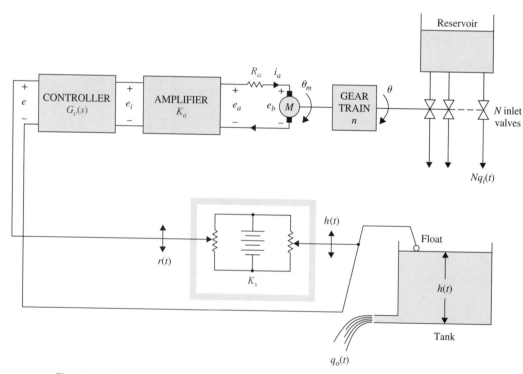

Figure 6P-13

(a) Draw a functional block diagram for the overall system, showing the functional relationship between the transfer functions.

(b) Let the transfer function of the controller, $G_c(s)$, be unity. Find the open-loop transfer function $G(s) = H(s)/E(s)$ and the closed-loop transfer function $M(s) = H(s)/R(s)$. Find the characteristic equation of the closed-loop system. Note that all the system parameters are specified except for N, the number of inlets.

(c) Apply the Routh-Hurwitz criterion to the characteristic equation and determine the maximum number of inlets (positive integer) so that the system is asymptotically stable. Set $G_c(s) = 1$.

• Stability of liquid-level control system

6-14. The block diagram shown in Fig. 6P-14 represents the liquid-level control system described in Problem 6-13. The following data are given: $K_a = 50$, $K_I = 50$, $K_b = 0.0706$, $J = 0.006$, $R_a = 10$, $K_i = 10$, and $n = 1/100$, the values of A, N, and K_o will be assigned in the following.

(a) Let the number of inlets, N, and the tank area, A, be the variable parameters, and $K_o = 40$. Find the ranges of N and A so that the closed-loop system is asymptotically stable. Indicate the region in the N-versus-A plane in which the system is stable. Use the fact that N is an integer. If the cross-sectional area of the tank, A, were infinitely large, what would be the maximum value of N (integer) for stability?

(b) It is useful to investigate the relationship between the gear ratio n and K_o, which control the liquid inflow and outflow, respectively. Find the stable region in the N-versus-K_o plane, for K_o up to 100. Let $A = 50$.

(c) Let $N = 10$ and $A = 50$. K_I and K_o are variable parameters. Find the stable region in the K_I-versus-K_o plane.

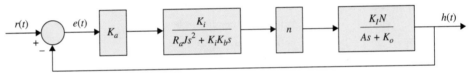

Figure 6P-14

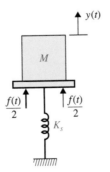

Stability of space-shuttle pointing control system

6-15. The payload of a space-shuttle-pointing control system is modeled as a pure mass M. The payload is suspended by magnetic bearings so that no friction is encountered in the control. The attitude of the payload in the y direction is controlled by magnetic actuators located at the base. The total force produced by the magnetic actuators is $f(t)$. The controls of the other degrees of motion are independent, and are not considered here. Since there are experiments located on the payload, electric power must be brought to the payload through cables. The linear spring with spring constant K_s is used to model the cable attachment. The dynamic system model for the control of the y-axis motion is shown in Fig. 6P-15. The force equation of motion in the y-direction is

$$f(t) = K_s y(t) + M\frac{d^2 y(t)}{dt^2}$$

where $K_s = 0.5$ N-m/m, and $M = 500$ kg.

Figure 6P-15

The magnetic actuators are controlled through state feedback, so that

$$f(t) = -K_P y(t) - K_D \frac{dy(t)}{dt}$$

(a) Draw a functional block diagram for the system.

(b) Find the characteristic equation of the closed-loop system.

(c) Find the region in the K_D-versus-K_P plane in which the system is asymptotically stable.

• Stability of inventory control system

6-16. An inventory-control system is modeled by the following differential equations:

$$\frac{dx_1(t)}{dt} = -x_2(t) + u(t)$$

$$\frac{dx_2(t)}{dt} = -Ku(t)$$

where $x_1(t)$ is the level of inventory; $x_2(t)$, the rate of sales of product; $u(t)$, the production rate; and K, a real constant. Let the output of the system by $y(t) = x_1(t)$, and $r(t)$ be the reference set point for the desired inventory level. Let $u(t) = r(t) - y(t)$. Determine the constraint on K so that the closed-loop system is asymptotically stable.

Additional Computer Problem

6-17. Use a root-finding computer program to find the roots of the following characteristic equations of linear continuous-data systems and determine the stability condition of the systems.

(a) $s^3 + 10s^2 + 10s + 130 = 0$

(b) $s^4 + 12s^3 + s^2 + 2s + 10 = 0$

(c) $s^4 + 12s^3 + 10s^2 + 10s + 10 = 0$

(d) $s^4 + 12s^3 + s^2 + 10s + 1 = 0$

(e) $s^6 + 6s^5 + 125s^4 + 100s^3 + 100s^2 + 20s + 10 = 0$

(f) $s^5 + 125s^4 + 100s^3 + 100s^2 + 20s + 10 = 0$

Answers to Review Questions

1. (a) No (b) No   2. (b)   3. (F)   4. (T)   5. (T)   6. (T)   7. (T)

CHAPTER 7

Time-Domain Analysis of Control Systems

▶ 7-1 TIME RESPONSE OF CONTINUOUS-DATA SYSTEMS: INTRODUCTION

Since time is used as an independent variable in most control systems, it is usually of interest to evaluate the state and output responses with respect to time, or simply, the **time response**. In the analysis problem, a reference input signal is applied to a system, and the performance of the system is evaluated by studying the system response in the time domain. For instance, if the objective of the control system is to have the output variable track the input signal, starting at some initial time and initial condition, it is necessary to compare the input and output responses as functions of time. Therefore, in most control-system problems, the final evaluation of the performance of the system is based on the time responses.

The time response of a control system is usually divided into two parts: the **transient response** and the **steady-state response**. Let $y(t)$ denote the time response of a continuous-data system; then, in general, it can be written as

• Transient response is defined as the part of the time response that goes to zero as time goes to infinity.

$$y(t) = y_t(t) + y_{ss}(t) \tag{7-1}$$

where $y_t(t)$ denotes the transient response, and $y_{ss}(t)$ denotes the steady-state response.

In control systems, *transient response* is defined as the part of the time response that goes to zero as time becomes very large. Thus, $y_t(t)$ has the property:

$$\lim_{t \to \infty} y_t(t) = 0 \tag{7-2}$$

• Steady-state response is the part of the total response that remains after the transient has died out.

The steady-state response is simply the part of the total response that remains after the transient has died out. Thus, the steady-state response can still vary in a fixed pattern, such as a sine wave, or a ramp function that increases with time.

All real, stable control systems exhibit transient phenomena to some extent before the steady state is reached. Since inertia, mass, and inductance are unavoidable in physical systems, the response of a typical control system cannot follow sudden changes in the input instantaneously, and transients are usually observed. Therefore, the control of the transient response is necessarily important, because it is a significant part of the dynamic behavior of the system; and the deviation between the output response and the input or the desired response, before the steady state is reached, must be closely controlled.

The steady-state response of a control system is also very important, since it indicates where the system output ends up at when time becomes large. For a position-control system, the steady-state response when compared with the desired reference position gives an indication of the final accuracy of the system. In general, if the steady-state response of the output does not agree with the desired reference exactly, the system is said to have a **steady-state error**.

The study of a control system in the time domain essentially involves the evaluation of the transient and the steady-state responses of the system. In the design problem, specifications are usually given in terms of the transient and the steady-state performances, and controllers are designed so that the specifications are all met by the designed system.

▶ 7-2 TYPICAL TEST SIGNALS FOR THE TIME RESPONSE OF CONTROL SYSTEMS

Unlike electric networks and communication systems, the inputs to many practical control systems are not exactly known ahead of time. In many cases, the actual inputs of a control system may vary in random fashion with respect to time. For instance, in a radar-tracking system for antiaircraft missiles, the position and speed of the target to be tracked may vary in an unpredictable manner, so that they cannot be predetermined. This poses a problem for the designer, since it is difficult to design a control system so that it will perform satisfactorily to all possible forms of input signals. For the purpose of analysis and design, it is necessary to assume some basic types of test inputs so that the performance of a system can be evaluated. By selecting these basic test signals properly, not only is the mathematical treatment of the problem systematized, but the response due to these inputs allows the prediction of the system's performance to other more complex inputs. In the design problem, performance criteria may be specified with respect to these test signals so that the system may be designed to meet the criteria. This approach is particularly useful for linear systems, since the response to complex signals can be determined by superposing those due to simple test signals.

When the response of a linear time-invariant system is analyzed in the frequency domain, a sinusoidal input with variable frequency is used. When the input frequency is swept from zero to beyond the significant range of the system characteristics, curves in terms of the amplitude ratio and phase between the input and the output are drawn as functions of frequency. It is possible to predict the time-domain behavior of the system from its frequency-domain characteristics.

To facilitate the time-domain analysis, the following deterministic test signals are used.

Step-Function Input: The step-function input represents an instantaneous change in the reference input. For example, if the input is an angular position of a mechanical shaft, a step input represents the sudden rotation of the shaft. The mathematical representation of a step function or magnitude R is

$$r(t) = R \qquad t \geq 0$$
$$= 0 \qquad t < 0 \tag{7-3}$$

where R is a real constant. Or,

$$r(t) = Ru_s(t) \tag{7-4}$$

where $u_s(t)$ is the unit-step function. The step function as a function of time is shown in Fig. 7-1(a). The step function is very useful as a test signal since its initial instantaneous jump in amplitude reveals a great deal about a system's quickness in responding to inputs

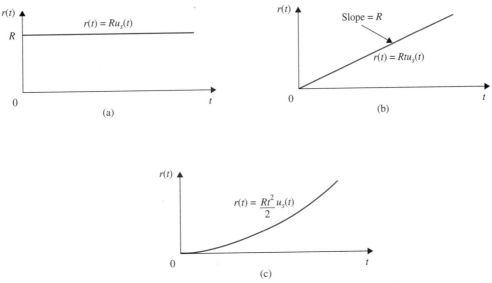

Figure 7-1   Basic time-domain test signals for control systems. (a) Step function.
(b) Ramp-function. (c) Parabolic function.

with abrupt changes. Also, since the step function contains, in principle, a wide band of
frequencies in its spectrum, as a result of the jump discontinuity, it is equivalent to the
application of numerous sinusoidal signals with a wide range of frequencies.

Ramp-Function Input: The ramp function is a signal that changes constantly with time.
Mathematically, a ramp function is represented by

$$r(t) = Rtu_s(t) \tag{7-5}$$

where R is a real constant. The ramp function is shown in Fig. 7-1(b). If the input variable represents the angular displacement of a shaft, the ramp input denotes the constant-
speed rotation of the shaft. The ramp function has the ability to test how the system would
respond to a signal that changes linearly with time.

Parabolic-Function Input: The parabolic function represents a signal that is one order
faster than the ramp function. Mathematically, it is represented as

$$r(t) = \frac{Rt^2}{2} u_s(t) \tag{7-6}$$

where R is a real constant, and the factor $^1/_2$ is added for mathematical convenience since
the Laplace transform of $r(t)$ is simply R/s^3. The graphical representation of the parabolic
function is shown in Fig. 7-1(c).

These signals all have the common feature that they are simple to describe
mathematically. From the step function to the parabolic function, the signals become pro-
gressively faster with respect to time. In theory we can define signals with still higher
rates, such as t^3, which is called the *jerk function*, and so forth. However, in reality, we
seldom find it necessary or feasible to use a test signal faster than a parabolic function.
This is because, as we shall see later, in order to track a high-order input accurately, the
system must have high-order integrations in the loop, which usually leads to serious stability problems.

▶ 7-3 THE UNIT-STEP RESPONSE AND TIME-DOMAIN SPECIFICATIONS

As defined earlier, the transient portion of the time response is the part that goes to zero as time becomes large. Nevertheless, the transient response of a control system is necessarily important, since both the amplitude and the time duration of the transient response must be kept within tolerable or prescribed limits. For example, in the automobile idle-speed control system described in Chapter 1, in addition to striving for a desirable idle speed in the steady state, the transient drop in engine speed must not be excessive, and the recovery in speed should be made as quickly as possible. For linear control systems, the characterization of the transient response is often done by use of the unit-step function $u_s(t)$ as the input. The response of a control system when the input is a unit-step function is called the **unit-step response**. Figure 7-2 illustrates a typical unit-step response of a linear control system. With reference to the unit-step response, performance criteria commonly used for the characterization of linear control systems in the time domain are defined as follows:

1. ***Maximum Overshoot.*** Let $y(t)$ be the unit-step response. Let y_{max} denote the maximum value of $y(t)$; and y_{ss}, the steady-state value of $y(t)$; and $y_{max} \geq y_{ss}$. The maximum overshoot of $y(t)$ is defined as

$$\text{maximum overshoot} = y_{max} - y_{ss} \tag{7-7}$$

The maximum overshoot is often represented as a percentage of the final value of the step response; that is,

$$\text{percent maximum overshoot} = \frac{\text{maximum overshoot}}{y_{ss}} \times 100\% \tag{7-8}$$

The maximum overshoot is often used to measure the relative stability of a control system. A system with a large overshoot is usually undesirable. For design purposes, the maximum overshoot is often given as a time-domain specification. The

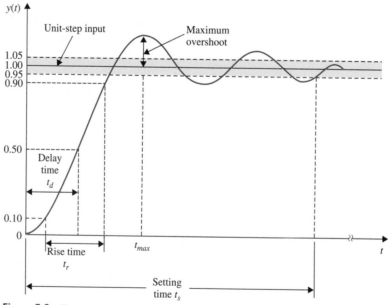

Figure 7-2 Typical unit-step response of a control system illustrating the time-domain specifications.

unit-step illustrated in Fig. 7-2 shows that the maximum overshoot occurs at the first overshoot. For some systems, the maximum overshoot may occur at a later peak, and if the system transfer function has an odd number of zeros in the right-half *s*-plane, a negative undershoot may even occur [3, 4] (Problem 7-23).

2. **Delay Time.**   The delay time t_d is defined as the time required for the step response to reach 50 percent of its final value. This is shown in Fig. 7-2.

3. **Rise Time.**   The rise time t_r is defined as the time required for the step response to rise from 10 to 90 percent of its final value, as shown in Fig. 7-2. An alternative measure is to represent the rise time as the reciprocal of the slope of the step response at the instant that the response is equal to 50 percent of its final value.

4. **Settling Time.**   The settling time t_s is defined as the time required for the step response to decrease and stay within a specified percentage of its final value. A frequently used figure is 5 percent.

   The four quantities just defined give a direct measure of the transient characteristics of a control system in terms of the unit-step response. These time-domain specifications are relatively easy to measure when the step response is well defined, as shown in Fig. 7-2. Analytically, these quantities are difficult to establish, except for simple systems lower than the third order.

5. **Steady-State Error.**   The steady-state error of a system response is defined as the discrepancy between the output and the reference input when the steady state $(t \rightarrow \infty)$ is reached.

   It should be pointed out that the steady-state error may be defined for any test signal such as a step-function, ramp-function, parabolic-function, or even a sinusoidal input, although Fig. 7-2 only shows the error for a step input.

▶ 7-4 STEADY-STATE ERROR

One of the objectives of most control systems is that the system output response follows a specific reference signal accurately in the steady state. The difference between the output and the reference in the steady state was defined earlier as the **steady-state error**. In the real world, because of friction and other imperfections, and the natural composition of the system, the steady state of the output response seldom agrees exactly with the reference. Therefore, steady-state errors in control systems are almost unavoidable. In a design problem, one of the objectives is to keep the steady-state error to a minimum, or below a certain tolerable value, and at the same time the transient response must satisfy a certain set of specifications.

The accuracy requirement on control systems depends to a great extent on the control objectives of the system. For instance, the final position accuracy of an elevator would be far less stringent than the pointing accuracy on the control of the Large Space Telescope, which is a telescope mounted on board of a space shuttle. The accuracy of position control of such a system is often measured in microradians.

7-4-1   Steady-State Error of Linear Continuous-Data Control Systems

Linear control systems are subject to steady-state errors for somewhat different causes than nonlinear systems, although the reason is still that the system no longer "sees" the error, and no corrective effort is exerted. In general, the steady-state errors of linear control systems depend on the type of the reference signal and the type of the system.

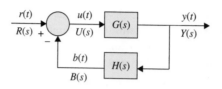

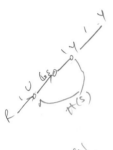

Figure 7-3 Nonunity feedback control system.

Definition of the Steady-State Error with Respect to System Configuration

Before embarking on the steady-state error analysis, we must first clarify what is meant by system error. In general, we can regard the error as a signal that should be quickly reduced to zero, if possible. Let us refer to the closed-loop system shown in Fig. 7-3, where $r(t)$ is the input, $u(t)$ the actuating signal, $b(t)$ the feedback signal, and $y(t)$ is the output. The error of the system may be defined as

$$e(t) = \text{reference signal} - y(t) \tag{7-9}$$

where the reference signal is the signal that the output $y(t)$ is to track. When the system has unity feedback, that is, $H(s) = 1$, then the input $r(t)$ is the reference signal, and the error is simply

$$e(t) = r(t) - y(t) \tag{7-10}$$

The steady-state error is defined as

$$e_{ss} = \lim_{s \to \infty} e(t) \tag{7-11}$$

When $H(s)$ is not unity, the actuating signal $u(t)$ in Fig. 7-2 may or may not be the error, depending on the form and the purpose of $H(s)$. Let us assume that the objective of the system in Fig. 7-3 is to have the output $y(t)$ track the input $r(t)$ as closely as possible, and the system transfer functions are

$$G(s) = \frac{1}{s^2(s + 12)} \qquad H(s) = \frac{5(s + 1)}{(s + 5)} \tag{7-12}$$

We can show that if $H(s) = 1$, the characteristic equation is

$$s^3 + 12s^2 + 1 = 0 \tag{7-13}$$

which has roots in the right-half s-plane, and the closed-loop system is unstable. We can show that the $H(s)$ given in Eq. (7-10) stabilizes the system and the characteristic equation becomes

$$s^4 + 17s^3 + 60s^2 + 5s + 5 = 0 \tag{7-14}$$

In this case, the system error may still be defined as in Eq.(7-10).

However, consider a velocity control system in which a step input is used to control the system output that contains a ramp in the steady state. The system transfer functions may be of the form

$$G(s) = \frac{1}{s^2(s + 12)} \qquad H(s) = K_t s \tag{7-15}$$

where $H(s)$ is the transfer function of an electromechanical or electronic tachometer, and K_t is the tachometer constant. The system error should be defined as in Eq. (7-9), where the reference signal is the *desired velocity*, and not $r(t)$. In this case, since $r(t)$ and $y(t)$ are not of the same dimension, it would be meaningless to define the error as in Eq. (7-10). To illustrate the system further, let $K_t = 10$ volts/rad/sec. This means that for a unit-step input of 1 volt, the desired velocity in the steady state is $1/10$ or 0.1 rad/sec, since when this is achieved, the output voltage of the tachometer would be 1 volt, and the steady-state error would be zero. The closed-loop transfer function of the system is

$$M(s) = \frac{Y(s)}{R(s)} = \frac{G(s)}{1 + G(s)H(s)} = \frac{1}{s(s^2 + 12s + 10)} \qquad (7\text{-}16)$$

For a unit-step function input, $R(s) = 1/s$. The output time response is

$$y(t) = 0.1t - 0.12 - 0.000796e^{-11.1t} + 0.1208e^{-0.901t} \qquad t \geq 0 \qquad (7\text{-}17)$$

Since the exponential terms of $y(t)$ in Eq. (7-17) all diminish as $t \to \infty$, the steady-state part of $y(t)$ is $0.1t - 0.12$. Thus, the steady-state error of the system is

$$e_{ss} = \lim_{t \to \infty}[0.1t - y(t)] = 0.12 \qquad (7\text{-}18)$$

More explanations will be given on how to define the reference signal when $H(s) \neq 1$ later when the general discussion on the steady-state error of nonunity feedback systems is given.

Not all system errors are defined with respect to the response due to the input. Figure 7-4 shows a system with a disturbance $d(t)$, in addition to the input $r(t)$. The output due to $d(t)$ acting alone may also be considered as an error.

Because of these reasons, the definition of system error has not been unified in the literature. To establish a systematic study of the steady-state error for linear systems, we shall classify three types of systems and treat these separately.

1. Systems with unity feedback; $H(s) = 1$.
2. Systems with nonunity feedback, but $H(0) = K_H = $ constant.
3. Systems with nonunity feedback, and $H(s)$ has zeros at $s = 0$ of order N.

The objective here is to establish a definition of the error with respect to one basic system configuration so that some fundamental relationships can be determined between the steady-state error and the system parameters.

Type of Control Systems: Unity Feedback Systems

Consider that a control system with unity feedback can be represented by or simplified to the block diagram with $H(s) = 1$ in Fig. 7-3. The steady-state error of the system is written

$$e_{ss} = \lim_{t \to \infty} e(t) = \lim_{s \to 0} sE(s)$$

- System type is defined here only for unity-feedback systems.

$$= \lim_{s \to 0} \frac{sR(s)}{1 + G(s)} \qquad (7\text{-}19)$$

Clearly, e_{ss} depends on the characteristics of $G(s)$. More specifically, we can show that e_{ss} depends on the number of poles $G(s)$ has at $s = 0$. This number is known as the **type** of the control system, or simply, system **type**.

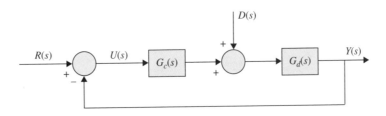

Figure 7-4 System with disturbance input.

We can show that the steady-state error e_{ss} depends on the **type** of the control system. Let us formalize the system **type** by referring to the form of the forward-path transfer function $G(s)$. In general, $G(s)$ can be expressed for convenience as

$$G(s) = \frac{K(1 + T_1 s)(1 + T_2 s) \cdots (1 + T_{m1} s + T_{m2} s^2)}{s^j (1 + T_a s)(1 + T_b s) \cdots (1 + T_{n1} s + T_{n2} s^2)} e^{-T_d s} \qquad (7\text{-}20)$$

where K and all the T's are real constants. The system **type** refers to the **order** of the pole of $G(s)$ at $s = 0$. Thus, the closed-loop system having the forward-path transfer function of Eq. (7-20) is type j, where $j = 0, 1, 2, \ldots$. The total number of terms in the numerator and the denominator and the values of the coefficients are not important to the system type, as system type refers only to the number of poles $G(s)$ has at $s = 0$. The following example illustrates the system type with reference to the form of $G(s)$.

▶ **EXAMPLE 7-1**

$$G(s) = \frac{K(1 + 0.5s)}{s(1 + s)(1 + 2s)(1 + s + s^2)} \qquad \text{type 1} \qquad (7\text{-}21)$$

$$G(s) = \frac{K(1 + 2s)}{s^3} \qquad \text{type 3} \qquad (7\text{-}22)$$

◀

Now let us investigate the effects of the types of inputs on the steady-state error. We shall consider only the step, ramp, and parabolic inputs.

Steady-State Error of System with a Step-Function Input

When the input $r(t)$ to the control system with $H(s) = 1$ of Fig. 7-3 is a step function with magnitude R, $R(s) = R/s$, the steady-state error is written from Eq. (7-19),

$$e_{ss} = \lim_{s \to 0} \frac{sR(s)}{1 + G(s)} = \lim_{s \to 0} \frac{R}{1 + G(s)} = \frac{R}{1 + \lim_{s \to 0} G(s)} \qquad (7\text{-}23)$$

• The **step-error constant** K_p is defined only when the system is subject to a step input.

For convenience, we define

$$K_p = \lim_{s \to 0} G(s) \qquad (7\text{-}24)$$

as the **step-error constant**. Then Eq. (7-23) becomes

$$e_{ss} = \frac{R}{1 + K_p} \qquad (7\text{-}25)$$

A typical e_{ss} due to a step input when K_p is finite and nonzero is shown in Fig. 7-5. We see from Eq. (7-25) that for e_{ss} to be zero, when the input is a step function, K_p must be

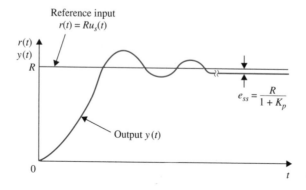

Figure 7-5 Typical steady-state error due to a step input.

infinite. If $G(s)$ is described by Eq. (7-20), we see that for K_p to be infinite, j must be at least equal to unity; that is, $G(s)$ must have at least one pole at $s = 0$. Therefore, we can summarize the steady-state error due to a step function input as follows:

Type 0 system: $\qquad\qquad e_{ss} = \dfrac{R}{1 + K_p} = \text{constant}$

Type 1 or higher system: $\quad e_{ss} = 0$

Steady-State Error of System with a Ramp-Function Input

When the input to the control system $[H(s) = 1]$ of Fig. 7-3 is a ramp-function with magnitude R,

$$r(t) = Rtu_s(t) \tag{7-26}$$

where R is a real constant, the Laplace transform of $r(t)$ is

$$R(s) = \frac{R}{s^2} \tag{7-27}$$

The steady-state error is written using Eq. (7-19),

$$e_{ss} = \lim_{s \to 0} \frac{R}{s + sG(s)} = \frac{R}{\lim_{s \to 0} sG(s)} \tag{7-28}$$

• The **ramp-error constant** K_v is defined only when the system is subject to a ramp input.

We define the **ramp-error constant** as

$$K_v = \lim_{s \to 0} sG(s) \tag{7-29}$$

Then, Eq. (7-26) becomes

$$e_{ss} = \frac{R}{K_v} \tag{7-30}$$

which is the steady-state error when the input is a ramp function. A typical e_{ss} due to a ramp input when K_v is finite and nonzero is illustrated in Fig. 7-6.

Equation (7-30) shows that for e_{ss} to be zero when the input is a ramp function, K_v must be infinite. Using Eqs. (7-20) and (7-29), we obtain

$$K_v = \lim_{s \to 0} sG(s) = \lim_{s \to 0} \frac{K}{s^{j-1}} \qquad j = 0, 1, 2, \ldots \tag{7-31}$$

Thus, in order for K_v to be infinite, j must be at least equal to 2, or the system must be of type 2 or higher. The following conclusions may be stated with regard to the steady-

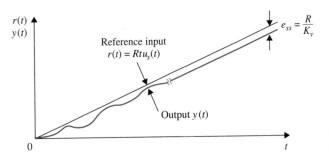

Figure 7-6 Typical steady-state error due to a ramp-function input.

state error of a system with ramp input:

$$\text{Type 0 system:} \qquad e_{ss} = \infty$$

$$\text{Type 1 system:} \qquad e_{ss} = \frac{R}{K_v} = \text{constant}$$

$$\text{Type 2 system:} \qquad e_{ss} = 0$$

Steady-State Error of System with a Parabolic-Function Input
When the input is described by the standard parabolic form,

$$r(t) = \frac{Rt^2}{2}u_s(t) \tag{7-32}$$

the Laplace transform of $r(t)$ is

$$R(s) = \frac{R}{s^3} \tag{7-33}$$

The steady-state error of the system in Fig. 7-3 with $H(s) = 1$ is

$$e_{ss} = \frac{R}{\lim\limits_{s \to 0} s^2 G(s)} \tag{7-34}$$

• The **parabolic error constant** K_a is defined only if the system is subject to a parabolic input.

A typical e_{ss} of a system with a nonzero and finite K_a due to a parabolic function input is shown in Fig. 7-7.

Defining the **parabolic-error constant** as

$$K_a = \lim\limits_{s \to 0} s^2 G(s) \tag{7-35}$$

the steady-state error becomes

$$e_{ss} = \frac{R}{K_a} \tag{7-36}$$

Following the pattern set with the step and ramp inputs, the steady-state error due to the parabolic input is zero if the system is of type 3 or greater. The following conclusions are made with regard to the steady-state error of a system with parabolic input:

$$\text{Type 0 system:} \qquad e_{ss} = \infty$$
$$\text{Type 1 system:} \qquad e_{ss} = \infty$$
$$\text{Type 2 system:} \qquad e_{ss} = \frac{R}{K_a} = \text{constant}$$
$$\text{Type 3 or higher system:} \qquad e_{ss} = 0$$

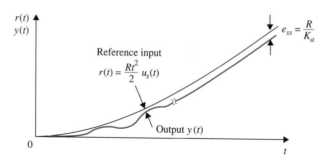

Figure 7-7 Typical steady-state error due to a parabolic-function input.

TABLE 7-1 Summary of the Steady-State Errors Due to Step-, Ramp-, and Parabolic-Function Inputs for Unity-Feedback Systems

Type of System j	Error Constants K_p	K_v	K_a	Steady-State Error e_{ss} Step Input $\dfrac{R}{1 + K_p}$	Ramp Input $\dfrac{R}{K_v}$	Parabolic $\dfrac{R}{K_a}$
0	K	0	0	$\dfrac{R}{1 + K}$	∞	∞
1	∞	K	0	0	$\dfrac{R}{K}$	∞
2	∞	∞	K	0	0	$\dfrac{R}{K}$
3	∞	∞	∞	0	0	0

We cannot emphasize often enough that in order for these results to be valid, the closed-loop system must be stable.

By using the method described, the steady-state error of any linear closed-loop system subject to an input with order higher than the parabolic function can also be derived if necessary. As a summary of the error analysis, Table 7-1 shows the relations among the error constants, the types of systems with reference to Eq. (7-20), and the input types.

As a summary, the following points should be noted when applying the error-constant analysis just presented.

1. The step-, ramp-, or parabolic-error constants is significant for the error analysis only when the input signal is a step function, ramp function, or parabolic function, respectively.

2. Since the error constants are defined with respect to the forward-path transfer function $G(s)$, the method is applicable to only the system configuration shown in Fig. 7-3 with $H(s) = 1$. Since the error analysis relies on the use of the final-value theorem of the Laplace transform, it is important to check first to see if $sE(s)$ has any poles on the $j\omega$-axis or in the right-half s-plane.

3. The steady-state error properties summarized in Table 7-1 are for systems with unity feedback only.

4. The steady-state error of a system with an input that is a linear combination of the three basic types of inputs can be determined by superimposing the errors due to each input component.

5. When the system configuration differs from that of Fig. 7-3 with $H(s) = 1$, we can either simplify the system to the form of Fig. 7-3, or establish the error signal and apply the final-value theorem. The error constants defined here may or may not apply, depending on the individual situation.

When the steady-state error is infinite, that is, when the error increases continuously with time, the error-constant method does not indicate how the error varies with time. This is one of the disadvantages of the error-constant method. The error-constant method also does not apply to systems with inputs that are sinusoidal, since the final-value theorem cannot be applied. The following examples illustrate the utility of the error constants and their values in the determination of the steady-state errors of linear control systems with unity feedback.

▶ **EXAMPLE 7-2** Consider that the system shown in Fig. 7-3 with $H(s) = 1$ has the following transfer functions. The error constants and steady-state errors are calculated for the three basic types of inputs using the error constants.

(a) $G(s) = \dfrac{K(s + 3.15)}{s(s + 1.5)(s + 0.5)}$ $\qquad H(s) = 1$ $\qquad$ Type 1 system

$\qquad$ Step input: $\qquad$ Step error constant $K_p = \infty$ $\qquad\qquad$ $e_{ss} = \dfrac{R}{1 + K_p} = 0$

$\qquad$ Ramp input: $\qquad$ Ramp error constant $K_v = 4.2K$ $\qquad$ $e_{ss} = \dfrac{R}{K_v} = \dfrac{R}{4.2K}$

$\qquad$ Parabolic input: $\quad$ Parabolic error constant $K_a = 0$ $\qquad$ $e_{ss} = \dfrac{R}{K_a} = \infty$

These results are valid only if the value of K stays within the range that corresponds to a stable closed-loop system, which is $0 < K < 1.304$.

(b) $G(s) = \dfrac{K}{s^2(s + 12)}$ $\qquad H(s) = 1$ $\qquad$ Type 2 system

The closed-loop system is unstable for all values of K, and error analysis is meaningless.

(c) $G(s) = \dfrac{5(s + 1)}{s^2(s + 12)(s + 5)}$ $\qquad H(s) = 1$ $\qquad$ Type 2 system

We can show that the closed-loop system is stable. The steady-state errors are calculated for the three basic types of inputs.

$\qquad$ Step input: $\qquad$ Step-error constant: $K_p = \infty$ $\qquad\qquad$ $e_{ss} = \dfrac{R}{1 + K_p} = 0$

$\qquad$ Ramp input: $\qquad$ Ramp-error constant: $K_v = \infty$ $\qquad\qquad$ $e_{ss} = \dfrac{R}{K_v} = 0$

$\qquad$ Parabolic input: $\quad$ Parabolic-error constant: $K_a = 1/12$ $\qquad$ $e_{ss} = \dfrac{R}{K_a} = 12R$ $\qquad$ ◀

Relationship between Steady-State Error and the Closed-Loop Transfer Function: Unity and Nonunity Feedback Systems

In the last section, the steady-state error of a closed-loop system is related to the forward-path transfer function $G(s)$ of the system, which is usually known. Often, the closed-loop transfer function is derived in the analysis process, and it would be of interest to establish the relationships between the steady-state error and the coefficients of the closed-loop transfer function. As it turns out, the closed-loop transfer function can be used to find the steady-state error of systems with unity as well as nonunity feedback. For the present discussion let us impose the following condition:

$$\lim_{s \to 0} H(s) = H(0) = K_H = \text{constant} \qquad (7\text{-}37)$$

which means that H(s) cannot have poles at $s = 0$. Since the signal that is fed back to be compared with the input in the steady state is K_H times the steady-state output, when this feedback signal equal the input, the steady-state error would be zero. Thus, we can define the reference signal as $r(t)/K_H$ and the error signal as

$$e(t) = \frac{1}{K_H} r(t) - y(t) \qquad (7\text{-}38)$$

or in the transform domain,

$$E(s) = \frac{1}{K_H} R(s) - Y(s) = \frac{1}{K_H}[1 - K_H M(s)]R(s) \qquad (7\text{-}39)$$

where $M(s)$ is the closed-loop transfer function, $Y(s)/R(s)$. Notice that the above development includes the unity-feedback case for which $K_H = 1$. Let us assume that $M(s)$ does not have any poles at $s = 0$, and is of the form:

$$M(s) = \frac{Y(s)}{R(s)} = \frac{b_m s^m + b_{m-1} s^{m-1} + \cdots + b_1 s + b_0}{s^n + a_{n-1} s^{n-1} + \cdots + a_1 s + a_0} \tag{7-40}$$

where $n > m$. We further require that all the poles of $M(s)$ are in the left-half s-plane, which means that the system is stable. The steady-state error of the system is written

$$e_{ss} = \lim_{s \to 0} sE(s) = \lim_{s \to 0} \frac{1}{K_H} [1 - K_H M(s)] sR(s) \tag{7-41}$$

Substituting Eq. (7-40) into the last equation and simplifying, we get

$$e_{ss} = \frac{1}{K_H} \lim_{s \to 0} \frac{s^n + \cdots + (a_1 - b_1 K_H)s + (a_0 - b_0 K_H)}{s^n + a_{n-1} s^{n-1} + \cdots + a_1 s + a_0} sR(s) \tag{7-42}$$

We consider the three basic types of inputs for $r(t)$.

1. **Step-Function Input.** $R(s) = R/s$.

   For a step-function input, the steady-state error in Eq. (7-42) becomes

   $$e_{ss} = \frac{1}{K_H} \left(\frac{a_0 - b_0 K_H}{a_0} \right) R \tag{7-43}$$

   Thus, the steady-state error due to a step input can be zero only if

   $$a_0 - b_0 K_H = 0 \tag{7-44}$$

   or

   $$M(0) = \frac{b_0}{a_0} = \frac{1}{K_H} \tag{7-45}$$

   This means that *for a unity-feedback system $K_H = 1$, the constant terms of the numerator and the denominator of $M(s)$ must be equal, that is, $b_0 = a_0$, in order for the steady-state error to be zero.*

   • For a system with unity feedback, the steady-state error due to a step input is zero only if the constant terms in the numerator and the denominator of the closed-loop transfer function are equal.

2. **Ramp-Function Input.** $R(s) = R/s^2$.

   For a ramp-function input, the steady-state error in Eq. (7-42) becomes

   $$e_{ss} = \frac{1}{K_H} \lim_{s \to 0} \frac{s^n + \cdots + (a_1 - b_1 K_H)s + (a_0 - b_0 K_H)}{s(s^n + a_{n-1} s^{n-1} + \cdots + a_1 s + a_0)} R \tag{7-46}$$

   The following values of e_{ss} are possible:

   $$e_{ss} = 0 \qquad \text{if} \quad a_0 - b_0 K_H = 0 \quad \text{and} \quad a_1 - b_1 K_H = 0 \tag{7-47}$$

   $$e_{ss} = \frac{a_1 - b_1 K_H}{a_0 K_H} R = \text{constant} \quad \text{if} \quad a_0 - b_0 K_H = 0 \quad \text{and} \quad a_1 - b_1 K_H \neq 0 \tag{7-48}$$

   $$e_{ss} = \infty \qquad \text{if} \quad a_0 - b_0 K_H \neq 0 \tag{7-49}$$

3. **Parabolic-Function Input.** $R(s) = R/s^3$.

   For a parabolic input, the steady-state error in Eq. (7-42) becomes

   $$e_{ss} = \frac{1}{K_H} \lim_{s \to 0} \frac{s^n + \cdots + (a_2 - b_2 K_H)s^2 + (a_1 - b_1 K_H)s + (a_0 - b_0 K_H)}{s^2(s^n + a_{n-1} s^{n-1} + \cdots + a_1 s + a_0)} R \tag{7-50}$$

The following values of e_{ss} are possible:

$$e_{ss} = 0 \qquad\qquad\qquad \text{if} \quad a_i - b_i K_H = 0 \quad \text{for} \quad i = 0, 1, \text{and } 2 \qquad (7\text{-}51)$$

$$e_{ss} = \frac{a_2 - b_2 K_H}{a_0 K_H} R = \text{constant} \quad \text{if} \quad a_i - b_i K_H = 0 \quad \text{for} \quad i = 0 \text{ and } 1 \qquad (7\text{-}52)$$

$$e_{ss} = \infty \qquad\qquad\qquad \text{if} \quad a_i - b_i K_H \neq 0 \quad \text{for} \quad i = 0 \text{ and } 1 \qquad (7\text{-}53)$$

▶ **EXAMPLE 7-3** The forward-path and closed-loop transfer functions of the system shown in Fig. 7-3 are given next. The system is assumed to have unity feedback, so $H(s) = 1$, and thus $K_H = H(0) = 1$.

$$G(s) = \frac{5(s+1)}{s^2(s+12)(s+5)} \qquad M(s) = \frac{5(s+1)}{s^4 + 17s^3 + 60s^2 + 5s + 5} \qquad (7\text{-}54)$$

The poles of $M(s)$ are all in the left-half s-plane. Thus, the system is stable. The steady-state errors due to the three basic types of inputs are evaluated as follows:

Step input: $\quad e_{ss} = 0 \quad$ since $a_0 = b_0 \, (= 5)$

Ramp input: $\quad e_{ss} = 0 \quad$ since $a_0 = b_0 \, (= 5)$ and $a_1 = b_1 \, (= 5)$

Parabolic input: $\quad e_{ss} = \dfrac{a_2 - b_2 K_H}{a_0 K_H} R = \dfrac{60}{5} R = 12R$

Since this is a Type 2 system with unity feedback, the same results are obtained with the error-constant method. ◀

▶ **EXAMPLE 7-4** Consider the system shown in Fig. 7-3 which has the following transfer functions:

$$G(s) = \frac{1}{s^2(s+12)} \qquad H(s) = \frac{5(s+1)}{s+5} \qquad (7\text{-}55)$$

Then, $K_H = H(0) = 1$. The closed-loop transfer function is

$$M(s) = \frac{Y(s)}{R(s)} = \frac{G(s)}{1 + G(s)H(s)} = \frac{s+5}{s^4 + 17s^3 + 60s^2 + 5s + 5} \qquad (7\text{-}56)$$

Comparing the last equation with Eq. (7-40), we have $a_0 = 5$, $a_1 = 5$, $a_2 = 60$, $b_0 = 5$, $b_1 = 1$, $b_2 = 0$. The steady-state errors of the system are calculated for the three basic types of inputs.

Unit-step input, $r(t) = u_s(t)$: $\qquad e_{ss} = \dfrac{a_0 - b_0 K_H}{a_0} = 0$

Unit-ramp input, $r(t) = tu_s(t)$: $\qquad e_{ss} = \dfrac{a_1 - b_1 K_H}{a_0 K_H} = \dfrac{5-1}{5} = 0.8$

Unit-parabolic input, $r(t) = tu_s(t)/2$: $\quad e_{ss} = \infty \quad$ since $\quad a_1 - b_1 K_H \neq 0$

It would be illuminating to calculate the steady-state errors of the system from the difference between the input and the output, and compare them with the results just obtained.

Applying the unit-step, unit-ramp, and unit-parabolic inputs to the system described by Eq. (7-56), and taking the inverse Laplace transform of $Y(s)$, the outputs are

Unit-step input: $y(t) = 1 - 0.00056e^{-12.05t} - 0.0001381e^{-4.886t}$

$\qquad\qquad - 0.9993e^{-0.0302t} \cos 0.2898t - 0.1301e^{-0.0302t} \sin 0.2898t \qquad t \geq 0$ $\qquad (7\text{-}57)$

Thus, the steady-state value of $y(t)$ is unity, and the steady-state error is zero.

Unit-ramp input: $y(t) = t - 0.8 + 4.682 \times 10^{-5}e^{-12.05t} + 2.826 \times 10^{-5}e^{-4.886t}$

$\qquad\qquad + 0.8e^{-0.0302t} \cos 0.2898t - 3.365e^{-0.0302t} \sin 0.2898t \qquad t \geq 0$ $\qquad (7\text{-}58)$

Thus, the steady-state portion of $y(t)$ is $t - 0.8$, and the steady-state error to a unit ramp is 0.8.

Unit-parabolic input: $y(t) = 0.5t^2 - 0.8t - 11.2 - 3.8842 \times 10^{-6}e^{-12.05t} - 5.784 \times 10^{-6}e^{-4.886t}$
$$+ 11.2e^{-0.0302t}\cos 0.2898t + 3.9289e^{-0.0302t}\sin 0.2898t \qquad t \geq 0$$
$$(7\text{-}59)$$

The steady-state portion of $y(t)$ is $0.5t^2 - 0.8t - 11.2$. Thus, the steady-state error is $0.8t + 11.2$, which becomes infinite as time goes to infinity. ◀

▶ **EXAMPLE 7-5** Consider that the system shown in Fig. 7-3 has the following transfer functions:

$$G(s) = \frac{1}{s^2(s + 12)} \qquad H(s) = \frac{10(s + 1)}{s + 5} \qquad (7\text{-}60)$$

Thus,

$$K_H = \lim_{s \to 0} H(s) = 2 \qquad (7\text{-}61)$$

The closed-loop transfer function is

$$M(s) = \frac{Y(s)}{R(s)} = \frac{G(s)}{1 + G(s)H(s)} = \frac{s + 5}{s^4 + 17s^3 + 60s^2 + 10s + 10} \qquad (7\text{-}62)$$

The steady-state errors of the system due to the three basic types of inputs are calculated as follows:

Unit-step input, $r(t) = u_s(t)$: $e_{ss} = \dfrac{1}{K_H}\left(\dfrac{a_0 - b_0 K_H}{a_0}\right) = \dfrac{1}{2}\left(\dfrac{10 - 5 \times 2}{10}\right) = 0 \qquad (7\text{-}63)$

Solving for the output using the $M(s)$ in Eq. (7-62), we get

$$y(t) = 0.5u_s(t) + \text{transient terms} \qquad (7\text{-}64)$$

Thus, the steady-state value of $y(t)$ is 0.5, and since $K_H = 2$, the steady-state error due to a unit-step input is zero.

Unit-ramp input, $r(t) = tu_s(t)$: $e_{ss} = \dfrac{1}{K_H}\left(\dfrac{a_1 - b_1 K_H}{a_0}\right) = \dfrac{1}{2}\left(\dfrac{10 - 1 \times 2}{10}\right) = 0.4 \qquad (7\text{-}65)$

The unit-ramp response of the system is written

$$y(t) = [0.5t - 0.4]u_s(t) + \text{transient terms} \qquad (7\text{-}66)$$

Thus, using Eq. (7-38), the steady-state error is calculated as

$$e(t) = \frac{1}{K_H}r(t) - y(t) = 0.4u_s(t) - \text{transient terms} \qquad (7\text{-}67)$$

Since the transient terms will die out as t approached infinity, the steady-state error due to a unit-ramp input is 0.4, as calculated in Eq. (7-66).

Unit-parabolic input, $r(t) = t^2 u_s(t)/2$: $e_{ss} = \infty$ since $a_1 - b_1 K_H \neq 0$

The unit-parabolic input is

$$y(t) = [0.25t^2 - 0.4t - 2.6]u_s(t) + \text{transient terms} \qquad (7\text{-}68)$$

The error due to the unit-parabolic input is

$$e(t) = \frac{1}{K_H}r(t) - y(t) = (0.4t - 2.6)u_s(t) - \text{transient terms} \qquad (7\text{-}69)$$

Thus, the steady-state error is $0.4t + 2.6$ which increases with time. ◀

Steady-State Error of Nonunity Feedback: $H(s)$ Has Nth-Order Zero at $s = 0$.

When $H(s)$ has an Nth-order zero at $s = 0$, this corresponds to desired output be proportional to the Nth-order derivative of the input in the steady state. In the real world, this corresponds to applying a tachometer or rate feedback. Thus, for the steady-state error analysis, the reference signal can be defined as $R(s)/K_H s^N$, and the error signal in the transform domain may be defined as

$$E(s) = \frac{1}{K_H s^N} R(s) - Y(s) \tag{7-70}$$

where

$$K_H = \lim_{s \to 0} \frac{H(s)}{s^N} \tag{7-71}$$

We shall derive only the results for $N = 1$ here. In this case, the transfer function of $M(s)$ in Eq. (7-40) will have a pole at $s = 0$, or $a_0 = 0$. The steady-state error is written from Eq. (7-70),

$$e_{ss} = \frac{1}{K_H} \lim_{s \to 0} \left[\frac{s^{n-1} + \cdots + (a_2 - b_1 K_H)s + (a_1 - b_0 K_H)}{s^n + a_{n-1}s^{n-1} + \cdots + a_1 s} \right] sR(s) \tag{7-72}$$

For a step input of magnitude R, the last equation is written

$$e_{ss} = \frac{1}{K_H} \lim_{s \to 0} \left[\frac{s^{n-1} + \cdots + (a_2 - b_1 K_H)s + (a_1 - b_0 K_H)}{s^n + a_{n-1}s^{n-1} + \cdots + a_1 s} \right] R \tag{7-73}$$

Thus, the steady-state error is

$$e_{ss} = 0 \qquad\qquad \text{if} \qquad a_2 - b_1 K_H = 0 \quad \text{and} \quad a_1 - b_0 K_H = 0 \tag{7-74}$$

$$e_{ss} = \frac{a_2 - b_1 K_H}{a_1 K_H} R = \text{constant} \qquad \text{if} \qquad a_1 - b_0 K_H = 0 \quad \text{but} \quad a_2 - b_1 K_H \neq 0 \tag{7-75}$$

$$e_{ss} = \infty \qquad\qquad \text{if} \qquad a_1 - b_0 K_H \neq 0 \tag{7-76}$$

We shall use the following example to illustrate these results.

▶ **EXAMPLE 7-6** Consider that the transfer functions of the system shown in Fig. 7-3 has the following transfer functions:

$$G(s) = \frac{1}{s^2(s + 12)} \qquad H(s) = \frac{10s}{s + 5} \tag{7-77}$$

Thus,

$$K_H = \lim_{s \to 0} \frac{H(s)}{s} = 2 \tag{7-78}$$

The closed-loop transfer function is

$$M(s) = \frac{Y(s)}{R(s)} = \frac{s + 5}{s^4 + 17s^3 + 60s^2 + 10s} \tag{7-79}$$

The velocity control system is considered to be stable as defined in Chapter 5, although $M(s)$ has a pole at $s = 0$, since the objective is to control velocity with a step input. The coefficients are identified to be $a_0 = 0$, $a_1 = 10$, $a_2 = 60$, $b_0 = 5$, and $b_1 = 1$.

For a unit-step input, the steady-state error, from Eq. (7-75), is

$$e_{ss} = \frac{1}{K_H}\left(\frac{a_2 - b_1 K_H}{a_1}\right) = \frac{1}{2}\left(\frac{60 - 1 \times 2}{10}\right) = 2.9 \qquad (7\text{-}80)$$

To verify this result, we find the unit-step response using the closed-loop transfer function in Eq. (7-79). The result is

$$y(t) = (0.5t - 2.9)u_s(t) + \text{transient terms} \qquad (7\text{-}81)$$

From the discussion that leads to Eq. (7-70), the reference signal is considered to be $tu_s(t)/K_H = 0.5tu_s(t)$ in the steady state; thus, the steady-state error is 2.9. Of course, it should be pointed out that if $H(s)$ were a constant for this Type 2 system, the closed-loop system would be unstable. So, the derivative control in the feedback path also has a stabilizing effect. ◀

7-4-2 Steady-State Error Caused by Nonlinear System Elements

In many instances, steady-state errors of control systems are attributed to some nonlinear system characteristics such as nonlinear friction or dead zone. For instance, if an amplifier used in a control system has the input-output characteristics shown in Fig. 7-8, then, when the amplitude of the amplifier input signal falls within the dead zone, the output of the amplifier would be zero, and the control would not be able to correct the error if any exists. Dead-zone nonlinearity characteristics shown in Fig. 7-8 are not limited to amplifiers. The flux-to-current relation of the magnetic field of an electric motor may exhibit a similar characteristic. As the current of the motor falls below the dead zone D, no magnetic flux, and, thus, no torque will be produced by the motor to move the load.

The output signals of digital components used in control systems, such as a microprocessor, can take on only discrete or quantized levels. This property is illustrated by the quantization characteristics shown in Fig. 7-9. When the input to the quantizer is within $\pm q/2$, the output is zero, and the system may generate an error in the output whose magnitude is related to $\pm q/2$. This type of error is also known as the **quantization error** in digital control systems.

When the control of physical objects is involved, friction is almost always present. Coulomb friction is a common cause of steady-state position errors in control systems. Figure 7-10 shows a restoring-torque-versus-position curve of a control system. The torque curve typically could be generated by a step motor or a switched-reluctance motor, or from a closed-loop system with a position encoder. Point 0 designates a stable equilibrium point on the torque curve, as well as the other periodic intersecting points along the axis where the slope on the torque curve is negative. The torque on either side of point 0 represents a restoring torque that tends to return the output to the equilibrium point when some angular-displacement disturbance takes place. When there is no friction, the position error should be zero, since there is always a restoring torque so long as the position

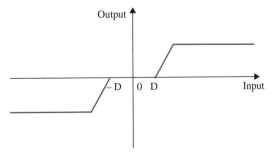

Figure 7-8 Typical input-output characteristics of an amplifier with dead zone and saturation.

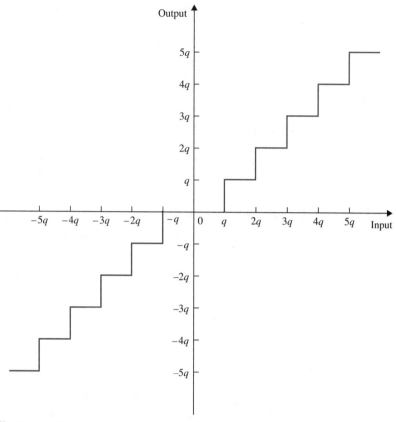

Figure 7-9 Typical input-output characteristics of a quantizer.

is not at the stable equilibrium point. If the rotor of the motor sees a Coulomb friction torque T_F, then the motor torque must first overcome this frictional torque before producing any motion. Thus, as the motor torque falls below T_F as the rotor position approaches the stable equilibrium point, it may stop at any position inside the error band bounded by $\pm\theta_e$, as shown in Fig. 7-10.

Although it is relatively simple to comprehend the effects of nonlinearities on errors, and to establish maximum upper bounds on the error magnitudes, it is difficult to

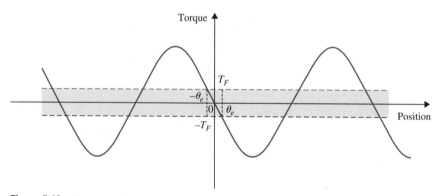

Figure 7-10 Torque-angle curve of a motor or closed-loop system with Coulomb friction.

establish general and closed-form solutions for nonlinear systems. Usually, exact and detailed analysis of errors in nonlinear control systems can be carried out only by computer simulations.

Therefore, we must realize that there are no error-free control systems in the real world, and since all physical systems have nonlinear characteristics of one form or another, steady-state errors can be reduced but never completely eliminated.

▶ 7-5 TIME RESPONSE OF A FIRST-ORDER SYSTEM

7-5-1 Speed Control of a DC Motor

Let us consider the dc motor system discussed in Chapter 4. We can assume that the armature inductance and the motor friction are all negligible. These assumptions are seldom justified, but for the sake of illustrating a first-order system, we are making these drastic assumptions. Then, the motor equations are simplified to

Torque: $\qquad T_m(t) = K_i i_a(t)$ (7-82)

Back emf: $\qquad e_b(t) = K_b \omega_m(t)$ (7-83)

Motor current: $\qquad i_a(t) = \dfrac{e_a(t) - e_b(t)}{R_a}$ (7-84)

Motor speed: $\qquad \omega_m(t) = \dfrac{1}{J_m} T_m(t)$ (7-85)

$\qquad e_a(t) = K_a e(t)$ (7-86)

where $e(t)$ is the applied voltage; K_a, the amplifier gain; K_b, the back-emf constant; K_i, the torque constant; and J_m and R_a, the motor inertia and armature resistance, respectively.

Since this is to be a speed-control problem, we are interested only in $\omega_m(t)$ and treat it as the output variable. The state diagram of the system is shown in Fig. 7-11, from which we obtain the closed-loop transfer function

$$\frac{\Omega_m(s)}{E(s)} = \frac{K}{s + a}$$ (7-87)

where

$$K = \frac{K_a K_i}{R_a J_m}$$ (7-88)

$$a = \frac{K_i K_b}{R_a J_m}$$ (7-89)

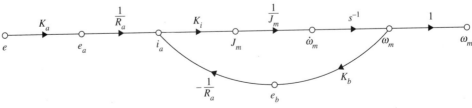

Figure 7-11 State diagram of the speed control of a first-order dc motor system.

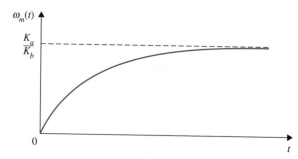

Figure 7-12 Unit-step responses of a first-order dc motor speed-control system.

For a unit-step input, $e(t) = u_s(t)$, and $E(s) = 1/s$. The motor speed is given by

$$\Omega_m(s) = \frac{K}{s(s + a)} \tag{7-90}$$

Taking the inverse Laplace transform on both sides of Eq. (7-90), we get

$$\omega_m(t) = \frac{K}{a}(1 - e^{-at}) = \frac{K_a}{K_b}(1 - e^{-at}) \qquad t \geq 0 \tag{7-91}$$

Figure 7-12 shows typical unit-step responses of $\omega_m(t)$ for several relative values of a We can make the following observations for this first-order system:

1. The step response will not have any overshoot for any combination of system parameters.

2. Figure 7-13 shows the location of the pole at $s = -a$ in the s-plane of the system transfer function in Eq. (7-84). Since the motor parameters are all positive, the value of a is always positive. The pole at $s = -a$ will always stay in the left-half s-plane, and the system is always stable.

3. The magnitude of the steady-state speed for a unit-step input voltage e_a is proportional to the amplifier gain K_a and inversely proportional to the back-emf constant K_b.

4. The rate of change of the motor speed in reaching its steady-state value depends on the value of a, which is given by Eq. (7-86). As the value of a increases, the motor accelerates faster. This is not difficult to comprehend, since a is proportional to the amplifier gain K_a and inversely proportional to the motor resistance R_a and inertial J_m.

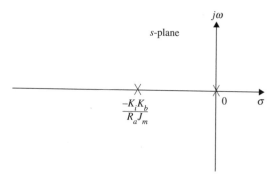

Figure 7-13 Pole configuration of the transfer function of a first-order dc motor speed-control system.

▶ 7-6 TRANSIENT RESPONSE OF A PROTOTYPE SECOND-ORDER SYSTEM

Although true second-order control systems are rare in practice, their analysis generally helps to form a basis for the understanding of analysis and design of higher-order systems, especially the ones that can be approximated by second-order systems.

Consider that a second-order control system with unity feedback is represented by the block diagram shown in Fig. 7-14. The open-loop transfer function of the system is

$$G(s) = \frac{Y(s)}{E(s)} = \frac{\omega_n^2}{s(s + 2\zeta\omega_n)} \tag{7-92}$$

where ζ and ω_n are real constants. The closed-loop transfer function of the system is

$$\frac{Y(s)}{R(s)} = \frac{\omega_n^2}{s^2 + 2\zeta\omega_n s + \omega_n^2} \tag{7-93}$$

The system in Fig. 7-14 with the transfer functions given by Eqs. (7-92) and (7-93) is defined as the **prototype second-order system**.

The characteristic equation of the prototype second-order system is obtained by setting the denominator of Eq. (7-93) to zero:

$$\Delta(s) = s^2 + 2\zeta\omega_n s + \omega_n^2 = 0 \tag{7-94}$$

For a unit-step function input, $R(s) = 1/s$, the output response of the system is obtained by taking the inverse Laplace transform of the output transform

$$Y(s) = \frac{\omega_n^2}{s(s^2 + 2\zeta\omega_n s + \omega_n^2)} \tag{7-95}$$

This can be done by referring to the Laplace-transform table in Appendix D. The result is

$$y(t) = 1 - \frac{e^{-\zeta\omega_n t}}{\sqrt{1 - \zeta^2}} \sin\left(\omega_n\sqrt{1 - \zeta^2}\, t + \cos^{-1}\zeta\right) \qquad t \geq 0 \tag{7-96}$$

Figure 7-15 shows the unit-step responses of Eq. (7-96) plotted as functions of the normalized time $\omega_n t$ for various values of ζ. As seen, the response becomes more oscillatory with larger overshoot as ζ decreases. When $\zeta \geq 1$, the step response does not exhibit any overshoot; that is, $y(t)$ never exceeds its final value during the transient. The responses also show that ω_n has a direct effect on the rise time, delay time, and settling time, but does not affect the overshoot. These will be studied in more detail in the following sections.

7-6-1 Damping Ratio and Damping Factor

The effects of the system parameters ζ and ω_n on the step response $y(t)$ of the prototype second-order system can be studied by referring to the roots of the characteristic equation in Eq. (7-94).

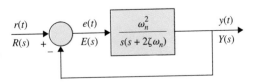

Figure 7-14 Prototype second-order control system.

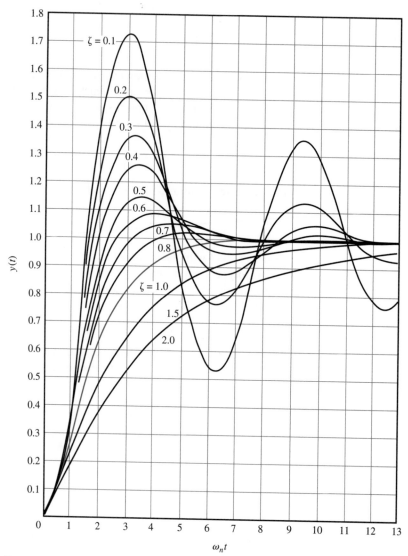

Figure 7-15 Unit-step responses of the prototype second-order system with various damping ratios.

The two roots can be expressed as

$$s_1, s_2 = -\zeta\omega_n \pm j\omega_n\sqrt{1 - \zeta^2} \tag{7-97}$$
$$= -\alpha \pm j\omega$$

where

$$\alpha = \zeta\omega_n \tag{7-98}$$

and

$$\omega = \omega_n\sqrt{1 - \zeta^2} \tag{7-99}$$

• When the damping is critical, $\zeta = 1$.

The physical significance of ζ and α is now investigated. As seen from Eqs. (7-96) and (7-98), α appears as the constant that is multiplied to t in the exponential term of

$y(t)$. Therefore, α controls the rate of rise or decay of the unit-step response $y(t)$. In other words, α controls the "damping" of the system, and is called the **damping factor**, or the **damping constant**. The inverse of α, $1/\alpha$, is proportional to the time constant of the system.

When the two roots of the characteristic equation are real and equal, we called the system **critically damped**. From Eq. (7-97), we see that critical damping occurs when $\zeta = 1$. Under this condition, the damping factor is simply $\alpha = \omega_n$. Thus, we can regard ζ as the damping ratio, that is,

$$\zeta = \text{damping ratio} = \frac{\alpha}{\omega_n} = \frac{\text{actual damping factor}}{\text{damping factor at critical damping}} \quad (7\text{-}100)$$

7-6-2 Natural Undamped Frequency

The parameter ω_n is defined as the **natural undamped frequency**. As seen from Eq. (7-97), when $\zeta = 0$, the damping is zero, the roots of the characteristic equation are imaginary, and Eq. (7-96) shows that the unit-step response is purely sinusoidal. Therefore, ω_n corresponds to the frequency of the undamped sinusoidal response. Equation (7-97) shows that when $0 < \zeta < 1$, the imaginary part of the roots has the magnitude of ω. Since when $\zeta \neq 0$, the response of $y(t)$ is not a periodic function, and ω defined in Eq. (7-99) is not a frequency. For the purpose of reference, ω is sometimes defined as the **conditional frequency**, or the **damped frequency**.

Figure 7-16 illustrates the relationships between the location of the characteristic equation roots and α, ζ, ω_n, and ω. For the complex-conjugate roots shown,

- ω_n is the radial distance from the roots to the origin of the s-plane.
- α is the real part of the roots.
- ω is the imaginary part of the roots.
- ζ is the cosine of the angle between the radial line to the roots and the negative axis when the roots are in the left-half s-plane, or

$$\zeta = \cos\theta \quad (7\text{-}101)$$

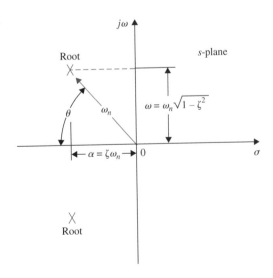

Figure 7-16 Relationship between the characteristic-equation roots of the prototype second-order system and α, ζ, ω_n, and ω.

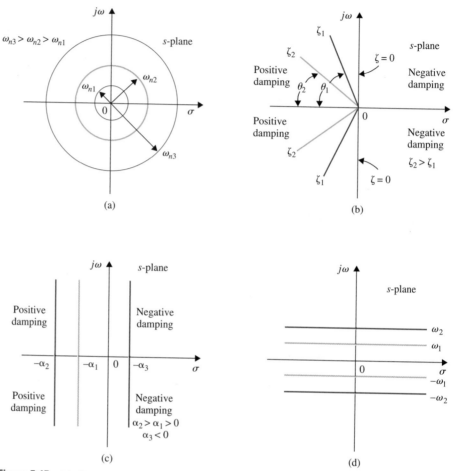

Figure 7-17 (a) Constant-natural-undamped-frequency loci. (b) Constant-damping-ratio loci. (c) Constant-damping-factor loci. (d) Constant-conditional-frequency loci.

Figure 7-17 shows in the s-plane, (a) the constant-ω_n loci, (b) the constant-ζ loci, (c) the constant-α loci, and (d) the constant-ω loci. Regions in the s-plane are identified with the system damping as follows:

- The left-half s-plane corresponds to positive damping, that is, the damping factor or damping ratio is positive. Positive damping causes the unit-step response to settle to a constant final value in the steady state due to the negative exponent of $\exp(-\zeta\omega_n t)$. The system is stable.

- The right-half s-plane corresponds to negative damping. Negative damping gives a response that grows in magnitude without bound with time, and the system is unstable.

- The imaginary axis corresponds to zero damping ($\alpha = 0$ or $\zeta = 0$). Zero damping results in a sustained oscillation response, and the system is marginally stable or marginally unstable.

Thus, we have demonstrated with the help of the simple prototype second-order system that the location of the characteristic equation roots plays an important role in the transient response of the system.

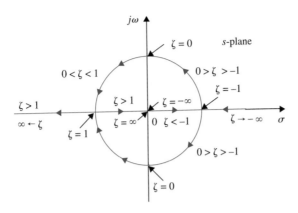

Figure 7-18 Locus of roots of the characteristic equation of the prototype second-order system, Eq. (7-94), when ω_n is held constant while the damping ratio is varied from $-\infty$ to ∞.

The effect of the characteristic equation roots on the damping of the second-order system is further illustrated by Figs. 7-18 and 7-19. In Fig. 7-18, ω_n is held constant while the damping ratio ζ is varied from $-\infty$ to $+\infty$. The following classification of the system dynamics with respect to the value of ζ is made:

$$0 < \zeta < 1: \quad s_1, s_2 = -\zeta\omega_n \pm j\omega_n\sqrt{1 - \zeta^2} \quad (-\zeta\omega_n < 0) \qquad \textit{underdamped}$$
$$\zeta = 1: \quad s_1, s_2 = -\omega_n \qquad\qquad\qquad\qquad\qquad \textit{critically damped}$$
$$\zeta > 1: \quad s_1, s_2 = -\zeta\omega_n \pm \omega_n\sqrt{\zeta^2 - 1} \qquad\qquad \textit{overdamped}$$
$$\zeta = 0: \quad s_1, s_2 = \pm j\omega_n \qquad\qquad\qquad\qquad\quad \textit{undamped}$$
$$\zeta < 0: \quad s_1, s_2 = -\zeta\omega_n \pm j\omega_n\sqrt{1 - \zeta^2} \quad (-\zeta\omega_n > 0) \qquad \textit{negatively damped}$$

Figure 7-19 illustrates typical unit-step responses that correspond to the various root locations already shown.

In practical applications, only stable systems that correspond to $\zeta > 0$ are of interest. Figure 7-15 gives the unit-step responses of Eq. (7-96) plotted as functions of the normalized time $\omega_n t$ for various values of the damping ratio ζ. As seen, the response becomes more oscillatory as ζ decreases in value. When $\zeta \geq 1$, the step response does not exhibit any overshoot; that is, $y(t)$ never exceeds its final value during the transient.

7-6-3 Maximum Overshoot

The exact relation between the damping ratio and the amount of overshoot can be obtained by taking the derivative of Eq. (7-96) with respect to t and setting the result to zero. Thus,

$$\frac{dy(t)}{dt} = \frac{\omega_n e^{-\zeta\omega_n t}}{\sqrt{1 - \zeta^2}}\left[\zeta \sin(\omega t + \theta) - \sqrt{1 - \zeta^2}\cos(\omega t + \theta)\right] \qquad t \geq 0 \quad (7\text{-}102)$$

• Maximum overshoot is defined only for a step function input.

where ω and θ are defined in Eqs. (7-99) and (7-101), respectively. We can show that the quantity inside the square bracket in Eq. (7-102) can be reduced to $\sin \omega t$. Thus, Eq. (7-102) is simplified to

$$\frac{dy(t)}{dt} = \frac{\omega_n}{\sqrt{1 - \zeta^2}}e^{-\zeta\omega_n t}\sin \omega_n\sqrt{1 - \zeta^2}t \qquad t \geq 0 \qquad\qquad (7\text{-}103)$$

Setting $dy(t)/dt$ to zero, we have the solutions: $t = \infty$, and

$$\omega_n\sqrt{1 - \zeta^2}t = n\pi \qquad n = 0, 1, 2, \ldots \qquad\qquad (7\text{-}104)$$

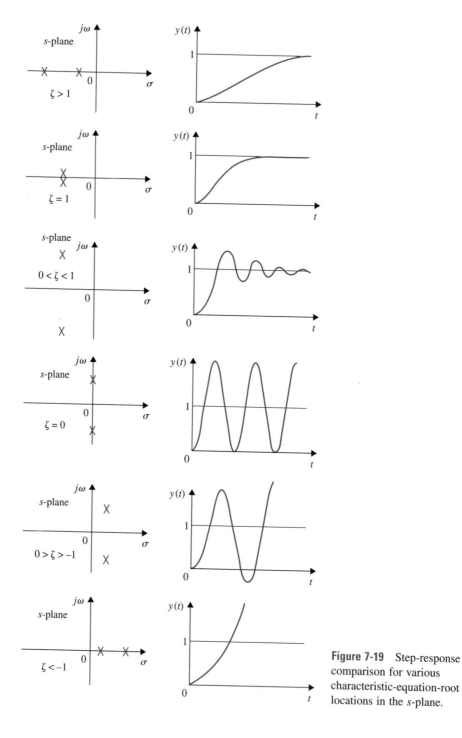

Figure 7-19 Step-response comparison for various characteristic-equation-root locations in the s-plane.

from which we get

$$t = \frac{n\pi}{\omega_n\sqrt{1 - \zeta^2}} \qquad n = 0, 1, 2, \ldots \qquad (7\text{-}105)$$

The solution at $t = \infty$ is the maximum of $y(t)$ only when $\zeta \geq 1$. For the unit-step responses shown in Fig. 7-14, the first overshoot is the maximum overshoot. This corresponds

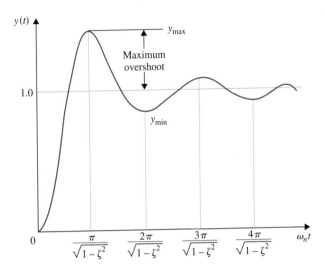

Figure 7-20 Unit-step response illustrating that the maxima and minima occur at periodic intervals.

to $n = 1$ in Eq. (7-105). Thus, the time at which the maximum overshoot occurs is

$$t_{max} = \frac{\pi}{\omega_n \sqrt{1 - \zeta^2}} \qquad (7\text{-}106)$$

With reference to Fig. 7-14, the overshoots occur at odd values of n, that is, $n = 1$, $3, 5, \ldots$, and the undershoots occur at even values of n. Whether the extremum is an overshoot or an undershoot, the time at which it occurs is given by Eq. (7-105). It should be noted that *although the unit-step response for $\zeta \neq 0$ is not periodic, the overshoots and the undershoots of the response do occur at periodic intervals, as shown in Fig. 7-20.*

The magnitudes of the overshoots and the undershoots can be determined by substituting Eq. (7-105) into Eq. (7-96). The result is

$$y(t)\Big|_{max\ or\ min} = 1 - \frac{e^{-n\pi\zeta/\sqrt{1-\zeta^2}}}{\sqrt{1 - \zeta^2}} \sin(n\pi + \theta) \qquad n = 1, 2, \ldots \qquad (7\text{-}107)$$

or

$$y(t)\Big|_{max\ or\ min} = 1 + (-1)^{n-1} e^{-n\pi\zeta/\sqrt{1-\zeta^2}} \qquad n = 1, 2, \ldots \qquad (7\text{-}108)$$

The maximum overshoot is obtained by letting $n = 1$ in Eq. (7-108). Therefore,

$$\text{maximum overshoot} = y_{max} - 1 = e^{-\pi\zeta/\sqrt{1-\zeta^2}} \qquad (7\text{-}109)$$

and

$$\text{percent maximum overshoot} = 100e^{-\pi\zeta/\sqrt{1-\zeta^2}} \qquad (7\text{-}110)$$

• The maximum overshoot of the prototype second-order system depends only on ζ.

Equation (7-109) shows that the maximum overshoot of the step response of the prototype second-order system is a function of only the damping ratio ζ. The relationship between the percent maximum overshoot and the damping ratio given in Eq. (7-110) is plotted in Fig. 7-21. The time t_{max} in Eq. (7-106) is a function of both ζ and ω_n.

7-6-4 Delay Time and Rise Time

It is more difficult to determine the exact analytical expressions of the delay time t_d, rise time t_r, and settling time t_s, even for just the simple prototype second-order system. For

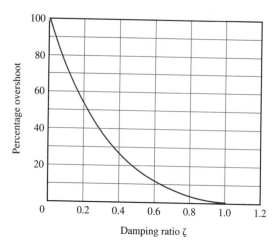

Figure 7-21 Percent overshoot as a function of damping ratio for the step response of the prototype second-order system.

instance, for the delay time, we would have to set $y(t) = 0.5$ in Eq. (7-96) and solve for t. An easier way would be to plot $\omega_n t_d$ versus ζ, as shown in Fig. 7-22, and then approximate the curve by a straight line or a curve over the range of $0 < \zeta < 1$. From Fig. 7-22, the delay time for the prototype second-order system is approximated as

$$t_d \cong \frac{1 + 0.7\zeta}{\omega_n} \qquad 0 < \zeta < 1.0 \qquad (7\text{-}111)$$

We can obtain a better approximation by using a second-order equation for t_d:

$$t_d \cong \frac{1.1 + 0.125\zeta + 0.469\zeta^2}{\omega_n} \qquad 0 < \zeta < 1.0 \qquad (7\text{-}112)$$

For the rise time t_r, which is the time for the step response to reach from 10 to 90 percent of its final value, the exact value can be determined directly from the responses of Fig. 7-15. The plot of $\omega_n t_r$ versus ζ is shown in Fig. 7-23. In this case, the relation can again

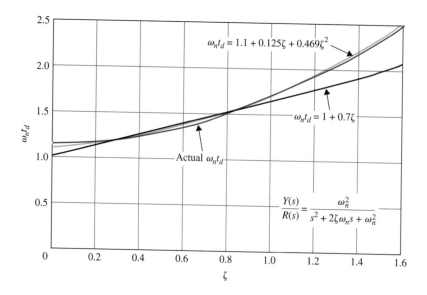

Figure 7-22 Normalized delay time versus ζ for the prototype second-order system.

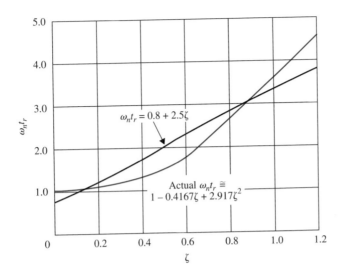

Figure 7-23 Normalized rise time versus ζ for the prototype second-order system.

be approximated by a straight line over a limited range of ζ:

$$t_r = \frac{0.8 + 2.5\zeta}{\omega_n} \qquad 0 < \zeta < 1 \tag{7-113}$$

A better approximation can be obtained by using a second-order equation:

$$t_r = \frac{1 - 0.4167\zeta + 2.917\zeta^2}{\omega_n} \qquad 0 < \zeta < 1 \tag{7-114}$$

From this discussion, the following conclusions can be made on the rise time and delay time of the prototype second-order system:

- t_r and t_d are proportional to ζ and inversely proportional to ω_n.
- Increasing (decreasing) the natural undamped frequency ω_n will reduce (increase) t_r and t_d.

7-6-5 Settling Time

From Fig. 7-15 we see that when $0 < \zeta < 0.69$, the unit-step response has a maximum overshoot greater than 5 percent, and the response can enter the band between 0.95 and 1.05 for the last time from either the top or the bottom. When ζ is greater than 0.69, the overshoot is less than 5 percent, and the response can enter the band between 0.95 and 1.05 only from the bottom. Figures 7-24(a) and (b) show the two different situations. Thus, the settling time has a discontinuity at $\zeta = 0.69$. The exact analytical description of the settling time t_s is difficult to obtain. We can obtain an approximation for t_s for $0 < \zeta < 0.69$ by using the envelope of the damped sinusoid of $y(t)$, as shown in Fig. 7-24(a) for a 5-percent requirement. In general, when the settling time corresponds to an intersection with the upper envelope of $y(t)$, the following relation is obtained:

$$1 + \frac{1}{\sqrt{1 - \zeta^2}} e^{-\zeta\omega_n t_s} = \text{upper bound of unit-step response} \tag{7-115}$$

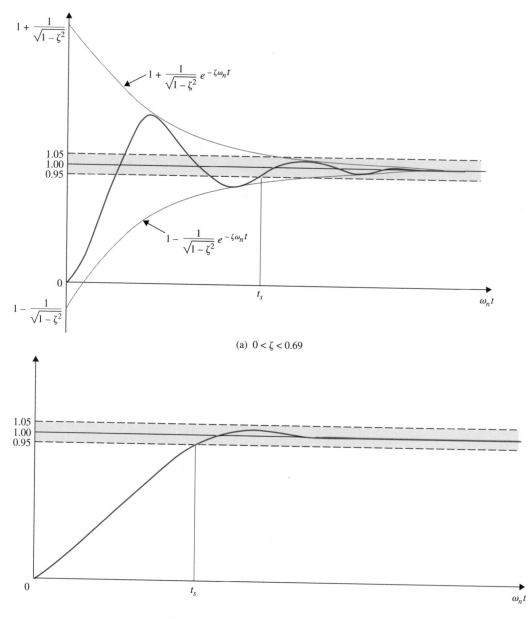

Figure 7-24 Settling time of the unit-step response.

When the settling time corresponds to an intersection with the bottom envelope of $y(t)$, t_s must satisfy the following condition:

$$1 - \frac{1}{\sqrt{1 - \zeta^2}} e^{-\zeta \omega_n t_s} = \text{lower-bound of unit-step response} \qquad (7\text{-}116)$$

For the 5-percent requirement on settling time, the right-hand side of Eq. (7-115) would be 1.05, and that of Eq. (7-116) would be 0.95. It is easily verified that the same result for t_s is obtained using either Eq. (7-115) or Eq. (7-116).

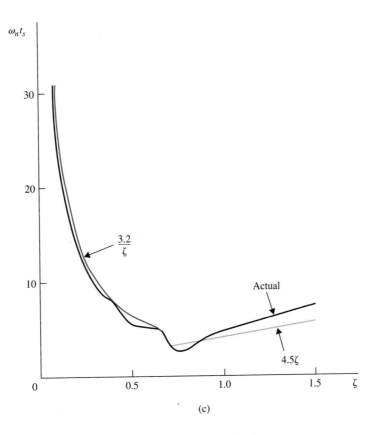

Figure 7-24 (*continued*)

(c)

Solving Eq. (7-115) for $\omega_n t_s$, we have

$$\omega_n t_s = -\frac{1}{\zeta} \ln\left(c_{ts}\sqrt{1 - \zeta^2}\right) \qquad (7\text{-}117)$$

where c_{ts} is the percentage set for the settling time. For example, if the threshold is 5 percent, the $c_{ts} = 0.05$. Thus, for a 5-percent settling time, the right-hand side of Eq. (7-117) varies between 3.0 and 3.32 as ζ varies from 0 to 0.69. We can approximate the settling time for the prototype second-order system as

$$t_s \cong \frac{3.2}{\zeta\omega_n} \qquad 0 < \zeta < 0.69 \qquad (7\text{-}118)$$

The approximation will be poor for small values of $\zeta(< 0.3)$.

When the damping ratio ζ is greater than 0.69, the unit-step response will always enter the band between 0.95 and 1.05 from below. We can show by observing the responses in Fig. 7-15 that the value of $\omega_n t_s$ is almost directly proportional to ζ. The following approximation is used for t_s for $\zeta > 0.69$.

$$t_s = \frac{4.5\zeta}{\omega_n} \qquad \zeta > 0.69 \qquad (7\text{-}119)$$

Figure 7-24(c) shows the actual values of $\omega_n t_s$ versus ζ for the prototype second-order system described by Eq. (7-93), along with the approximations using Eqs. (7-118) and (7-119) for their respective effective ranges. The numerical values are tabulated in Table 7-2.

TABLE 7-2   Comparison of Settling Times of Prototype Second-Order System, $\omega_n t_s$

ζ	Actual	$\dfrac{3.2}{\zeta}$	4.5ζ
0.10	28.7	30.2	
0.20	13.7	16.0	
0.30	10.0	10.7	
0.40	7.5	8.0	
0.50	5.2	6.4	
0.60	5.2	5.3	
0.62	5.16	5.16	
0.64	5.00	5.00	
0.65	5.03	4.92	
0.68	4.71	4.71	
0.69	4.35	4.64	
0.70	2.86		3.15
0.80	3.33		3.60
0.90	4.00		4.05
1.00	4.73		4.50
1.10	5.50		4.95
1.20	6.21		5.40
1.50	8.20		6.75

We can summarize the relationships between t_s and the system parameters as follows:

- For $\zeta < 0.69$, the settling time is inversely proportional to ζ and ω_n. A practical way of reducing the settling time is to increase ω_n while holding ζ constant. Although the response will be more oscillatory, the maximum overshoot depends only on ζ, and can be controlled independently.

- For $\zeta > 0.69$, the settling time is proportional to ζ and inversely proportional to ω_n. Again, t_s can be reduced by increasing ω_n.

It should be commented that the settling time for $\zeta > 0.69$ is truly a measure of how fast the step response rises to its final value. It seems that for this case, the rise and delay times should be adequate to describe the response behavior. As the name implies, settling time should be used to measure how fast the step response settles to its final value. It should also be pointed out that the 5-percent threshold is by no means a number cast in stone. More stringent design problems may require the system response to settle in any number less than 5 percent.

Keep in mind that while the definitions on y_{max}, t_{max}, t_d, t_r, and t_s, apply to a system of any order, the damping ratio ζ and the natural undamped frequency ω_n strictly apply only to a second-order system whose closed-loop transfer function is given in Eq. (7-93). Naturally, the relationships among t_d, t_r, and t_s and ζ and ω_n are valid only for the same second-order system model. However, these relationships can be used to measure the performance of higher-order systems that can be approximated by second-order ones, under the stipulation that some of the higher-order poles can be neglected.

▶ 7-7 TIME-DOMAIN ANALYSIS OF A POSITION-CONTROL SYSTEM

In this section we shall analyze the performance of a system using the time-domain criteria established in the preceding section. The purpose of the system considered here is to control the positions of the fins of a modern airship. Due to the requirements of improved response and reliability, the surfaces of modern aircraft are controlled by electric actuators with electronic controls. Gone are the days when the ailerons, rudder, and elevators of the aircraft were all linked to the cockpit through mechanical linkages. The so-called fly-by-wire control system used in modern aircraft implies that the attitude of aircrafts is no longer controlled entirely be mechanical linkages. Figure 7-25 illustrates the controlled surfaces and the block diagram of one axis of such a position-control system. Figure 7-26 shows the analytical block diagram of the system using the dc-motor model given in Fig. 4-49. The system is simplified to the extent that saturation of the amplifier gain and motor torque, gear backlash, and shaft compliances have all been neglected. (When you get into the real world, some of these nonlinear effects should be incorporated into the mathematical model to come up with a better controller design that works in reality. The reader should refer to Chapter 11 on the Virtual Lab, where these topics are discussed in more detail.)

The objective of the system is to have the output of the system, $\theta_y(t)$, follow the input $\theta_r(t)$. The following system parameters are given initially:

Gain or encoder	$K_s = 1$ V/rad
Gain of preamplifier	$K = $ adjustable
Gain of power amplifier	$K_1 = 10$ V/V
Gain of current feedback	$K_2 = 0.5$ V/A
Gain of tachometer feedback	$K_t = 0$ V/rad/sec
Armature resistance of motor	$R_a = 5.0$ Ω
Armature inductance of motor	$L_a = 0.003$ H
Torque constant of motor	$K_i = 9.0$ oz-in./A
Back-emf constant of motor	$K_b = 0.0636$ V/rad/sec
Inertia of motor rotor	$J_m = 0.0001$ oz-in.-sec^2
Inertia of load	$J_L = 0.01$ oz-in.-sec^2

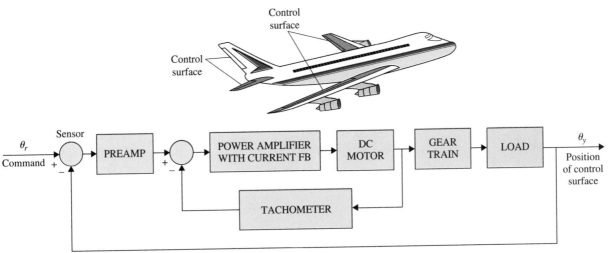

Figure 7-25 Block diagram of an attitude-control system of an aircraft.

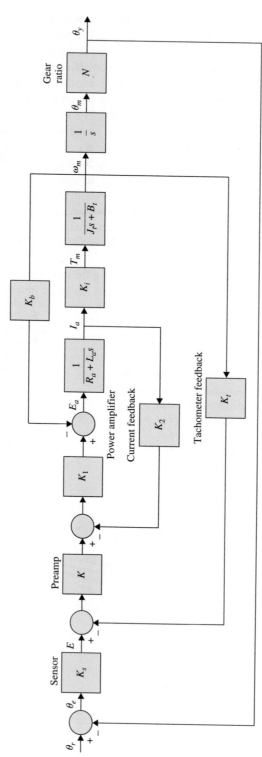

Figure 7-26 Transfer-function block diagram of the system shown in Fig. 7-25.

Viscous-friction coefficient of motor $B_m = 0.005$ oz-in.-sec
Viscous-friction coefficient of load $B_L = 1.0$ oz-in.-sec
Gear-train ratio between motor and load $N = \theta_y/\theta_m = 1/10$

Since the motor shaft is coupled to the load through a gear train with a gear ratio of N, $\theta_y = N\theta_m$, the total inertia and viscous-friction coefficient seen by the motor are

$$J_t = J_m + N^2 J_L = 0.0001 + 0.01/100 = 0.0002 \quad \text{oz-in.-sec}^2 \qquad (7\text{-}120)$$
$$B_t = B_m + N^2 B_L = 0.005 + 1/100 = 0.015 \quad \text{oz-in.-sec} \qquad (7\text{-}121)$$

respectively. The forward-path transfer function of the unity-feedback system is written from Fig. 7-26 by applying the SFG gain formula:

$$
\begin{aligned}
G(s) &= \frac{\Theta_y(s)}{\Theta_e(s)} \\
&= \frac{K_s K_1 K_i KN}{s[L_a J_t s^2 + (R_a J_t + L_a B_t + K_1 K_2 J_t)s + R_a B_t + K_1 K_2 B_t + K_i K_b + KK_1 K_t K_i]}
\end{aligned}
\qquad (7\text{-}122)
$$

The system is of the third order, since the highest-order term in $G(s)$ is s^3. The electrical time constant of the amplifier-motor system is

$$\tau_a = \frac{L_a}{R_a + K_1 K_2} = \frac{0.003}{5 + 5} = 0.0003 \quad \text{sec} \qquad (7\text{-}123)$$

The mechanical time constant of the motor-load system is

$$\tau_t = \frac{J_t}{B_t} = \frac{0.0002}{0.015} = 0.01333 \quad \text{sec} \qquad (7\text{-}124)$$

Since the electrical time constant is much smaller than the mechanical time constant, on account of the low inductance of the motor, we can perform an initial approximation by neglecting the armature inductance L_a. The result is a second-order approximation of the third-order system. Later we will show that this is not the best way of approximating a high-order system by a low-order one. The forward-path transfer function is now

$$
\begin{aligned}
G(s) &= \frac{K_s K_1 K_i KN}{s[(R_a J_t + K_1 K_2 J_t)s + R_a B_t + K_1 K_2 B_t + K_i K_b + KK_1 K_i K_t]} \\
&= \frac{\dfrac{K_s K_1 K_i KN}{R_a J_t + K_1 K_2 J_t}}{s\left(s + \dfrac{R_a B_t + K_1 K_2 B_t + K_i K_b + KK_1 K_i K_t}{R_a J_t + K_1 K_2 J_t}\right)}
\end{aligned}
\qquad (7\text{-}125)
$$

Substituting the system parameters in the last equation, we get

$$G(s) = \frac{4500K}{s(s + 361.2)} \qquad (7\text{-}126)$$

Comparing Eq. (7-125) and (7-126) with the prototype second-order transfer function of Eq. (7-92), we have

$$\text{natural undamped frequency } \omega_n = \pm\sqrt{\frac{K_s K_1 K_i KN}{R_a J_t + K_1 K_2 J_t}} = \pm\sqrt{4500K} \quad \text{rad/sec} \qquad (7\text{-}127)$$

$$\text{damping ratio } \zeta = \frac{R_a B_t + K_1 K_2 B_t + K_i K_b + KK_1 K_i K_t}{2\sqrt{K_s K_1 K_i KN \,(R_a J_t + K_1 K_2 J_t)}} = \frac{2.692}{\sqrt{K}} \qquad (7\text{-}128)$$

Thus we see that the natural undamped frequency ω_n is proportional to the square root of the amplifier gain K, whereas the damping ratio ζ is inversely proportional to $\sqrt{K}$.

The characteristic equation of the unity-feedback control system is

$$s^2 + 361.2s + 4500K = 0 \tag{7-129}$$

7-7-1 Unit-Step Transient Response

For time-domain analysis, it is informative to analyze the system performance by applying the unit-step input with zero initial conditions. In this way, it is possible to characterize the system performance in terms of the maximum overshoot and some of the other measures, such as rise time, delay time, and settling time, if necessary.

Let the reference input be a unit-step function $\theta_r(t) = u_s(t)$ rad; then $\Theta(s) = 1/s$. The output of the system, with zero initial conditions, is

$$\theta_y(t) = \mathcal{L}^{-1}\left[\frac{4500K}{s(s^2 + 361.2s + 4500K)}\right] \tag{7-130}$$

The inverse Laplace transform of the right-hand side of the last equation is carried out using the Laplace transform table in Appendix D, or using Eq. (7-96) directly. The following results are obtained for the three values of K indicated.

$K = 7.248(\zeta \cong 1.0)$:

$$\theta_y(t) = (1 - 151e^{-180t} + 150e^{-181.2t})u_s(t) \tag{7-131}$$

$K = 14.5(\zeta = 0.707)$:

$$\theta_y(t) = (1 - e^{-180.6t}\cos 180.6t - 0.9997e^{-180.6t}\sin 180.6t)u_s(t) \tag{7-132}$$

$K = 181.17(\zeta = 0.2)$:

$$\theta_y(t) = (1 - e^{-180.6t}\cos 884.7t - 0.2041e^{-180.6t}\sin 884.7t)u_s(t) \tag{7-133}$$

The three responses are plotted as shown in Fig. 7-27. Table 7-3 gives the comparison of the characteristics of the three unit-step responses for the three values of K used. When $K = 181.17$, $\zeta = 0.2$, the system is lightly damped, and the maximum overshoot is 52.7 percent, which is excessive. When the value of K is set at 7.248, ζ is very close to 1.0, and the system is almost critically damped. The unit-step response does not have any overshoot or oscillation. When K is set at 14.5, the damping ratio is 0.707, and the overshoot is 4.3 percent. It should be pointed out that in practice, it would be time consuming, even with the aid of a computer, to compute the time response for each change of a system parameter for either analysis or design purposes. Indeed, one of the main objectives of studying

TABLE 7-3   Comparison of the Performance of the Second-Order Position-Control System with the Gain K Varies

Gain K	ζ	ω_n (rad/sec)	% Max overshoot	t_d (sec)	t_r (sec)	t_s (sec)	t_{max} (sec)
7.24808	1.000	180.62	0	0.00929	0.0186	0.0259	—
14.50	0.707	255.44	4.3	0.00560	0.0084	0.0114	0.01735
181.17	0.200	903.00	52.2	0.00125	0.00136	0.0150	0.00369

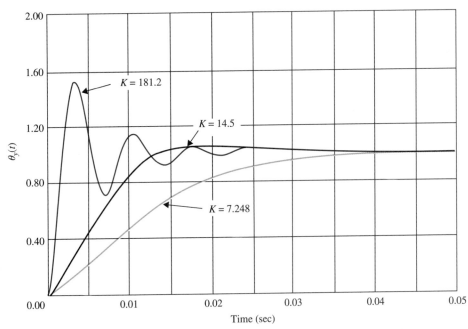

Figure 7-27   Unit-step responses of the attitude-control system in Fig. 7-26; $L_a = 0$.

control systems theory, using either the conventional or modern approach, is to establish methods so that the total reliance on computer simulation can be reduced. The motivation behind this discussion is to show that the performance of some control systems can be predicted by investigating the roots of the characteristic equation of the sytem. For the characteristic equation of Eq. (7-129), the roots are

$$s_1 = -180.6 + \sqrt{32616 - 4500K} \tag{7-134}$$
$$s_2 = -180.6 - \sqrt{32616 - 4500K} \tag{7-135}$$

For $K = 7.24808$, 14.5, and 181.2, the roots of the characteristic equation are tabulated as follows:

$K = 7.24808$: $s_1 = s_2 = -180.6$

$K = 14.5$: $s_1 = -180.6 + j180.6$ $s_2 = -180.6 - j180.6$

$K = 181.2$ $s_1 = -180.6 + j884.7$ $s_2 = -180.6 + j884.7$

These roots are marked as shown in Fig. 7-28. The trajectories of the two characteristic equation roots when K varies continuously from $-\infty$ to ∞ are also shown in Fig. 7-28. These root trajectories are called the **root loci** (see Chapter 8) of Eq. (7-129), and are used extensively for the analysis and design of linear control systems.

From Eqs. (7-134) and (7-135), we see that the two roots are real and negative for values of K between 0 and 7.24808. This means that the system is overdamped, and the step response will have no overshoot for this range of K. For values of K greater than 7.24808, the natural undamped frequency will increase with $\sqrt{K}$. When K is negative, one of the roots is positive, which corresponds to a time response that increases monotonically with time, and the system is unstable. The dynamic characteristics of

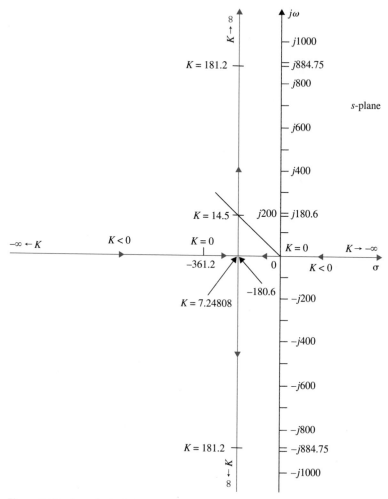

Figure 7-28 Root loci of the characteristic equation in Eq. (7-129) as K varies.

the transient step response as determined from the root loci of Fig. 7-28 are summarized as follows:

Amplifier Gain Dynamics	Characteristic Equation Roots	System
$0 < K < 7.24808$	Two negative distinct real roots	Overdamped ($\zeta > 1$)
$K = 7.24808$	Two negative equal real roots	Critically damped ($\zeta = 1$)
$7.24808 < K < \infty$	Two complex-conjugate roots with negative real parts	Underdamped ($\zeta < 1$)
$-\infty < K < 0$	Two distinct real roots, one positive and one negative	Unstable system ($\zeta < 0$)

7-7-2   The Steady-State Response

Since the forward-path transfer function in Eq. (7-126) has a simple pole at $s = 0$, the system is of type 1. This means that the steady-state error of the system is zero for all positive values of K when the input is a step function. Substituting Eq. (7-126) into Eq. (7-24), the step-error constant is

$$K_p = \lim_{s \to 0} \frac{4500K}{s(s + 361.2)} = \infty \tag{7-136}$$

Thus, the steady-state error of the system due to a step input, as given by Eq. (7-25), is zero. The unit-step responses in Fig. 7-27 verify this result. The zero-steady-state condition is achieved because only viscous friction is considered in the simplified system model. In the practical case, Coulomb friction is almost always present, so the steady-state positioning accuracy of the system can never be perfect.

7-7-3   Time Response to a Unit-Ramp Input

The control of position may be affected by the control of the profile of the output, rather than just by applying a step input. In other words, the system may be designed to follow a reference profile that represents the desired trajectory. It may be necessary to investigate the ability of the position-control system to follow a ramp-function input.

For a unit-ramp input, $\theta_r(t) = tu_s(t)$. The output response of the system in Fig. 7-26 is

$$\theta_y(t) = \mathcal{L}^{-1} \left[\frac{4500K}{s^2(s^2 + 361.2s + 4500K)} \right] \tag{7-137}$$

which can be solved by using the Laplace-transform table in Appendix D. The result is

$$\theta_y(t) = \left[t - \frac{2\zeta}{\omega_n} + \frac{e^{-\zeta\omega_n t}}{\omega_n\sqrt{1 - \zeta^2}} \sin(\omega_n\sqrt{1 - \zeta^2}t + \theta) \right] u_s(t) \tag{7-138}$$

where

$$\theta = \cos^{-1}(2\zeta^2 - 1) \qquad (\zeta < 1) \tag{7-139}$$

The values of ζ and ω_n are given in Eqs. (7-128) and (7-127), respectively. The ramp responses of the system for the three values of K are presented in the following equations.

$K = 7.248$:

$$\theta_y(t) = (t - 0.01107 - 0.8278e^{-181.2t} + 0.8389e^{-180t})u_s(t) \tag{7-140}$$

$K = 14.5$

$$\theta_y(t) = (t - 0.005536 + 0.005536e^{-180.6t}\cos 180.6t$$
$$-5.467 \times 10^{-7}e^{-180.6t}\sin 180.6t)u_s(t) \tag{7-141}$$

$K = 181.2$

$$\theta_y(t) = (t - 0.000443 + 0.000443e^{-180.6t}\cos 884.7t$$
$$-0.00104e^{-180.6t}\sin 884.7t)u_s(t) \tag{7-142}$$

These ramp responses are plotted as shown in Fig. 7-29. Notice that the steady-state error of the ramp response is not zero. The last term in Eq. (7-138) is the transient response. The steady-state portion of the unit-ramp response is

$$\lim_{t \to \infty} \theta_y(t) = \lim_{t \to \infty} \left[\left(t - \frac{2\zeta}{\omega_n} \right) u_s(t) \right] \tag{7-143}$$

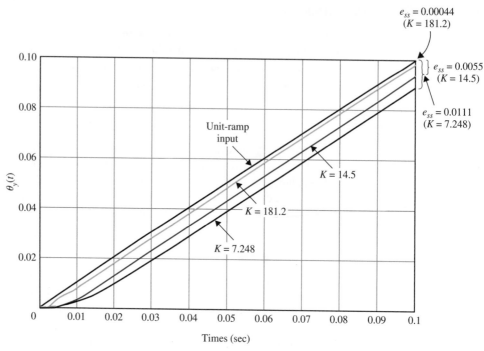

Figure 7-29 Unit-ramp responses of the attitude-control system in Fig. 7-26; $L_a = 0$.

Thus, the steady-state error of the system due to a unit-ramp input is

$$e_{ss} = \frac{2\zeta}{\omega_n} = \frac{0.0803}{K} \tag{7-144}$$

which is a constant.

A more direct method of determining the steady-state error due to a ramp input is to use the ramp-error constant K_v. From Eq. (7-31),

$$K_v = \lim_{s \to 0} sG(s) = \lim_{s \to 0} \frac{4500K}{s + 361.2} = 12.46K \tag{7-145}$$

Thus, the steady-state error is

$$e_{ss} = \frac{1}{K_v} = \frac{0.0803}{K} \tag{7-146}$$

which agrees with the result in Eq. (7-144).

The result in Eq. (7-146) shows that the steady-state error is inversely proportional to K. For $K = 14.5$, which corresponds to a damping ratio of 0.707, the steady-state error is 0.0055 rad, or more appropriately, 0.55 percent of the ramp-input magnitude. Apparently, if we attempt to improve the steady-state accuracy of the system due to ramp inputs by increasing the value of K, the transient step response will become more oscillatory and have a higher overshoot. This phenomenon is rather typical in all control systems. For higher-order systems, if the loop gain of the system is too high, the system can become unstable. Thus, by using the controller in the system loop, the transient and the steady-state error can be improved simultaneously.

7-7-4 Time Response of a Third-Order System

• All second-order systems with positive coefficients in the characteristic equations are stable.

In the preceding section, we have shown that the prototype second-order system, obtained by neglecting the armature inductance, is always stable for all positive values of K. It is not difficult to prove that, in general, all second-order systems with positive coefficients in the characteristic equations are stable.

Let us investigate the performance of the position-control system with the armature inductance $L_a = 0.003$ H. The forward-path transfer function of Eq. (7-122) becomes

$$G(s) = \frac{1.5 \times 10^7 K}{s(s^2 + 3408.3s + 1,204,000)}$$

$$= \frac{1.5 \times 10^7 K}{s(s + 400.26)(s + 3008)} \qquad (7\text{-}147)$$

The closed-loop transfer function is

$$\frac{\Theta_y(s)}{\Theta_r(s)} = \frac{1.5 \times 10^7 K}{s^3 + 3408.3s^2 + 1,204,000s + 1.5 \times 10^7 K} \qquad (7\text{-}148)$$

The system is now of the third order, and the characteristic equation is

$$s^3 + 3408.3s^2 + 1,204,000s + 1.5 \times 10^7 K = 0 \qquad (7\text{-}149)$$

Transient Response

The roots of the characteristic equation are tabulated for the three values of K used earlier for the second-order system:

$K = 7.248$:	$s_1 = -156.21$	$s_2 = -230.33$	$s_3 = -3021.8$
$K = 14.5$	$s_1 = -186.53 + j192$	$s_2 = -186.53 - j192$	$s_3 = -3035.2$
$K = 181.2$	$s_1 = -57.49 + j906.6$	$s_2 = -57.49 - j906.6$	$s_3 = -3293.3$

Comparing these results with those of the approximating second-order system, we see that when $K = 7.428$, the second-order system is critically damped, whereas the third-order system has three distinct real roots, and the system is slightly overdamped. The root at -3021.8 corresponds to a time constant of 0.33 millisecond, which is over 13 times faster than the next fastest time constant because of the pole at -230.33. Thus, the transient response due to the pole at -3021.8 decays rapidly, and the pole can be neglected from the transient standpoint. The output transient response is dominated by the two roots at -156.21 and -230.33. This analysis is verified by writing the transformed output response as

$$\Theta_y(s) = \frac{10.87 \times 10^7}{s(s + 156.21)(s + 230.33)(s + 3021.8)} \qquad (7\text{-}150)$$

Taking the inverse Laplace transform of the last equation, we get

$$\theta_y(t) = (1 - 3.28e^{-156.21t} + 2.28e^{-230.33t} - 0.0045e^{-3021.8t})u_s(t) \qquad (7\text{-}151)$$

• The dominant roots of a system are the roots that are closest to the $j\omega$ axis in the left-half s-plane.

The last term in Eq. (7-151), which is due to the root at -3021.8, decays to zero very rapidly. Furthermore, the magnitude of the term at $t = 0$ is very small compared to the other two transient terms. This simply demonstrates that, in general, the contribution of roots that lie relatively far to the left in the s-plane to the transient response will be small. The roots that are closer to the imaginary axis will dominate the transient response, and these are defined as the **dominant roots** of the characteristic equation or of the system.

When $K = 14.5$, the second-order system has a damping ratio of 0.707, since the real and imaginary parts of the two characteristic equation roots are identical. For the third-order system, recall that damping ratio is strictly not defined. However, since the effect on transient of the root at -3021.8 is negligible, the two roots that dominate the transient response correspond to a damping ratio of 0.697. Thus, for $K = 14.5$, the second-order approximation by setting L_a to zero is not a bad one. It should be noted, however, that the fact that the second-order approximation is justified for $K = 14.5$ does not mean that the approximation is valid for all values of K.

When $K = 181.2$, the two complex-conjugate roots of the third-order system again dominate the transient response, and the equivalent damping ratio due to the two roots is only 0.0633, which is much smaller than the value of 0.2 for the second-order system. Thus, we see that the justification and accuracy of the second-order approximation

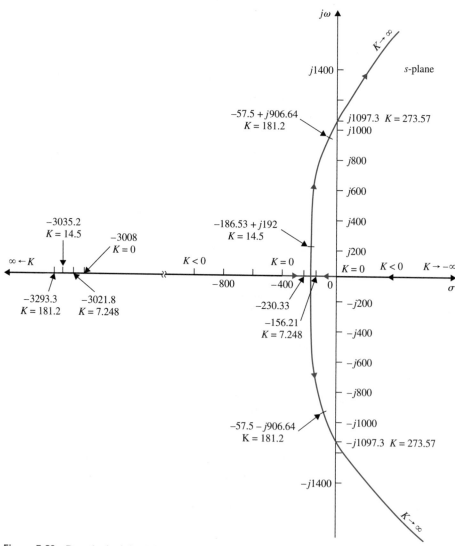

Figure 7-30 Root loci of the third-order attitude-control system.

diminish as the value of K is increased. Figure 7-28 illustrates the root loci of the third-order characteristic equation of Eq. (7-149) as K varies. When $K = 181.2$, the real root at -3293.3 still contributes little to the transient response, but the two complex-conjugate roots at $-57.49 \pm j906.6$ are much closer to the $j\omega$-axis than those of the second-order system for the same K which are at $-180.6 \pm j884.75$. This explains why the third-order system is a great deal less stable than the second-order system when $K = 181.2$.

By using the Routh-Hurwitz criterion, the marginal value of K for stability is found to be 273.57. With this critical value of K, the closed-loop transfer function becomes

$$\frac{\Theta_y(s)}{\Theta_r(s)} = \frac{1.0872 \times 10^8}{(s + 3408.3)(s^2 + 1.204 \times 10^6)} \tag{7-152}$$

The roots of the characteristic equation are at $s = -3408.3$, $-j1097.3$, and $j1097.3$. These points are shown on the root loci in Fig. 7-30.

The unit-step response of the system when $K = 273.57$ is

$$\theta_y(t) = [1 - 0.094e^{-3408.3t} - 0.952\sin(1097.3t + 72.16°)]u_s(t) \tag{7-153}$$

The steady-state response is an undamped sinusoid with a frequency of 1097.3 rad/sec, and the system is said to be marginally stable. When K is greater than 273.57, the two complex-conjugate roots will have positive real parts, the sinusoidal component of the time response will increase with time, and the system is unstable. Thus we see that the third-order system is capable of being unstable, whereas the second-order system obtained with $L_a = 0$ is stable for all finite positive values of K.

Figure 7-31 shows the unit-step responses of the third-order system for the three values of K used. The responses for $K = 7.248$ and $K = 14.5$ are very close to those of the second-order system with the same values of K that are shown in Fig. 7-27. However, the two responses for $K = 181.2$ are quite different.

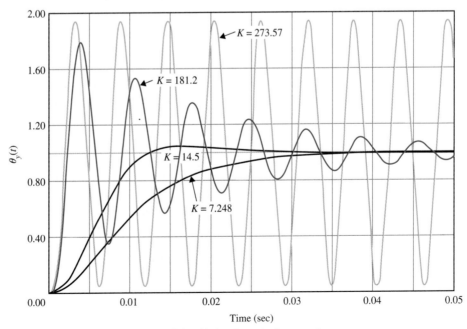

Figure 7-31 Unit-step responses of the third-order attitude-control system.

Steady-State Response

From Eq. (7-147), we see that when the inductance is restored, the third-order system is still of type 1. The value of K_v is still the same as that given in Eq. (7-145). Thus, the inductance of the motor does not affect the steady-state performance of the system, provided that the system is stable. This is expected, since L_a affects only the rate of change and not the final value of the motor current. A good engineer should always try to interpret the analytical results with the physical system.

▶ 7-8 EFFECTS OF ADDING POLES AND ZEROS TO TRANSFER FUNCTIONS

The position-control system discussed in the preceding section reveals important properties of the time responses of typical second- and third-order closed-loop systems. Specifically, the effects on the transient response relative to the location of the roots of the characteristic equation are demonstrated. However, in practice, successful design of a control system cannot depend only on choosing values of the system parameters so that the characteristic equation roots are properly placed. We shall show that, although the roots of the characteristic equation, which are the poles of the closed-loop transfer function, affect the transient response of linear time-invariant control systems, particularly the stability, the zeros of the transfer function, if there are any, are also important. Thus, the addition of poles and zeros and/or cancellation of undesirable poles and zeros of the transfer function often are necessary in achieving satisfactory time-domain performance of control systems.

In this section, we shall show that the addition of poles and zeros to loop and closed-loop transfer functions has varying effects on the transient response of the closed-loop system.

7-8-1   Addition of a Pole to the Forward-Path Transfer Function: Unity-Feedback Systems

For the position-control system described in Section 7-6, when the motor inductance is neglected, the system is of the second order, and the forward-path transfer function is of the prototype given in Eq. (7-126). When the motor inductance is restored, the system is of the third order, and the forward-path transfer function is given in Eq. (7-147). Comparing the two transfer functions of Eqs. (7-126) and (7-147), we see that the effect of the motor inductance is equivalent to adding a pole at $s = -3008$ to the forward-path transfer function of Eq. (7-126) while shifting the pole at -361.2 to -400.26, and the proportional constant is also increased. The apparent effect of adding a pole to the forward-path transfer function is that the third-order system can now become unstable if the value of the amplifier gain K exceeds 273.57. As shown by the root-locus diagrams of Figs. 7-28 and 7-30, the new pole of $G(s)$ at $s = -3008$ essentially "pushes" and "bends" the complex-conjugate portion of the root loci of the second-order system toward the right-half s-plane. Actually, because of the specific value of the inductance chosen, the additional pole of the third-order system is far to the left of the pole at -400.26, so its effect is small except when the value of K is relatively large.

To study the general effect of the addition of a pole, and its relative location, to a forward-path transfer function of a unity-feedback system, consider the transfer function

$$G(s) = \frac{\omega_n^2}{s(s + 2\zeta\omega_n)(1 + T_p s)} \qquad (7\text{-}154)$$

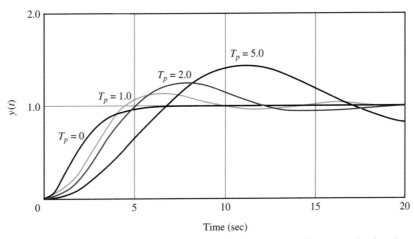

Figure 7-32 Unit-step responses of the system with the closed-loop transfer function in Eq. (7-155): $\zeta = 1$, $\omega_n = 1$, and $T_p = 0, 1, 2$, and 5.

The pole at $s = -1/T_p$ is considered to be added to the prototype second-order transfer function. The closed-loop transfer function is written

$$M(s) = \frac{Y(s)}{R(s)} = \frac{G(s)}{1 + G(s)} = \frac{\omega_n^2}{T_p s^3 + (1 + 2\zeta\omega_n T_p)s^2 + 2\zeta\omega_n s + \omega_n^2} \quad (7\text{-}155)$$

Figure 7-32 illustrates the unit-step responses of the closed-loop system when $\omega_n = 1$, $\zeta = 1$, and $T_p = 0, 1, 2$, and 5. These responses again show that the addition of a pole to the forward-path transfer function generally has the effect of increasing the maximum overshoot of the closed-loop system.

As the value of T_p increases, the pole at $-1/T_p$ moves closer to the origin in the s-plane, and the maximum overshoot increases. These responses also show that the added pole increases the rise time of the step response. This is not surprising, since the additional pole has the effect of reducing the bandwidth (see Chapter 9) of the system, thus cutting out the high-frequency components of the signal transmitted through the system.

The same conclusion can be drawn from the unit-step responses of Fig. 7-33, which are obtained with $\omega_n = 1$, $\zeta = 0.25$, and $T_p = 0, 0.2, 0.667$, and 1.0. In this case, when T_p is greater than 0.667, the amplitude of the unit-step response increases with time, and the system is unstable.

7-8-2 Addition of a Pole to the Closed-Loop Transfer Function

Since the poles of the closed-loop transfer function are roots of the characteristic equation, they control the transient response of the system directly. Consider the closed-loop transfer function

$$M(s) = \frac{Y(s)}{R(s)} = \frac{\omega_n^2}{(s^2 + 2\zeta\omega_n s + \omega_n^2)(1 + T_p s)} \quad (7\text{-}156)$$

where the term $(1 + T_p s)$ is added to a prototype second-order transfer function. Figure 7-34 illustrates the unit-step response of the system with $\omega_n = 1$, $\zeta = 0.5$, and $T_p = 0$,

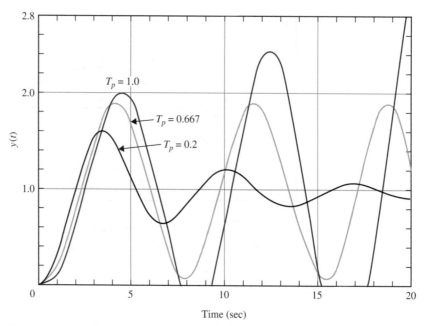

Figure 7-33 Unit-step responses of the system with the closed-loop transfer function in Eq. (7-155): $\zeta = 0.25$, $\omega_n = 1$, and $T_p = 0$, 0.2, 0.667, and 1.0.

0.5, 1, 2, and 4. As the pole at $s = -1/T_p$ is moved toward the origin in the s-plane, the rise time increases, and the maximum overshoot decreases. Thus, as far as the overshoot is concerned, adding a pole to the closed-loop transfer function has just the opposite effect to that of adding a pole to the forward-path transfer function.

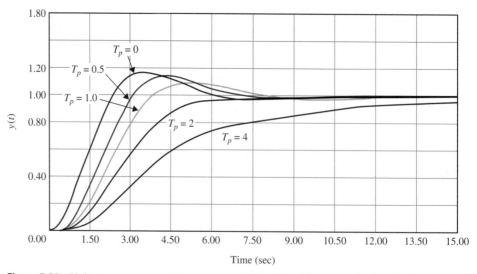

Figure 7-34 Unit-step responses of the system with the closed-loop transfer function in Eq. (7-156): $\zeta = 0.5$, $\omega_n = 1$, and $T_p = 0$, 0.5, 1.0, 2.0, and 4.0.

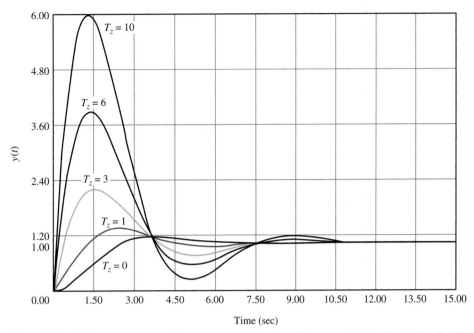

Figure 7-35 Unit-step responses of the system with the closed-loop transfer function in Eq. (7-157): $T_z = 0, 1, 2, 3, 6,$ and 10.

7-8-3 Addition of a Zero to the Closed-loop Transfer Function

Figure 7-35 shows the unit-step responses of the closed-loop system with the transfer function

$$M(s) = \frac{Y(s)}{R(s)} = \frac{\omega_n^2(1 + T_z s)}{(s^2 + 2\zeta\omega_n s + \omega_n^2)} \tag{7-157}$$

where $\omega_n = 1$, $\zeta = 0.5$, and $T_z = 0, 1, 3, 6,$ and 10. In this case, we see that adding a zero to the closed-loop transfer function decreases the rise time, and increases the maximum overshoot of the step response.

We can analyze the general case by writing Eq. (7-157) as

$$M(s) = \frac{Y(s)}{R(s)} = \frac{\omega_n^2}{s^2 + 2\zeta\omega_n s + \omega_n^2} + \frac{T_z\omega_n^2 s}{s^2 + 2\zeta\omega_n s + \omega_n^2} \tag{7-158}$$

For a unit-step input, let the output response that corresponds to the first term of the right-hand side of Eq. (7-158) be $y_1(t)$ Then, the total unit-step response is

$$y(t) = y_1(t) + T_z\frac{dy_1(t)}{dt} \tag{7-159}$$

Figure 7-36 shows why the addition of the zero at $s = -1/T_z$ reduces the rise time and increases the maximum overshoot, according to Eq. (7-159). In fact, as T_z approaches infinity, the maximum overshoot also approaches infinity, and yet the system is still stable as long as the overshoot is finite and ζ is positive.

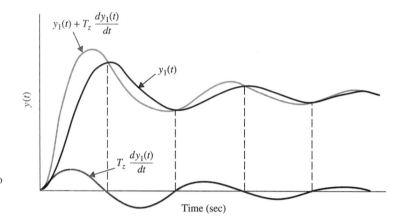

Figure 7-36 Unit-step responses showing the effect of adding a zero to the closed-loop transfer function.

7-8-4 Addition of a Zero to the Forward-Path Transfer Function: Unity-Feedback Systems

Let us consider that a zero at $-1/T_z$ is added to the forward-path transfer function of a third-order system, so

$$G(s) = \frac{6(1 + T_z s)}{s(s + 1)(s + 2)} \tag{7-160}$$

The closed-loop transfer function is

$$M(s) = \frac{Y(s)}{R(s)} = \frac{6(1 + T_z s)}{s^3 + 3s^2 + (2 + 6T_z)s + 6} \tag{7-161}$$

The difference between this case and that of adding a zero to the closed-loop transfer function is that in the present case, not only the term $(1 + T_z s)$ appears in the numerator of $M(s)$, but the denominator of $M(s)$ also contains T_z. The term $(1 + T_z s)$ in the numerator of $M(s)$ increases the maximum overshoot, but T_z appears in the coefficient of the s term in the denominator, which has the effect of improving damping, or reducing the maximum overshoot. Figure 7-37 illustrates the unit-step responses when $T_z = 0, 0.2, 0.5, 2,$

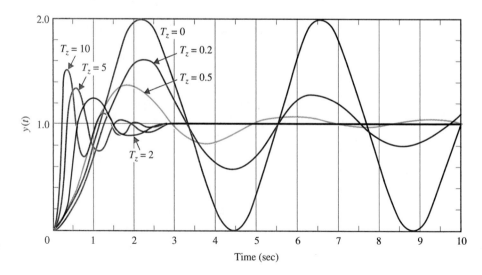

Figure 7-37 Unit-step responses of the system with the closed-loop transfer function in Eq. (7-161); $T_z = 0, 0.2, 0.5, 2, 5,$ and 10.

5, and 10. Notice that when $T_z = 0$, the closed-loop system is on the verge of becoming unstable. When $T_z = 0.2$ and 0.5, the maximum overshoots are reduced, mainly because of the improved damping. As T_z increases beyond 2, although the damping is still further improved, the $(1 + T_z s)$ term in the numerator becomes more dominant, so the maximum overshoot actually becomes greater as T_z is increased further.

An important finding from these discussions is that although the characteristic equation roots are generally used to study the relative damping and relative stability of linear control systems, the zeros of the transfer function should not be overlooked in their effects on the transient performance of the system.

► 7-9 DOMINANT POLES OF TRANSFER FUNCTIONS

From the discussions given in the preceding sections, it becomes apparent that the location of the poles of a transfer function in the s-plane greatly affects the transient response of the system. For analysis and design purposes, it is important to sort out the poles that have a dominant effect on the transient response and call these the **dominant poles**.

Since most control systems in practice are of orders higher than two, it would be useful to establish guidelines on the approximation of high-order systems by lower-order ones insofar as the transient response is concerned. In design, we can use the dominant poles to control the dynamic performance of the system, whereas the **insignificant poles** are used for the purpose of ensuring that the controller transfer function can be realized by physical components.

For all practical purposes, we can divide the s-plane into regions in which the dominant and insignificant poles can lie, as shown in Fig. 7-38. We intentionally do not assign specific values to the coordinates, since these are all relative to a given system.

The poles that are close to the imaginary axis in the left-half s-plane give rise to transient responses that will decay relatively slowly, whereas the poles that are far away from the axis (relative to the dominant poles) correspond to fast-decaying time responses. The distance D between the dominant region and the least significant region shown in Fig. 7-38 will be subject to discussion. The question is: How large a pole is considered to be really large? It has been recognized in practice and in the literature that if the magnitude of the real part of a pole is at least 5 to 10 times that of a dominant pole or a pair of complex dominant poles, then the pole may be regarded as insignificant insofar as the transient response is concerned.

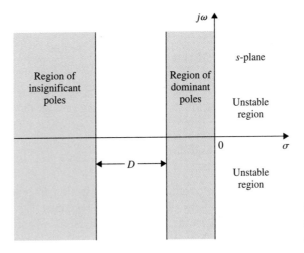

Figure 7-38 Regions of dominant and insignificant poles in the s-plane.

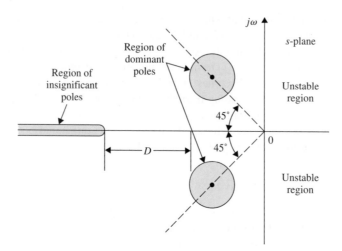

Figure 7-39 Regions of dominant and insignificant poles in the s-plane for design purposes.

We must point out that the regions shown in Fig. 7-38 are selected merely for the definitions of dominant and insignificant poles. For design purposes, such as in pole-placement design, the dominant poles and the insignificant poles should most likely be located in the tinted regions in Fig. 7-39. Again, we do not show any absolute coordinates, except that the desired region of the dominant poles is centered around the line that corresponds to $\zeta = 0.707$. It should also be noted that, while designing, we cannot place the insignificant poles arbitrarily far to the left in the s-plane or these may require unrealistic system parameter values when the pencil-and-paper design is implemented by physical components.

7-9-1 The Relative Damping Ratio

When a system is higher than the second order we can no longer strictly use the damping ratio ζ and the natural undamped frequency ω_n, which are defined for the prototype second-order systems. However, if the system dynamics can be accurately represented by a pair of complex-conjugate dominant poles, then we can still use ζ and ω_n to indicate the dynamics of the transient response, and the damping ratio in this case is referred to as the **relative damping ratio** of the system. For example, consider the closed-loop transfer function

$$M(s) = \frac{Y(s)}{R(s)} = \frac{20}{(s + 10)(s^2 + 2s + 2)} \qquad (7\text{-}162)$$

The pole at $s = -10$ is 10 times the real part of the complex conjugate poles, which are at $-1 \pm j1$. We can refer to the relative damping ratio of the system as 0.707.

7-9-2 The Proper Way of Neglecting the Insignificant Poles with Consideration of the Steady-State Response

Thus far we have provided guidelines of neglecting insignificant poles of a transfer function from the standpoint of the transient response. However, going through with the mechanics, the steady-state performance must also be considered. Let us consider the transfer function in Eq. (7-162); the pole at -10 can be neglected from the transient standpoint. To do this, we should first express Eq. (7-162) as

$$M(s) = \frac{20}{10(s/10 + 1)(s^2 + 2s + 2)} \tag{7-163}$$

Then we reason that $|s/10| \ll 1$ when the absolute value of s is much smaller than 10, because of the dominant nature of the complex poles. The term $s/10$ can be neglected when compared with 1. Then, Eq. (7-163) is approximated by

$$M(s) \cong \frac{20}{10(s^2 + 2s + 2)} \tag{7-164}$$

This way, the steady-state performance of the third-order system will not be affected by the approximation. In other words, the third-order system described by Eq. (7-162) and the second-order system approximated by Eq. (7-164) all have a final value of unity when a unit-step input is applied. On the other hand, if we simply throw away the term $(s + 10)$ in Eq. (7-162), the approximating second-order system will have a steady-state value of 5 when a unit-step input is applied.

▶ 7-10 APPROXIMATION OF HIGH-ORDER SYSTEMS BY LOW-ORDER SYSTEMS: THE FORMAL APPROACH

In the last section, we have shown that a high-order control system often contains less important poles that have little effect on the system response. Thus, given a high-order system, it is desirable to find a low-order approximating system, if possible, so that the analysis and design effort can be reduced. Given a high-order transfer function $M_H(s)$, how can we find a low-order transfer function $M_L(s)$ as an approximation, in the sense that the responses of the two systems are similar according to some prescribed criterion?

In the last section, we set up practical but nonrigorous guidelines for neglecting the poles that are far to the left in the s-plane relative to the dominant poles. However, in general, the transfer function may not have the so-called dominant poles or the latter may not be obviously defined, so a more scientific method may be necessary to arrive at a low-order equivalent.

In this section, we shall introduce a method [2] of approximating a high-order system by a low-order one in the sense that the frequency responses of the two systems are similar.

Let the high-order system transfer function be represented by

$$M_H(s) = K\frac{1 + b_1s + b_2s^2 + \cdots + b_ms^m}{1 + a_1s + a_2s^2 + \cdots + a_ns^n} \tag{7-165}$$

where $n \geq m$. Let the transfer function of the approximating low-order system be

$$M_L(s) = K\frac{1 + c_1s + c_2s^2 + \cdots + c_qs^q}{1 + d_1 + d_2s^2 + \cdots + d_ps^p} \tag{7-166}$$

where $n \geq p \geq q$. Notice that the zero-frequency ($s = 0$) gain K of the two transfer functions is the same. This will ensure that the steady-state behavior of the high-order system is preserved in the low-order system. Furthermore, we assume that the poles of $M_H(s)$ and $M_L(s)$ are all in the left-half s-plane since we are not interested in unstable systems.

The transfer functions $M_L(s)$ and $M_L(s)$ generally refer to the closed-loop transfer function, but if necessary, they can be treated as the loop transfer functions.

7-10-1 Approximation Criterion

The criterion of finding the low-order $M_L(s)$, given $M_H(s)$, is that the following relation should be satisfied as closely as possible:

$$\frac{|M_H(j\omega)|^2}{|M_L(j\omega)|^2} = 1 \qquad \text{for } 0 \leq \omega \leq \infty \qquad (7\text{-}167)$$

The last condition implies that the amplitude characteristics of the two systems in the frequency domain ($s = j\omega$) are similar. It is hoped that this will lead to similar time responses for the two systems.

The approximation procedure involves the following two steps:

1. Choose the appropriate orders of the numerator polynomial, q, and the denominator polynomial, p, of $M_L(s)$.

2. Determine the coefficients c_i, $i = 1, 2, \ldots, q$, and d_j, $j = 1, 2, \ldots p$ so that the condition in Eq. (7-167) is approached.

By using Eqs. (7-165) and (7-166), the ratio of $M_H(s)$ to $M_L(s)$ is

$$\begin{aligned}
\frac{M_H(s)}{M_L(s)} &= \frac{(1 + b_1 s + b_2 s^2 + \cdots + b_m s^m)(1 + d_1 s + d_2 s^2 + \cdots + d_p s^p)}{(1 + a_1 s + a_2 s^2 + \cdots + a_n s^n)(1 + c_1 s + c_2 s^2 + \cdots + c_q s^q)} \\
&= \frac{1 + m_1 s + m_2 s^2 + \cdots + m_u s^u}{1 + l_1 s + l_2 + \cdots + l_v s^v}
\end{aligned} \qquad (7\text{-}168)$$

where $u = m + p$ and $v = n + q$.

Equation (7-167) is written as

$$\frac{|M_H(j\omega)|^2}{|M_L(j\omega)|^2} = \frac{M_H(s)M_H(-s)}{M_L(s)M_L(s)}\bigg|_{s=j\omega} \qquad (7\text{-}169)$$

where $M_H(s)M_H(-s)$ and $M_L(s)M_L(-s)$ are even polynomials of s; that is, they contain only even powers of s. Thus, Eq. (7-169) can be written as

$$\frac{|M_H(j\omega)|^2}{|M_L(j\omega)|^2} = \frac{1 + e_2 s^2 + e_4 s^4 + \cdots + e_{2u} s^{2u}}{1 + f_2 s^2 + f_4 s^4 + \cdots + f_{2v} s^{2v}}\bigg|_{s=j\omega} \qquad (7\text{-}170)$$

Dividing the numerator by the denominator once on the right-hand side of the last equation, we have

$$\frac{|M_H(j\omega)|^2}{|M_L(j\omega)|^2} = 1 + \frac{(e_2 - f_2)s^2 + (e_4 - f_4)s^4 + \cdots}{1 + f_2 s^2 + f_4 s^4 + \cdots + f_{2v} s^{2v}}\bigg|_{s=j\omega} \qquad (7\text{-}171)$$

If $u = v$, the last term in the numerator of Eq. (7-171) will be $(e_{2u} - f_{2u})s^{2u}$. However, if $u < v$, as in most practical cases, then beyond the term $(e_{2u} - f_{2u})s^{2u}$, in addition, there will be

$$-f_{2(u+1)}s^{2(u+1)} - \cdots - f_{2v}s^{2v}$$

in the numerator of Eq. (7-171).

We see that to satisfy the condition of Eq. (7-167), one possible set of approximating solutions is obtained from Eq. (7-171):

$$e_2 = f_2$$
$$e_4 = f_4$$
$$e_6 = f_6$$
$$\vdots$$
$$e_{2u} = f_{2u}$$

(7-172)

if $u = v$. If $u < v$, then the error generated by the low-order model is

$$|\varepsilon| = \frac{|M_H(j\omega)|^2}{|M_L(j\omega)|^2} - 1 = \left. \frac{-f_{2(u+1)}s^{2(u+1)} - \cdots - f_{2v}s^{2v}}{1 + f_2 s^2 + f_4 s^4 + \cdots + f_{2v}s^{2v}} \right|_{s=j\omega}$$

(7-173)

The conditions in Eq. (7-172) are used to solve for the unknown coefficients in $M_L(s)$ once $M_H(s)$ is given. This is done by writing

$$
\frac{M_H(s)M_H(-s)}{M_L(s)M_L(-s)}
$$
$$
= \frac{[1 + m_1 s + m_2 s^2 + \cdots + m_u s^u][1 - m_1 s + m_2 s^2 - \cdots + (-1)^u m_u s^u]}{[1 + l_1 s + l_2 s^2 + \cdots + l_v s^v][1 - l_1 s + l_2 s^2 - \cdots + (-1)^v l_v s^v]}
$$
$$
= \frac{1 + e_2 s^2 + e_4 s^4 + \cdots + e_{2u} s^{2u}}{1 + f_2 s^2 + f_4 s^4 + \cdots + f_{2v} s^{2v}}
$$

(7-174)

Equating both sides of Eq. (7-174), we can express $e_2, e_4, \ldots, e_{2u}$ in terms of m_1, $m_2, \ldots, m_u$. Similar relationships can be obtained for $f_2, f_4, \ldots, f_{2v}$ in terms of $l_1, l_2, \ldots, l_v$.

As an illustrative example, for $u = 8$, Eq. (7-174) gives

$$e_2 = 2m_2 - m_1^2$$
$$e_4 = 2m_4 - 2m_1 m_3 + m_2^2$$
$$e_6 = 2m_6 - 2m_1 m_5 + 2m_2 m_4 - m_3^2$$
$$e_8 = 2m_8 - 2m_1 m_7 + 2m_2 m_6 - 2m_3 m_5 + m_4^2$$
$$e_{10} = 2m_2 m_8 - 2m_3 m_7 + 2m_4 m_6 - m_5^2$$
$$e_{12} = 2m_4 m_8 - 2m_3 m_7 + m_6^2$$
$$e_{14} = 2m_6 m_8 - m_7^2$$
$$e_{16} = m_8^2$$

(7-175)

In general,

$$e_{2x} = \sum_{i=0}^{x-1} (-1)^i 2m_i m_{2x-i} + (-1)^x m_x^2$$

(7-176)

for $x = 1, 2, \ldots, u$, and $m_0 = 1$. Similarly,

$$f_{2y} = \sum_{i=0}^{y-1} (-1)^i 2l_i l_{2y-i} + (-1)^y l_y^2$$

(7-177)

for $y = 1, 2, \ldots, v$, and $l_0 = 1$.

The following examples illustrate the method of simplification of linear systems outlined in this section.

▶ **EXAMPLE 7-9** Consider that the forward-path transfer function of a unity-feedback control system is given as

$$G(s) = \frac{8}{s(s^2 + 6s + 12)} \tag{7-178}$$

The closed-loop transfer function is

$$M_H(s) = \frac{Y(s)}{R(s)} = \frac{8}{s^3 + 6s^2 + 12s + 8}$$

$$= \frac{1}{1 + 1.5s + 0.75s^2 + 0.125s^3} \tag{7-179}$$

The poles of the closed-loop transfer function are at $s = -2$, -2, and -2. The low-order approximating system is considered for the following two cases.

Case 1

Consider that the simplified low-order system is of the second order, and the transfer function is of the form:

$$M_L(s) = \frac{1}{1 + d_1 s + d_2 s^2} \tag{7-180}$$

Equation (7-168) gives

$$\frac{M_H(s)}{M_L(s)} = \frac{1 + d_1 s + d_2 s^2}{1 + 1.5s + 0.75s^2 + 0.125s^3}$$

$$= \frac{1 + m_1 s + m_2 s^2}{1 + l_1 s + l_2 s^2 + l_3 s^3} \tag{7-181}$$

Thus,

$$l_1 = 1.5 \qquad l_2 = 0.75 \qquad l_3 = 0.125 \qquad d_1 = m_1 \qquad d_2 = m_2$$

Using Eq. (7-174), we have

$$\frac{M_H(s) M_H(-s)}{M_L(s) M_L(-s)} = \frac{1 + e_2 s^2 + e_4 s^4}{1 + f_2 s^2 + f_4 s^4 + f_6 s^6} \tag{7-182}$$

Now using Eqs. (7-172) and (7-175), the following nonlinear equations are obtained:

$$e_2 = f_2 = 2m_2 - m_1^2 = 2d_2 - d_1^2$$
$$e_4 = f_4 = 2m_4 - 2m_4 - 2m_1 m_3 = m_2^2 = d_2^2$$

Similarly,

$$f_2 = 2l_2 - l_1^2 = 1.5 - (1.5)^2 = -0.75$$
$$f_4 = 2l_4 - 2l_1 l_3 + l_2^2 = 0.1875$$
$$f_6 = 2l_6 - 2l_1 l_5 + 2l_2 l_4 - l_3^2 = -l_3^2 = -0.156$$

Solving the preceding equations, we have

$$d_1^2 = 1.616 \qquad d_2^2 = 0.1875$$

Thus,

$$d_1 = 1.271 \qquad d_2 = 0.433$$

where we have taken the positive values so that $M_L(s)$ will represent a stable system. Thus, the second-order simplified system transfer function is

$$M_H(s) = \frac{1}{1 + 1.271s + 0.433s^2} = \frac{2.31}{s^2 + 2.936s + 2.31} \tag{7-183}$$

The poles of $M_L(s)$ are at $s = -1.468 + j0.384$ and $-1.468 - j0.384$. Thus, the third-order system with the transfer function M_H, which has real poles at -2, -2, and -2, is approximated by the second-order transfer function with two complex-conjugate poles. The forward path transfer function of the second-order system is

$$G_L(s) = \frac{2.31}{s(s + 2.936)} \tag{7-184}$$

Viewed from the open-loop standpoint, the third-order system with open-loop poles at $s = 0$, $-3 + j1.732$, and $-3 - j1.732$ and a gain of 8 is approximated by a second-order system with the open-loop poles at $s = 0$ and $s = -2.936$ and a gain of 2.31.

Figure 7-40 shows the unit-step responses of the systems with the closed-loop transfer functions $M_H(s)$ and $M_L(s)$. Notice that the second-order approximating system has a faster rise time. This is due to the two complex-conjugate poles of $M_L(s)$.

Case 2

Now let us use a first-order system to approximate the closed-loop transfer function of Eq. (7-179). We choose

$$M_L(s) = \frac{1}{1 + d_1 s} \tag{7-185}$$

Then

$$\frac{M_H(s)}{M_L(s)} = \frac{1 + d_1 s}{1 + 1.5s + 0.75s^2 + 0.125s^3} = \frac{1 + m_1 s}{1 + l_1 s + l_2 s^2 + l_3 s^3} \tag{7-186}$$

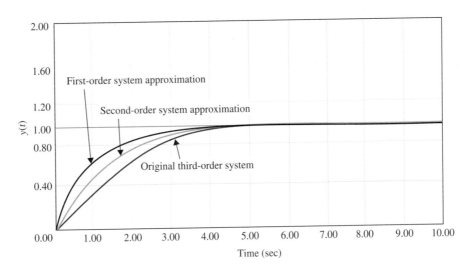

Original system: $\quad M_H(s) = \dfrac{1}{1 + 1.5s + 0.75s^2 + 0.125s^3}$

Second-order system: $\quad M_L(s) = \dfrac{1}{1 + 1.271s + 0.433s^2}$

First-order system: $\quad M_L(s) = \dfrac{1}{1 + 0.866s}$

Figure 7-40 Approximation of a third-order system by a first-order system and a second-order system.

Thus,

$$l_1 = 1.5 \qquad l_2 = 0.75 \qquad l_3 = 0.125 \qquad d_1 = m_1$$

and

$$\frac{M_H(s)\,M_H(-s)}{M_L(s)\,M_L(-s)} = \frac{1 + e_2 s^2}{1 + f_2 s^2 + f_4 s^4 + f_6 s^6} \tag{7-187}$$

Thus,

$$e_2 = f_2 = 2d_2 - d_1^2 = -d_1^2 = -0.75$$
$$f_4 = 0.1875$$
$$f_6 = -0.156$$

This gives $d_1 = 0.866$. The transfer function of $M_L(s)$ is

$$M_L(s) = \frac{1}{1 + 0.866s} = \frac{1.1547}{s + 1.1547} \tag{7-188}$$

The forward-path transfer function of the first-order system is

$$G_L(s) = \frac{1.1546}{s} \tag{7-189}$$

The unit-step response of the first-order system is shown in Fig. 7-40. As expected, the first-order system gives a more inferior approximation to the third-order system than the second-order system. ◀

▶ **EXAMPLE 7-10** Consider the following closed-loop transfer function:

$$M_H(s) = \frac{1}{(1 + s + 0.5s^2)(1 + Ts)} = \frac{1}{1 + (1 + T)s + (0.5 + T)s^2 + 0.5s^3} \tag{7-190}$$

where T is a positive variable parameter.

We shall investigate the approximation of $M_H(s)$ by a second-order model when T takes on various values.

Let the second-order system be modeled by the transfer function

$$M_L(s) = \frac{1}{1 + d_1 s + d_2 s^2} \tag{7-191}$$

Substituting Eqs. (7-190) and (7-191) into Eq. (7-168), we get

$$\frac{M_H(s)}{M_L(s)} = \frac{1 + d_1 s + d_2 s^2}{1 + (1 + T)s + (0.5 + T)s^2 + 0.5Ts^3} = \frac{1 + m_1 s + m_2 s^2}{1 + l_1 s + l_2 s^2 + l_3 s^3} \tag{7-192}$$

Then,

$$l_1 = 1 + T \qquad l_2 = 0.5 + T \qquad l_3 = 0.5T \tag{7-193}$$

From Eq. (7-176), we get

$$e_2 = f_2 = 2d_2 - d_1^2 \qquad e_4 = f_4 = m_2^2 = d_2^2 \tag{7-194}$$

From Eq. (7-177), we get

$$f_2 = 2l_2 - l_1^2 = 2(0.5 + T) - (1 + T)^2 = -T^2$$
$$f_4 = 2l_4 - 2l_1 l_3 + l_2^2 = -2(1 + T)(0.5T) + (0.5 + T)^2 = 0.25 \tag{7-195}$$
$$f_6 = -l_3^2 = -(0.5T)^2 = -0.25T^2$$

From the equations in Eqs. (7-193) through (7-195), we solve for the values of d_1 and d_2, and the results are

$$d_1 = \sqrt{1 + T^2} \qquad d_2 = 0.5$$

Thus, the transfer function of the second-order approximating system is

$$M_L(s) = \frac{1}{1 + \sqrt{1 + T^2}s + 0.5s^2} \tag{7-196}$$

The two poles of $M_L(s)$ are at

$$s = -\sqrt{1 + T^2} + j\sqrt{1 - T^2} \quad \text{and} \quad s = -\sqrt{1 + T^2} - j\sqrt{1 - T^2}$$

The poles of $M_H(s)$ are at $s = -1 + j, -1 - j$, and $-1/T$. The two poles of $M_L(s)$ are tabulated for $T = 0$ to $T = 6$:

$T = 0$	$s = -1.000 + j$	$s = -1.000 - j$
$T = 0.1$	$s = -1.005 + j0.995$	$s = -1.005 - j0.995$
$T = 0.25$	$s = -1.031 + j0.968$	$s = -1.031 - j0.968$
$T = 0.5$	$s = -1.118 + j0.866$	$s = -1.118 - j0.866$
$T = 0.75$	$s = -1.250 + j0.661$	$s = -1.250 - j0.661$
$T = 1.0$	$s = -1.414$	$s = -1.414$
$T = 2.0$	$s = -3.968$	$s = -0.504$
$T = 4.0$	$s = -7.996$	$s = -0.250$
$T = 6.0$	$s = -11.999$	$s = -0.167$

Notice that as the value of T decreases, the original system approaches a second-order system as the pole at $-1/T$ moves toward $-\infty$. Thus, when $T = 0$, $M_L(s)$ is identical to $M_H(s)$, and the approximate solution is exact. As the value of T increases, the pole at $-1/T$ moves toward the origin and becomes more dominant; the poles of $M_L(s)$ move toward the real axis, and eventually when $T \geq 1$, the poles become real. As T increases, one of the real poles of the second-order system moves toward $-\infty$, and the other approaches $-1/T$. Thus, the second-order approximation is good for small and large values of T.

Figures 7-41 through 7-44 illustrate the unit-step responses of the third-order system and the second-order approximating system when $T = 0.1, 0.5, 1.0$, and 6.0, respectively. When the

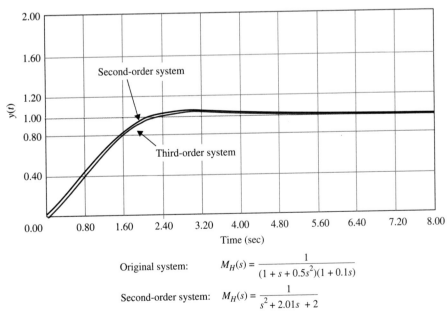

Original system: $\quad M_H(s) = \dfrac{1}{(1 + s + 0.5s^2)(1 + 0.1s)}$

Second-order system: $\quad M_H(s) = \dfrac{1}{s^2 + 2.01s + 2}$

Figure 7-41 Approximation of a third-order system by a second-order system.

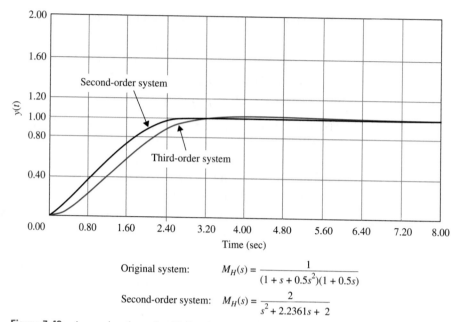

Original system: $M_H(s) = \dfrac{1}{(1 + s + 0.5s^2)(1 + 0.5s)}$

Second-order system: $M_H(s) = \dfrac{2}{s^2 + 2.2361s + 2}$

Figure 7-42 Approximation of a third-order system by a second-order system.

value of T is very small, the unit-step responses of the third-order and the second-order systems are very close. The error between the two responses increases as T is increased to 0.5 and 1.0. Figure 7-44 shows that the responses of the two systems again move closer to each other when $T = 6$. As the value of T increases, the real pole at $-1/T$ becomes dominant relative to the two

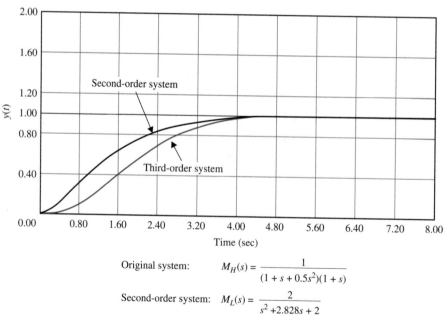

Original system: $M_H(s) = \dfrac{1}{(1 + s + 0.5s^2)(1 + s)}$

Second-order system: $M_L(s) = \dfrac{2}{s^2 + 2.828s + 2}$

Figure 7-43 Approximation of a third-order system by a second-order system.

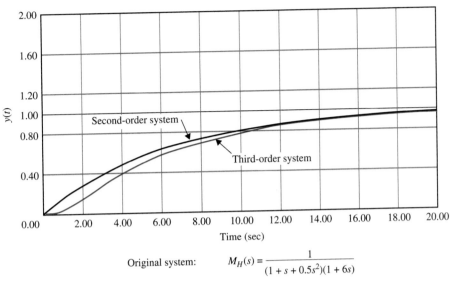

Original system: $\qquad M_H(s) = \dfrac{1}{(1 + s + 0.5s^2)(1 + 6s)}$

Second-order system: $\quad M_L(s) = \dfrac{2}{s^2 + 12.17s + 2}$

Figure 7-44 Approximation of a third-order system by a second-order system.

complex-conjugate poles, and the second-order model for $M_L(s)$ in Eq. (7-181) would give a better approximation to the third-order system. ◄

▶ **EXAMPLE 7-11** The third-order position-control system described in Section 7-7 was approximated by a second-order system by neglecting the small motor inductance. The second-order model arrived at in Eq. (7-126) was intended for all values of K. It was shown that the approximation was relatively good for small values of K but inferior for large values of K. Now we shall use the approximating method presented in this section in the hope of arriving at a better second-order approximation for a wider range of K.

The closed-loop transfer function of the system is given in Eq. (7-148) and is repeated here:

$$M_H(s) = \frac{1.5 \times 10^7 K}{s^3 + 3408.3s^2 + 1,204,000s + 1.5 \times 10^7 K} \qquad (7\text{-}197)$$

For a given value of K, this third-order transfer function is to be approximated by a second-order transfer function of the form

$$M_L(s) = \frac{1}{1 + d_1 s + d_2 s^2} \qquad (7\text{-}198)$$

Once $M_L(s)$ is determined, the forward-path transfer function of the approximating second-order system is written

$$G_L(s) = \frac{d_1/d_2}{s(s + d_1/d_2)} \qquad (7\text{-}199)$$

The coefficients d_1 and d_2 are determined for $K = 7.248$, 14.5, and 100, using the method described in this section. The results are tabulated as follows, where the first column of $G(s)$ denotes the approximation with $L_a = 0$, and the second column gives the corresponding $G_L(s)$ with the values of d_1 and d_2 indicated.

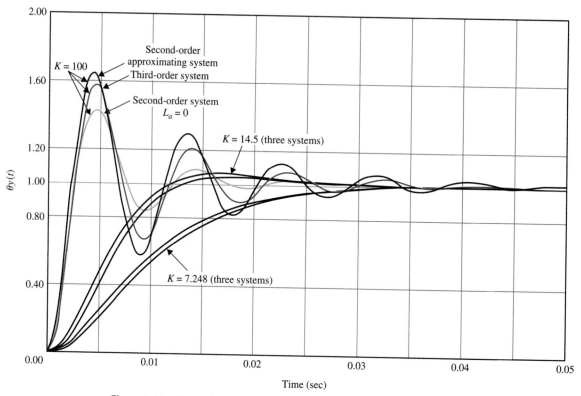

Figure 7-45 Comparison of the unit-step responses of the system in Example 7-11.

K	$G(s)$	$G_L(s)$
7.248	$\dfrac{326166}{s(s + 361.2)}$	$\dfrac{35829.5}{s(s + 385.5)}$
14.5	$\dfrac{65{,}250}{s(s + 361.2)}$	$\dfrac{71{,}684.6}{s(s + 378.2)}$
100.0	$\dfrac{450{,}000}{s(s + 361.2)}$	$\dfrac{494{,}300}{s(s + 188.84)}$

Comparisons of the unit-step responses of the third-order system and of the second-order systems with $L_a = 0$ and with the approximation method are shown in Fig. 7-45 for the three values of K indicated. When the value of K is relatively small, both second-order approximations are quite good. For $K = 100$, the second-order system with $L_a = 0$ has a lower maximum overshoot and is less oscillatory than those of the true third-order system. The second-order system from the approximation method has a step response that is closer to the actual response, although the maximum overshoot is slightly larger. ◀

One of the problems that surfaces from this investigation is that the low-order approximation method described in this section provides a system that is less damped than the high-order system. This means that when the original system is lightly damped, it is possible that the dominant roots of the low-order system may end up in the right-half s-plane. For the system treated in Example 7-11 we run into this difficulty when $K = 181.2$, one of the values chosen in Section 7-6. Since d_1 and d_2 are solved from $d_1^2 = 2d_2 - f_2$, which comes from Eqs. (7-172) and (7-175), then if f_2 is greater than $2d_2$, no real solution could be obtained for d_1. For the system in Example 7-11, we can show that when $K = 129.3$, the two poles of the second-order approximating system will be on the $j\omega$-axis. To approximate the third-order system when K is large would require that the second-order transfer function $M_L(s)$ contain a zero.

In conclusion, we would like to point out that the problem of reduced-order approximation of a high-order system is a complex subject. Most of the existing methods have flaws of one type or another. The serious reader should refer to the literature [1, 2] for more in-depth treatment of the subject.

▶ 7-11 MATLAB TOOLS AND CASE STUDIES

The Time Response Analysis Tool (timetool) consists of a number of m-files and GUIs for the time analysis of control engineering transfer functions. The timetool can be invoked either from the MATLAB command line by simply typing **timetool** or from the Automatic Control Systems launch applet (**ACSYS**) by clicking on the appropriate button. This software allows the user to conduct the following tasks:

- Enter the transfer function values in polynomial form (user must use the tftool discussed in Chapter 2 to convert the transfer function from pole, zero, gain form into the polynomial form).
- Obtain the step, impulse, parabolic, and ramp time responses.
- Investigate the effect of poles on the time response by relating the s-plane and the system root locus to the system time response.
- Understand the effect of adding zeros and poles to the closed-loop or open-loop transfer functions.
- Compare higher-order transfer functions to their approximations.
- Study position and speed-control problems.

To better illustrate how to use timetool, let us go through the steps involved in solving the earlier example involving position control in Section 7-6. Rewriting the second-order closed-loop transfer function, we have

$$\frac{\Theta_y(s)}{\Theta_r(s)} = \frac{4500K}{s^2 + 361.2s + 4500K} \tag{7-200}$$

The loop transfer function in this case is

$$G(s) = \frac{4500K}{s(s + 361.2)} \tag{7-201}$$

In order to examine the time response, invoke the Time Response Analysis Tool by typing timetool at the MATLAB prompt in the MATLAB command window. Figure 7-46 will appear on the computer screen.

In order to define the transfer function, click on the "Enter Transfer Function" popup button to go to the input module. In the timetool, the user can only use the **polynomial** approach to enter the transfer functions. In order to convert a transfer

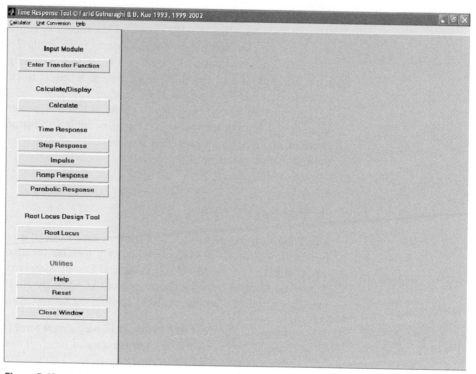

Figure 7-46   Main GUI for Time Response Tool.

function from pole-zero form to a polynomial form, **tftool** must be used. **Note:** The value of transfer function G may directly be transferred from tftool to timetool, if tftool is activated after timetool (do not close timetool). The steps you require to take in this case are

- Activate tftool.
- Enter the poles and zeros of G.
- Calculate the transfer function (conversion to polynomial form).
- Close tftool.
- Activate the Enter Transfer Function button in timetool (**caution:** if this button is activated prior to running tftool, the value of G will not be transferred).
- The coefficients of G in polynomial form will be shown upon clicking the G button.
- Continue entering the other transfer functions.

To enter the transfer function Eq. (7-190), click the Enter Transfer Function button. Next, the window in Fig. 7-47a appears. Press block G to enter its properties, and follow the instructions to enter the system parameters (the coefficient must be separated by a space), as shown in Fig. 7-47b. We have entered the values using Eq. (7-201), excluding the value of K. Next, enter the values of H and G_c the same way. The default values of H and G_c are set to 1. So you need to enter a value for K in transfer function G_c. We have used $K = 1$ for simplicity (the default value of G_c). Use the APPLY pushbutton to exit this window once all values are properly entered.

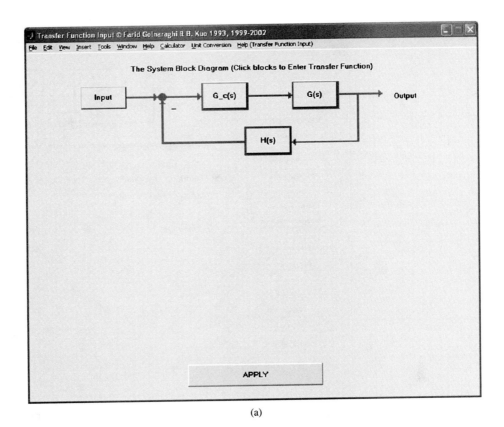

(a)

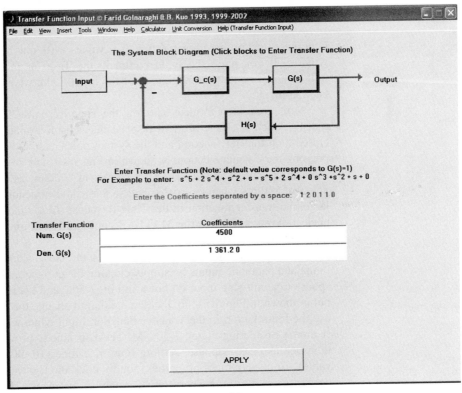

Figure 7-47 (a) Transfer Function Input window. (b) Entering transfer function *G* after clicking the *G* button.

(b)

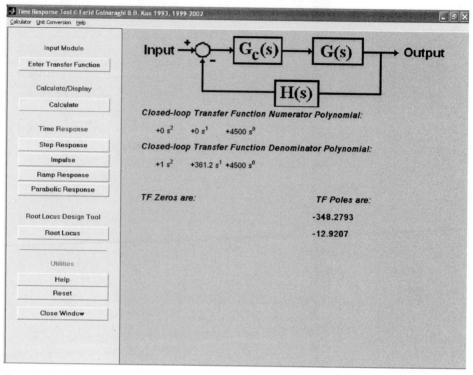

Figure 7-48   Closed-loop system transfer function in polynomial form and system poles-zeros form.

Otherwise the proper transfer function values will not be transferred to the main window. If you press the Input button in the Transfer Function Input window, you will be able to enter the magnitude of the input. Otherwise a default value of one has been used.

Next, press the Calculate button in the Time Response Tool window to evaluate the transfer function. There are two display mechanisms available. First, you may see the closed-loop transfer function and the system poles and zeros on the same window after pressing the Calculate button, as shown in Fig. 7-48. The second and more detailed representation appears in the MATLAB command window, as shown in Fig. 7-49. Please note that at times when the coefficients in numerator or denominator polynomials are too large, the transfer function in the Time Response Tool window may appear messy and unclear. **Always refer to the MATLAB command window for an accurate representation of transfer functions**.

You can now find the time response of the closed-loop system to step, impulse, ramp, and parabolic inputs by simply clicking the corresponding button. The system response to a unit-step input is shown in Fig. 7-50a, and the response for a unit-ramp input is shown in Figure 7-50b. Please note that to change the input magnitude; you must use the Input button in the Transfer Function Input window as described earlier in this section. Considering Figure 7-50a, MATLAB is able to provide the system characteristics such as peak response, settling time, rise time (10–90%), and steady state (final value). To display these properties, simply click the right mouse button and select the desired property in the Characteristics item in the menu. This action will result in the

```
G=

Transfer function:
     4500
----------------
s^2 + 361.2 s

Gc=

Transfer function:
I

H=

Transfer function:
I

G*G_c==>open loop

Transfer function:
     4500
----------------
s^2 + 361.2 s

G*G_c*H==>loop
Transfer function:
     4500
----------------
s^2 + 361.2 s

G*G_c/(1+G*G_c*H)==>closed loop

Transfer function:
        4500
----------------------
s^2 + 361.2 s + 4500

system coefficients are
Closed loop TF in zero/pole format

Zero/pole/gain:
           4500
------------------------
(s+348.3) (s+12.92)

System zeros are

zero TF =

   Empty matrix: 1-by-0

System poles are

poleTF=

-348.2793 -12.9207
```

Figure 7-49 Detailed system-transfer function and pole-zero value in MATLAB command window.

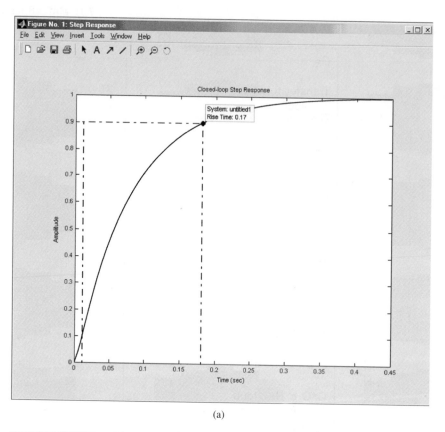

(a)

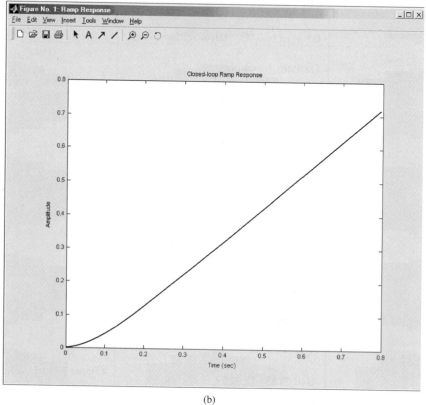

Figure 7-50 (a) Unit-step response of the closed-loop system, (b) Unit-ramp response.

(b)

298

appearance of a blue dot on the plot. The blue dot reflects the 10 to 90% rise time. The numerical values of these properties may be obtained by clicking the left mouse button while the mouse arrow is on the blue dot, as shown in Fig. 7-50a. In this case, the rise time is 0.17 second. You may change the plot's various properties, using the Figure menu.

The root locus of the system may be obtained by clicking the corresponding button in the Time Response window, which will activate the MATLAB SISO design tool. Fig. 7-51a shows root locus of the system. Root locus construction and properties will be discussed in depth in Chapter 7, but at this point it is important to show how the poles change in Eq. (7-200) as K varies from 0 to infinity. The limiting value of $K = 0$ corresponds to the poles and zeros of Eq. (7-201). In order to see the poles and zeros of G and H, go to View menu and select System Data, or alternatively double click the blocks G or H in the top right corner of the block diagram. You should see the System Data window as shown in Fig. 7-51b.

The red squares in Fig. 7-51b correspond to the closed-loop system poles for $K = 1$. In order to see the closed-loop system time response to a unit-step input, you must select the Tools menu, then Loop Responses ..., and select the Closed-Loop Step. You should see Fig. 7-52a appear. You may also obtain the closed-loop system poles by selecting the Closed-Loop Poles from the View menu in the SISO Design Tool window. Recall that the poles of the closed-loop system Eq. (7-200) are

$$s_{1,2} = -180.6 \pm \sqrt{32616 - 4500K} \tag{7-202}$$

So, by varying K, the pole locations change. In the Root Locus window, you may see the red squares that represent the closed-loop poles move if you vary the value of $C(s)$. Note that $C(s) = 1$ is a scaling factor for K. Hence, if $C(s)$ is reduced, the effective K value is decreased, and in this example, the closed-loop poles move away from each other. For $C(s) > 1$ the poles move toward each other and ultimately become a complex conjugate. For $C(s) = 181.2$, the closed-loop poles become complex, as shown in Fig. 7-53a and 7-53b. The step response of the system shows a damped oscillatory behavior, as expected. You may display the time response characteristics by right-clicking the mouse and selecting the appropriate characteristic. In this case, as in Fig. 7-54a, we have shown the peak value, rise time, and settling time. The properties associated with these values may be adjusted using the LTI Viewer Preferences from the figure Edit menu as shown in the Fig. 7-54b.

As discussed in Chapter 2, to obtain an analytical expression for the time response for different K values, you may use the Laplace-transform tables in tftool, or use the Transfer Function Symbolic (tfsym) tool. Alternatively, it is very easy to go to the MATLAB command window and use the *ilaplace* command as shown in Fig. 7-55 for value of $K = 181.2$. The command "syms s" makes s a symbolic variable. Since the inverse Laplace is in symbolic form, you must further simplify it to a numeric form. The easiest way is shown in Fig. 7-53. Othewise, refer to MATLAB help on the Symbolic Toolbox to find a more proper approach.

You can use the SISO Design Tool to investigate the effect of adding poles and zeros to the transfer function, by changing the properties of $C(s)$. For example, to add a pole to Eq. (7-200) when $K = 181.2$, click on the compensator $C(s)$ box in the block diagram located in the right corner of the SISO Design Tool window. Enter a value for a compensator pole, say, a real pole at $s = -10^6$, as shown in Fig. 7-56.

The root locus will change to reflect a third-order system, as shown in Fig. 7-57a. The closed-loop poles for $K = 181.2$ are shown in Fig. 7-57b, and are slightly different than those of the second-order system shown earlier in Fig. 7-53. The step response of

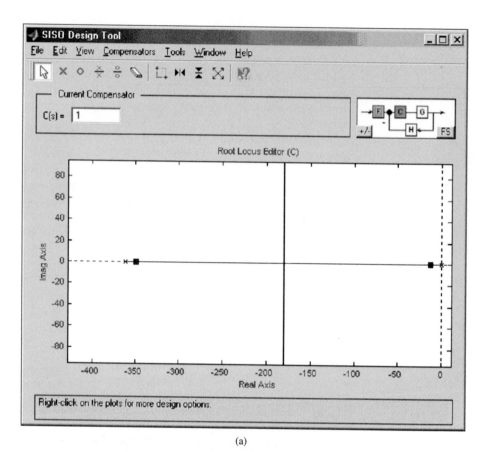

Figure 7-51 (a) Root locus
of Eq. (7-201), (b) *G* and
H poles and zeros.

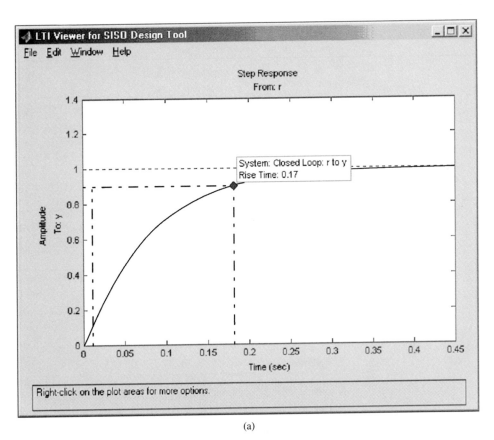

(a)

(b)

Figure 7-52 (a) Unit-step response of Eq. (7-201) for $K = 1$; (b) poles and zeros of Eq. (7-201) for $K = 1$.

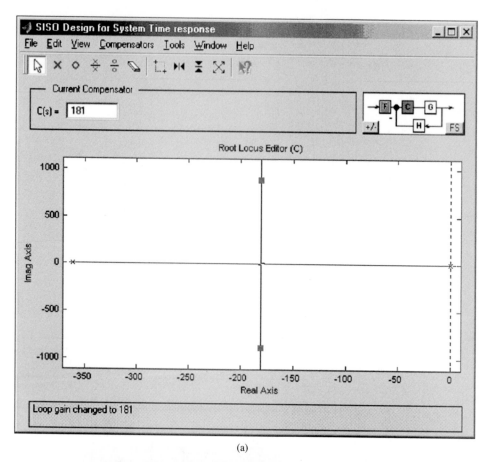

(a)

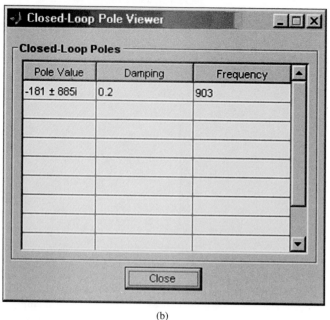

(b)

Figure 7-53 (a) Root locus of Eq. (7-201); (b) poles and zeros of (7-201) for $K = 181.2$.

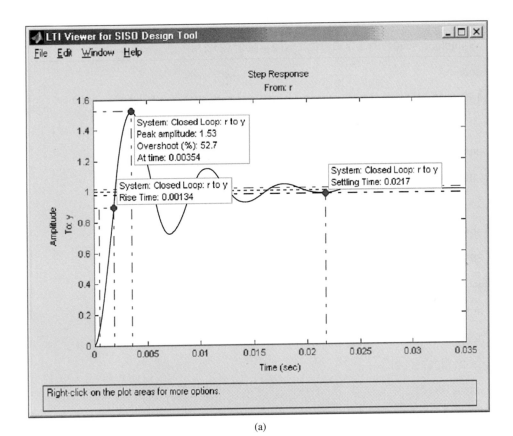

(a)

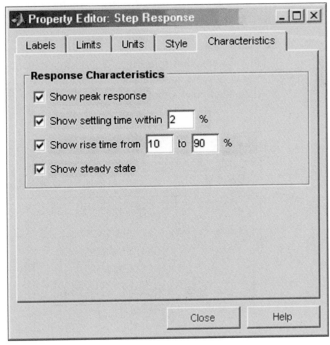

(b)

Figure 7-54 (a) The unit-step response of Eq. (7-200) for $K = 180.2$, (b) The Property Editor window for the response characteristics.

```
>> syms s
>> ilaplace(4500*181.2/(s^3+361.2*s^2+4500*181.2*s))

ans =

1-exp(-903/5*t)*cos(3/5*2174399^(1/2*)*t)-301/2174399*exp(-903/5*t)*2174399^(1/2)*sin(3/5*2174399^(1/2)*t)

>> -903/5

ans =

-180.6000

>> 3/5*2174399^(1/2)

ans =

884.7506

>> 301/2174399

ans =

1.3843e-004

>> 2174399^(1/2)

ans =

1.4746e+003
```

Figure 7-55 The inverse Laplace transform of the closed-loop system of Eq. (7-200) for $K = 181.2$ and a unit-step input.

this system, shown in Fig. 7-58, is almost identical to the second-order system shown in Fig. 7-54a. As stated earlier in Sections 7-7-4 and 7-8, the added pole is insignificant. If the added pole is moved closer to the two dominant poles of the system, for example, at

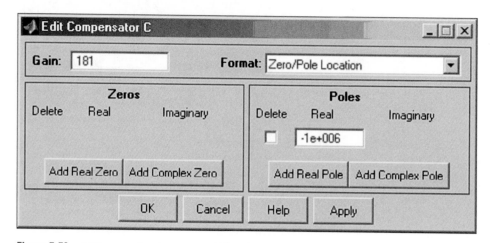

Figure 7-56 Adding a real pole at $s = -10^6$ to the open-loop system.

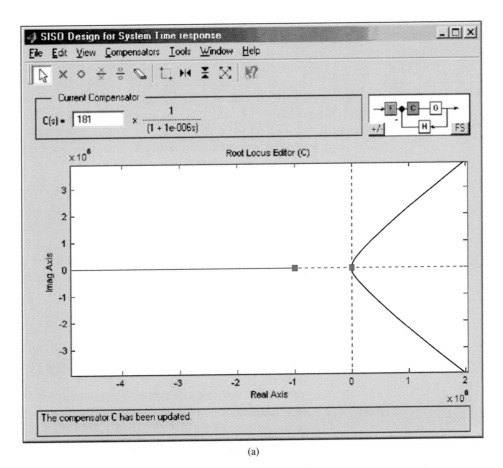

(a)

(b)

Figure 7-57 (a) Root locus of the third-order system with an added pole at $s = -10^6$; (b) poles and zeros for $K = 181.2$.

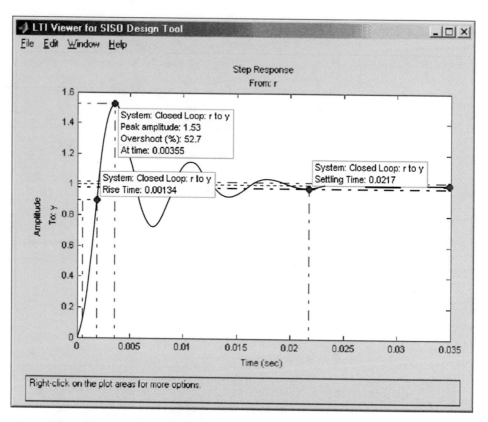

Figure 7-58 Unit-step response of the third-order system with an added pole at $s = -10^6$.

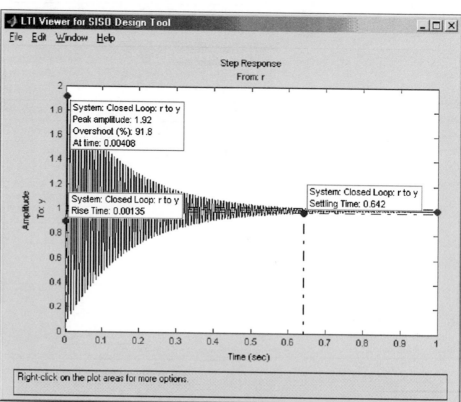

Figure 7-59 Unit-step response of the third-order system with an added pole at $s = -2000$.

$s = -2000$, the response of the system would drastically change to oscillator response with a higher natural frequency and faster rise time, as shown in Fig. 7-59.

The MATLAB SISO tool is very powerful and has many useful features. Students are encouraged to spend time with this important tool and apply it to all the problems discussed in this chapter.

▶ 7-12 SUMMARY

This chapter was devoted to the time-domain analysis of linear continuous-data control systems. The time response of control systems is divided into the transient and the steady-state responses. The steady-state error is a measure of the accuracy of the system as time approaches infinity. When the system has unity feedback for the step, ramp, and parabolic inputs, the steady-state error is characterized by the error constants K_p, K_v, and K_a, respectively, as well as the system **type**. When applying the steady-state error analysis, the final-value theorem of the Laplace transform is the basis; it should be ascertained that the closed-loop system is stable or the error analysis will be invalid. The error constants are not defined for systems with nonunity feedback. For nonunity-feedback systems, a method of determining the steady-state error was introduced by using the closed-loop transfer function.

The transient response is characterized by such criteria as the **maximum overshoot**, **rise time**, **delay time**, and **settling time**, and such parameters as **damping ratio, natural undamped frequency**, and **time constant**. The analytical expressions of these parameters can all be related to the system parameters simply if the transfer function is of the second-order prototype. For second-order systems that are not of the prototype, and for higher-order systems, the analytical relationships between the transient parameters and the system constants are difficult to determine. Computer simulations are recommended for these systems.

Time-domain analysis of a position-control system was conducted. The transient and steady-state analyses were carried out first by approximating the system as a second-order system. The effect of varying the amplifier gain K on the transient and steady-state performance was demonstrated. The concept of the root-locus technique was introduced, and the system was then analyzed as a third-order system, and it was shown that the second-order approximation was accurate only for low values of K.

The effects of adding poles and zeros to the forward-path and closed-loop transfer functions were demonstrated. The dominant poles of transfer functions were also discussed. This established the significance of the location of the poles of the transfer function in the s-plane, and under what conditions the insignificant poles (and zeros) could be neglected with regard to the transient response.

In Section 7-9, a method of approximating a high-order system by an equivalent low-order system was introduced. The low-order system tracks the high-order one in the sense that the frequency responses of the two systems are similar.

Finally, the last portion of this chapter was devoted to the Time Response Analysis tool (**timetool**), which may be used to solve most of the problems discussed in this chapter.

▶ REVIEW QUESTIONS

1. Give the definitions of the error constants, K_p, K_v, and K_a.

2. Specify the type of input to which the error constant K_p is dedicated.

3. Specify the type of input to which the error constant K_v is dedicated.

4. Specify the type of input to which the error constant K_a is dedicated.

5. Define an error constant if the input to a unity-feedback control system is described by $r(t) = t^3 u_s(t)/6$.

6. Give the definition of the system **type** of a linear time-invariant system.

7. If a unity-feedback control system type is 2, then it is certain that the steady-state error of the system to a step input or a ramp input will be zero. **(T)** **(F)**

8. Linear and nonlinear frictions will generally degrade the steady-state error of a control system. **(T)** **(F)**

9. The maximum overshoot of a unit-step response of the second-order prototype system will never exceed 100 percent when the damping ratio ζ and the natural undamped frequency ω_n are all positive.

10. For the second-order prototype system, when the undamped natural frequency ω_n increases, the maximum overshoot of the output stays the same.

11. The maximum overshoot of the following system will never exceed 100 percent when ζ, ω_n, and T are all positive.

$$\frac{Y(s)}{R(s)} = \frac{\omega_n^2(1 + Ts)}{s^2 + 2\zeta\omega_n s + \omega_n^2}$$ **(T)** **(F)**

12. Increasing the undamped natural frequency will generally reduce the rise time of the step response. **(T)** **(F)**

13. Increasing the undamped natural frequency will generally reduce the settling time of the step response. **(T)** **(F)**

14. Adding a zero to the forward-path transfer function will generally improve the system damping, and thus will always reduce the maximum overshoot of the system. **(T)** **(F)**

15. Given the following characteristic equation of a linear control system, increasing the value of K will increase the frequency of oscillation of the system

$$s^3 + 3s^2 + 5s + K = 0$$ **(T)** **(F)**

16. For the characteristic equation given in Review Question 15, increasing the coefficient of the s^2 term will generally improve the damping of the system. **(T)** **(F)**

17. The location of the roots of the characteristic equation in the s-plane will give a definite indication on the maximum overshoot of the transient response of the system. **(T)** **(F)**

18. The following transfer function $G(s)$ can be approximated by $G_L(s)$ since the pole at -20 is much larger than the dominant pole at $s = -1$.

$$G(s) = \frac{10}{s(s + 1)(s + 20)} \qquad G_L(s) = \frac{10}{s(s + 1)}$$ **(T)** **(F)**

Answers to the true-and-false questions will be given at the end of the Problems section.

▶ REFERENCES

Simplification of Linear Systems

1. E. J. Davison, "A Method for Simplifying Linear Dynamic Systems," *IEEE Trans. Automatic Control,* Vol. AC-11, pp. 93–101, Jan. 1966.
2. T. C. Hsia, "On the Simplification of Linear Systems," *IEEE Trans. Automatic Control,* Vol. AC-17, pp. 372–374, June 1972.

Undershoot in Step Response

3. M. Vidyasagar, "On Undershoot and Nonminimum Phase Zeros," *IEEE Trans. Automatic Control,* Vol. AC-31, p. 440, May 1986.
4. T. Norimatsu and M. Ito, "On the Zero Non-regular Control System," *J. Inst. Elec. Eng. Japan,* Vol. 81, pp. 567–575, 1961.

You may use timetool (**ACSYS**) to solve most of the following problems.

▶ PROBLEMS

• s-domain specifications

7-1. A pair of complex-conjugate poles in the s-plane is required to meet the various specifications that follow. For each specification, sketch the region in the s-plane in which the poles should be located.

(a) $\zeta \geq 0.707$ $\omega_n \geq 2$ rad/sec   (positive damping)

(b) $0 \leq \zeta \leq 0.707$ $\omega_n \leq 2$ rad/sec   (positive damping)

(c) $\zeta \leq 0.5$ $1 \leq \omega_n \leq 5$ rad/sec   (positive damping)

(d) $0.5 \leq \zeta \leq 0.707$ $\omega_n \leq 5$ rad/sec   (positive and negative damping)

• System type

7-2. Determine the type of the following unity-feedback systems for which the forward-path transfer functions are given.

(a) $G(s) = \dfrac{K}{(1 + s)(1 + 10s)(1 + 20s)}$ (b) $G(s) = \dfrac{10e^{-0.2s}}{(1 + s)(1 + 10s)(1 + 20s)}$

(c) $G(s) = \dfrac{10(s + 1)}{s(s + 5)(s + 6)}$ (d) $G(s) = \dfrac{100(s - 1)}{s^2(s + 5)(s + 6)^2}$

(e) $G(s) = \dfrac{10(s + 1)}{s^3(s^2 + 5s + 5)}$ (f) $G(s) = \dfrac{100}{s^3(s + 2)^2}$

• Error constants

7-3. Determine the step, ramp, and parabolic error constants of the following unity-feedback control systems. The forward-path transfer functions are given.

(a) $G(s) = \dfrac{1000}{(1 + 0.1s)(1 + 10s)}$ (b) $G(s) = \dfrac{100}{s(s^2 + 10s + 100)}$

(c) $G(s) = \dfrac{K}{s(1 + 0.1s)(1 + 0.5s)}$ (d) $G(s) = \dfrac{100}{s^2(s^2 + 10s + 100)}$

(e) $G(s) = \dfrac{1000}{s(s + 10)(s + 100)}$ (f) $G(s) = \dfrac{K(1 + 2s)(1 + 4s)}{s^2(s^2 + s + 1)}$

• Steady-state error of
unity-feedback systems

7-4. For the unity-feedback control systems described in Problem 7-2, determine the steady-state error for a unit-step input, a unit-ramp input, and a parabolic input, $(t^2/2)u_s(t)$. Check the stability of the system before applying the final-value theorem.

• Steady-state error of
nonunity-feedback systems

7-5. The following transfer functions are given for a single-loop nonunity-feedback control system. Find the steady-state errors due to a unit-step input, a unit-ramp input, and a parabolic input, $(t^2/2)u_s(t)$.

(a) $G(s) = \dfrac{1}{(s^2 + s + 2)}$ $H(s) = \dfrac{1}{(s + 1)}$

(b) $G(s) = \dfrac{1}{s(s + 5)}$ $H(s) = 5$

(c) $G(s) = \dfrac{1}{s^2(s + 10)}$ $H(s) = \dfrac{s + 1}{s + 5}$

(d) $G(s) = \dfrac{1}{s^2(s + 12)}$ $H(s) = 5(s + 2)$

• Steady-state error of
single-loop systems

7-6. Find the steady-state errors of the following single-loop control systems for a unit-step input, a unit-ramp input, and a parabolic input, $(t^2/2)u_s(t)$. For systems that include a parameter K, find its value so that the answers are valid.

(a) $M(s) = \dfrac{s + 4}{s^4 + 16s^3 + 48s^2 + 4s + 4}, K_H = 1$

(b) $M(s) = \dfrac{K(s + 3)}{s^3 + 3s^2 + (K + 2)s + 3K}, K_H = 1$

(c) $M(s) = \dfrac{s + 5}{s^4 + 15s^3 + 50s^2 + 10s}, H(s) = \dfrac{10s}{s + 5}$

(d) $M(s) = \dfrac{K(s + 5)}{s^4 + 17s^3 + 60s^2 + 5Ks + 5K}, K_H = 1$

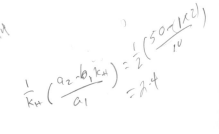

• Error constants and
steady-state errors

7-7. The block diagram of a control system is shown in Fig. 7P-7. Find the step-, ramp-, and parabolic-error constants. The error signal is defined to be $e(t)$. Find the steady-state errors in terms of K and K_t when the following inputs are applied. Assume that the system is stable.

(a) $r(t) = u_s(t)$ (b) $r(t) = tu_s(t)$ (c) $r(t) = (t^2/2)u_s(t)$

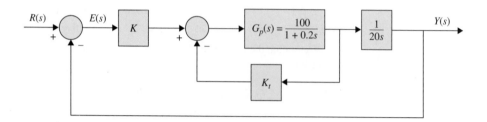

Figure 7P-7

• Error constants and
steady-state errors

7-8. Repeat Problem 7-7 when the transfer function of the process is, instead,

$$G_p(s) = \frac{100}{(1 + 0.1s)(1 + 0.5s)}$$

What constraints must be made, if any, on the values of K and K_t so that the answers are valid? Determine the minimum steady-state error that can be achieved with a unit-ramp input by varying the values of K and K_t.

• Steady-state error

7-9. For the position-control system shown in Fig. 3P-18, determine the following.

(a) Find the steady-state value of the error signal $\theta_e(t)$ in terms of the system parameters when the input is a unit-step function.

(b) Repeat part (a) when the input is a unit-ramp function. Assume that the system is stable.

• Steady-state error

7-10. The block diagram of a feedback control system is shown in Fig. 7P-10. The error signal is defined to be $e(t)$.

(a) Find the steady-state error of the system in terms of K and K_t when the input is a unit-ramp function. Give the constraints on the values of K and K_t so that the answer is valid. Let $n(t) = 0$ for this part.

(b) Find the steady-state value of $y(t)$ when $n(t)$ is a unit-step function. Let $r(t) = 0$. Assume that the system is stable.

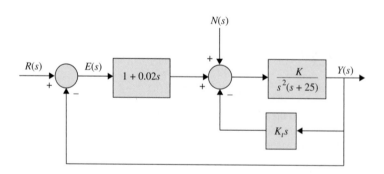

Figure 7P-10

• Steady-state error due to
input and disturbance

7-11. The block diagram of a linear control system is shown in Fig. 7P-11, where $r(t)$ is the reference input and $n(t)$ is the disturbance.

(a) Find the steady-state value of $e(t)$ when $n(t) = 0$ and $r(t) = tu_s(t)$. Find the conditions on the values of α and K so that the solution is valid.

(b) Find the steady-state value of $y(t)$ when $r(t) = 0$ and $n(t) = u_s(t)$.

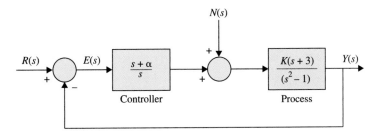

Figure 7P-11

• Transfer function of
second-order prototype
system

7-12. The unit-step response of a linear control system is shown in Fig. 7P-12. Find the transfer function of a second-order prototype system to model the system.

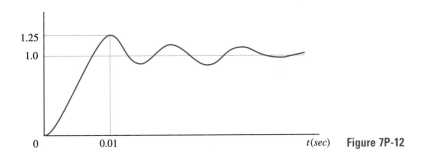

Figure 7P-12

• Maximum overshoot and
rise time

7-13. For the control system shown in Fig. 7P-7, find the values of K and K_t so that the maximum overshoot of the output is approximately 4.3 percent and the rise time t_r is approximately 0.2 sec. Use Eq. (7-104) for the rise-time relationship. Simulate the system with any time-response simulation program to check the accuracy of your solutions.

• Maximum overshoot and
rise time

7-14. Repeat Problem 7-13 with a maximum overshoot of 10 percent and a rise time of 0.1 sec.

7-15. Repeat Problem 7-13 with a maximum overshoot of 20 percent and a rise time of 0.05 sec.

• Maximum overshoot and
delay time

7-16. For the control system shown in Fig. 7P-7, find the values of K and K_t so that the maximum overshoot of the output is approximately 4.3 percent and the delay time t_d is approximately 0.1 sec. Use Eq. (7-102) for the delay-time relationship. Simulate the system with a computer program to check the accuracy of your solutions.

• Maximum orershoot and
delay time

7-17. Repeat Problem 7-16 with a maximum overshoot of 10 percent and a delay time of 0.05 sec.

7-18. Repeat Problem 7-16 with a maximum overshoot of 20 percent and a delay time of 0.01 sec.

• Damping ratio and
settling time

7-19. For the control system shown in Fig. 7P-7, find the values of K and K_t so that the damping ratio of the system is 0.6 and the settling time of the unit-step response is 0.1 sec. Use Eq. (7-108) for the settling time relationship. Simulate the system with a computer program to check on the accuracy of your results.

• Maximum overshoot and
settling time

7-20. (a) Repeat Problem 7-19 with a maximum overshoot of 10 percent and a settling time of 0.05 sec.

(b) Repeat Problem 7-19 with a maximum overshoot of 20 percent and a settling time of 0.01 sec.

• Damping ratio and
settling time

7-21. Repeat Problem 7-19 with a damping ratio of 0.707 and a settling time of 0.1 sec. Use Eq. (7-109) for the settling time relationship.

• Damping ratio, rise time, steady-state error

7-22. The forward-path transfer function of a control system with unity feedback is

$$G(s) = \frac{K}{s(s + a)(s + 30)}$$

where a and K are real constants.

(a) Find the values of a and K so that the relative damping ratio of the complex roots of the characteristic equation is 0.5 and the rise time of the unit-step response is approximately 1 sec. Use Eq. (7-104) as an approximation of the rise time. With the values of a and K found, determine the actual rise time using computer simulation.

(b) With the values of a and K found in part (a), find the steady-state errors of the system when the reference input is (i) a unit-step function, and (ii) a unit-ramp function.

• Time response

7-23. The block diagram of a linear control system is shown in Fig. 7P-23.

(a) By means of trial and error, find the value of K so that the characteristic equation has two equal real roots and the system is stable. You may use any root-finding computer program to solve this problem.

(b) Find the unit-step response of the system when K has the value found in part (a). Use any computer simulation program for this. Set all the initial conditions to zero.

(c) Repeat part (b) when $K = -1$. What is peculiar about the step response for small t, and what may have caused it?

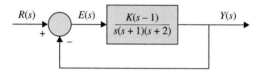

Figure 7P-23

• State feedback and parameter plane

7-24. A controlled process is represented by the following dynamic equations:

$$\frac{dx_1(t)}{dt} = -x_1(t) + 5x_2(t)$$

$$\frac{dx_2(t)}{dt} = -6x_1(t) + u(t)$$

$$y(t) = x_1(t)$$

The control is obtained through state feedback with

$$u(t) = -k_1 x_1(t) - k_2 x_2(t) + r(t)$$

where k_1 and k_2 are real constants, and $r(t)$ is the reference input.

(a) Find the locus in the k_1-versus-k_2 plane (k_1 = vertical axis) on which the overall system has a natural undamped frequency of 10 rad/sec.

(b) Find the locus in the k_1-versus-k_2 plane on which the overall system has a damping ratio of 0.707.

(c) Find the values of k_1 and k_2 such that $\zeta = 0.707$ and $\omega_n = 10$ rad/sec.

(d) Let the error signal be defined as $e(t) = r(t) - y(t)$. Find the steady-state error when $r(t) = u_s(t)$ and k_1 and k_2 are at the values found in part (c).

(e) Find the locus in the k_1-versus-k_2 plane on which the steady-state error due to a unit-step input is zero.

- Parameter plane and special trajectories

7-25 The block diagram of a linear control system is shown in Fig. 7P-25. Construct a parameter plane of K_p versus K_d (K_p is the vertical axis) and show the following trajectories or regions in the plane.

(a) Unstable and stable regions.

(b) Trajectories on which the damping is critical ($\zeta = 1$).

(c) Region in which the system is overdamped ($\zeta > 1$).

(d) Region in which the system is underdamped ($\zeta < 1$).

(e) Trajectory on which the parabolic-error constant K_a is 1000 sec^{-2}.

(f) Trajectory on which the natural undamped frequency ω_n is 50 rad/sec.

(g) Trajectory on which the system is either uncontrollable or unobservable. (Hint: look for pole-zero cancellation.)

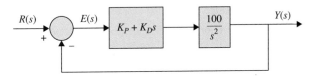

Figure 7P-25

- Steady-state performances

7-26. The block diagram of a linear control system is shown in Fig. 7P-26. The fixed parameters of the system are given as $T = 0.1$, $J = 0.01$, and $K_i = 10$.

(a) When $r(t) = tu_s(t)$ and $T_d(t) = 0$, determine how the values of K and K_t affect the steady-state value of $e(t)$. Find the restrictions on K and K_t so that the system is stable.

(b) Let $r(t) = 0$. Determine how the values of K and K_t affect the steady-state value of $y(t)$ when the disturbance input $T_d(t) = u_s(t)$.

(c) Let $K_t = 0.01$ and $r(t) = 0$. Find the minimum steady-state value of $y(t)$ that can be obtained by varying K, when $T_d(t)$ is a unit-step function. Find the value of this K. From the transient standpoint, would you operate the system at this value of K? Explain.

(d) Assume that it is desired to operate the system with the value of K as selected in part (c). Find the value of K_t so that the complex roots of the characteristic equation will have a real part of -2.5. Find all the three roots of the characteristic equation.

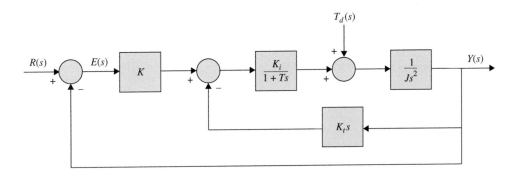

Figure 7P-26

Printwheel control systrem

7-27. The dc-motor control system for controlling a printwheel described in Problem 4-14 has the forward-path transfer function

$$G(s) = \frac{\Theta_o(s)}{\Theta_e(s)} = \frac{nK_sK_iK_LK}{\Delta(s)}$$

where $\Delta(s) = s[L_a J_m J_L s^4 + J_L(R_a J_m + B_m L_a)s^3 + (n^2 K_L L_a J_L + K_L L_a J_m$
$+ K_i K_b J_L + R_a B_m J_L)s^2 + (n^2 R_a K_L J_L + R_a K_L J_m + B_m K_L L_a)s$
$+ R_a B_m K_L + K_i K_b K_L]$

where $K_i = 9$ oz-in./A, $K_b = 0.636$ V/rad/sec, $R_a = 5\ \Omega$, $L_a = 1$ mH, $K_s = 1$ V/rad. $n = 1/10$, $J_m = J_L = 0.001$ oz-in.-sec^2, and $B_m \cong 0$. The characteristic equation of the closed-loop system is

$$\Delta(s) + nK_s K_i K_L K = 0$$

(a) Let $K_L = 10{,}000$ oz-in./rad. Write the forward-path transfer function $G(s)$ and find the poles of $G(s)$. Find the critical value of K for the closed-loop system to be stable. Find the roots of the characteristic equation of the closed-loop system when K is at marginal stability.

(b) Repeat part (a) when $K_L = 1000$ oz-in./rad.

(c) Repeat part (a) when $K_L = \infty$; that is, the motor shaft is rigid.

(d) Compare the results of parts (a), (b), and (c), and comment on the effects of the values of K_L on the poles of $G(s)$ and the roots of the characteristic equation.

• Guided-missile control system

7-28. The block diagram of the guided-missile attitude-control system described in Problem 4-13 is shown in Fig. 7P-28. The command input is $r(t)$, and $d(t)$ represents disturbance input. The objective of this problem is to study the effect of the controller $G_c(s)$ on the steady-state and transient responses of the system.

(a) Let $G_c(s) = 1$. Find the steady-state error of the system when $r(t)$ is a unit-step function. Set $d(t) = 0$.

(b) Let $G_c(s) = (s + \alpha)/s$. Find the steady-state error when $r(t)$ is a unit-step function.

(c) Obtain the unit-step response of the system for $0 \le t \le 0.5$ sec. with $G_c(s)$ as given in part (b), and $\alpha = 5$, 50, and 500. Assume zero initial conditions. Record the maximum overshoot of $y(t)$ for each case. Use any available computer simulation program. Comment on the effect of varying the value of α of the controller on the transient response.

(d) Set $r(t) = 0$, and $G_c(s) = 1$. Find the steady-state value of $y(t)$ when $d(t) = u_s(t)$.

(e) Let $G_c(s) = (s + \alpha)/s$. Find the steady-state value of $y(t)$ when $d(t) = u_s(t)$.

(f) Obtain the output response for $0 \le t \le 0.5$ sec, with $G_c(s)$ as given in part (e) when $r(t) = 0$ and $d(t) = u_s(t)$; $\alpha = 5$, 50, and 500. Use zero initial conditions.

(g) Comment on the effect of varying the value of α of the controller on the transient response of $y(t)$ and $d(t)$.

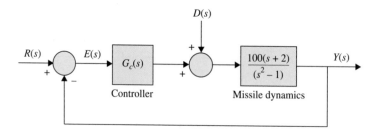

$D(s)$

$R(s)$ $E(s)$ $G_c(s)$ Controller $\dfrac{100(s + 2)}{(s^2 - 1)}$ Missile dynamics $Y(s)$

Figure 7P-28

• Reduced-order liquid-level control system

7-29. The block diagram shown in Fig. 7P-29 represents the liquid-level control system described in Problem 7-13. The liquid level is represented by $h(t)$, and N denotes the number of inlets.

(a) Since one of the poles of the open-loop transfer function is relatively far to the left on the real axis of the s-plane at $s = -10$, it is suggested that this pole can be neglected. Approximate the system by a second-order system be neglecting the pole of $G(s)$ at $s = -10$. The approximation should be valid for both the transient and the steady-state responses. Apply the formulas for the maximum overshoot and the peak time t_{max} to the second-order model for $N = 1$ and $N = 10$.

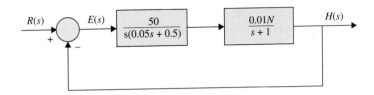

Figure 7P-29

(b) Obtain the unit-step response (with zero initial conditions) of the original third-order system with $N = 1$ and then with $N = 10$. Compare the responses of the original system with those of the second-order approximating system. Comment on the accuracy of the approximation as a function of N.

• Effects of moving open-loop zeros

7-30. The forward-path transfer function of a unity-feedback control system is

$$G(s) = \frac{1 + T_z s}{s(s + 1)^2}$$

Compute and plot the unit-step responses of the closed-loop system for $T_z = 0, 0.5, 1, 10,$ and 50. Assume zero initial conditions. Use any computer simulation program that is available. Comment on the effects of the various values of T_z on the step response.

• Effect of moving open-loop pole

7-31. The forward-path transfer function of a unity-feedback control system is

$$G(s) = \frac{1}{s(s + 1)^2(1 + T_p s)}$$

Compute and plot the unit-step responses of the closed-loop system for $T_p = 0, 0.5,$ and 0.707. Assume zero initial conditions. Use any computer simulation program. Find the critical value of T_p so that the closed-loop system is marginally stable. Comment on the effects of the pole at $s = -1/T_p$ in $G(s)$.

• Reduced order of liquid-level control system

7-32. This problem is devoted to the approximation of a third-order system by a second-order one, using the method described in Section 7-9. The liquid-level control system described in Problems 7-13 and 7-29 has the forward-path transfer function

$$G(s) = \frac{10N}{s(s + 1)(s + 10)}$$

(a) For $N = 1$, find the transfer function $G_L(s)$ of a second-order system that approximates the third-order system. Compute the unit-step responses of the systems, and observe the closeness of the approximation. Compare the roots of the characteristic equations of the two closed-loop systems.

(b) Repeat part (a) with $N = 2$.

(c) Repeat part (a) with $N = 3$.

(d) Repeat part (a) with $N = 4$.

(e) Repeat part (a) with $N = 5$.

Reduced-order printwheel control system

7-33. The dc-motor control system for controlling a printwheel described in Problem 4-14 has the following forward-path transfer function when $K_L = \infty$.

$$G(s) = \frac{891,100K}{s(s^2 + 5000s + 566,700)}$$

The system is to be approximated by a second-order system using the method described in Section 7-9.

(a) For $K = 1$, find the second-order transfer function $G_L(s)$ that approximates $G(s)$. Compute and plot the unit-step responses of the systems, and evaluate the closeness of the approximation. Compare the roots of the characteristic equations of the two closed-loop systems.

(b) Repeat part (a) with $K = 100$.

(c) Repeat part (a) with $K = 1000$.

• Reduced-order
approximation

7-34. Approximate the control system shown in Fig. 7P-23 when $K = -1$ by a second-order system with the closed-loop transfer function

$$M(s) = \frac{Y(s)}{R(s)} = \frac{1 + c_1 s}{1 + d_1 s + d_2 s^2}$$

where c_1, d_1, and d_2 are real constants. Use the method outlined in Section 7-9. Compute and plot the unit-step responses of the original third-order system ($K = -1$) and the second-order approximating system. Compare the roots of the characteristic equations of the two systems.

7-35. A unity-feedback control system has the forward-path transfer function

$$G(s) = \frac{K}{s(s + 1)(s + 2)(s + 20)}$$

(a) For $K = 10$, the fourth-order system is to be approximated by a second-order system with the closed-loop transfer function

$$L(s) = \frac{G_L(s)}{1 + G_L(s)} = \frac{1}{1 + d_1 s + d_2 s^2}$$

Find d_1 and d_2 using the method outlined in Section 7-9. Compute and plot the unit-step responses of the fourth-order and the second-order systems and compare. Find the roots of the characteristic equations of the two systems.

(b) For $K = 10$, approximate the system by a third-order model with

$$L(s) = \frac{1}{1 + d_1 s + d_2 s^2 + d_3 s^3}$$

Carry out the tasks specified in part (a).

(c) Repeat parts (a) and (b) with $K = 40$.

▶ **ADDITIONAL
COMPUTER
PROBLEMS**

MATLAB

Compare and plot the unit-step responses of the unity-feedback closed-loop systems with the forward-path transfer functions given. Assume zero initial conditions. Use timetool program.

7-36. (a) For $T_z = 0, 1, 5, 20$,

$$G(s) = \frac{1 + T_z s}{s(s + 0.55)(s + 1.5)}$$

(b) For $T_z = 0, 1, 5, 20$,

$$G(s) = \frac{1 + T_z s}{(s^2 + 2s + 2)}$$

7-37. (a) For $T_p = 0, 0.5, 1.0$,

$$G(s) = \frac{2}{(s^2 + 2s + 2)(1 + T_p s)}$$

(b) For $T_p = 0, 0.5, 1.0$,

$$G(s) = \frac{10}{s(s + 5)(1 + T_p s)}$$

7-38. $G(s) = \dfrac{K}{s(s + 1.25)(s^2 + 2.5s + 10)}$

(a) For $K = 5$.   (b) For $K = 10$.   (c) For $K = 30$.

7-39. $G(s) = \dfrac{K(s + 2.5)}{s(s + 1.25)(s^2 + 2.5s + 10)}$

(a) For $K = 5$.   (b) For $K = 10$.   (c) For $K = 30$.

► ANSWERS TO
TRUE-AND-FALSE
REVIEW
QUESTIONS

7. (F) 8. (T) 9. (T) 11. (F) 12. (T) 13. (T) 14. (F) 15. (T) 16. (T) 17. (F)
18. (F) 21. (T)

CHAPTER 8

Root-Locus Technique

▶ 8-1 INTRODUCTION

• Root loci are trajectories of roots of characteristic equation when a system parameter varies.

In the preceding chapters, we have demonstrated the importance of the poles and zeros of the closed-loop transfer function of a linear control system on the dynamic performance of the system. The roots of the characteristic equation, which are the poles of the closed-loop transfer function, determine the absolute and the relative stability of linear SISO systems. Keep in mind that the transient properties of the system also depend on the zeros of the closed-loop transfer function.

An important study in linear control systems is the investigation of the trajectories of the roots of the characteristic equation—or, simply, the **root loci**—when a certain system parameter varies. In Chapter 6, several examples already illustrated the usefulness of the root loci of the characteristic equation in the study of linear control systems.

The basic properties and the systematic construction of the root loci are first due to W. R. Evans [1, 3]. In general, root loci may be sketched by following some simple rules and properties. For plotting the root loci accurately the MATLAB root-locus tool in the Control Systems Toolbox (**controls**) or in the Time Response Analysis Tool (timetool) of **ACSYS**, can be used. As a design engineer, it may be sufficient for us to learn how to use these computer tools to generate the root loci for design purposes. However, it is important to learn the basics of the root loci and their properties, as well as how to interpret the data provided by the root loci for analysis and design purposes. As a proficient engineer, we must also know if the data provided by the root loci are indeed correct and be able to derive vital information from the root loci. The material in this text is prepared with these objectives in mind; details on the properties and construction of the root loci are presented in Appendix F.

The root-locus technique is not confined only to the study of control systems. In general, the method can be applied to study the behavior of roots of any algebraic equation with one or more variable parameters. The general root-locus problem can be formulated by referring to the following algebraic equation of the complex variable, say, s:

$$F(s) = P(s) + KQ(s) = 0 \qquad (8\text{-}1)$$

where $P(s)$ is an nth-order polynomial of s,

$$P(s) = s^n + a_{n-1}s^{n-1} + \cdots + a_1s + a_0 \qquad (8\text{-}2)$$

and $Q(s)$ is an mth-order polynomial of s; n and m are positive integers.

$$Q(s) = s^m + b_{m-1}s^{m-1} + \cdots + b_1s + b_0 \qquad (8\text{-}3)$$

For the present, we do not place any limitations on the relative magnitudes between n and m. K is a real constant that can vary from $-\infty$ to $+\infty$.

The coefficients $a_1, a_2, \ldots, a_n, b_1, b_2, \ldots, b_m$ are considered to be real and fixed.

• Root loci of multiple-variable parameters are called root contours.

Root loci of multiple variable parameters can be treated by varying one parameter at a time. The resultant loci are called the **root contours**, and the subject is treated in Section 8-5. By replacing s with z in Eqs. (8-1) through (8-3), the root loci of the characteristic equation of a linear discrete-data system can be constructed in a similar fashion (Appendix I).

For the purpose of identification in this text, we define the following categories of root loci based on the values of K:

1. **Root Loci (RL):** refers to the entire root loci for $-\infty < K < \infty$.

2. **Root Contours (RC):** contour of roots when more than one parameter varies.

In general, for most control-system applications, the values of K are positive. Under unusual conditions when a system has positive feedback or the loop gain is negative, then we have the situation that K is negative. Although we should be aware of this possibility, we need to place the emphasis only on positive values of K in developing the root-locus techniques.

▶ 8-2 BASIC PROPERTIES OF THE ROOT LOCI (RL)

Since our main interest is control systems, let us consider the closed-loop transfer function of a single-loop control system:

$$\frac{Y(s)}{R(s)} = \frac{G(s)}{1 + G(s)H(s)} \tag{8-4}$$

keeping in mind that the transfer function of multiple-loop SISO systems can also be expressed in a similar form. The characteristic equation of the closed-loop system is obtained by setting the denominator polynomial of $Y(s)/R(s)$ to zero. Thus, the roots of the characteristic equation must satisfy

$$1 + G(s)H(s) = 0 \tag{8-5}$$

Suppose that $G(s)H(s)$ contains a real variable parameter K as a multiplying factor, such that the rational function can be written as

$$G(s)H(s) = \frac{KQ(s)}{P(s)} \tag{8-6}$$

where $P(s)$ and $Q(s)$ are polynomials as defined in Eqs. (8-2) and (8-3), respectively. Equation (8-5) is written

$$1 + \frac{KQ(s)}{P(s)} = \frac{P(s) + KQ(s)}{P(s)} = 0 \tag{8-7}$$

The numerator polynomial of Eq. (8-7) is identical to Eq. (8-1). Thus, by considering that the loop transfer function $G(s)H(s)$ can be written in the form of Eq. (8-6), we have identified the RL of a control system with the general root-locus problem.

When the variable parameter K does not appear as a multiplying factor of $G(s)H(s)$, we can always condition the functions in the form of Eq. (8-1). As an illustrative example, consider that the characteristic equation of a control system is

$$s(s + 1)(s + 2) + s^2 + (3 + 2K)s + 5 = 0 \tag{8-8}$$

To express the last equation in the form of Eq. (8-7), *we divide both sides of the equation by the terms that do not contain K*, and we get

$$1 + \frac{2Ks}{s(s+1)(s+2) + s^2 + 3s + 5} = 0 \tag{8-9}$$

Comparing the last equation with Eq. (8-7), we get

$$\frac{Q(s)}{P(s)} = \frac{2s}{s^3 + 4s^2 + 5s + 5} \tag{8-10}$$

Now K is isolated as a multiplying factor to the function $Q(s)/P(s)$.

We shall show that the RL of Eq. (8-5) can be constructed based on the properties of $Q(s)/P(s)$. In the case where $G(s)H(s) = KQ(s)/P(s)$, the root-locus problem is another example in which the characteristics of the closed-loop system, in this case represented by the roots of the characteristic equation, are determined from the knowledge of the loop transfer function $G(s)H(s)$.

Now we are ready to investigate the conditions under which Eq. (8-5) or Eq. (8-7) is satisfied.

Let us express $G(s)H(s)$ as

$$G(s)H(s) = KG_1(s)H_1(s) \tag{8-11}$$

where $G_1(s)H_1(s)$ does not contain the variable parameter K. Then, Eq. (8-5) is written

$$G_1(s)H_1(s) = -\frac{1}{K} \tag{8-12}$$

To satisfy Eq. (8-12), the following conditions must be satisfied simultaneously:

Condition on Magnitude

$$|G_1(s)H_1(s)| = \frac{1}{|K|} \qquad -\infty < K < \infty \tag{8-13}$$

Condition on Angles

$$\angle G_1(s)H_1(s) = (2i+1)\pi \qquad K \geq 0$$
$$= \text{odd multiples of } \pi \text{ radians or } 180° \tag{8-14}$$
$$\angle G_1(s)H_1(s) = 2i\pi \qquad K \leq 0$$
$$= \text{even multiples of } \pi \text{ radians or } 180° \tag{8-15}$$

where $i = 0, \pm1, \pm2, \ldots$ (any integer).

In practice, the conditions stated in Eqs. (8-13) through (8-15) play different roles in the construction of the root loci.

- The conditions on angles in Eq. (8-14) or Eq. (8-15) are used to determine the trajectories of the root loci in the s-plane.
- Once the root loci are drawn, the values of K on the loci are determined by using the condition on magnitude in Eq. (8-13).

The construction of the root loci is basically a graphical problem, although some of the properties are derived analytically. The graphical construction of the RL is based on the knowledge of the poles and zeros of the function $G(s)H(s)$. In other words, $G(s)H(s)$

must first be written as

$$G(s)H(s) = KG_1(s)H_1(s) = \frac{K(s + z_1)(s + z_2)\cdots(s + z_m)}{(s + p_1)(s + p_2)\cdots(s + p_n)} \tag{8-16}$$

where the zeros and poles of $G(s)H(s)$ are real or in complex-conjugate pairs.

Applying the conditions in Eqs. (8-13), (8-14), and (8-15) to Eq. (8-16), we have

$$|G_1(s)H_1(s)| = \frac{\displaystyle\prod_{i=1}^{m}|s + z_i|}{\displaystyle\prod_{k=1}^{n}|s + p_k|} = \frac{1}{|K|} \qquad -\infty < K < \infty \tag{8-17}$$

For $0 \leq K < \infty$:

$$\angle G_1(s)H_1(s) = \sum_{k=1}^{m}\angle(s + z_k) - \sum_{j=1}^{n}\angle(s + p_j) = (2i + 1) \times 180° \tag{8-18}$$

For $-\infty < K \leq 0$:

$$\angle G_1(s)H_1(s) = \sum_{k=1}^{m}\angle(s + z_k) - \sum_{j=1}^{n}\angle(s + p_j) = 2i \times 180° \tag{8-19}$$

where $i = 0, \pm 1, \pm 2, \ldots$.

The graphical interpretation of Eq. (8-18) is that any point s_1 on the RL that corresponds to a positive value of K must satisfy the condition:

The difference between the sums of the angles of the vectors drawn from the zeros and those from the poles of $G(s)H(s)$ to s_1 is an odd multiple of 180 degrees.

For negative values of K, any point s_1 on the RL must satisfy the condition:

The difference between the sums of the angles of the vectors drawn from the zeros and those from the poles of $G(s)H(s)$ to s_1 is an even multiple of 180 degrees, including zero degrees.

Once the root loci are constructed, the values of K along the loci can be determined by writing Eq. (8-17) as

$$|K| = \frac{\displaystyle\prod_{j=1}^{n}|s + p_j|}{\displaystyle\prod_{i=1}^{m}|s + z_i|} \tag{8-20}$$

The value of K at any point s_1 on the RL is obtained from Eq. (8-20) by substituting the value of s_1 into the equation. Graphically, the numerator of Eq. (8-20) represents the product of the lengths of the vectors drawn from the poles of $G(s)H(s)$ to s_1, and the denominator represents the product of lengths of the vectors drawn from the zeros of $G(s)H(s)$ to s_1.

To illustrate the use of Eqs. (8-18) to (8-20) for the construction of the root loci, let us consider the function

$$G(s)H(s) = \frac{K(s + z_1)}{s(s + p_2)(s + p_3)} \tag{8-21}$$

The location of the poles and zero of $G(s)H(s)$ are arbitrarily assigned as shown in Fig. 8-1. Let us select an arbitrary trial point s_1 in the s-plane and draw vectors directing from the poles and zeros of $G(s)H(s)$ to the point. If s_1 is indeed a point on

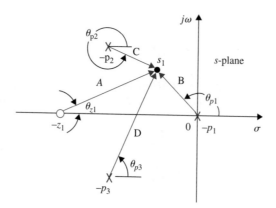

Figure 8-1 Pole-zero configuration of $G(s)H(s) = K(s + z_1)/[s(s + p_2)(s + p_3)]$

the RL for positive K, it must satisfy Eq. (8-18); that is, the angles of the vectors shown in Fig. 8-1 must satisfy

$$\angle(s_1 + z_1) - \angle s_1 - \angle(s_1 + p_2) - \angle(s_1 + p_3)$$
$$= \theta_{z1} - \theta_{p1} - \theta_{p2} - \theta_{p3} = (2i + 1) \times 180° \qquad (8\text{-}22)$$

where $i = 0, \pm 1, \pm 2, \ldots$. As shown in Fig. 8-1, the angles of the vectors are measured with the positive real axis as reference. Similarly, if s_1 is a point on the RL for negative values of K, it must satisfy Eq. (8-19); that is,

$$\angle(s_1 + z_1) - \angle s_1 - \angle(s_1 + p_2) - \angle(s_1 + p_3)$$
$$= \theta_{z1} - \theta_{p1} - \theta_{p2} - \theta_{p3} = 2i \times 180° \qquad (8\text{-}23)$$

where $i = 0, \pm 1, \pm 2, \ldots$.

If s_1 is found to satisfy either Eq. (8-22) or Eq. (8-23), Eq. (8-20) is used to find the magnitude of K at the point. As shown in Fig. 8-1, the lengths of the vectors are represented by A, B, C, and D. The magnitude of K is

$$|K| = \frac{|s_1||s_1 + p_2||s_1 + p_3|}{|s_1 + z_1|} = \frac{BCD}{A} \qquad (8\text{-}24)$$

The sign of K depends on whether s_1 satisfies Eq. (8-22) ($K \geq 0$) or Eq. (8-23) ($K \leq 0$). Thus, given the function $G(s)H(s)$ with K as a multiplying factor and the poles and zeros are known, the construction of the RL of the zeros of $1 + G(s)H(s)$ involves the following two steps:

1. A search for all the s_1 points in the s-plane that satisfy Eq. (8-18) for positive K. If the RL for negative values of K are desired, then Eq. (8-19) must be satisfied.
2. Use Eq. (8-20) to find the magnitude of K on the RL.

We have established the basic conditions on the construction of the root-locus diagram. However, if we were to use the trial-and-error method just described, the search for all the root-locus points in the s-plane that satisfy Eq. (8-18) or Eq. (8-19) and Eq. (8-20) would be a very tedious task. Years ago, when Evans [1, 2] first invented the root-locus technique, digital computer technology was still at its infancy; he had to devise a special tool, called the **Spirule**, which can be used to assist in adding and subtracting angles of vectors quickly, according to Eq. (8-18) or Eq. (8-19). Even with the Spirule, for the device to be effective, the user still has to first know the general proximity of the roots in the s-plane.

With the availability of digital computers and efficient root-finding subroutines, the Spirule and the trial-and-error method have long become obsolete. Nevertheless, even with a high-speed computer and an effective root-locus program, the analyst should still have an understanding of the properties of the root loci in order to be able to manually sketch the root loci of simple and moderately complex systems, if necessary, and interpret the computer results correctly, when applying the root loci for analysis and design of control systems.

▶ 8-3 PROPERTIES OF THE ROOT LOCI

The following properties of the root loci are useful for the purpose of constructing the root loci manually and for the understanding of the root loci. The properties are developed based on the relation between the poles and zeros of $G(s)H(s)$ and the zeros of $1 + G(s)H(s)$, which are the roots of the characteristic equation. We shall limit the discussion only to the properties, but leave the details of the proofs and the applications of the properties to the construction of the root loci in Appendix F.

8-3-1 $K = 0$ and $K = \pm\infty$ Points

The $K = 0$ points on the root loci are at the poles of $G(s)H(s)$.
The $K = \pm\infty$ points on the root loci are at the zeros of $G(s)H(s)$.

The poles and zeros referred to here include those at infinity, if any. The reason for these properties are seen from the condition of the root loci given by Eq. (8-12), which is,

$$G_1(s)H_1(s) = -\frac{1}{K} \tag{8-25}$$

As the magnitude of K approaches zero, $G_1(s)H_1(s)$ approaches infinity, so s must approach the poles of $G_1(s)H_1(s)$ or of $G(s)H(s)$. Similarly, as the magnitude of K approaches infinity, s must approach the zeros of $G(s)H(s)$.

▶ **EXAMPLE 8-1** Consider the equation

$$s(s + 2)(s + 3) + K(s + 1) = 0 \tag{8-26}$$

When $K = 0$, the three roots of the equation are at $s = 0$, -2, and -3. When the magnitude of K is infinite, the three roots of the equation are at $s = -1$, ∞ and ∞. It is useful to consider that infinity in the s-plane is a point concept. We can visualize that the finite s-plane is only a small portion of a sphere with an infinite radius. Then, infinity in the s-plane is a point on the opposite side of the sphere that we face.

Dividing both sides of Eq. (8-26) by the terms that do not contain K, we get

$$1 + G(s)H(s) = 1 + \frac{K(s + 1)}{s(s + 2)(s + 3)} = 0 \tag{8-27}$$

which gives

$$G(s)H(s) = \frac{K(s + 1)}{s(s + 2)(s + 3)} \tag{8-28}$$

Thus, the three roots of Eq. (8-26) when $K = 0$ are the same as the poles of the function $G(s)H(s)$. The three roots of Eq. (8-26) when $K = \pm\infty$ are at the three zeros of $G(s)H(s)$, including those at

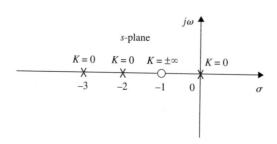

Figure 8-2 Points at which $K = 0$ and $K = \pm\infty$ on the *RL* of $s(s + 2)(s + 3) + K(s + 1) = 0$

infinity. In this case, one finite zero is at $s = -1$, but there are two zeros at infinity. The three points on the root loci at which $K = 0$ and those at which $K = \pm\infty$ are shown in Fig. 8-2. ◀

8-3-2 Number of Branches on the Root Loci

• It is important to keep track of the total number of branches of the root loci.

A branch of the RL is the locus of one root when K varies between $-\infty$ and ∞. The following property of the RL results, since the number of branches of the RL must equal the number of roots of the equation.

The number of branches of the RL of Eq. (8-1) or Eq. (8-5) is equal to the order of the polynomial.

For example, the number of branches of the root loci of Eq. (8-26) when K varies from $-\infty$ to ∞ is three, since the equation has three roots.

Keeping track of the individual branches and the total number of branches of the root-locus diagram is important in making certain that the plot is done correctly. This is particularly true when the root-locus plot is done by a computer, since unless each root locus branch is coded by a different color, it is up to the user to make the distinctions.

8-3-3 Symmetry of the RL

• It is important to pay attention to the symmetry of the root loci.

The RL are symmetrical with respect to the real axis of the s-plane. In general, the RL are symmetrical with respect to the axes of symmetry of the pole-zero configuration of G(s)H(s).

The reason behind this property is because for real coefficient in Eq. (8-1), the roots must be real or in complex-conjugate pairs.

8-3-4 Angles of Asymptotes of the RL: Behavior of the RL at $|s| = \infty$

• Asymptotes of root loci refers to behavior of root loci at $s \to \infty$.

When n, the order of $P(s)$, is not equal to m, the order of $Q(s)$, some of the loci will approach infinity in the s-plane. The properties of the RL near infinity in the s-plane are described by the **asymptotes** of the loci when $|s| \to \infty$. In general when $n \neq m$, there will be $2|n - m|$ asymptotes that describe the behavior of the RL at $|s| = \infty$. The angles of the asymptotes and their intersect with the real axis of the s-plane are described as follows.

For large values of s, the RL for $K \geq 0$ are asymptotic to asymptotes with angles given by

$$\theta_i = \frac{(2i + 1)}{|n - m|} \times 180° \qquad n \neq m \qquad (8\text{-}29)$$

where $i = 0, 1, 2, \ldots, |n - m| - 1$; n and m are the number of finite poles and zeros of G(s)H(s), respectively.

The asymptotes of the root loci for $K \leq 0$ are simply the extensions of the asymptotes for $K \geq 0$.

8-3-5 Intersect of the Asymptotes (Centroid)

The intersect of the $2|n - m|$ asymptotes of the RL lies on the real axis of the s-plane, at

$$\sigma_1 = \frac{\sum \text{finite poles of } G(s)H(s) - \sum \text{finite zeros of } G(s)H(s)}{n - m} \tag{8-30}$$

where n is the number of finite poles and m is the number of finite zeros of $G(s)H(s)$, respectively. The intersect of the asymptotes σ_1 represents the center of gravity of the root loci, and is always a real number, or

$$\sigma_1 = \frac{\sum \text{real parts of poles of } G(s)H(s) - \sum \text{real parts of zeros of } G(s)H(s)}{n - m} \tag{8-31}$$

The root loci and their asymptotes for Eq. (8-26) for $-\infty \leq K \leq \infty$ are shown in Fig. 8-3. More examples on root-loci asymptotes and constructions are found in Appendix F.

8-3-6 Root Loci on the Real Axis

• The entire real axis of the s-plane is occupied by root loci.

The entire real axis of the s-plane is occupied by the RL for all values K. On a given section of the real axis, RL for $K \geq 0$ are found in the section only if the total number of poles and zeros of $G(s)H(s)$ to the right of the section is odd. Note that the remaining sections of the real axis are occupied by the RL for $K \leq 0$. Complex poles and zeros of $G(s)H(s)$ do not affect the type of RL found on the real axis.

8-3-7 Angles of Departure and Angles of Arrival of the RL

The angle of departure or arrival of a root locus at a pole or zero, respectively, of $G(s)H(s)$ denotes the angle of the tangent to the locus near the point.

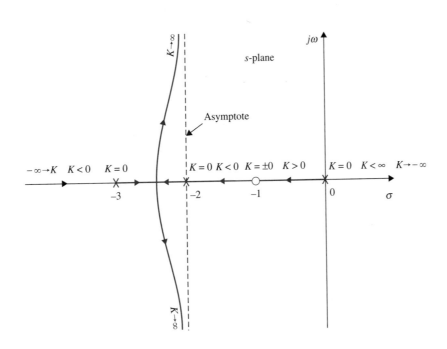

Figure 8-3 Root loci and asymptotes of $s(s + 2)(s + 3) + K(s + 1) = 0$ for $-\infty \leq K \leq \infty$.

8-3-8 Intersection of the RL with the Imaginary Axis

• Routh-Hurwitz criterion may be used to find the intersects of the RL on the imaginary axis.

The points where the RL intersect the imaginary axis of the s-plane, and the corresponding values of K, may be determined by means of the Routh-Hurwitz criterion. For complex situations, when the RL have multiple number of intersections on the imaginary axis, the intersects and the critical values of K can be determined with the help of the root-locus computer program. The Bode diagram method in Chapter 9, associated with the frequency response, can also be used for this purpose.

8-3-9 Breakaway Points (Saddle Points) on the RL

• A root locus plot may have more than one breakaway points.

• Breakaway points may be complex conjugates in the s-plane.

Breakaway points on the RL of an equation correspond to multiple-order roots of the equation.

The breakaway points on the RL of $1 + KG_1(s)H_1(s) = 0$ must satisfy

$$\frac{dG_1(s)H_1(s)}{ds} = 0 \qquad (8\text{-}32)$$

It is important to point out that the condition for the breakaway point given in Eq. (8-32) is *necessary* but *not sufficient*. In other words, all breakaway points on the root loci must satisfy Eq. (8-32), but not all solutions of Eq. (8-32) are breakaway points. To be a breakaway point, the solution of Eq. (8-32) must also satisfy Eq. (8-5), that is, must also be a point on the root loci for some real K.

If we take the derivatives on both sides of Eq. (8-12) with respect to s, we get

$$\frac{dK}{ds} = \frac{dG_1(s)H_1(s)/ds}{[G_1(s)H_1(s)]^2} \qquad (8\text{-}33)$$

Thus, the condition in Eq. (8-32) is equivalent to

$$\frac{dK}{ds} = 0 \qquad (8\text{-}34)$$

In summary, except for extremely complex cases, the properties on the root loci just presented should be adequate for making a reasonably accurate sketch of the root-locus diagram short of plotting it point by point. The computer program can be used to solve for the exact root locations, the breakaway points, and some of the other specific details of the root loci, including the plotting of the final loci. However, one cannot rely on the computer solution completely, since the user still has to decide on the range and resolution of K so that the root-locus plot has a reasonable appearance. For quick reference, the important properties described are summarized in Table 8-1 and the details are given in Appendix F.

8-3-10 The Root Sensitivity [17, 18, 19]

• The root sensitivity at the breakaway points is infinite.

The condition on the breakaway points on the RL in Eq. (8-34) leads to the **root sensitivity** of the characteristic equation. The sensitivity of the roots of the characteristic equation when K varies is defined as the **root sensitivity**, and is given by:

$$S_K = \frac{ds/s}{dK/K} = \frac{K}{s}\frac{ds}{dK} \qquad (8\text{-}35)$$

Thus, Eq. (8-34) shows that *the root sensitivity at the breakaway points is infinite*. From the root-sensitivity standpoint, we should avoid selecting the value of K to operate at

TABLE 8-1   Properties of the Root Loci of $1 + KG_1(s)H_1(s) = 0$

1. $K = 0$ points	The $K = 0$ points are at the poles of $G(s)H(s)$, including those at $s = \infty$.
2. $K = \pm\infty$ points	The $K = \infty$ points are at the zeros of $G(s)H(s)$, including those at $s = \infty$.
3. Number of separate root loci	The total number of root loci is equal to the order of the equation $F(s) = 0$.
4. Symmetry of root loci	The root loci are symmetrical about the axes of symmetry the of pole-zero configuration of $G(s)H(s)$.
5. Asymptotes of root loci as $s \to \infty$	For large values of s, the RL ($K > 0$) are asymptotic to asymptotes with angles given by $$\theta_i = \frac{2i + 1}{|n - m|} \times 180°$$ For $K < 0$, the RL are asymptotic to $$\theta_i = \frac{2i}{|n - m|} \times 180°$$ where $i = 0, 1, 2, ..., |n - m| - 1$, n = number of finite poles of $G(s)H(s)$, and m = number of finite zeros of $G(s)H(s)$.
6. Intersection of the asymptotes	(a) The intersection of the asymptotes lies only on the real axis in the s-plane. (b) The point of intersection of the asymptotes is given by $$\sigma_1 = \frac{\Sigma \text{real parts of poles of } G(s)H(s) - \Sigma \text{real parts of zeros of } G(s)H(s)}{n - m}$$
7. Root loci on the real axis	RL for $K \geq 0$ are found in a section of the real axis only if the total number of real poles and zeros of $G(s)H(s)$ to the **right** of the section is **odd**. If the total number of real poles and zeros to the right of a given section is **even**, RL for $K \leq 0$ are found.
8. Angles of departure	The angle of departure or arrival of the RL from a pole or a zero of $G(s)H(s)$ can be determined by assuming a point s_1 that is very close to the pole, or zero, and applying the equation, $$\angle G(s_1)H(s_1) = \sum_{k=1}^{m} \angle(s_1 + z_k) - \sum_{j=1}^{n} \angle(s_1 + p_j)$$ $$= 2(i + 1)180° \qquad K \geq 0$$ $$= 2i \times 180° \qquad K \leq 0$$ where $i = 0, \pm1, \pm2, ...$
9. Intersection of the root loci with the imaginary axis	The crossing points of the root loci on the imaginary axis and the corresponding values of K may be found by use of the Routh-Hurwitz criterion.
10. Breakaway points	The breakaway points on the root loci are determined by finding the roots of $dK/ds = 0$, or $dG(s)H(s)/ds = 0$. These are necessary conditions only.
11. Calculation of the values of K	The absolute value of K at any point s_1 on the root loci is on the root loci determined from the equation $$|K| = \frac{1}{|G_1(s_1)H_1(s_1)|}$$

the breakaway points, which correspond to multiple-order roots of the characteristic equation. In the design of control systems, not only it is important to arrive at a system that has the desired characteristics, but, just as important, the system should be insensitive to parameter variations. For instance, a system may perform satisfactorily at a certain K, but if it is very sensitive to the variation of K, it may get into the undesirable performance region or become unstable if K varies by only a small amount. In formal control-system terminology, a system that is insensitive to parameter variations is called a **robust system**. Thus, the root-locus study of control systems must involve not only the shape of the root loci with respect to the variable parameter K, but also how the roots along the loci vary with the variation of K.

► **EXAMPLE 8-2** Figure 8-4 shows the root locus diagram of

$$s(s + 1) + K = 0 \qquad (8\text{-}36)$$

with K incremented uniformly over 100 values from -20 to 20. The RL are computed and plotted digitally. Each dot on the root-locus plot represents one root for a distinct value of K. Thus, we see that the root sensitivity is low when the magnitude of K is large. As the magnitude of K decreases, the movements of the roots become larger for the same incremental change in K. At the breakaway point, $s = -0.5$, the root sensitivity is infinite.

Figure 8-5 shows the RL of

$$s^2(s + 1)^2 + K(s + 2) = 0 \qquad (8\text{-}37)$$

with K incremented uniformly over 200 values from -40 to 50. Again, the loci show that the root sensitivity increases as the roots approach the breakaway points at $s = 0$, -0.543, -1.0, and -2.457. We can investigate the root sensitivity further by using the expression in Eq. (8-34). For the second-order equation in Eq. (8-36),

$$\frac{dK}{ds} = -2s - 1 \qquad (8\text{-}38)$$

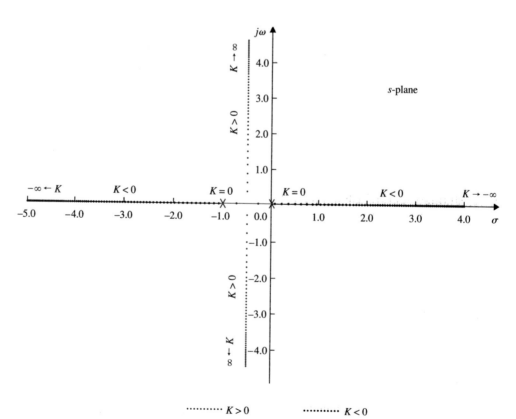

Figure 8-4   RL of $s(s + 1) + K = 0$ showing the root sensitivity with respect to K.

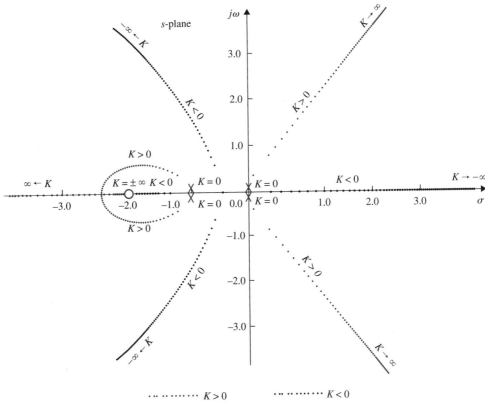

Figure 8-5 RL of $s^2(s + 1)^2 + K(s + 2) = 0$, showing the root sensitivity with respect to K.

From Eq. (8-36), $K = -s(s + 1)$; the root sensitivity becomes

$$S_K = \frac{ds}{dK}\frac{K}{s} = \frac{s + 1}{2s + 1} \tag{8-39}$$

where $s = \sigma + j\omega$, and s must take on the values of the roots of Eq. (8-39). For the roots on the real axis, $\omega = 0$. Thus, Eq. (8-39) leads to

$$|S_K|_{\omega=0} = \left|\frac{\sigma + 1}{2\sigma + 1}\right| \tag{8-40}$$

When the two roots are complex, $\sigma = -0.5$ for all values of ω; Eq. (8-39) gives

$$|S_K|_{\sigma=-0.5} = \left(\frac{0.25 + \omega^2}{4\omega^2}\right)^{1/2} \tag{8-41}$$

From Eq. (8-41), it is apparent that the sensitivities of the pair of complex-conjugate roots are the same, since ω appears only as ω^2 in the equation. Equation (8-40) indicates that the sensitivities of the two real roots are different for a given value of K. Table 8-2 gives the magnitudes of the sensitivities of the two roots of Eq. (8-36) for several values of K, where $|S_{K1}|$ denotes the root sensitivity of the first root, and $|S_{K2}|$ denotes that of the second root. These values indicate that although the two real roots reach $\sigma = -0.5$ for the same value of $K = 0.25$, and each root travels the same distance from $s = 0$ and $s = -1$, respectively, the sensitivities of the two real roots are not the same.

TABLE 8-2 Root Sensitivity

K	ROOT 1	$|S_{K1}|$	ROOT 2	$|S_{K2}|$
0	0	1.000	-1.000	0
0.04	-0.042	1.045	-0.958	0.454
0.16	-0.200	1.333	-0.800	0.333
0.24	-0.400	3.000	-0.600	2.000
0.25	-0.500	∞	-0.500	∞
0.28	$-0.5 + j0.173$	1.527	$-0.5 - j0.173$	1.527
0.40	$-0.5 + j0.387$	0.817	$-0.5 - j0.387$	0.817
1.20	$-0.5 + j0.975$	0.562	$-0.5 - j0.975$	0.562
4.00	$-0.5 + j1.937$	0.516	$-0.5 - j1.937$	0.516
∞	$-0.5 + j\infty$	0.500	$-0.5 - j\infty$	0.500

◀

▸ 8-4 DESIGN ASPECTS OF THE ROOT LOCI

One of the important aspects of the root-locus technique is that for most control systems with moderate complexity, the analyst or designer can obtain vital information on the performance of the system by making a quick sketch of the RL using some or all of the properties of the root loci. It is of importance to understand all the properties of the RL even when the diagram is to be plotted with the help of a digital computer program. From the design standpoint, it is useful to learn the effects on the RL when poles and zeros of $G(s)H(s)$ are added or moved around in the s-plane. Some of these properties are helpful in the construction of the root locus diagram. The design of the PI, PID, phase-lead, phase-lag, and the lead-lag controllers discussed in Chapter 10 all have implications of adding poles and zeros to the loop transfer function in the s-plane.

8-4-1 Effects of Adding Poles and Zeros to $G(s)H(s)$

• Adding a pole to $G(s)H(s)$ has the effect of pushing the root loci to the right.

The general problem of controller design in control systems may be treated as an investigation of the effects to the root loci when poles and zeros are added to the loop transfer function $G(s)H(s)$.

Addition of Poles of $G(s)H(s)$
Adding a pole to $G(s)H(s)$ has the effect of pushing the root loci toward the right-half. The effect of adding a zero to $G(s)H(s)$ can be illustrated with several examples.

▸ **EXAMPLE 8-3** Consider the function

$$G(s)H(s) = \frac{K}{s(s + a)} \qquad a > 0 \tag{8-42}$$

The RL of $1 + G(s)H(s) = 0$ are shown in Fig. 8-6(a). These RL are constructed based on the poles of $G(s)H(s)$, which are at $s = 0$ and $-a$. Now let us introduce a pole at $s = -b$, with $b > a$. The function $G(s)H(s)$ now becomes

$$G(s)H(s) = \frac{K}{s(s + a)(s + b)} \tag{8-43}$$

Figure 8-6(b) shows that the pole at $s = -b$ causes the complex part of the root loci to bend toward the right-half s-plane. The angles of the asymptotes for the complex roots are changed from $\pm 90°$ to $\pm 60°$. The intersect of the asymptotes is also moved from $-a/2$ to $-(a + b)/2$ on the real axis.

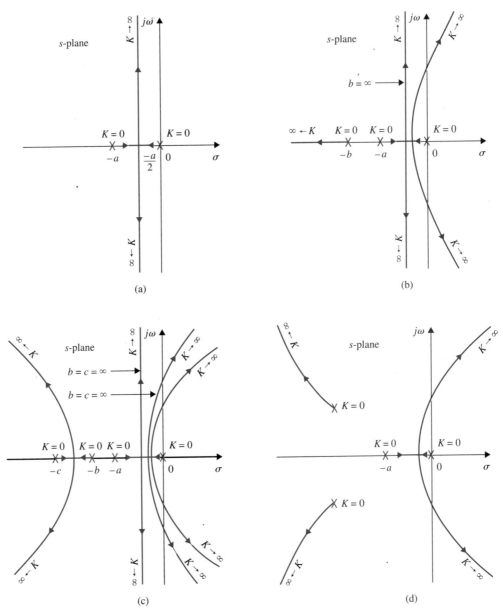

Figure 8-6 Root-locus diagrams that show the effects of adding poles to $G(s)H(s)$.

If $G(s)H(s)$ represents the loop transfer function of a control system, the system with the root loci in Fig. 8-6(b) may become unstable if the value of K exceeds the critical value for stability, whereas the system represented by the root loci in Fig. 8-6(a) is always stable for $K > 0$. Figure 8-6(c) shows the root loci when another pole is added to $G(s)H(s)$ at $s = -c$, $c > b$. The system is now of the fourth order, and the two complex root loci are bent farther to the right. The angles of asymptotes of these two complex loci are now $\pm 45°$. The stability condition of the fourth-order system is even more acute than that of the third-order system. Figure 8-6(d) illustrates that the addition of a pair of complex-conjugate poles to the transfer function of Eq. (8-42) will result in a similar effect. Therefore, we may draw a general conclusion that the addition of poles to $G(s)H(s)$ has the effect of moving the dominant portion of the root loci toward the right-half s-plane. ◀

Addition of Zeros to $G(s)H(s)$

Adding left-half plane zeros to the function $G(s)H(s)$ generally has the effect of moving and bending the root loci toward the left-half s-plane.

The following example illustrates the effect of adding a zero and zeros to $G(s)H(s)$ on the RL.

• Adding zeros to $G(s)H(s)$ tends to push the root loci to the left.

▶ **EXAMPLE 8-4** Figure 8-7(a) shows the RL of the $G(s)H(s)$ in Eq. (8-42) with a zero added at $s = -b$ ($b > a$). The complex-conjugate part of the RL of the original system is bent toward the left and forms a circle. Thus, if $G(s)H(s)$ is the loop transfer function of a control system, the relative stability of

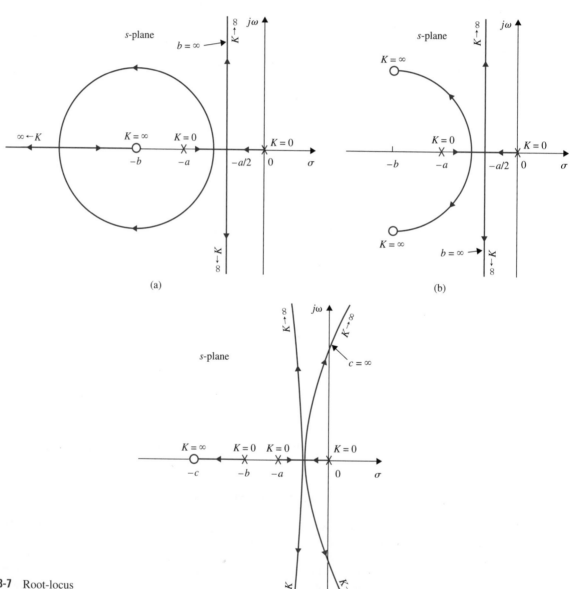

(a)

(b)

(c)

Figure 8-7 Root-locus diagrams that show the effects of adding a zero to $G(s)H(s)$.

the system is improved by the addition of the zero. Figure 8-7(b) shows that a similar effect will result if a pair of complex-conjugate zeros is added to the function of Eq. (8-42). Figure 8-7(c) shows the RL when a zero at $s = -c$ is added to the transfer function of Eq. (8-43).   ◄

▶ **EXAMPLE 8-5** Consider the equation

$$s^2(s + a) + K(s + b) = 0 \tag{8-44}$$

Dividing both sides of Eq. (8-44) by the terms that do not contain K, we have the loop transfer function,

$$G(s)H(s) = \frac{K(s + b)}{s^2(s + a)} \tag{8-45}$$

It can be shown that the nonzero breakaway points depend on the value of a, and are

$$s = -\frac{a + 3}{4} \pm \frac{1}{4}\sqrt{a^2 - 10a + 9} \tag{8-46}$$

Figure 8-8 shows the RL of Eq. (8-44) with $b = 1$ and several values of a. The results are summarized as follows:

Figure 8-8(a): $a = 10$. Breakaway points: $s = -2.5$ and -4.0.

Figure 8-8(b): $a = 9$. The two breakaway points given by Eq. (8-46) converge to one point at $s = -3$. Note the change in the RL when the pole at $-a$ is moved from -10 to -9.

For values of a less than 9, the values of s as given by Eq. (8-46) no longer satisfy Eq. (8-44), which means that there are no finite, nonzero, breakaway points.

Figure 8-8(c): $a = 8$. No breakaway point on RL.

As the pole at $s = -a$ is moved further to the right along the real axis, the complex portion of the RL is pushed further toward the right-half plane.

Figure 8-8(d): $a = 3$.

Figure 8-8(e): $a = b = 1$. The pole at $s = -a$ and the zero at $-b$ cancel each other, and the RL degenerate into a second-order case and lie entirely on the $j\omega$-axis.

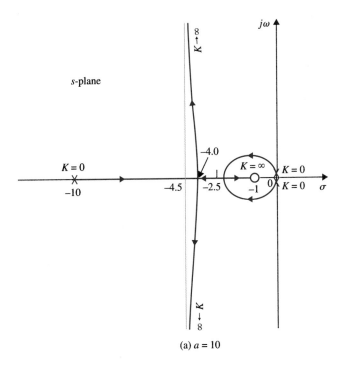

(a) $a = 10$

Figure 8-8 Root-locus diagrams that show the effects of moving a pole of $G(s)H(s)$. $G(s)H(s) = K(s + 1)/[s^2(s + a)]$.

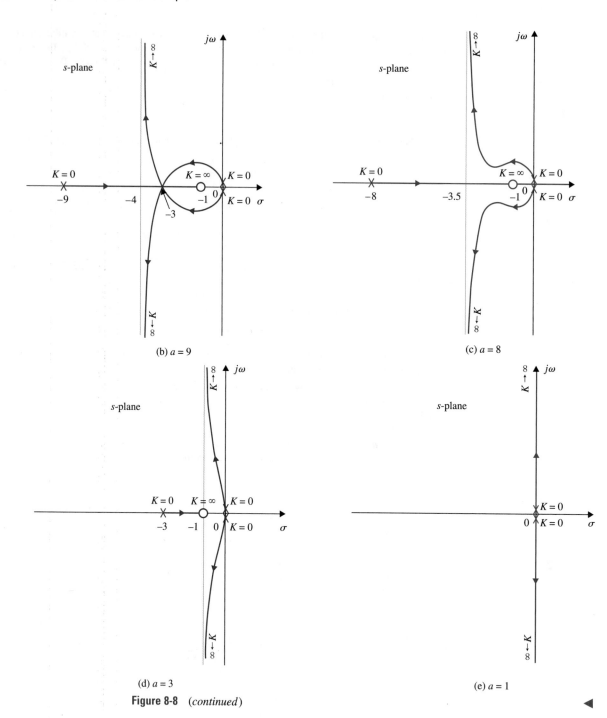

(b) $a = 9$

(c) $a = 8$

(d) $a = 3$

(e) $a = 1$

Figure 8-8 (*continued*)

► **EXAMPLE 8-6** Consider the equation

$$s(s^2 + 2s + a) + K(s + 2) = 0 \qquad (8\text{-}47)$$

which leads to the equivalent $G(s)H(s)$ as

$$G(s)H(s) = \frac{K(s + 2)}{s(s^2 + 2s + a)} \qquad (8\text{-}48)$$

The objective is to study the RL for various values of a (> 0). The breakaway point equation of the RL is determined as

$$s^3 + 4s^2 + 4s + a = 0 \qquad (8\text{-}49)$$

Figure 8-9 shows the RL of Eq. (8-47) under the following conditions.

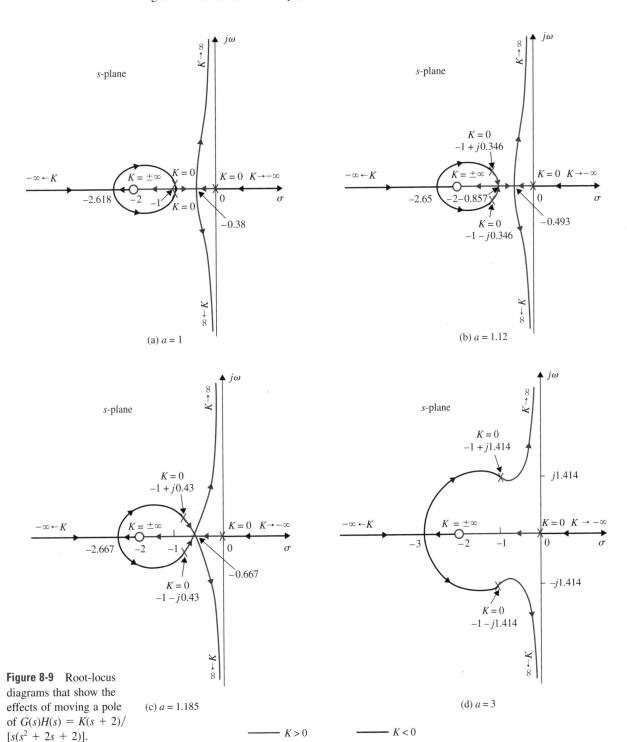

Figure 8-9 Root-locus diagrams that show the effects of moving a pole of $G(s)H(s) = K(s + 2)/[s(s^2 + 2s + 2)]$.

Figure 8-9(a): $a = 1$. Breakaway points: $s = -0.38$, -1.0, and -2.618, with the last point being on the RL for $K \geq 0$.

As the value of a is increased from unity, the two double poles of $G(s)H(s)$ at $s = -1$ will move vertically up and down with the real parts equal to -1. The breakaway points at $s = -0.38$ and $s = -2.618$ will move to the left, whereas the breakaway point at $s = -1$ will move to the right.

Figure 8-9(b): $a = 1.12$. Breakaway points: $s = -0.493$, -0.857, and -2.65. Since the real parts of the poles and zeros of $G(s)H(s)$ are not affected by the value of a, the intersect of the asymptotes is always at $s = 0$.

Figure 8-9(c): $a = 1.185$. Breakaway points: $s = -0.667$, -0.667, and -2.667. The two breakaway points of the RL that lie between $s = 0$ and -1 converge to a point.

Figure 8-9(d): $a = 3$. Breakaway point: $s = -3$. When a is greater than 1.185, Eq. (8-49) yields only one solution for the breakaway point.

The reader may investigate the difference between the RL in Figs. 8-9(c) and 8-9(d), and fill in the evolution of the loci when the value of a is gradually changed from 1.185 to 3 and beyond. ◄

► 8-5 ROOT CONTOURS (RC): MULTIPLE-PARAMETER VARIATION

The root-locus technique discussed thus far is limited to only one variable parameter in K. In many control-systems problem, the effects of varying several parameters should be investigated. For example, when designing a controller that is represented by a transfer function with poles and zeros, it would be useful to investigate the effects on the characteristic equation roots when these poles and zeros take on various values. In Section 8-4, the root loci of equations with two variable parameters are studied by fixing one parameter and assigning different values to the other. In this section, the multiparameter problem is investigated through a more systematic method of embedding. When more than one parameter varies continuously from $-\infty$ to ∞, the root loci are referred to as the **root contours (RC)**. It will be shown that the root contours still possess the same properties as the single-parameter root loci, so that the methods of construction discussed thus far are all applicable.

The principle of root contour can be described by considering the equation

$$P(s) + K_1 Q_1(s) + K_2 Q_2(s) = 0 \tag{8-50}$$

where K_1 and K_2 are the variable parameters, and $P(s)$, $Q_1(s)$, and $Q_2(s)$ are polynomials of s. The first step involves setting the value of one of the parameters to zero. Let us set K_2 to zero. Then, Eq. (8-50) becomes

$$P(s) + K_1 Q_1(s) = 0 \tag{8-51}$$

which now has only one variable parameter in K_1. The root loci of Eq. (8-51) may be determined by dividing both sides of the equation by $P(s)$. Thus,

$$1 + \frac{K_1 Q_1(s)}{P(s)} = 0 \tag{8-52}$$

Equation (8-52) is of the form of $1 + K_1 G_1(s) H_1(s) = 0$, so we can construct the RL of the equation based on the pole-zero configuration of $G_1(s)H_1(s)$. Next, we restore the value of K_2, while considering the value of K_1 fixed, and divide both sides of Eq. (8-50) by the terms that do not contain K_2. We have

$$1 + \frac{K_2 Q_2(s)}{P(s) + K_1 Q_1(s)} = 0 \tag{8-53}$$

which is of the form of $1 + K_2 G_2(s)H_2(s) = 0$. The root contours of Eq. (8-50) when K_2 varies (while K_1 is fixed) are constructed based on the pole-zero configuration of

$$G_2(s)H_2(s) = \frac{Q_2(s)}{P(s) + K_1 Q_1(s)} \tag{8-54}$$

It is important to note that the poles of $G_2(s)H_2(s)$ are identical to the roots of Eq. (8-51). Thus, the root contours of Eq. (8-50) when K_2 varies must all start ($K_2 = 0$) at the points that lie on the root loci of Eq. (8-51). This is the reason why one root-contour problem is considered to be embedded in another. The same procedure may be extended to more than two variable parameters. The following examples illustrate the construction of RCs when multiparameter-variation situations exist.

▶ **EXAMPLE 8-7** Consider the equation

$$s^3 + K_2 s^2 + K_1 s + K_1 = 0 \tag{8-55}$$

where K_1 and K_2 are the variable parameters, which vary from 0 to ∞.

As the first step, we let $K_2 = 0$, and Eq. (8-55) becomes

$$s^3 + K_1 s + K_1 = 0 \tag{8-56}$$

Dividing both sides of the last equation by s^3, which is the term that does not contain K_1, we have

$$1 + \frac{K_1(s + 1)}{s^3} = 0 \tag{8-57}$$

The root contours of Eq. (8-56) are drawn based on the pole-zero configuration of

$$G_1(s)H_1(s) = \frac{s + 1}{s^3} \tag{8-58}$$

as shown in Fig. 8-10(a). Next, we let K_2 vary between 0 and ∞ while holding K_1 at a constant nonzero value. Dividing both sides of Eq. (8-55) by the terms that do not contain K_2, we have

$$1 + \frac{K_2 s^2}{s^3 + K_1 s + K_1} = 0 \tag{8-59}$$

Thus, the root contours of Eq. (8-55) when K_2 varies may be drawn from the pole-zero configuration of

$$G_2(s)H_2(s) = \frac{s^2}{s^3 + K_1 s + K_1} \tag{8-60}$$

The zeros of $G_2(s)H_2(s)$ are at $s = 0, 0$; but the poles are at the zeros of $1 + K_1 G_1(s)H_1(s)$, which are found on the RL of Fig. 8-10(a). Thus, for fixed K_1, the RC when K_2 varies must all emanate from the root contours of Eq. 8-10(a). Figure 8-10(b) shows the root contours of Eq. (8-55) when K_2 varies from 0 to ∞, for $K_1 = 0.0184$, 0.25, and 2.56. ◀

▶ **EXAMPLE 8-8** Consider the loop transfer function

$$G(s)H(s) = \frac{K}{s(1 + Ts)(s^2 + 2s + 2)} \tag{8-61}$$

of a closed-loop control system. It is desired to construct the root contours of the characteristic equation with K and T as variable parameters. The characteristic equation of the system is

$$s(1 + Ts)(s^2 + 2s + 2) + K = 0 \tag{8-62}$$

First, we set the value of T to zero. The characteristic equation becomes

$$s(s^2 + 2s + 2) + K = 0 \tag{8-63}$$

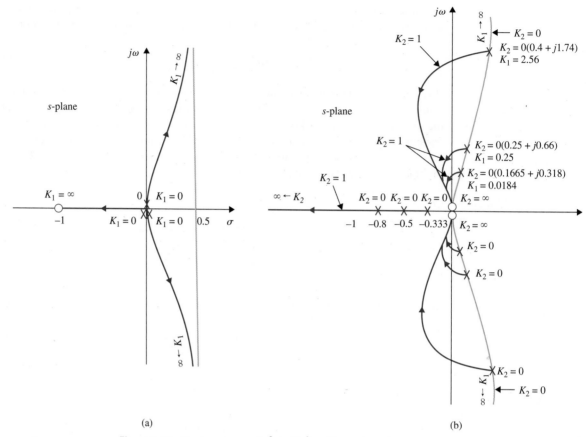

Figure 8-10 Root contours of $s^2 + K_2 s^2 + K_1 s + K_1 = 0$. (a) $K_2 = 0$. (b) K_2 varies and K_1 is a constant.

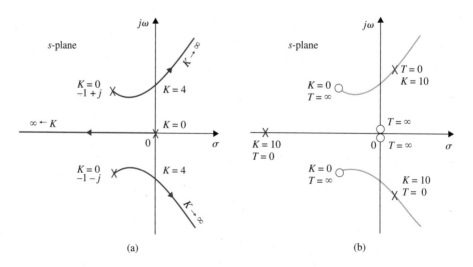

Figure 8-11 (a) RL for $s(s^2 + 2s + 2) + K = 0$. (b) Pole-zero configuration of $G_2(s)H_2(s) = Ts^2(s^2 + 2s + 2)/[s(s^2 + 2s + 2) + K]$.

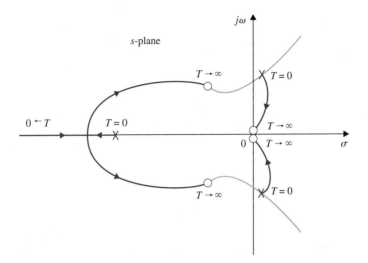

Figure 8-12 Root contours for $s(1 + Ts)$ $(s^2 + 2s + 2) + K = 0.$ $K > 4.$

The root contours of this equation when K varies are drawn based on the pole-zero configuration of

$$G_1(s)H_1(s) = \frac{1}{s(s^2 + 2s + 2)} \qquad (8\text{-}64)$$

as shown in Fig. 8-11(a). Next, we let K be fixed, and consider that T is the variable parameter. Dividing both sides of Eq. (8-62) by the terms that do not contain T, we get

$$1 + TG_2(s)H_2(s) = 1 + \frac{Ts^2(s^2 + 2s + 2)}{s(s^2 + 2s + 2) + K} = 0 \qquad (8\text{-}65)$$

The root contours when T varies are constructed based on the pole-zero configuration of $G_2(s)H_2(s)$. When $T = 0$, the points on the root contours are at the poles of $G_2(s)H_2(s)$, which are on the root contours of Eq. (8-63). When $T = \infty$, the roots of Eq. (8-62) are at the zeros of $G_2(s)H_2(s)$, which are at $s = 0, 0, -1 + j$, and $-1 - j$. Figure 8-11(b) shows the pole-zero configuration of $G_2(s)H_2(s)$ for $K = 10$. Notice that $G_2(s)H_2(s)$ has three finite poles and four finite zeros. The root contours for Eq. (8-62) when T varies are shown in Figs. 8-12, 8-13, and 8-14 for three different values of K.

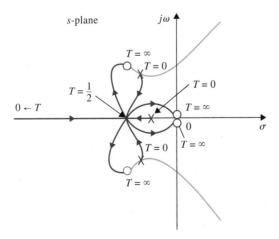

Figure 8-13 Root contours for $s(1 + Ts)(s^2 + 2s + 2) + K = 0.$ $K = 0.5.$

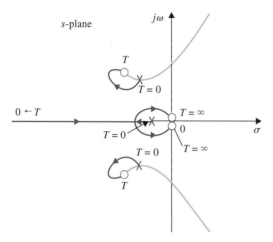

Figure 8-14 Root contours for $s(1 + Ts)(s^2 + 2s + 2) + K = 0$. $K < 0.5$.

The root contours in Fig. 8-13 show that when $K = 0.5$ and $T = 0.5$, the characteristic equation in Eq. (8-62) has a quadruple root at $s = -1$. ◀

▶ **EXAMPLE 8-9** As an example to illustrate the effect of the variation of a zero of $G(s)H(s)$, consider the function

$$G(s)H(s) = \frac{K(1 + Ts)}{s(s + 1)(s + 2)} \tag{8-66}$$

The characteristic equation is

$$s(s + 1)(s + 2) + K(1 + Ts) = 0 \tag{8-67}$$

Let us first set T to zero and consider the effect of varying K. Equation (8-67) becomes

$$s(s + 1)(s + 2) + K = 0 \tag{8-68}$$

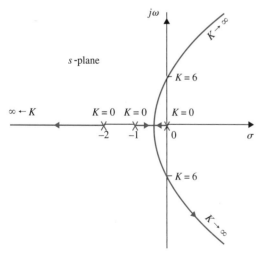

Figure 8-15 Root loci for $s(s + 1)(s + 2) + K = 0$.

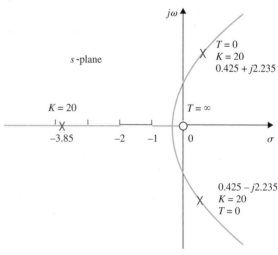

Figure 8-16 Pole-zero configuration of $G_2(s)H_2(s) = TKs[s(s + 1)(s + 2) + K]$. $K = 20$.

This leads to

$$G_1(s)H_1(s) = \frac{1}{s(s + 1)(s + 2)} \tag{8-69}$$

The root contours of Eq. (8-68) are drawn based on the pole-zero configuration of Eq. (8-69), and are shown in Fig. 8-15.

When the K is fixed and nonzero, we divide both sides of Eq. (8-67) by the terms that do not contain T, and we get

$$1 + TG_2(s)H_2(s) = 1 + \frac{TKs}{s(s + 1)(s + 2) + K} = 0 \tag{8-70}$$

The points that correspond to $T = 0$ on the root contours are at the poles of $G_2(s)H_2(s)$ or the zeros of $s(s + 1)(s + 2) + K$, whose root contours are sketched as shown in Fig. 8-15 when K varies. If we choose $K = 20$ just as an illustration, the pole-zero configuration of $G_2(s)H_2(s)$ is shown in Fig. 8-16. The root contours of Eq. (8-67) for $0 \le T < \infty$ are shown in Fig. 8-17 for three different

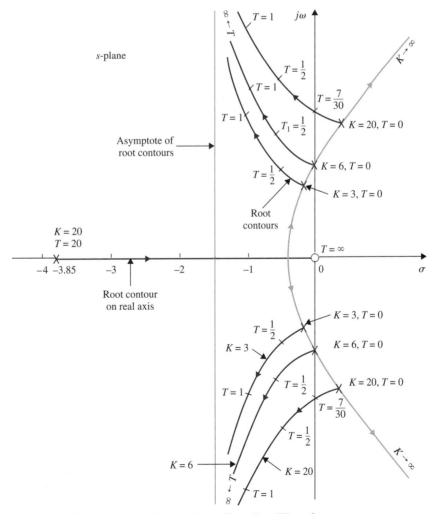

Figure 8-17 Root contours of $s(s + 1)(s + 2) + K + KTs = 0$.

values of K. Since $G_2(s)H_2(s)$ has three poles and one zero, the angles of the asymptotes of the root contours when T varies are at $90°$ and $-90°$. We can show that the intersection of the asymptotes is always at $s = 1.5$. This is because the sum of the poles of $G_2(s)H_2(s)$, which is given by the negative of the coefficient of the s^2 term in the denominator polynomial of Eq. (8-70), is 3, the sum of the zeros of $G_2(s)H_2(s)$ is 0, and $n - m$ in Eq. (8-30) is 2.

The root contours in Fig. 8-17 show that adding a zero to the loop transfer function generally improves the relative stability of the closed-loop system by moving the characteristic equation roots toward the left in the s-plane. As shown in Fig. 8-17, for $K = 20$, the system is stabilized for all values of T greater than 0.2333. However, the largest relative damping ratio that the system can have by increasing T is only approximately 30 percent. ◀

▶ 8-6 ROOT LOCUS WITH THE MATLAB TOOLBOX

In Chapter 7 we introduced the Time Response Analysis toolbox, **timetool**, for obtaining the time response of control systems, and further introduced the root-locus design tool. The timetool may also be used to solve any of the problems in this chapter involving construction of the root locus. You should use the **tftool** to convert the transfer function from the zero-pole format into polynomial form (refer to Chapter 7 for detailed description).

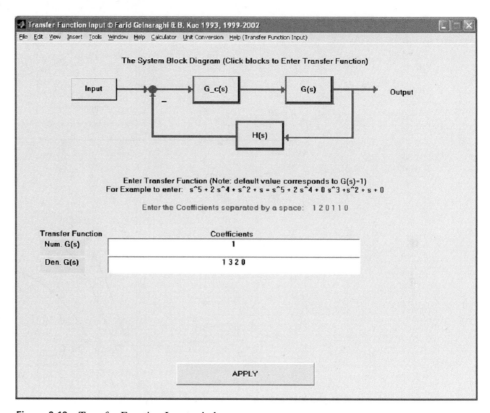

Figure 8-18 Transfer Function Input window.

Consider the loop transfer function of a single-loop control system:

$$G(s)H(s) = \frac{K}{s(s + 1)(s + 2)} = \frac{K}{s^3 + 3s^2 + 2s} \tag{8-71}$$

In order to find the root locus, activate the timetool and click the "Enter Transfer Function" button. In the resulting Transfer Function Input window, enter the values of $G(s)$ and $H(s)$. In this case, you may assign $H(s)$ a numerical value corresponding to K, say, $K = 1$. In this case the plant $G(s)$ is

$$G(s) = \frac{1}{s^3 + 3s^2 + 2s} \tag{8-72}$$

Figure 8-18 shows the transfer function input module after entering the $G(s)$ values. Note the default values for $G\_c(s)$ and $H(s)$ are set to 1. Next use the "APPLY" button to go back to the Time Response Analysis window, and press "Calculate" to evaluate the transfer function. As shown in Fig. 8-19, the loop transfer function for $G\_c(s) = 1$ and $H(s) = 1$ is the same as Eq. (8-72).

The root locus is found by pressing the "Root Locus" button as shown in Fig. 8-20a. The red squares define the closed-loop poles, which may also be obtained by activating the Closed-Loop Pole Viewer from the edit menu, as shown in Fig. 8-20b.

Students are encouraged to try the root locus function for all the examples in this chapter and those in Appendix F.

```
G=
Transfer function:
       1
---------------
s^3 + 3 s^2 + 2 s

Gc=

Transfer function:
1
H=

Transfer function:
1

G*G_c ==>open loop

Transfer function:
       1
---------------
s^3 + 3 s^2 + 2 s

G*G_c*H ==>loop

Transfer function:
       1
---------------
s^3 + 3 s^2 + 2 s
```

```
G*G_c/(1+G*G_c*H) ==> closed loop

Transfer function:
          1
-----------------------
s^3 + 3 s^2 + 2 s + 1

system coefficients are
Closed loop TF in zero/pole format

Zero/pole/gain:
              1
---------------------------------
(s+2.325) (s^2 + 0.6753s + 0.4302)

System zeros are

zeroTF =

  Empty matrix: 1-by-0

System poles are

poleTF =

-2.3247    -0.3376 - 0.5623i -0.3376 + 0.5623i
```

Figure 8-19 MATLAB command window representation of the transfer function.

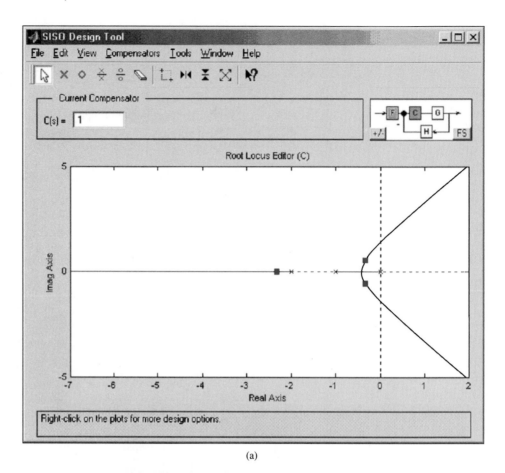

(a)

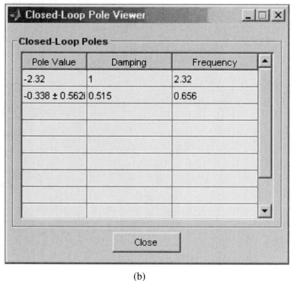

(b)

Figure 8-20 (a) The root locus of Eq. (8-107), (b) Poles of the closed-loop system for $K = 1$.

► 8-7 SUMMARY

In this chapter we introduced the root-locus technique for linear continuous-data control systems. The technique represents a graphical method of investigating the roots of the characteristic equation of a linear time-invariant system when one or more parameters vary. In Chapter 10 the root-locus method will be used heavily for the design of control systems. However, keep in mind that the characteristic equation roots give exact indication on the absolute stability of linear SISO systems but give only qualitative information on the relative stability, since the zeros of the closed-loop transfer function, if any, also play an important role on the dynamic performance of the system.

The root-locus technique can also be applied to discrete-data systems with the characteristic equation expressed in the z-transform. As will be shown in Appendix I, the properties and construction of the root loci in the z-plane are essentially the same as those of the continuous-data systems in the s-plane, except that the interpretation of the root location to system performance must be made with respect to the unit circle $|z| = 1$ and the significance of the regions in the z-plane.

The majority of the material in this chapter is designed to provide the basics of constructing the root loci. Computer programs can be used to plot the root loci and provide details of the plot. The final section of this chapter deals with the root-locus tools of MAT-LAB. However, the authors believe that a computer program can be used only as a tool, and the intelligent investigator should have a thorough understanding of the fundamentals of the subject.

The root-locus technique can also be applied to linear systems with pure time delay in the system loop. The subject is not treated here, since systems with pure time delays are more easily treated with the frequency-domain methods discussed in Chapter 9.

► REVIEW QUESTIONS

The following questions and true-and-false problems all refer to the equation $P(s) + KQ(s) = 0$, where $P(s)$ and $Q(s)$ are polynomials of s with constant coefficients.

1. Give the condition from which the root loci are constructed.

2. Determine the points on the complete root loci at which $K = 0$, with reference to the poles and zeros of $Q(s)/P(s)$.

3. Determine the points on the root loci at which $K = \pm\infty$, with reference to the poles and zeros of $Q(s)/P(s)$.

4. Give the significance of the breakaway points with respect to the roots of $P(s) + KQ(s) = 0$.

5. Give the equation of intersect of the asymptotes.

6. The asymptotes of the root loci refer to the angles of the root loci when $K = \pm\infty$.   **(T)**   **(F)**

7. There is only one intersect of the asymptotes of the complete root loci.   **(T)**   **(F)**

8. The intersect of the asymptotes must always be on the real axis.   **(T)**   **(F)**

9. The breakaway points of the root loci must always be on the real axis.   **(T)**   **(F)**

10. Given the equation: $1 + KG_1(s)H_1(s) = 0$, where $G_1(s)H_1(s)$ is a rational function of s, and does not contain K, the roots of $dG_1(s)H_1(s)/ds$ are all breakaway points on the root loci ($-\infty < K < \infty$).   **(T)**   **(F)**

11. At the breakaway points on the root loci, the root sensitivity is infinite.   **(T)**   **(F)**

12. Without modification, all the rules and properties for the construction of root loci in the s-plane can be applied to the construction of root loci of discrete-data systems in the z-plane.   **(T)**   **(F)**

13. The determination of the intersections of the root loci in the *s*-plane with the *jω*-axis can be made by solving the auxiliary equation of Routh's tabulation of the equation. **(T)** **(F)**

14. Adding a pole to $Q(s)/P(s)$ has the general effect of pushing the root loci to the right, whereas adding a zero pushes the loci to the left. **(T)** **(F)**

Answers to these true-and-false questions are found after the Problems section.

▶ REFERENCES

General Subjects

1. W. R. Evans, "Graphical Analysis of Control Systems," *Trans. AIEE*, Vol. 67, pp. 548–551,1948.
2. W. R. Evans, "Control System Synthesis by Root Locus Method," *Trans. AIEE*, Vol. 69, pp. 66–69, 1950.
3. W. R. Evans, *Control System Dynamics*, McGraw-Hill Book Company, New York, 1954.

Construction and Properties of Root Loci

4. C. C. MacDuff, *Theory of Equations*, pp. 29–104, John Wiley & Sons, New York, 1954.
5. C. S. Lorens and R. C. Titsworth, "Properties of Root Locus Asymptotes," *IRE Trans. Automatic Control*, AC-5, pp. 71–72, Jan. 1960.
6. C. A. Stapleton, "On Root Locus Breakaway Points," *IRE Trans. Automatic Control*, Vol. AC-7, pp. 88–89, April 1962.
7. M. J. Remec, "Saddle-Points of a Complete Root Locus and an Algorithm for Their Easy Location in the Complex Frequency Plane," *Proc. Natl. Electronics Conf.*, Vol. 21, pp. 605–608, 1965.
8. C. F. Chen, "A New Rule for Finding Breaking Points of Root Loci Involving Complex Roots," *IEEE Trans. Automatic Control*, AC-10, pp. 373–374, July 1965.
9. V. Krishran, "Semi-analytic Approach to Root Locus," *IEEE Trans. Automatic Control*, Vol. AC-11, pp. 102–108, Jan. 1966.
10. R. H. Labounty and C. H. Houpis, "Root Locus Analysis of a High-Grain Linear System with Variable Coefficients; Application of Horowitz's Method," *IEEE Trans. Automatic Control*, Vol. AC-11, pp. 255–263, April 1966.
11. A. Fregosi and J. Feinstein, "Some Exclusive Properties of the Negative Root Locus," *IEEE Trans. Automatic Control*, Vol. AC-14, pp. 304–305, June 1969.

Analytical Representation of Root Loci

12. G. A. Bendrikov and K. F. Teodorchik, "The Analytic Theory of Constructing Root Loci," *Automation and Remote Control*, pp. 340–344, March 1959.
13. K. Steiglitz, "Analytical Approach to Root Loci," *IRE Trans. Automatic Control*, Vol. AC-6, pp. 326–332, Sept. 1961.
14. C. Wojcik, "Analytical Representation of Root Locus," *Trans. ASME*, J. Basic Engineering, Ser. D. Vol. 86, March 1964.
15. C. S. Chang, "An Analytical Method for Obtaining the Root Locus with Positive and Negative Gain," *IEEE Trans. Automatic Control*, Vol. AC-10, pp. 92–94, Jan. 1965.
16. B. P. Bhattacharyya, "Root Locus Equations of the Fourth Degree," *Interna. J. Control*, Vol. 1, No. 6, pp. 533–556, 1965.

Root Sensitivity

17. J. G. Truxal and M. Horowitz, "Sensitivity Consideration in Active Network Synthesis," *Proc. Second Midwest Symposium on Circuit Theory*, East Lansing, MI, 1956.
18. R. Y. Huang, "The Sensitivity of the Poles of Linear Closed-Loop Systems," *IEEE Trans. Appl. Ind.*, Vol. 77, Part 2, pp. 182–187, Sept. 1958.
19. H. Ur, "Root Locus Properties and Sensitivity Relations in Control Systems," *IRE Trans. Automatic Control*, Vol. AC-5, pp. 58–65, Jan. 1960.

▶ PROBLEMS

The MATLAB toolbox provided with this text (timetool), can be used to solve the problems in this chapter. Do the problems without relying on the computer first, and then carry out the computer solutions. Determine all the vital information and mark on the root loci, including the points at which the variable parameter, e.g., *K*, equals zero and $\pm\infty$, asymptotic properties, intersections on the $j\omega$-axis and the corresponding parameter value, and the breakaway points. Indicate the direction of increase of the variable parameter on the loci with arrows.

• Asymptotes

8-1. Find the angles of the asymptotes and the intersect of the asymptotes of the root loci of the following equations when K varies from $-\infty$ to ∞.

(a) $s^4 + 4s^3 + 4s^2 + (K + 8)s + K = 0$ (b) $s^3 + 5s^2 + (K + 1)s + K = 0$

(c) $s^2 + K(s^3 + 3s^2 + 2s + 8) = 0$ (d) $s^3 + 2s^2 + 3s + K(s^2 - 1)(s + 3) = 0$

(e) $s^5 + 2s^4 + 3s^3 + K(s^2 + 3s + 5) = 0$ (f) $s^4 + 2s^2 + 10 + K(s + 5) = 0$

• Angles of departure and arrival

8-2 For the loop transfer functions that follow, find the angle of departure or arrival of the root loci at the designated pole or zero.

(a) $G(s)H(s) = \dfrac{Ks}{(s + 1)(s^2 + 1)}$

Angle of arrival $(K < 0)$ and angle of departure $(K > 0)$ at $s = j$.

(b) $G(s)H(s) = \dfrac{Ks}{(s - 1)(s^2 + 1)}$

Angle of arrival $(K < 0)$ and angle of departure $(K > 0)$ at $s = j$.

(c) $G(s)H(s) = \dfrac{K}{s(s + 2)(s^2 + 2s + 2)}$

Angle of departure $(K > 0)$ at $s = -1 + j$.

(d) $G(s)H(s) = \dfrac{K}{s^2(s^2 + 2s + 2)}$

Angle of departure $(K > 0)$ at $s = -1 + j$.

(e) $G(s)H(s) = \dfrac{K(s^2 + 2s + 2)}{s^2(s + 2)(s + 3)}$

Angle of arrival $(K > 0)$ at $s = -1 + j$.

• $K = 0$ and $K = \pm\infty$ points and RL and CRL on real axis

8-3. Mark the $K = 0$ and $K = \pm\infty$ points and the RL and CRL on the real axis for the pole-zero configurations shown in Fig. 8P-3. Add arrows on the root loci on the real axis in the direction of increasing K.

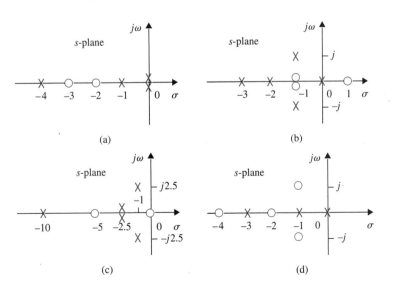

Figure 8P-3

• Breakaway points

8-4. Find all the breakaway points of the root loci of the systems described by the pole-zero configurations shown in Fig. 8P-3.

• Root-loci construction.

8-5. Construct the root-locus diagram for each of the following control systems for which the poles and zeros of $G(s)H(s)$ are given. The characteristic equation is obtained by equating the numerator of $1 + G(s)H(s)$ to zero.

(a) Poles at 0, −5, −6; zero at −8.

(b) Poles at 0, −1, −3, −4; no finite zeros.

(c) Poles at 0, 0, −2, −2; zero at −4.

(d) Poles at 0, −1 +j, −1 −j; zero at −2.

(e) Poles at 0, −1 +j, −1 −j; zero at −5.

(f) Poles at 0, −1 +j, −1 −j; no finite zeros.

(g) Poles at 0, 0, −8, −8; zeros at −4, −4.

(h) Poles at 0, 0, −8, −8; no finite zeros.

(i) Poles at 0, 0, −8, −8; zeros at −4 + $j2$, −4 −$j2$.

(j) Poles at −2, 2; zeros at 0, 0.

(k) Poles at j, −j, $j2$, −$j2$; zeros at −2, 2.

(l) Poles at j, −j, $j2$, −$j2$; zeros at −1, 1.

(m) Poles at 0, 0, 0, 1; zeros at −1, −2, −3.

(n) Poles at 0, 0, 0, −100, −200; zeros at −5, −40.

(o) Poles at 0, −1, −2; zero at 1.

• Root-loci construction

8-6. The characteristic equation of linear control systems are given as follows. Construct the root loci for $K \geq 0$.

(a) $s^3 + 3s^2 + (K + 2)s + 5K = 0$

(b) $s^3 + s^2 + (K + 2)s + 3K = 0$

(c) $s^3 + 5Ks^2 + 10 = 0$

(d) $s^4 + (K + 3)s^3 + (K + 1)s^2 + (2K + 5) + 10 = 0$

(e) $s^3 + 2s^2 + 2s + K(s^2 − 1)(s + 2) = 0$

(f) $s^3 − 2s + K(s + 4)(s + 1) = 0$

(g) $s^4 + 6s^3 + 9s^2 + K(s^2 + 4s + 5) = 0$

(h) $s^3 + 2s^2 + 2s + K(s^2 − 2)(s + 4) = 0$

(i) $s(s^2 − 1) + K(s + 2)(s + 0.5) = 0$

(j) $s^4 + 2s^3 + 2s^2 + 2Ks + 5K = 0$

(k) $s^5 + 2s^4 + 3s^3 + 2s^2 + s + K = 0$

• Root-loci construction

8-7. The forward-path transfer functions of a unity-feedback control systems are given in the following.

(a) $G(s) = \dfrac{K(s + 3)}{s(s^2 + 4s + 4)(s + 5)(s + 6)}$

(b) $G(s) = \dfrac{K}{s(s + 2)(s + 4)(s + 10)}$

(c) $G(s) = \dfrac{K(s^2 + 2s + 8)}{s(s + 5)(s + 10)}$

(d) $G(s) = \dfrac{K(s^2 + 4)}{(s + 2)^2(s + 5)(s + 6)}$

(e) $G(s) = \dfrac{K(s + 10)}{s^2(s + 2.5)(s^2 + 2s + 2)}$

Construct the root loci for $K \geq 0$. Find the value of K that makes the relative damping ratio of the closed-loop system (measured by the dominant complex characteristic equation roots) equal to 0.707 if such solution exists.

• Root loci, breakaway points

8-8. A unity-feedback control system has the forward-path transfer function given in the following. Construct the root-locus diagram for $K \geq 0$. Find the values of K at all the breakaway points.

(a) $G(s) = \dfrac{K}{s(s + 10)(s + 20)}$

(b) $G(s) = \dfrac{K}{s(s + 1)(s + 3)(s + 5)}$

• Root-loci construction

8-9. The forward-path transfer function of a unity-feedback control system is

$$G(s) = \frac{K}{(s + 4)^n}$$

Construct the root loci of the characteristic equation of the closed-loop system for $K \geq \infty$, with
(a) $n = 1$, (b) $n = 2$, (c) $n = 3$, (d) $n = 4$, and (e) $n = 5$.

• Root loci of system with tachometer feedback

8-10. The characteristic equation of the control system shown in Fig. 7P-10 on p. 310 when $K = 100$ is

$$s^3 + 25s^2 + (100K_t + 2)s + 100 = 0$$

Construct the root loci of the equation for $K_t \geq 0$.

• Root loci of system with tachometer feedback

8-11. The block diagram of a control system with tachometer feedback is shown in Fig. 8P-11.
(a) Construct the root loci of the characteristic equation for $K \geq 0$ when $K_t = 0$.
(b) Set $K = 10$. Construct the root loci of the characteristic equation for $K_t \geq 0$.

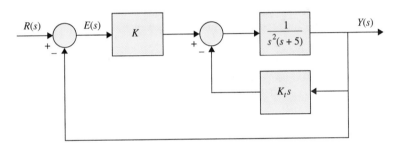

Figure 8P-11

• Root loci of dc-motor control system

8-12. The characteristic equation of the dc-motor control system described in Problem 7-27 can be approximated as

$$2.05J_L s^3 + (1 + 10.25J_L)s^2 + 116.84s + 1843 = 0$$

when $K_L = \infty$, and the load inertia J_L is considered as a variable parameter. Construct the root loci of the characteristic equation for $J_L \geq 0$.

• Root loci of control system

8-13. The foward-path transfer function of the control system shown in Fig. 8P-11 is

$$G(s) = \frac{K(s + \alpha)(s + 3)}{s(s^2 - 1)}$$

(a) Construct the root loci for $K \geq 0$ with $\alpha = 5$.
(b) Construct the root loci for $\alpha \geq 0$ with $K = 10$.

• Root loci of liquid-level control system

8-14. The characteristic equation of the liquid-level control system described in Problem 6P-13 is written

$$0.06s(s + 12.5)(As + K_o) + 250N = 0$$

(a) For $A = K_o = 50$, construct the root loci of the characteristic equation as N varies from 0 to ∞.
(b) For $N = 10$ and $K_o = 50$, construct the root loci of the characteristic equation for $A \geq 0$.
(c) For $A = 50$ and $N = 20$, construct the root loci for $K_o \geq 0$.

• Root loci of liquid-level control system

8-15. Repeat Problem 8-14 for the following cases.
(a) $A = K_o = 100$.
(b) $N = 20$ and $K_o = 50$.
(c) $A = 100$ and $N = 20$.

8-16. The transfer functions of a single-feedback-loop control system are

$$G(s) = \frac{K}{s^2(s + 1)(s + 5)} \qquad H(s) = 1$$

(a) Construct the loci of the zeros of $1 + G(s)$ for $K \geq 0$.

(b) Repeat part (a) when $H(s) = 1 + 5s$.

8-17. The transfer functions of a single-feedback-loop control system are

$$G(s) = \frac{10}{s^2(s + 1)(s + 5)} \qquad H(s) = 1 + T_d s$$

Construct the root loci of the characteristic equation for $T_d \geq 0$.

8-18. For the dc-motor control system described in Problem 7-27, it is of interest to study the effects of the motor-shaft compliance K_L on the system performance.

(a) Let $K = 1$, with the other system parameters as given in Problem 7-27. Find an equivalent $G(s)H(s)$ with K_L as the gain factor. Construct the root loci of the characteristic equation for $K_L \geq 0$. The system can be approximated as a fourth-order system by canceling the large negative pole and zero of $G(s)H(s)$ that are very close to each other.

(b) Repeat part (a) with $K = 1000$.

8-19. The characteristic equation of the dc-motor control system described in Problem 7-27 is given in the following when the motor shaft is considered to be rigid ($K_L = \infty$). Let $K = 1$, $J_m = 0.001$, $L_a = 0.001$, $n = 0.1$, $R_a = 5$, $K_i = 9$, $K_b = 0.0636$, $B_m = 0$, and $K_s = 1$.

$$L_a(J_m + n^2 J_L)s^3 + (R_a J_m + n^2 R_a J_L + B_m L_a)s^2 + (R_a B_m + K_i K_b)s + n K_s K_i K = 0$$

Construct the root loci for $J_L \geq 0$ to show the effects of variation of the load inertia on system performance.

8-20. Given the equation

$$s^3 + \alpha s^2 + Ks + K = 0$$

it is desired to investigate the root loci of this equation for $-\infty < K < \infty$ and for various values of α.

(a) Construct the root loci for $-\infty < K < \infty$ when $\alpha = 12$.

(b) Repeat part (a) when $\alpha = 4$.

(c) Determine the value of α so that there is only one nonzero breakaway point on the entire root loci for $-\infty < K < \infty$. Construct the root loci.

8-21. The forward-path transfer function of a unity-feedback control system is

$$G(s) = \frac{K(s + \alpha)}{s^2(s + 3)}$$

Determine the values of α so that the root loci ($-\infty < K < \infty$) will have zero, one, and two breakaway points, respectively, not including the one at $s = 0$. Construct the root loci for $-\infty < K < \infty$ for all three cases.

8-22. The pole-zero configuration of $G(s)H(s)$ of a single-feedback-loop control system is shown in Fig. 8P-22(a). Without actually plotting, apply the angle-of-departure (and -arrival) property of the root loci to determine which root-locus diagram shown is the correct one.

Answers to True-and-False Review Questions

6. (F)   7. (T)   8. (T)   9. (F)   10. (F)   11. (T)   12. (T)   13. (T)   14. (T)

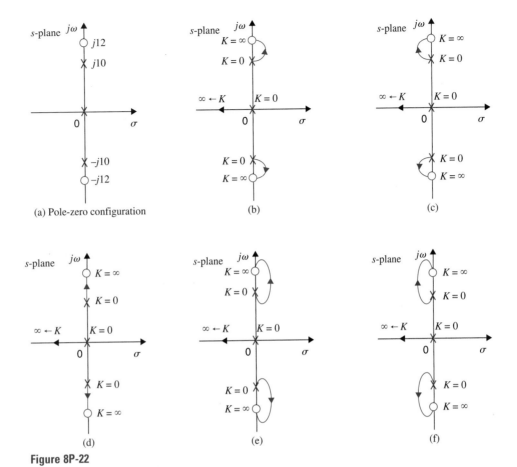

(a) Pole-zero configuration

(b)

(c)

(d)

(e)

(f)

Figure 8P-22

CHAPTER 9

Frequency-Domain Analysis

▶ 9-1 INTRODUCTION

In practice the performance of a control system is more realistically measured by its time-domain characteristics. The reason is that the performance of most control systems is judged based on the time responses due to certain test signals. This is in contrast to the analysis and design of communication systems for which the frequency response is of more importance, since most of the signals to be processed are either sinusoidal or composed of sinusoidal components. We learned in Chapter 7 that the time response of a control system is usually more difficult to determine analytically, especially for high-order systems. In design problems, there are no unified methods of arriving at a designed system that meets the time-domain performance specifications, such as maximum overshoot, rise time, delay time, settling time, and so on. On the other hand, in the frequency domain there is a wealth of graphical methods available that are not limited to low-order systems. It is important to realize that there are correlating relations between the frequency-domain and the time-domain performances in a linear system, so the time-domain properties of the system can be predicted based on the frequency-domain characteristics. The frequency domain is also more convenient for measurements of system sensitivity to noise and parameter variations. With these concepts in mind, we consider the primary motivation for conducting control systems analysis and design in the frequency domain to be convenience and the availability of the existing analytical tools. Another reason is that it presents an alternative point of view to control-system problems, which often provides valuable or crucial information in the complex analysis and design of control systems. Therefore, to conduct a frequency-domain analysis of a linear control system does not imply that the system will only be subject to a sinusoidal inputs. It may never be. Rather, from the frequency-response studies, we will be able to project the time-domain performance of the system.

The starting point for frequency-domain analysis of a linear system is its transfer function. It is well known from linear system theory that when the input to a linear time-invariant system is sinusoidal with amplitude R and frequency ω_0,

$$r(t) = R \sin \omega_0 t \tag{9-1}$$

the steady-state output of the system, $y(t)$, will be a sinusoid with the same frequency ω_0, but possibly with different amplitude and phase; that is,

$$y(t) = Y \sin(\omega_0 t + \phi) \tag{9-2}$$

where Y is the amplitude of the output sine wave, and ϕ is the phase shift in degrees or radians. Let the transfer function of a linear SISO system be $M(s)$; then the Laplace transforms of the input and the output are related through

$$Y(s) = M(s)R(s) \tag{9-3}$$

For sinusoidal steady-state analysis, we replace s by $j\omega$, and the last equation becomes

$$Y(j\omega) = M(j\omega)R(j\omega) \tag{9-4}$$

By writing the function $Y(j\omega)$ as

$$Y(j\omega) = |Y(j\omega)| \angle Y(j\omega) \tag{9-5}$$

with similar definitions for $M(j\omega)$ and $R(j\omega)$, Eq. (9-4) leads to the magnitude relation between the input and the output:

$$|Y(j\omega)| = |M(j\omega)||R(j\omega)| \tag{9-6}$$

and the phase relation:

$$\angle Y(j\omega) = \angle M(j\omega) + \angle R(j\omega) \tag{9-7}$$

Thus, for the input and output signals described by Eqs. (9-1) and (9-2), respectively, the amplitude of the output sinusoid is

$$Y = R|M(j\omega_0)| \tag{9-8}$$

and the phase of the output is

$$\phi = \angle M(j\omega_0) \tag{9-9}$$

Thus, by knowing the transfer function $M(s)$ of a linear system, the magnitude characteristic, $|M(j\omega)|$, and the phase characteristic, $\angle M(j\omega)$, completely describe the steady-state performance when the input is a sinusoid. The crux of frequency-domain analysis is that the amplitude and phase characteristics of a closed-loop system can be used to predict both time-domain transient and steady-state system performances.

9-1-1 Frequency Response of Closed-Loop Systems

For the single-loop control-system configuration studied in the preceding chapters, the closed-loop transfer function is

$$M(s) = \frac{Y(s)}{R(s)} = \frac{G(s)}{1 + G(s)H(s)} \tag{9-10}$$

Under the sinusoidal steady state, $s = j\omega$, Eq. (9-10) becomes

$$M(j\omega) = \frac{Y(j\omega)}{R(j\omega)} = \frac{G(j\omega)}{1 + G(j\omega)H(j\omega)} \tag{9-11}$$

The sinusoidal steady-state transfer function $M(j\omega)$ may be expressed in terms of its magnitude and phase; that is,

$$M(j\omega) = |M(j\omega)| \angle M(j\omega) \tag{9-12}$$

Or $M(j\omega)$ can be expressed in terms of its real and imaginary parts:

$$M(j\omega) = \text{Re}[M(j\omega)] + j\,\text{Im}[M(j\omega)] \tag{9-13}$$

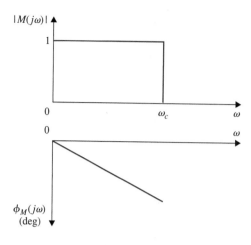

Figure 9-1 Gain-phase characteristics of an ideal low-pass filter.

The magnitude of $M(j\omega)$ is

$$|M(j\omega)| = \left|\frac{G(j\omega)}{1 + G(j\omega)H(j\omega)}\right| = \frac{|G(j\omega)|}{|1 + G(j\omega)H(j\omega)|} \qquad (9\text{-}14)$$

and the phase of $M(j\omega)$ is

$$\angle M(j\omega) = \phi_M(j\omega) = \angle G(j\omega) - \angle[1 + G(j\omega)H(j\omega)] \qquad (9\text{-}15)$$

If $M(s)$ represents the input-output transfer function of an electric filter, then the magnitude and phase of $M(j\omega)$ indicate the filtering characteristics on the input signal. Figure 9-1 shows the gain and phase characteristics of an ideal low-pass filter that has a sharp cutoff frequency at ω_c. It is well known that an ideal filter characteristic is physically unrealizable. In many ways, the design of control systems is quite similar to filter design, and the control system is regarded as a signal processor. In fact, if the ideal low-pass-filter characteristics shown in Fig. 9-1 were physically realizable, they would be highly desirable for a control system, since all signals would be passed without distortion below the frequency ω_c, and completely eliminated at frequencies above ω_c where noise may lie.

If ω_c is increased indefinitely, the output $Y(j\omega)$ would be identical to the input $R(j\omega)$ for all frequencies. Such a system would follow a step-function input in the time domain exactly. From Eq. (9-14), we see that for $|M(j\omega)|$ to be unity at all frequencies, the magnitude of $G(j\omega)$ must be infinite. An infinite magnitude of $G(j\omega)$ is, of course, impossible to achieve in practice, nor would it be desirable, since most control systems may become unstable when their loop gains become very high. Furthermore, all control systems are subject to noise during operation. Thus, in addition to responding to the input signal, the system should be able to reject and suppress noise and unwanted signals. For control systems with high-frequency noise, such as air-frame vibration of an aircraft, the frequency response should have a finite cutoff frequency ω_c.

The phase characteristics of the frequency response of a control system are also of importance, as we shall see that they affect the stability of the system.

Figure 9-2 illustrates typical gain and phase characteristics of a control system. As shown by Eqs. (9-14) and (9-15), the gain and phase of a closed-loop system can be determined from the forward-path and loop transfer functions. In practice, the frequency responses of $G(s)$ and $H(s)$ can often be determined by applying sine-wave inputs to the system and sweeping the frequency from 0 to a value beyond the frequency range of the system.

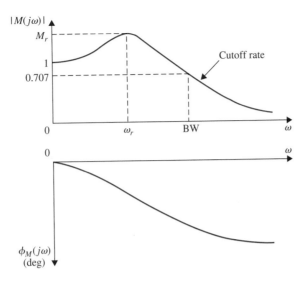

Figure 9-2 Typical gain-phase characteristics of a feedback control system.

9-1-2 Frequency-Domain Specifications

In the design of linear control systems using the frequency-domain methods, it is necessary to define a set of specifications so that the performance of the system can be identified. Specifications such as the maximum overshoot, damping ratio, and the like used in the time domain can no longer be used directly in the frequency domain. The following frequency-domain specifications are often used in practice.

Resonant Peak M_r

The resonant peak M_r is the maximum value of $|M(j\omega)|$.

- M_r indicates the relative stability of a stable closed-loop system.

In general, the magnitude of M_r gives indication on the relative stability of a stable closed-loop system. Normally, a large M_r corresponds to a large maximum overshoot of the step response. For most control systems, it is generally accepted in practice that the desirable value of M_r should be between 1.1 and 1.5.

Resonant Frequency ω_r

The resonant frequency ω_r is the frequency at which the peak resonance M_r occurs.

Bandwidth BW

- BW gives an indication of the transient response properties of a control system.

- BW gives an indication of the noise-filtering characteristics and robustness of the system.

The bandwidth BW is the frequency at which $|M(j\omega)|$ drops to 70.7 percent of, or 3 dB down from, its zero-frequency value.

In general, the bandwidth of a control system gives indication on the transient-response properties in the time domain. A large bandwidth corresponds to a faster rise time, since higher-frequency signals are more easily passed through the system. Conversely, if the bandwidth is small, only signals of relatively low frequencies are passed, and the time response will be slow and sluggish. Bandwidth also indicates the noise-filtering characteristics and the robustness of the system. The robustness represents a measure of the sensitivity of a system to parameter variations. A robust system is one that is insensitive to parameter variations.

Cutoff Rate

Often, bandwidth alone is inadequate to indicate the ability of a system in distinguishing signals from noise. Sometimes it may be necessary to look at the slope of $|M(j\omega)|$, which is called the cutoff rate of the frequency response, at high frequencies. Apparently, two systems can have the same bandwidth, but the cutoff rates may be different.

The performance criteria for the frequency-domain defined above are illustrated in Fig. 9-2. Other important criteria for the frequency domain will be defined in later sections of this chapter.

▶ 9-2 M_r, ω_r, AND BANDWIDTH OF THE PROTOTYPE SECOND-ORDER SYSTEM

9-2-1 Resonant Peak and Resonant Frequency

For the prototype second-order system defined in Section 7-6, the resonant peak M_r, the resonant frequency ω_r, and the bandwidth BW are all uniquely related to the damping ratio ζ and the natural undamped frequency ω_n of the system.

Consider the closed-loop transfer function of the prototype second-order system,

$$M(s) = \frac{Y(s)}{R(s)} = \frac{\omega_n^2}{s^2 + 2\zeta\omega_n s + \omega_n^2} \tag{9-16}$$

At sinusoidal steady state, $s = j\omega$, Eq. (9-16) becomes

$$M(j\omega) = \frac{Y(j\omega)}{R(j\omega)} = \frac{\omega_n^2}{(j\omega)^2 + 2\zeta\omega_n(j\omega) + \omega_n^2}$$

$$= \frac{1}{1 + j2(\omega/\omega_n)\zeta - (\omega/\omega_n)^2} \tag{9-17}$$

We can simplify Eq. (9-17) by letting $u = \omega/\omega_n$. Then, Eq. (9-17) becomes

$$M(ju) = \frac{1}{1 + j2u\zeta - u^2} \tag{9-18}$$

The magnitude and phase of $M(ju)$ are

$$|M(ju)| = \frac{1}{[(1 - u^2)^2 + (2\zeta u)^2]^{1/2}} \tag{9-19}$$

and

$$\angle M(ju) = \phi_M(ju) = -\tan^{-1}\frac{2\zeta u}{1 - u^2} \tag{9-20}$$

respectively. The resonant frequency is determined by setting the derivative of $|M(ju)|$ with respect to u to zero. Thus,

$$\frac{d|M(ju)|}{du} = -\frac{1}{2}[(1 - u^2)^2 + (2\zeta u)^2]^{-3/2}(4u^3 - 4u + 8u\zeta^2) = 0 \tag{9-21}$$

from which we get

$$4u^3 - 4u + 8u\zeta^2 = 4u(u^2 - 1 + 2\zeta^2) = 0 \tag{9-22}$$

In normalized frequency, the roots of Eq. (9-22) are $u_r = 0$ and

$$u_r = \sqrt{1 - 2\zeta^2} \tag{9-23}$$

The solution of $u_r = 0$ merely indicates that the slope of the $|M(ju)|$-versus-ω curve is zero at $\omega = 0$; it is not a true maximum if ζ is less than 0.707. Equation (9-23) gives the resonant frequency

$$\omega_r = \omega_n \sqrt{1 - 2\zeta^2} \tag{9-24}$$

Since frequency is a real quantity, Eq. (9-24) is meaningful only for $2\zeta^2 \le 1$, or $\zeta \le 0.707$. This means simply that for all values of ζ greater than 0.707, the resonant frequency is $\omega_r = 0$ and $M_r = 1$.

Substituting Eq. (9-23) in Eq. (9-20) for u and simplifying, we get

$$M_r = \frac{1}{2\zeta\sqrt{1 - \zeta^2}} \qquad \zeta \le 0.707 \tag{9-25}$$

• For the prototype second-order system, M_R is a function of ζ only.

It is important to note that for the prototype second-order system, M_r is a function of the damping ratio ζ only, and ω_r is a function of both ζ and ω_n. Furthermore, although taking the derivative of $|M(ju)|$ with respect to u is a valid method of determining M_r, and ω_r, for higher-order systems, this analytical method is quite tedious and is not recommended. Graphical methods to be discussed and computer methods are much more efficient for high-order systems.

• For the prototype second-order system, $M_r = 1$ and $\omega_r = 0$ when $\zeta \ge 0.707$.

Figure 9-3 illustrates the plots of $|M(ju)|$ of Eq. (9-19) versus u for various values of ζ. Notice that if the frequency scale were unnormalized, the value of $\omega_r = u_r\omega_n$ would

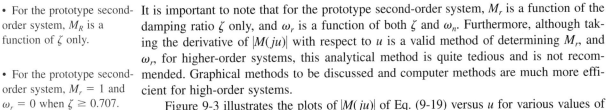

Figure 9-3 Magnification versus normalized frequency of the prototype second-order control system.

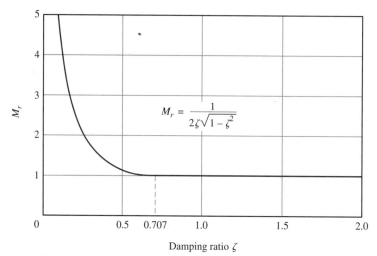

$$M_r = \frac{1}{2\zeta\sqrt{1-\zeta^2}}$$

Figure 9-4 M_r versus damping ratio for the prototype second-order system.

increase when ζ decreases, as indicated by Eq. (9-24). When $\zeta = 0$, $\omega_r = \omega_n$. Figures 9-4 and 9-5 illustrate the relationship between M_r and ζ, and u_r ($= \omega_r/\omega_n$) and ζ, respectively.

9-2-2 Bandwidth

In accordance with the definition of bandwidth, we set the value of $|M(ju)|$ to $1/\sqrt{2} \cong 0.707$.

$$|M(ju)| = \frac{1}{[(1 - u^2)^2 + (2\zeta u)^2]^{1/2}} = \frac{1}{\sqrt{2}} \qquad (9\text{-}26)$$

Thus,

$$[(1 - u^2)^2 + (2\zeta u)^2]^{1/2} = \sqrt{2} \qquad (9\text{-}27)$$

• BW/ω_n decreases monotonically as the damping ratio ζ decreases.

which leads to

$$u^2 = (1 - 2\zeta^2) \pm \sqrt{4\zeta^4 - 4\zeta^2 + 2} \qquad (9\text{-}28)$$

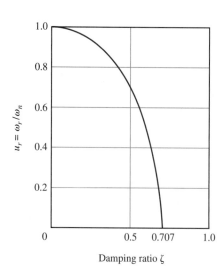

Figure 9-5 Normalized resonant frequency versus damping ratio for the prototype second-order system. $u_r = \sqrt{1 - 2\zeta^2}$.

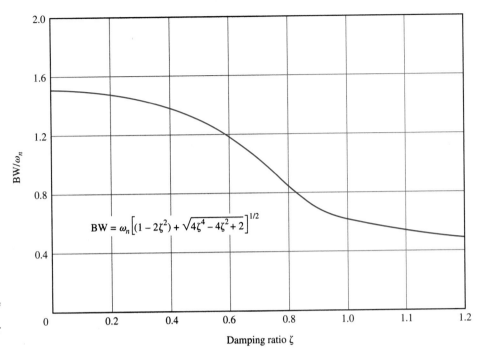

Figure 9-6 Bandwidth/ω_n versus damping ratio for the prototype second-order system.

The plus sign should be chosen in the last equation, since u must be a positive real quantity for any ζ. Therefore, the bandwidth of the prototype second-order system is determined from Eq. (9-28) as

$$BW = \omega_n[(1 - 2\zeta^2) + \sqrt{4\zeta^4 - 4\zeta^2 + 2}]^{1/2} \qquad (9\text{-}29)$$

• BW is directly proportional to ω_n.

Figure 9-6 shows a plot of BW/ω_n as a function of ζ. Notice that as ζ increases, BW/ω_n decreases monotonically. Even more important, Eq. (9-29) shows that BW is directly proportional to ω_n.

We have established some simple relationships between the time-domain response and the frequency-domain characteristics of the prototype second-order system. The summary of these relationships is as follows.

• When a system is unstable, M_r no longer has any meaning.

1. The resonant peak M_r of the closed-loop frequency response depends on ζ only [Eq. (9-25)]. When ζ is zero, M_r is infinite. When ζ is negative, the system is unstable, and the value of M_r ceases to have any meaning. As ζ increases, M_r decreases.

2. For $\zeta \geq 0.707$, $M_r = 1$ (see Fig. 9-4), and $\omega_r = 0$ (see Fig. 9-5). In comparison with the unit-step time response, the maximum overshoot in Eq. (7-109) also depends only on ζ. However, the maximum overshoot is zero when $\zeta \geq 1$.

• Bandwidth and rise time are inversely proportional to each other.

3. Bandwidth is directly proportional to ω_n [Eq. (9-29)]; that is, BW increases and decreases linearly with ω_n. BW also decreases with increase in ζ for a fixed ω_n (see Fig. 9-6). For the unit-step response, rise time increases as ω_n decreases, Eq. (7-114) and Fig. 7-22. Therefore, BW and rise time are inversely proportional to each other.

4. Bandwidth and M_r are proportional to each other for $0 \leq \zeta \leq 0.707$.

The correlation between pole locations, unit-step response, and the magnitude of the frequency response for the prototype second-order system are summarized in Fig. 9-7.

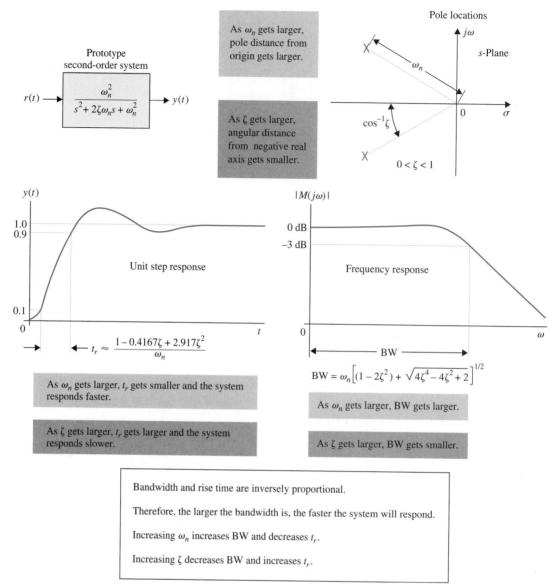

Figure 9-7 Correlation among pole locations, unit-step response, and the magnitude of frequency response of the prototype second-order system.

▷ 9-3 EFFECTS OF ADDING A ZERO TO THE FORWARD-PATH TRANSFER FUNCTION

The relationships between the time-domain and the frequency-domain responses arrived at in the preceding section apply only to the prototype second-order system described by Eq. (9-16). When other second-order or higher-order systems are involved, the relationships are different and may be more complex. It is of interest to consider the effects on the frequency-domain response when poles and zeros are added to the prototype second-order transfer function. It is simpler to study the effects of adding poles and zeros to the closed-loop transfer function; however, it is more realistic from a design standpoint to modify the forward-path transfer function.

The closed-loop transfer function of Eq. (9-16) may be considered as that of a unity-feedback control system with the prototype second-order forward-path transfer function

$$G(s) = \frac{\omega_n^2}{s(s + 2\zeta\omega_n)} \tag{9-30}$$

Let us add a zero at $s = -1/T$ to the transfer function so that Eq. (9-30) becomes

$$G(s) = \frac{\omega_n^2(1 + Ts)}{s(s + 2\zeta\omega_n)} \tag{9-31}$$

The closed-loop transfer function is

$$M(s) = \frac{\omega_n^2(1 + Ts)}{s^2 + (2\zeta\omega_n + T\omega_n^2)s + \omega_n^2} \tag{9-32}$$

• The general effect of adding a zero to the forward-path transfer function is to increase the BW of the closed-loop system.

In principle, M_r, ω_r and BW of the system can all be derived using the same steps used in the previous section. However, since there are now three paramters in ζ, ω_n, and T, the exact expression for M_r, ω_r and BW are difficult to obtain analytically even though the system is still second order. After a length derivation, the bandwidth of the system is found to be

$$BW = (-b + \tfrac{1}{2}\sqrt{b^2 + 4\omega_n^4})^{1/2} \tag{9-33}$$

where

$$b = 4\zeta^2\omega_n^2 + 4\zeta\omega_n^3 T - 2\omega_n^2 - \omega_n^4 T^2 \tag{9-34}$$

While it is difficult to see how each of the parameters in Eq. (9-33) affects the bandwidth, Fig. 9-8 shows the relationship between BW and T for $\zeta = 0.707$ and $\omega_n = 1$.

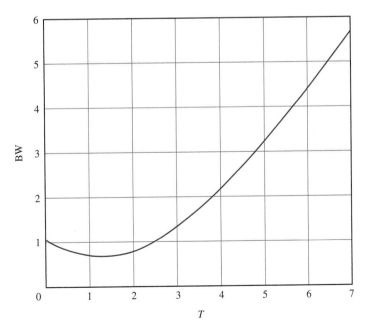

Figure 9-8 Bandwidth of a second-order system with open-loop transfer function $G(s) = (1 + Ts)/[s(s + 1.414)]$.

Notice that *the general effect of adding a zero to the forward-path transfer function is to increase the bandwidth of the closed-loop system.*

However, as shown in Fig. 9-8, over a range of small values of T, the bandwidth is actually decreased. Figures 9-9(a) and 9-9(b) gives the plots of $|M(j\omega)|$ of the closed-loop system that has the $G(s)$ of Eq. (9-31) as its forward-path transfer function; $\omega_n = 1$, $\zeta = 0.707$ and 0.2, respectively, and T takes on various values. These curves verify that the bandwidth

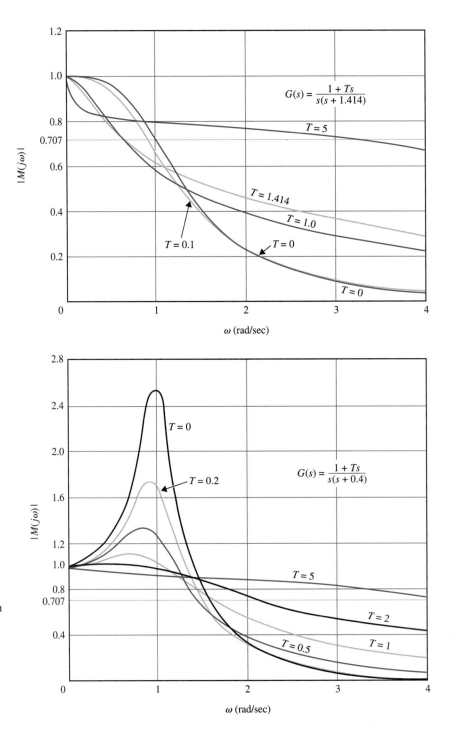

Figure 9-9 Magnification curves for the second-order system with the forward-path transfer function $G(s)$ in Eq. (9-32). (a) $\omega_n = 1$, $\zeta = 0.707$ (b) (a) $\omega_n = 1$, $\zeta = 0.2$.

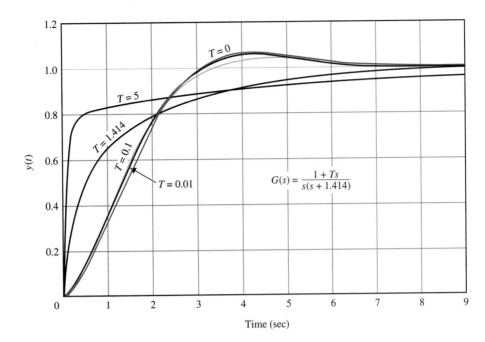

Figure 9-10 Unit-step responses of a second-order system with a forward-path transfer function $G(s)$.

$$G(s) = \frac{1 + Ts}{s(s + 1.414)}$$

generally increases with the increase of T by the addition of a zero to $G(s)$, but for a range of small values of T for which BW is actually decreased. Figures 9-10 and 9-11 show the corresponding unit-step responses of the closed-loop system. These curves show that a high bandwidth corresponds to a faster rise time. However, as T become very large, the zero of the closed-loop transfer function, which is at $s = -1/T$, moves very close to the origin, causing the system to have a large time constant. Thus, Fig. 9-10 illustrates the situation that the rise time is fast, but the large time constant of the zero near the origin of the s-plane causes the time response to drag out in reaching the final steady state (i.e., the settling time will be longer).

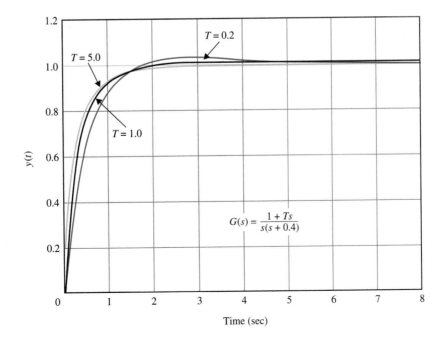

Figure 9-11 Unit-step responses of a second-order system with a forward-path transfer function $G(s)$.

$$G(s) = \frac{1 + Ts}{s(s + 0.4)}$$

▶ 9-4 EFFECTS OF ADDING A POLE TO THE FORWARD-PATH TRANSFER FUNCTION

Adding a pole at $s = -1/T$ to the forward-path transfer function of Eq. (9-30) leads to

$$G(s) = \frac{\omega_n^2}{s(s + 2\zeta\omega_n)(1 + Ts)} \qquad (9\text{-}35)$$

• Adding a pole to the forward-path transfer function makes the closed-loop system less stable, and decreases the bandwidth.

The derivation of the bandwidth of the closed-loop system with $G(s)$ given in Eq. (9-35) is quite tedious. We can obtain a qualitative indication on the bandwidth properties by referring to Fig. 9-12, which shows the plots of $|M(j\omega)|$ versus ω for $\omega_n = 1$, $\zeta = 0.707$, and various values of T. Since the system is now of the third order, it can be unstable for a certain set of system parameters. It can be shown that for $\omega_n = 1$ and $\zeta = 0.707$, the system is stable for all positive values of T. The $|M(j\omega)|$-versus-ω curves of Fig. 9-12 show that for small values of T, the bandwidth of the system is slightly increased by the addition of the pole, but M_r is also increased. When T becomes large, the pole added to $G(s)$ has the effect of decreasing the bandwidth but increasing M_r. Thus, we can conclude that, in general,

the effect of adding a pole to the forward-path transfer function is to make the closed-loop system less stable, while decreasing the bandwidth.

The unit-step responses of Fig. 9-13 show that for larger values of T, $T = 1$ and $T = 5$, the following relations are observed:

1. The rise time increases with the decrease of the bandwidth.

2. The larger values of M_r also correspond to a larger maximum overshoot in the unit-step responses.

• When $M_r = \infty$ the closed-loop system is marginally stable. When the system is unstable, M_r no longer has any meaning.

The correlation between M_r and the maximum overshoot of the step response is meaningful only when the system is stable. When $G(j\omega) = -1$, $|M(j\omega)|$ is infinite, and the closed-loop system is marginally stable. On the other hand, when the system is unstable, the value of $|M(j\omega)|$ is analytically finite, but it no longer has any significance.

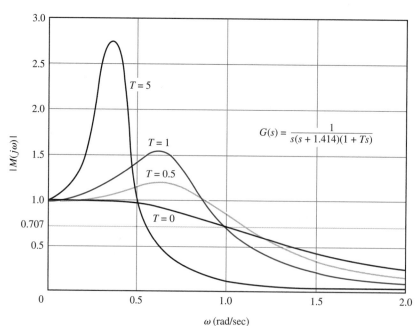

Figure 9-12 Magnification curves for a third-order system with a forward-path transfer function $G(s)$.

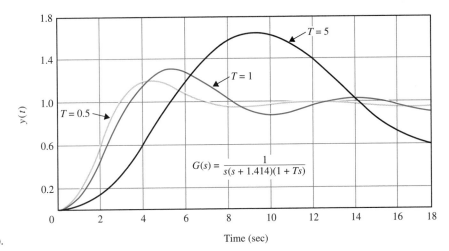

Figure 9-13 Unit-step responses of a third-order system with a forward-path transfer function $G(s)$.

The objective of these last two sections is to demonstrate the simple relationships between BW, M_r, and the time-domain response. Typical effects on BW of adding a pole and a zero to the foreward-path transfer function are investigated. No attempt is made to include all general cases.

► 9-5 NYQUIST STABILITY CRITERION: FUNDAMENTALS

• The Nyquist plot of $L(j\omega)$ is done in polar coordinates as ω varies from 0 to ∞.

Thus far we have presented two methods of determining the stability of linear SISO systems: the Routh-Hurwitz criterion and the root-locus method of determining stability by locating the roots of the characteristic equation in the s-plane. Of course, if the coefficients of the characteristic equation are all known, we can solve for the roots of the equation by use of a computer program.

The Nyquist criterion is a semigraphical method that determines the stability of a closed-loop system by investigating the properties of the frequency-domain plot, the **Nyquist plot**, of the loop transfer function $G(s)H(s)$, or $L(s)$. Specifically, the Nyquist plot of $L(s)$ is a plot of $L(j\omega)$ in the polar coordinates of $\text{Im}[L(j\omega)]$ versus $\text{Re}[L(j\omega)]$ as ω varies from 0 to ∞. This is another example of using the properties the loop transfer function to find the performance of the closed-loop system. The Nyquist criterion has the following features that make it an alternative method that is attractive for the analysis and design of control systems.

• Nyquist criterion also gives indication on relative stability.

1. In addition to providing the absolute stability, like the Routh-Hurwitz criterion, the Nyquist criterion also gives information on the relative stability of a stable system, and the degree of instability of an unstable system. It also gives indication on how the system stability may be improved, if needed.

2. The Nyquist plot of $G(s)H(s)$ or of $L(s)$ is very easy to obtain, especially with the aid of a computer.

3. The Nyquist plot of $G(s)H(s)$ gives information on the frequency-domain characteristics such as M_r, ω_r, BW, and others, with ease.

4. The Nyquist plot is useful for systems with pure time delay that cannot be treated with the Routh-Hurwitz criterion and are difficult to analyze with the root-locus method.

9-5-1 Stability Problem

The Nyquist criterion represents a method of determining the location of the characteristic equation roots with respect to the left half and the right half of the s-plane. Unlike the root-locus method, the Nyquist criterion does not give the exact location of the characteristic equation roots.

Let us consider that the closed-loop transfer function of a SISO system is

$$M(s) = \frac{G(s)}{1 + G(s)H(s)} \qquad (9\text{-}36)$$

where $G(s)H(s)$ can assume the following form:

$$G(s)H(s) = \frac{K(1 + T_1 s)(1 + T_2 s) \cdots (1 + T_m s)}{s^p(1 + T_a s)(1 + T_b s) \cdots (1 + T_n s)} e^{-T_d s} \qquad (9\text{-}37)$$

where the T's are real or complex-conjugate coefficients, and T_d is a real time delay.

Since the characteristic equation is obtained by setting the denominator polynomial of $M(s)$ to zero, the roots of the characteristic equation are also the zeros of $1 + G(s)H(s)$. Or, the characteristic equation roots must satisfy

$$\Delta(s) = 1 + G(s)H(s) = 0 \qquad (9\text{-}38)$$

In general, for a system with multiple number of loops, the denominator of $M(s)$ can be written as

$$\Delta(s) = 1 + L(s) = 0 \qquad (9\text{-}39)$$

where $L(s)$ is the loop transfer function, and is of the form of Eq. (9-37).

Before embarking on the details of the Nyquist criterion, it is useful to summarize the pole-zero relationships of the various system transfer functions.

Identification of Poles and Zeros

Loop transfer function zeros: zeros of $L(s)$

Loop transfer function poles: poles of $L(s)$

Closed-loop transfer function poles: zeros of $1 + L(s)$ = roots of the characteristic equation poles of $1 + L(s)$ = poles of $L(s)$.

Stability Conditions

We define two types of stability with respect to the system configuration.

Open-Loop Stability A system is said to be **open-loop stable** if the poles of the loop transfer function $L(s)$ are all in the left-half s-plane. For a single-loop system this is equivalent to the system being stable when the loop is opened at any point.

Closed-Loop Stability A system is said to be **closed-loop stable**, or simply stable, if the poles of the closed-loop transfer function or the zeros of $1 + L(s)$ are all in the left-half s-plane. Exceptions to the above definitions are systems with poles or zeros intentionally placed at $s = 0$.

9-5-2 Definition of Encircled and Enclosed

Since the Nyquist criterion is a graphical method, we need to establish the concepts of **encircled** and **enclosed**, which are used for the interpretation of the Nyquist plots for stability.

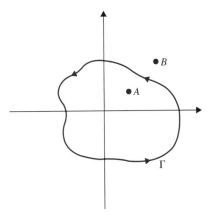

Figure 9-14 Definition of encirclement.

Encircled

A point or region in a complex function plane is said to be encircled by a closed path if it is found inside the path.

For example, point A in Fig. 9-14 is encircled by the closed path Γ, since A is *inside* the closed path. Point B is not encircled by the closed path Γ, since it is *outside* the path. Furthermore, when the closed path Γ has a direction assigned to it, the encirclement, if made, can be in the clockwise (CW) or the counterclockwise (CCW) direction. As shown in Fig. 9-14, point A is encircled by Γ in the CCW direction. We can say that the region *inside* the path is encircled in the prescribed direction, and the region *outside* the path is not encircled.

Enclosed

A point or region is said to be enclosed by a closed path if it is encircled in the CCW direction, or the point or region lies to the left of the path when the path is traversed in the prescribed direction.

The concept of enclosure is particularly useful if only a portion of the closed path is shown. For example, the shaded regions in Figs. 9-15(a) and (b) are considered to be *enclosed* by the closed path Γ. In other words, point A in Fig. 9-15(a) is *enclosed* by Γ, but point A in Fig. 9-15(b) is not. However, point B and all the points in the shaded region outside Γ in Fig. 9-15(b) are *enclosed*.

9-5-3 Number of Encirclements and Enclosures

When a point is encircled by a closed path Γ, a number N can be assigned to the number of times it is encircled. The magnitude of N can be determined by drawing an arrow from

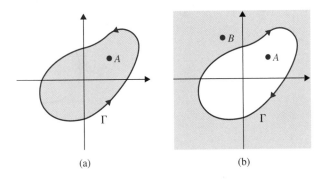

(a) (b)

Figure 9-15 Definition of enclosed points and regions. (a) Point A is enclosed by Γ. (b) Point A is not enclosed, but B is enclosed by the locus Γ.

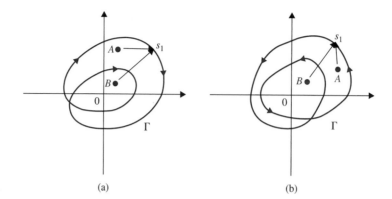

Figure 9-16 Definition of the number of encirclements and enclosures.

(a)   (b)

the point to any arbitrary point s_1 on the closed path Γ and then letting s_1 follow the path in the prescribed direction until it returns to the starting point. The total *net* number of revolutions traversed by this arrow is N, or the net angle is $2\pi N$ radians. For example, point A in Fig. 9-16(a) is *encircled once* or 2π radians by Γ, and point B is *encircled twice* or 4π radians, all in the CW direction. In Fig. 9-16(b), point A is *enclosed one*, and point B is *enclosed twice* by Γ. By definition, N is positive for CCW encirclement, and negative for CW encirclement.

9-5-4 Principle of the Argument

The Nyquist criterion was originated as an engineering application of the well-known "principle of the argument" concept in complex-variable theory. The principle is stated in the following in a heuristic manner.

Let $\Delta(s)$ be a single-valued function of the form of the right-hand side of Eq. (9-37), which has a finite number of poles in the s-plane. Single valued means that for each point in the s-plane, there is one and only one corresponding point, including infinity, in the complex $\Delta(s)$-plane. As defined in Chapter 8, infinity in the complex plane is interpreted as a point.

Suppose that a continuous closed path Γ_s is arbitrarily chosen in the s-plane, as shown in Fig. 9-17(a). If Γ_s does not go through any poles of $\Delta(s)$, then the trajectory Γ_Δ mapped by $\Delta(s)$ into the $\Delta(s)$-plane is also a closed one, as shown in Fig. 9-17(b). Starting from a point s_1, the Γ_s locus is traversed in the arbitrarily chosen direction (CW in the illustrated

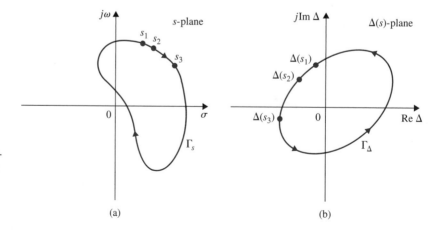

Figure 9-17 (a) Arbitrarily chosen closed path in the s-plane. (b) Corresponding locus Γ_s in the $\Delta(s)$-plane.

(a)   (b)

case), through the points s_2 and s_3, and then returning to s_1 after going through all the points on the Γ_s locus, as shown in Fig. 9-17(a), the corresponding Γ_Δ locus will start from the point $\Delta(s_1)$ and go through points $\Delta(s_2)$ and $\Delta(s_3)$, corresponding to s_1, s_2, and s_3, respectively, and finally return to the starting point, $\Delta(s_1)$. The direction of traverse of Γ_Δ can be either CW or CCW; that is, in the same direction or the opposite direction as that of Γ_s, depending on the function $\Delta(s)$. In Fig. 9-17(b), the direction of Γ_Δ is arbitrarily assigned, for illustration purposes, to be CCW.

• Do not attempt to relate $\Delta(s)$ with $L(s)$. They are not the same.

Although the mapping from the s-plane to the $\Delta(s)$-plane is single-valued, the reverse process is not a single-valued mapping. For example, consider the function

$$\Delta(s) = \frac{K}{s(s + 1)(s + 2)} \tag{9-40}$$

which has poles $s = 0$, -1, and -2 in the s-plane. For each point in the s-plane, there is only one corresponding point in the $\Delta(s)$-plane. However, for each point in the $\Delta(s)$-plane, the function maps into three corresponding points in the s-plane. The simplest way to illustrate this is to write Eq. (9-40) as

$$s(s + 1)(s + 2) - \frac{K}{\Delta(s)} = 0 \tag{9-41}$$

If $\Delta(s)$ is a real constant, which represents a point on the real axis in the $\Delta(s)$-plane, the third-order equation in Eq. (9-41) gives three roots in the s-plane. The reader should recognize the parallel of this situation to the root-locus diagram that essentially represents the mapping of $\Delta(s) = -1 + j0$ onto the loci of roots of the characteristic equation in the s-plane, for a given value of K. Thus, the root loci of Eq. (9-40) have three individual branches in the s-plane.

The principle of the argument can be stated:

Let $\Delta(s)$ be a single-valued function that has a finite number of poles in the s-plane. Suppose that an arbitrary closed path Γ_s is chosen in the s-plane so that the path does not go through any one of the poles or zeros of $\Delta(s)$; the corresponding Γ_Δ locus mapped in the $\Delta(s)$-plane will encircle the origin as many times as the difference between the number of zeros and poles of $\Delta(s)$ that are encircled by the s-plane locus Γ_s.

In equation form, the principle of the argument is stated as

$$N = Z - P \tag{9-42}$$

where

N = number of encirclements of the origin made by the $\Delta(s)$-plane locus Γ_Δ.

Z = number of zeros of $\Delta(s)$ encircled by the s-plane locus Γ_s in the s-plane.

P = number of poles of $\Delta(s)$ encircled by the s-plane locus Γ_s in the s-plane.

In general, N can be positive $(Z > P)$, zero $(Z = P)$, or negative $(Z < P)$. These three situations are described in more detail as follows.

1. $N > 0$ $(Z > P)$. If the s-plane locus encircles more zeros than poles of $\Delta(s)$ in a certain prescribed direction (CW or CCW), N is a positive integer. In this case, the $\Delta(s)$-plane locus Γ_Δ will encircle the origin of the $\Delta(s)$-plane N times in the same direction as that of Γ_s.

2. $N = 0$ $(Z = P)$. If the s-plane locus encircles as many poles as zeros, or no poles and zeros, of $\Delta(s)$, the $\Delta(s)$-plane locus Γ_Δ will not encircle the origin of the $\Delta(s)$-plane.

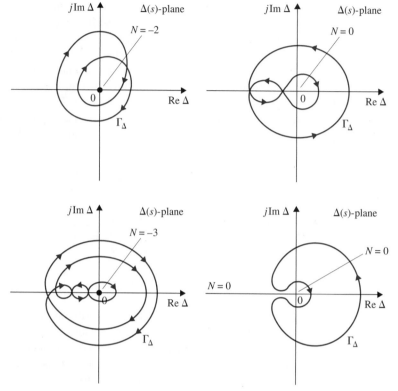

Figure 9-18 Examples of the determination of N in the $\Delta(s)$-plane.

3. $N < 0$ $(Z < P)$. If the s-plane locus encircles more poles than zeros of $\Delta(s)$ in a certain direction, N is a negative integer. In this case, the $\Delta(s)$-plane locus Γ_Δ will encircle the origin N times in the *opposite* direction as that of Γ_s.

A convenient way of determining N with respect to the origin (or any point) of the Δ-plane is to draw a line from the point in any direction to a point as far as necessary; the number of *net* intersections of this line with the $\Delta(s)$ locus gives the magnitude of N. Figure 9-18 gives several examples of this method of determining N. In these illustrated cases, it is assumed that the Γ_s locus has a CCW sense.

Critical Point

For convenience, we shall designate the origin of the Δ-plane as the **critical point** from which the value of N is determined. Later, we shall designate other points in the complex-function plane as critical points, dependent on the way the Nyquist criterion is applied.

A rigorous proof of the principle of the argument is not given here. The following illustrative example may be considered as a heuristic explanation of the principle.

Let us consider the function $\Delta(s)$ is of the form:

$$\Delta(s) = \frac{K(s + z_1)}{(s + p_1)(s + p_2)} \tag{9-43}$$

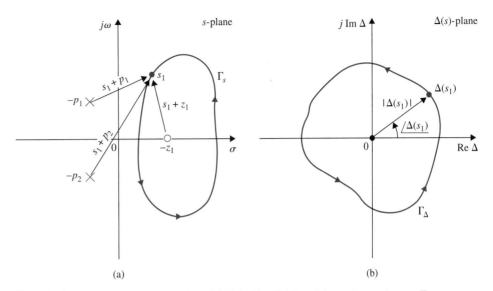

Figure 9-19 (a) Pole-zero configuration of $\Delta(s)$ in Eq. (9-44) and the s-plane trajectory Γ_s. (b) $\Delta(s)$-plane locus Γ_Δ, which corresponds to the Γ_s locus of (a) through the mapping of Eq. (9-44).

where K is a positive real number. The poles and zeros of $\Delta(s)$ are assumed to be as shown in Fig. 9-19(a). The function $\Delta(s)$ can be written as

$$\Delta(s) = |\Delta(s)| \angle \Delta(s)$$
$$= \frac{K|s + z_1|}{|s + p_1||s + p_2|}[\angle(s + z_1) - \angle(s + p_1) - \angle(s + p_2)]$$

(9-44)

Figure 9-19(a) shows an arbitrarily chosen trajectory Γ_s in the s-plane, with the arbitrary point s_1 on the path, and Γ_s does not pass through any of the poles and the zeros of $\Delta(s)$. The function $\Delta(s)$ evaluated at $s = s_1$ is

$$\Delta(s_1) = \frac{K(s + z_1)}{(s_1 + p_1)(s + p_2)}$$

(9-45)

• Z and P refer to only the zeros and poles, respectively, of $\Delta(s)$ that are encircled by Γ_s.

The term $(s_1 + z_1)$ can be represented graphically by the vector drawn from $-z_1$ to s_1. Similar vectors can be drawn for $(s_1 + p_1)$ and $(s + p_2)$. Thus, $\Delta(s_1)$ is represented by the vectors drawn from the finite poles and zeros of $\Delta(s)$ to the point s_1, as shown in Fig. 9-19(a). Now, if the point s_1 is moved along the locus Γ_s in the prescribed CCW direction until it returns to the starting point, the angles generated by the vectors drawn from the two poles that are not encircled by Γ_s when s_1 completes one round trip are zero; whereas the vector $(s_1 + z_1)$ drawn from the zero at $-z_1$, which is encircled by Γ_s, generates a positive angle (CCW) of 2π radians, which means that the corresponding $\Delta(s)$ plot must go around the origin 2π radians, or one revolution, in the CCW direction, as shown in Fig. 9-19(b). This is why only the poles and zeros of $\Delta(s)$ that are inside the Γ_s trajectory in the s-plane will contribute to the value of N of Eq. (9-42). Since the poles of $\Delta(s)$ contribute to a negative phase, and zeros contribute to positive phase, the value of N depends only on the difference between Z and P. For the case illustrated in Fig. 9-19(a), $Z = 1$ and $P = 0$.

TABLE 9-1   Summary of All Possible Outcomes of the Principle of the Argument

		$\Delta(s)$-Plane Locus	
$N = Z - P$	Sense of the s-plane Locus	Number of Encirclements of the Origin	Direction of Encirclement
$N > 0$	CW	N	CW
	CCW	N	CCW
$N < 0$	CW	N	CCW
	CCW	N	CW
$N = 0$	CW	0	No encirclement
	CCW	0	No encirclement

Thus,

$$N = Z - P = 1 \tag{9-46}$$

which means that the $\Delta(s)$-plane locus Γ_Δ should encircle the origin once in the same direction as that of the s-plane locus Γ_s. It should be kept in mind that Z and P refer only to the zeros and poles, respectively, of $\Delta(s)$ that are encircled by Γ_s, and not the total number of zeros and poles of $\Delta(s)$.

In general, the net angle traversed by the $\Delta(s)$-plane locus, as the s-plane locus is traversed once in any direction, is equal to

$$2\pi(Z - P) = 2\pi N \qquad \text{radians} \tag{9-47}$$

This equation implies that if there are N more zeros than poles of $\Delta(s)$, which are encircled by the s-plane locus Γ_s, in a prescribed direction, the $\Delta(s)$-plane locus will encircle the origin N times in the *same direction* as that of Γ_s. Conversely, if N more poles than zeros are encircled by Γ_s in a given direction, N in Eq. (9-47) will be negative, and the $\Delta(s)$-plane locus must encircle the origin N times in the *opposite direction* to that of Γ_s.

A summary of all the possible outcomes of the principle of the argument is given in Table 9-1.

9-5-5   Nyquist Path

Years ago when Nyquist was faced with the problem of solving the stability problem which involves determining if the function $\Delta(s) = 1 + L(s)$ has zeros in the right-half s-plane, he apparently discovered that the principle of the argument could be applied to solve the stability problem if the s-plane locus Γ_s is taken to be one that encircles the entire right-half of the s-plane. Of course, as an alternative, Γ_s can be chosen to encircle the entire left-half s-plane, as the solution is a relative one. Figure 9-20 illustrates a Γ_s locus with a CCW sense, that encircles the entire right half of the s-plane. This path is chosen to be the s-plane trajectory Γ_s for the Nyquist criterion, since in mathematics, CCW is traditionally defined to be the positive sense. The path Γ_s shown in Fig. 9-20 is defined to be the **Nyquist path**. Since the Nyquist path must not pass through any poles and zeros of $\Delta(s)$, the small semicircles shown along the $j\omega$-axis in Fig. 9-20 are used to indicate that the path should go around these poles and zeros if they fall on the $j\omega$-axis. It is apparent that if any pole or zero of $\Delta(s)$ lies inside the right-half s-plane, it will be encircled by the Nyquist path Γ_s.

• The Nyquist path is defined to encircle the entire right-half s-plane.

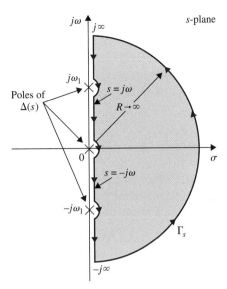

Figure 9-20 Nyquist path.

9-5-6 Nyquist Criterion and the *L*(*s*) or the *G*(*s*)*H*(*s*) Plot

The Nyquist criterion is a direct application of the principle of the argument when the *s*-plane locus is the Nyquist path of Fig. 9-20. In principle, once the Nyquist path is specified, the stability of a closed-loop system can be determined by plotting the $\Delta(s) = 1 + L(s)$ locus when *s* takes on values along the Nyquist path, and investigating the behavior of the $\Delta(s)$ plot with respect to the **critical point**, which in this case is the origin of the $\Delta(s)$-plane.

Since the function *L*(*s*) is generally known,

*it would be simpler to construct the **L**(**s**) plot that corresponds to the Nyquist path, and the same conclusion on the stability of the closed-loop system can be obtained by observing the behavior of the **L**(**s**) plot with respect to the* $(-1, j0)$ *point in the **L**(**s**)-plane.*

This is because the origin of the $\Delta(s) = 1 + L(s)$ plane corresponds to the $(-1, j0)$ point in the *L*(*s*)-plane. Thus the $(-1, j0)$ point in the *L*(*s*)-plane becomes the critical point for the determination of closed-loop stability.

For single-loop systems, $L(s) = G(s)H(s)$, the previous development leads to the determination of the closed-loop stability by investigating the behavior of the *G*(*s*)*H*(*s*) plot with respect to the $(-1, j0)$ point of the *G*(*s*)*H*(*s*)-plane. Thus, the Nyquist stability criterion is another example of the using the the loop transfer function properties to find the behavior of closed-loop systems.

Thus, given a control system that has the characteristic equation given by equating the numerator polynomial of $1 + L(s)$ to zero, where *L*(*s*) is the loop transfer function, the application of the Nyquist criterion to the stability problem involves the following steps.

1. The Nyquist path Γ_s is defined in the *s*-plane, as shown in Fig. 9-20.
2. The *L*(*s*) plot corresponding to the Nyquist path is constructed in the *L*(*s*)-plane.
3. The value of *N*, the number of encirclement of the $(-1, j0)$ point made by the *L*(*s*) plot is observed.
4. The Nyquist criterion follows from Eq. (9-42),

$$N = Z - P \qquad (9\text{-}48)$$

where

N = number of encirclements of the $(-1, j0)$ point made by the $L(s)$ plot.

Z = number of zeros of $1 + L(s)$ that are inside the Nyquist path, that is, the right-half s-plane.

P = number of poles of $1 + L(s)$ that are inside the Nyquist path, that is, the right-half s-plane. Notice that the poles of $1 + L(s)$ are the same as that of $L(s)$.

The stability requirements for the two types of stability defined earlier are interpreted in terms of Z and P.

For closed-loop stability, Z must equal zero.

For open-loop stability, P must equal zero.

Thus, the condition of stability according to the Nyquist criterion is stated as

$$N = -P \tag{9-49}$$

That is,

for a closed-loop system to be stable, the $L(s)$ plot must encircle the $(-1, j0)$ point as many times as the number of poles of $L(s)$ that are in the right-half s-plane, and the encirclement, if any, must be made in the clockwise direction (if Γ_s is defined in the CCW sense).

▶ 9-6 NYQUIST CRITERION FOR SYSTEMS WITH MINIMUM-PHASE TRANSFER FUNCTIONS

We shall first apply the Nyquist criterion to systems with $L(s)$ that are **minimum-phase transfer functions**. The properties of the minimum-phase transfer functions are described in Appendix A, and are summarized as follows:

1. A minimum-phase transfer function does not have poles or zeros in the right-half s-plane or on the $j\omega$-axis, excluding the origin.

2. For a minimum-phase transfer function $L(s)$ with m zeros and n poles, excluding the poles at $s = 0$, when $s = j\omega$ and as ω varies from ∞ to 0, the total phase variation of $L(j\omega)$ is $(n - m)\pi/2$ radians.

3. The value of a minimum-phase transfer function cannot become zero or infinity at any finite nonzero frequency.

4. A nonminimum-phase transfer function will always have a more positive phase shift as ω varies from ∞ to 0. Or equally true, it will always have a more negative phase shift as ω varies from 0 to ∞.

• A minimum-phase transfer function does not have poles or zeros in the right-half s-plane or on the $j\omega$-axis, except at $s = 0$.

Since a majority of the loop transfer functions encountered in the real world satisfy condition 1, and are of the minimum-phase type, it would be prudent to investigate the application of the Nyquist criterion to this class of systems. As if turns out, this is quite simple.

Since a minimum-phase $L(s)$ does not have any poles or zeros in the right-half s-plane or on the $j\omega$-axis, except at $s = 0$, $P = 0$, and the poles of $\Delta(s) = 1 + L(s)$ also have the same properties. Thus, the Nyquist criterion for a system with $L(s)$ being a minimum-phase transfer function is simplified to:

$$N = 0 \tag{9-50}$$

Thus, the Nyquist criterion can be stated:

For a closed-loop system with loop transfer function L(s) that is of minimum-phase type, the system is closed-loop stable, if the plot of L(s) that corresponds to the Nyquist path does not encircle the critical point(−1, j0) in the L(s)-plane.

Furthermore, if the system is unstable, $Z \neq 0$; N in Eq. (9-50) would be a positive integer, which means that the critical point $(-1, j0)$ is **enclosed** N times (corresponding to the direction of the Nyquist path defined here). Thus, the Nyquist criterion of stability for systems with minimum-phase loop transfer functions can be further simplified to:

For a closed-loop system with loop transfer function L(s) that is of minimum-phase type, the system is closed-loop stable, if the L(s) plot that corresponds to the Nyquist path does not enclose the (−1, j0) point. If the (−1, j0) is enclosed by the Nyquist plot, the system is unstable.

• For $L(s)$ that is minimum-phase type, Nyquist criterion can be checked by plotting the segment of $L(j\omega)$ from $\omega = \infty$ to 0.

Since the region that is enclosed by a trajectory is defined as the region that lies to the left when the trajectory is traversed in the prescribed direction, the *Nyquist criterion can be checked simply by plotting the segment of L(jω) from ω = ∞ to 0, or, points on the positive jω-axis.* This simplifies the procedure considerably, since the plot can be made easily on a computer. The only drawback to this method is that the Nyquist plot that corresponds to the $j\omega$-axis tells only whether the critical point is enclosed or not; and if it does, not how many times. Thus, if the system is found to be unstable, the enclosure property does not give information on how many roots of the characteristic equation are in the right-half s-plane. However, in practice, this information is not vital. From this point on, we shall define *the L(jω) plot that corresponds to the positive jω-axis of the s-plane as the Nyquist plot of L(s).*

9-6-1 Application of the Nyquist Criterion to Minimum-Phase Transfer Functions that Are Not Strictly Proper

Just as in the case of the root locus, it is often necessary in design to create an equivalent loop transfer function $L_{eq}(s)$ so that a variable parameter K will appear as a multiplying factor in $L_{eq}(s)$; that is, $L(s) = KL_{eq}(s)$. Since the equivalent loop transfer function does not correspond to any physical entity, it may not have more poles than zeros, and the transfer function is not strictly proper as defined in Chapter 3. In principle, there is no difficulty in constructing the Nyquist plot of a transfer function that is not strictly proper, and the Nyquist criterion can be applied for stability studies without any complications. However, some computer programs may not be prepared for handling improper transfer functions, and it may be necessary to reformulate the equation for compatibility with the computer program. To consider this case, consider that the characteristic equation of a system with a variable parameter K is conditioned to

$$1 + KL_{eq}(s) = 0 \qquad (9\text{-}51)$$

If $L_{eq}(s)$ does not have more poles than zeros, we can rewrite Eq. (9-51) as

$$1 + \frac{1}{KL_{eq}(s)} = 0 \qquad (9\text{-}52)$$

by dividing both sides of the equation by $KL_{eq}(s)$. Now we can plot the Nyquist plot of $1/L_{eq}(s)$, and the critical point is still $(-1, j0)$ for $K > 0$. The variable parameter on the Nyquist plot is now $1/K$. Thus, with this minor adjustment, the Nyquist criterion can still be applied.

The Nyquist criterion presented here is cumbersome when the loop transfer function is of the nonminimum-phase type, for example, when $L(s)$ has poles or/and zeros in the right-half s-plane. A generalized Nyquist criterion is presented in Appendix H that will take care of transfer functions of all types.

▶ 9-7 RELATION BETWEEN THE ROOT LOCI AND THE NYQUIST PLOT

Since both the root locus analysis and the Nyquist criterion deal with the location of the roots of the characteristic equation of a linear SISO system, the two analyses are closely related. Exploring the relationship between the two methods will enhance the understanding of both methods. Given the characteristic equation

$$1 + L(s) = 1 + KG_1(s)H_1(s) = 0 \tag{9-53}$$

the Nyquist plot of $L(s)$ in the $L(s)$-plane is the mapping of the Nyquist path in the s-plane. Since the root loci of Eq. (9-53) must satisfy the conditions:

$$\angle KG_1(s)H_1(s) = (2j + 1)\pi \qquad K \geq 0 \tag{9-54}$$
$$\angle KG_1(s)H_1(s) = 2j\pi \qquad K \leq 0 \tag{9-55}$$

for $j = 0, \pm 1, \pm 2, \ldots$, the root loci simply represent a mapping of the real axis of the $L(s)$-plane or the $G(s)H(s)$-plane onto the s-plane. In fact, for the RL, $K \geq 0$, the mapping points are on the negative real axis of the $L(s)$-plane; and for the RL, $K \leq 0$, the mapping points are on the positive real axis of the $L(s)$-plane. It was pointed out earlier that the mapping from the s-plane to the function plane for a rational function is single valued, but the reverse process is multivalued. As a simple illustration, the Nyquist plot of a type-1 third-order transfer function $G(s)H(s)$ that corresponds to points on the $j\omega$-axis of the s-plane is shown in Fig. 9-21. The root loci for the same system are shown in Fig. 9-22 as a mapping of the real axis of the $G(s)H(s)$-plane onto the s-plane. Note that in this case, each point of the $G(s)H(s)$-plane corresponds to three points in the s-plane. The $(-1, j0)$ point of the $G(s)H(s)$-plane corresponds to the two points where the root loci intersect the $j\omega$-axis and a point on the real axis.

The Nyquist plot and the root loci each represents the mapping of only a very limited portion of one domain to the other. In general, it would be useful to consider the mapping of points other than those on the $j\omega$-axis of the s-plane and on the real axis of the $G(s)H(s)$-plane. For instance, we may use the mapping of the constant-damping-ratio lines in the s-plane onto the $G(s)H(s)$-plane for the purpose of determining relative stability of the closed-loop system. Figure 9-23 illustrates the $G(s)H(s)$ plots that correspond to

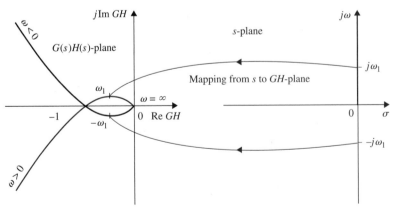

Figure 9-21 Polar plot of $G(s)H(s) = K/[s(s + a)(s + b)]$ interpreted as a mapping of the $j\omega$-axis of the s-plane onto the $G(s)H(s)$-plane.

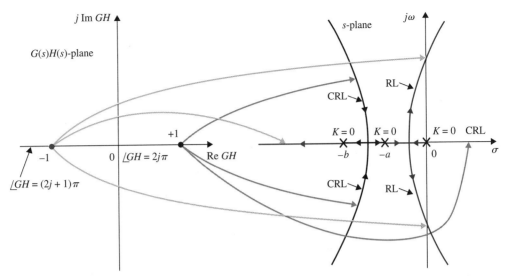

Figure 9-22 Root-locus diagram of $G(s)H(s) = K/[s(s + a)(s + b)]$ interpreted as a mapping of the real axis of the $G(s)H(s)$-plane onto the s-plane.

different constant-damping-ratio lines in the s-plane. As shown by curve (3) in Fig. 9-23, when the $G(s)H(s)$ curve passes through the $(-1, j0)$ point, it means that Eq. (9-52) is satisfied, and the corresponding trajectory in the s-plane passes through the root of the characteristic equation. Similarly, we can construct the root loci that correspond to the straight lines rotated at various angles from the real axis in the $G(s)H(s)$-plane, as shown in Fig. 9-24. Notice that these root loci now satisfy the condition of

$$\angle KG_1(s)H_1(s) = (2j + 1)\pi - \theta \qquad K \geq 0 \tag{9-56}$$

Or the root loci of Fig. 9-24 must satisfy the equation

$$1 + G(s)H(s)e^{j\theta} = 0 \tag{9-57}$$

for the various values of θ indicated.

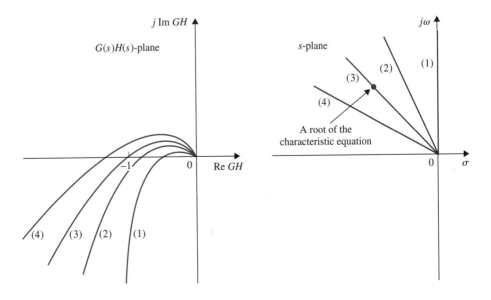

Figure 9-23 $G(s)H(s)$ plots that correspond to constant-damping-ratio lines in the s-plane.

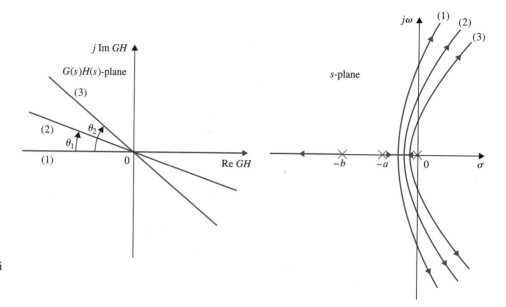

Figure 9-24 Root loci that correspond to different phase-angle loci in the $G(s)H(s)$-plane.

▶ 9-8 ILLUSTRATIVE EXAMPLES: NYQUIST CRITERION FOR MINIMUM-PHASE TRANSFER FUNCTIONS

The following examples serve to illustrate the application of the Nyquist criterion to systems with minimum-phase loop transfer functions. All examples in this chapter may also be solved using the **ACSYS freqtool**. For the description of the software, please refer to Section 9-17.

▶ **EXAMPLE 9-1** Consider that a single-loop feedback control system has the loop transfer function

$$L(s) = G(s)H(s) = \frac{K}{s(s+2)(s+10)} \quad (9\text{-}58)$$

which is of minimum-phase type. The stability of the closed-loop system can be conducted by investigating whether the Nyquist plot of $L(j\omega)/K$ for $\omega = \infty$ to 0 encloses the $(-1, j0)$ point. The Nyquist plot of $L(j\omega)/K$ may be plotted using freqtool. Figure 9-25 shows the Nyquist plot of $L(j\omega)/K$ for $\omega = \infty$ to 0. However, since we are interested only in whether the critical point is enclosed, in general, it is not necessary to produce an accurate Nyquist plot. Since the area that is enclosed by the Nyquist plot is to the left of the curve, traversed in the direction that corresponds to $\omega = \infty$ to 0 on the Nyquist path, all that is necessary to determine stability is to find the point or points at which the Nyquist plot crosses the real axis in the $L(j\omega)/K$-plane. In many cases, information on the intersection on the real axis and the properties of $L(j\omega)/K$ at $\omega = \infty$ and $\omega = 0$ would allow the sketching of the Nyquist plot without actual plotting. We can use the following steps to obtain a sketch of the Nyquist plot of $L(j\omega)/K$.

1. Substitute $s = j\omega$ in $L(s)$.

   Setting $s = j\omega$ in Eq. (9-58), we get

   $$L(j\omega)/K = \frac{1}{j\omega(j\omega + 2)(j\omega + 10)} \quad (9\text{-}59)$$

2. Substituting $\omega = 0$ in the last equation, we get the zero-frequency property of $L(j\omega)$,

   $$L(j0)/K = \infty \angle -90° \quad (9\text{-}60)$$

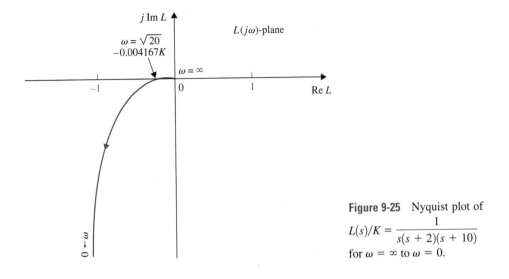

Figure 9-25 Nyquist plot of $L(s)/K = \dfrac{1}{s(s + 2)(s + 10)}$ for $\omega = \infty$ to $\omega = 0$.

3. Substituting $\omega = \infty$ in Eq. (9-59), the property of the Nyquist plot at infinite frequency is established.

$$L(j\infty)/K = 0\angle{-270°} \tag{9-61}$$

Apparently, these results are verified by the plot shown in Fig. 9-25.

4. To find the intersect(s) of the Nyquist plot with the real axis, if any, we rationalize $L(j\omega)/K$ by multiplying the numerator and the denominator of the equation by the complex conjugate of the denominator. Thus, Eq. (9-59) becomes

$$L(j\omega)/K = \frac{[-12\omega^2 - j\omega(20 - \omega^2)]}{[-12\omega^2 + j\omega(20 - \omega^2)][-12\omega^2 - j\omega(20 - \omega^2)]} \tag{9-62}$$
$$= \frac{[-12\omega - j(20 - \omega^2)]}{\omega[144\omega^2 + (20 - \omega^2)]}$$

5. To find the possible intersects on the real axis, we set the imaginary part of $L(j\omega)/K$ to zero. The result is

$$\mathrm{Im}[L(j\omega)/K] = \frac{-(20 - \omega^2)}{\omega[144\omega^2 + (20 - \omega^2)]} = 0 \tag{9-63}$$

The solutions of the last equation are: $\omega = \infty$, which is known to be a solution at $L(j\omega)/K = 0$, and

$$\omega = \pm \sqrt{20} \quad \text{rad/sec} \tag{9-64}$$

Since ω is positive, the correct answer is $\omega = \sqrt{20}$ rad/sec. Substituting this frequency into Eq. (9-62), we have the intersect on the real axis of the $L(j\omega)$-plane at

$$L(j\sqrt{20})/K = -\frac{12}{2880} = -0.004167 \tag{9-65}$$

The last five steps should lead to an adequate sketch of the Nyquist plot of $L(j\omega)/K$ short of plotting it. Thus, we see that if K is less than 240, the intersect of the $L(j\omega)$ locus on the real axis would be to the right of the critical point $(-1, j0)$; the latter is not enclosed, and the system is stable. If $K = 240$, the Nyquist plot of $L(j\omega)$ would intersect the real axis at the -1 point, and the system would be marginally stable. In this case, the characteristic equation would have two roots on the $j\omega$-axis in the s-plane at $s = \pm j\sqrt{20}$. If the gain is increased to a value beyond 240, the intersect would be to the left of the -1 point on the real axis, and the system would be unstable. When K

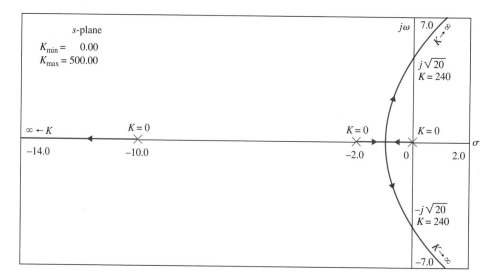

Figure 9-26   RL of
$$L(s) = \frac{K}{s(s + 2)(s + 10)}.$$

is negative, we can use the $(+1, j0)$ point in the $L(j\omega)$-plane as the critical point. Figure 9-25 shows that under this condition, the $+1$ point on the real axis would be enclosed for all negative values of K, and the system would always be unstable. Thus, the Nyquist criterion leads to the conclusion that the system is stable in the range of $0 < K < 240$. Note that application of the Routh-Hurwitz stability criterion leads to this same result.

Figure 9-26 shows the root loci of the characteristic equation of the system described by the loop transfer function in Eq. (9-58). The correlation between the Nyquist criterion and the root loci is easily observed.   ◀

▶ **EXAMPLE 9-2**   Consider the characteristic equation

$$Ks^3 + (2K + 1)s^2 + (2K + 5)s + 1 = 0 \qquad (9\text{-}66)$$

Dividing both sides of the last equation by the terms that do not contain K, we have

$$1 + KL_{eq}(s) = 1 + \frac{Ks(s^2 + 2s + 2)}{s^2 + 5s + 1} = 0 \qquad (9\text{-}67)$$

Thus,

$$L_{eq}(s) = \frac{s(s^2 + 2s + 2)}{s^2 + 5s + 1} \qquad (9\text{-}68)$$

which is an improper function. We can obtain the information to manually sketch the Nyquist plot of $L_{eq}(s)$ to determine the stability of the system. Setting $s = j\omega$ in Eq. (9-68), we get

$$\frac{L_{eq}(j\omega)}{K} = \frac{\omega[-2\omega + j(2 - \omega^2)]}{(1 - \omega^2) + 5j\omega} \qquad (9\text{-}69)$$

From the last equation we obtain the two end points of the Nyquist plot:

$$L_{eq}(j0) = 0\angle 90° \qquad \text{and} \qquad L_{eq}(j\infty) = \infty\angle 90° \qquad (9\text{-}70)$$

Rationalizing Eq. (9-69) by multiplying its numerator and denominator by the complex conjugate of the denominator, we get

$$\frac{L_{eq}(j\omega)}{K} = \frac{\omega^2[5(2 - \omega^2) - 2(1 - \omega^2)] + j\omega[10\omega^2 + (2 - \omega^2)(1 - \omega^2)]}{(1 - \omega^2)^2 + 25\omega^2} \qquad (9\text{-}71)$$

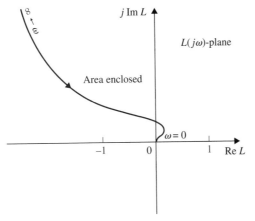

Figure 9-27   Nyquist plot of $\dfrac{L_{eq}(s)}{K} =$

$\dfrac{s(s^2 + 2s + 2)}{s^2 + 5s + 1}$ for $\omega = \infty$ to $\omega = 0$.

To find the possible intersects of the $L_{eq}(j\omega)/K$ plot on the real axis, we set the imaginary part of Eq. (9-71) to zero. We get $\omega = 0$ and

$$\omega^4 + 7\omega^2 + 2 = 0 \tag{9-72}$$

We can show that all the four roots of Eq. (9-72) are imaginary, which indicates that the $L_{eq}(j\omega)/K$ locus intersects the real axis only at $\omega = 0$. Using the information given by Eq. (9-70), and the fact that there are no other intersections on the real axis than at $\omega = 0$, the Nyquist plot of $L_{eq}(j\omega)/K$ is sketched as shown in Fig. 9-27. Notice that this plot is sketched without any detailed data computed on $L_{eq}(j\omega)/K$, and in fact, could be grossly inaccurate. However, the sketch is adequate to determine the stability of the system. Since the Nyquist plot in Fig. 9-27 does not enclose the $(-1, j0)$ point as ω varies from ∞ to 0, the system is stable for all finite positive values of K.

Figure 9-28 shows the Nyquist plot of Eq. (9-66), based on the poles and zeros of $L_{eq}(s)/K$ in Eq. (9-68). Notice that the RL stays in the left-half s-plane for all positive values of K, and the results confirm the Nyquist criterion results on system stability.

$$\frac{K}{L_{eq}(j\omega)} = \frac{(1 - \omega^2) + 5 j\omega}{[-2\omega^2 + j\omega(2 - \omega^2)]} \tag{9-73}$$

for $\omega = \infty$ to 0. The plot again does not enclose the $(-1, j0)$ point, and the system is again stable for all positive values of K by interpreting the Nyquist plot of $K/L_{eq}(j\omega)$.

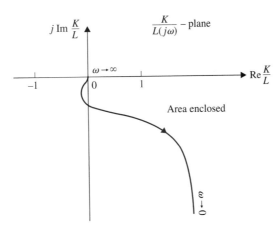

Figure 9-28   Nyquist plot of $K/L_{eq}(j\omega)$

for $\dfrac{L_{eq}(s)}{K} = \dfrac{s(s^2 + 2s + 2)}{s^2 + 5s + 1}$

for $\omega = \infty$ to $\omega = 0$.

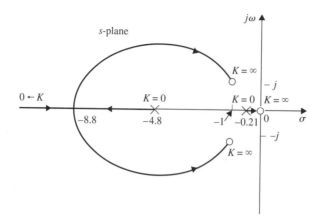

Figure 9-29   RL of
$$L(s) = \frac{Ks(s^2 + 2s + 2)}{s^2 + 5s + 1}.$$

Figure 9-29 shows the RL of Eq. (9-67) for $K > 0$, using the pole-zero configuration of $L_{eq}(s)$ of Eq. (9-68). Since the RL stays in the left-half s-plane for all positive values of K, the system is stable for $0 < K < \infty$, which agrees with the conclusion obtained with the Nyquist criterion.   ◀

▶ 9-9 EFFECTS OF ADDITION OF POLES AND ZEROS TO $L(s)$ ON THE SHAPE OF THE NYQUIST PLOT

Since the performance of a control system is often affected by adding and moving poles and zeros of the loop transfer function, it is important to investigate how the Nyquist plot is affected when poles and zeros are added to $L(s)$.

Let us begin with a first-order transfer function

$$L(s) = \frac{K}{1 + T_1 s} \tag{9-74}$$

where T_1 is a positive real constant. The Nyquist plot of $L(j\omega)$ for $0 \leq \omega \leq \infty$ is a semicircle, as shown in Fig. 9-30. The figure also shows the interpretation of the closed-loop stability with respect to the critical point for all values of K between $-\infty$ and ∞.

Addition of Poles at $s = 0$

Consider that a pole at $s = 0$ is added to the transfer function of Eq. (9-74); then

$$L(s) = \frac{K}{s(1 + T_1 s)} \tag{9-75}$$

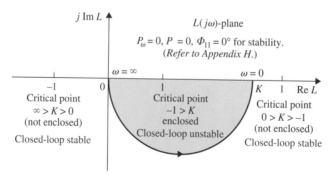

Figure 9-30   Nyquist plot of $L(s) = \dfrac{K}{s(1 + T_1 s)}.$

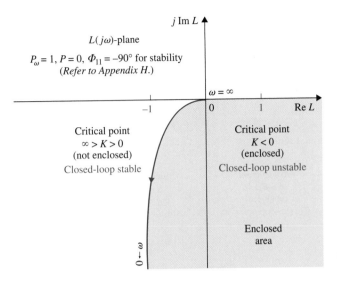

Figure 9-31 Nyquist plot of $L(s) = \dfrac{K}{s(1 + T_1 s)}$.

Since adding a pole at $s = 0$ is equivalent to dividing $L(s)$ by $j\omega$, the phase of $L(j\omega)$ is reduced by 90° at both zero and infinite frequencies. In addition, the magnitude of $L(j\omega)$ at $\omega = 0$ becomes infinite. Figure 9-31 illustrates the Nyquist plot of $L(j\omega)$ in Eq. (9-75) and the closed-loop stability interpretations with respect to the critical points for $-\infty < K < \infty$. In general, adding a pole of multiplicity p at $s = 0$ to the transfer function of Eq. (9-74) will give the following properties to the Nyquist plot of $L(j\omega)$:

$$\lim_{\omega \to \infty} \angle L(j\omega) = -(p + 1)90° \tag{9-76}$$

$$\lim_{\omega \to 0} \angle L(j\omega) = -p \times 90° \tag{9-77}$$

$$\lim_{\omega \to \infty} |L(j\omega)| = 0 \tag{9-78}$$

$$\lim_{\omega \to 0} |L(j\omega)| = \infty \tag{9-79}$$

The following example illustrates the effects of adding multiple-order poles to $L(s)$.

▶ **EXAMPLE 9-3** Figure 9-32 shows the Nyquist plot of

$$L(s) = \frac{K}{s^2(1 + T_1 s)} \tag{9-80}$$

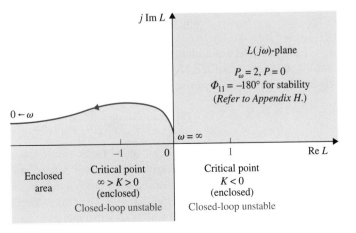

Figure 9-32 Nyquist plot of $L(s) = \dfrac{K}{s^2(1 + T_1 s)}$.

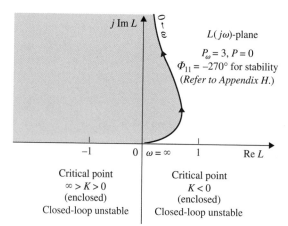

$j \operatorname{Im} L$

$L(j\omega)$-plane

$P_\omega = 3, P = 0$
$\Phi_{11} = -270°$ for stability
(*Refer to Appendix H.*)

$\omega = \infty$ Re L

Critical point
$\infty > K > 0$
(enclosed)
Closed-loop unstable

Critical point
$K < 0$
(enclosed)
Closed-loop unstable

Figure 9-33   Nyquist plot

of $L(s) = \dfrac{K}{s^3(1 + T_1 s)}$.

and the critical points, with stability interpretations. Figure 9-33 illustrates the same for

$$L(s) = \frac{K}{s^3(1 + T_1 s)} \tag{9-81}$$

◀

• Adding poles at $s = 0$ to a loop transfer function will reduce stability of the closed-loop system.

The conclusion from these illustrations is that the addition of poles at $s = 0$ to a loop transfer function will affect the stability of the closed-loop system adversely. A system that has a loop transfer function with more than one pole at $s = 0$ (type 2 or higher) is likely to be unstable or difficult to stabilize.

Addition of Finite Nonzero Poles

When a pole at $s = -1/T_2$ $(T_2 > 0)$ is added to the function $L(s)$ of Eq. (9-74), we have

$$L(s) = \frac{K}{(1 + T_1 s)(1 + T_2 s)} \tag{9-82}$$

The Nyquist plot of $L(j\omega)$ at $\omega = 0$ is not affected by the addition of the pole, since

$$\lim_{\omega \to 0} L(j\omega) = K \tag{9-83}$$

The value of $L(j\omega)$ at $\omega = \infty$ is

$$\lim_{\omega \to \infty} L(j\omega) = \lim_{\omega \to \infty} \frac{-K}{T_1 T_2 \omega^2} = 0 \angle -180° \tag{9-84}$$

Thus, the effect of adding a pole at $s = -1/T_2$ to the transfer function of Eq. (9-75) is to shift the phase of the Nyquist plot by $-90°$ at $\omega = \infty$, as shown in Fig. 9-34. The figure also shows the Nyquist plot of

$$L(s) = \frac{K}{(1 + T_1 s)(1 + T_2 s)(1 + T_3 s)} \tag{9-85}$$

• Adding nonzero poles to the loop transfer function also reduces stability of the closed-loop system.

where two nonzero poles have been added to the transfer function of Eq. (9-74) (T_1, T_2, T_3, > 0). In this case, the Nyquist plot at $\omega = \infty$ is rotated clockwise by another $90°$ from that of Eq. (9-82). *These examples show the adverse effects on closed-loop stability when poles are added to the loop transfer function.* The closed-loop systems with the loop transfer functions of Eqs. (9-74) and (9-82) are all stable as long as K is positive. The system

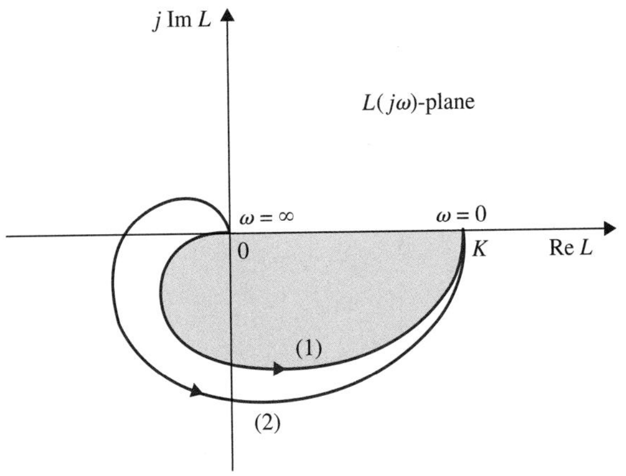

jIm L

$L(j\omega)$-plane

$\omega = \infty$ $\omega = 0$

0   K   Re L

(1)

(2)

Figure 9-34 Nyquist plots. Curve (1): $L(s) = \dfrac{K}{(1 + T_1 s)(1 + T_2 s)}$

Curve (2): $L(s) = \dfrac{K}{(1 + T_1 s)(1 + T_2 s)(1 + T_3 s)}$.

represented by Eq. (9-85) is unstable if the intersect of the Nyquist plot on the negative real axis is to the left of the $(-1, j0)$ point when K is positive.

Addition of Zeros

It was demonstrated in Chapter 6 that adding zeros to the loop transfer function has the effect of reducing the overshoot and the general effect of stabilization. In terms of the Nyquist criterion, this stabilization effect is easily demonstrated, since the multiplication of the term $(1 + T_d s)$ to the loop transfer function increases the phase of $L(s)$ by 90° at $\omega = \infty$. The following example illustrates the effect on stability of adding a zero at $-1/T_d$ to a loop transfer function.

▶ **EXAMPLE 9-4** Consider that the loop transfer function of a closed-loop control system is

$$L(s) = \frac{K}{s(1 + T_1 s)(1 + T_2 s)} \tag{9-86}$$

It can be shown that the closed-loop system is stable for

$$0 < K < \frac{T_1 + T_2}{T_1 T_2} \tag{9-87}$$

Suppose that a zero at $s = -1/T_d$ ($T_d > 0$) is added to the transfer function of Eq. (9-86), then,

$$L(s) = \frac{K(1 + T_d s)}{s(1 + T_1 s)(1 + T_2 s)} \tag{9-88}$$

• Adding zeros to the loop transfer function has the effect of stabilizing the closed-loop system.

The Nyquist plots of the two transfer functions of Eqs. (9-86) and (9-188) are shown in Fig. 9-35. The effect of the zero in Eq. (9-88) is to add 90° to the phase of the $L(j\omega)$ in Eq. (9-126) at $\omega = \infty$ while not affecting the value at $\omega = 0$. The intersect on the negative real axis of the $L(j\omega)$-plane is moved from $-KT_1 T_2/(T_1 + T_2)$ to $-K(T_1 T_2 - T_d T_1 - T_d T_2)/(T_1 + T_2)$. Thus, the system with the loop transfer function in Eq. (9-88) is stable for

$$0 < K < \frac{T_1 + T_2}{T_1 T_2 - T_d(T_1 + T_2)} \tag{9-89}$$

which for positive T_d and K has a higher upper bound than that of Eq. (9-87).

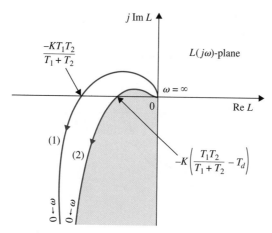

Figure 9-35 Nyquist plots. Curve (1): $L(s) = \dfrac{K}{s(1 + T_1 s)(1 + T_2 s)}$.

Curve (2): $L(s) = \dfrac{K(1 + T_d s)}{s(1 + T_1 s)(1 + T_2 s)}$; $\quad T_d < T_1; T_2$.

▶ 9-10 RELATIVE STABILITY: GAIN MARGIN AND PHASE MARGIN

We have demonstrated in Sections 9-2 through 9-4 the general relationship between the resonance peak M_p of the frequency response and the maximum overshoot of the time response. Comparisons and correlations between frequency-domain and time-domain parameters such as these are useful in the prediction of the performance of control systems. In general, we are interested not only in the absolute stability of a system, but also how stable it is. The latter is often called **relative stability**. In the time domain, relative stability is measured by parameters such as the maximum overshoot and the damping ratio. In the frequency domain, the resonance peak M_p can be used to indicate relative stability. Another way of measuring relative stability in the frequency domain is by how close the Nyquist plot of $L(j\omega)$ is to the $(-1, j0)$ point.

To demonstrate the concept of relative stability in the frequency domain, the Nyquist plots and the corresponding step responses and frequency responses of a typical third-order system are shown in Fig. 9-36 for four different values of loop gain K. It is assumed that the function $L(j\omega)$ is of minimum-phase type, so that the enclosure of the $(-1, j0)$ point is sufficient for stability analysis. The four cases are evaluated as follows.

- Relative stability is used to indicate how stable a system is.

1. *Figure 9-36(a); the loop gain K is low:* The Nyquist plot of $L(j\omega)$ intersects the negative real axis at a point that is quite far to the right of the $(-1, j0)$ point. The corresponding step response is quite well damped, and the value of M_r of the frequency response is low.

2. *Figure 9-36(b); K is increased:* The intersect is moved closer to the $(-1, j0)$ point; the system is still stable, since the critical point is not enclosed, but the step response has a larger maximum overshoot, and M_r is also larger.

3. *Figure 9-36(c); K is increased further:* The Nyquist plot now passes through the $(-1, j0)$ point, and the system is marginally stable. The step response becomes oscillatory with constant amplitude, and M_r becomes infinite.

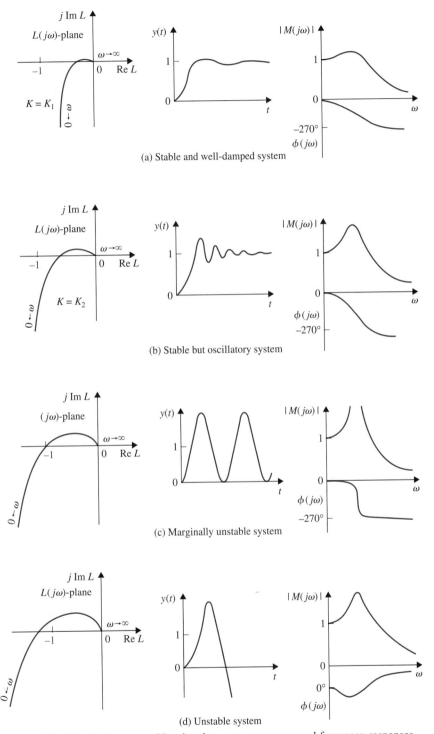

Figure 9-36 Correlation among Nyquist plots, step responses, and frequency responses.

• M_r ceases to have any meaning when the closed-loop system is unstable.

4. ***Figure 9-36(d); K relatively very large:*** The Nyquist plot now encloses the $(-1, j0)$ point, and the system is unstable. The step response becomes unbounded. The magnitude curve of $|M(j\omega)|$-versus-ω ceases to have any significance. In fact, for the unstable system, the value of M_r is still finite! In all the above analysis, the phase curve $\phi(j\omega)$ of the closed-loop frequency response also gives a qualitative information about stability. Notice that the negative slope of the phase curve becomes steeper as the relative stability decreases. When the system is unstable, the slope beyond the resonant frequency becomes positive. In practice, the phase characteristics of the closed-loop system are seldom used for analysis and design purposes.

9-10-1 Gain Margin (GM)

Gain Margin (GM) is one of the most frequently used criteria for measuring relative stability of control systems. In the frequency domain, gain margin is used to indicate the closeness of the intersection of the negative real axis made by the Nyquist plot of $L(j\omega)$ to the $(-1, j0)$ point. Before giving the definition of gain margin, let us first define the **phase crossover** on the Nyquist plot and the **phase-crossover frequency**.

Phase Crossover: A phase-crossover on the $L(j\omega)$ plot is a point at which the plot intersects the negative real axis.

• The definition of gain margin given here is for minimum-phase loop transfer functions.

Phase-Crossover Frequency: The **phase-crossover frequency** ω_p is the frequency at the phase crossover, or where

$$\angle L(j\omega_p) = 180° \tag{9-90}$$

• Gain margin is measured at the phase crossover.

The Nyquist plot of a loop transfer function $L(j\omega)$ that is of minimum-phase type is shown in Fig. 9-37. The phase-crossover frequency is denoted as ω_p, and the magnitude of $L(j\omega)$ at $\omega = \omega_p$ is designated as $|L(j\omega_p)|$. Then, the gain margin of the closed-loop system that has $L(s)$ as its loop transfer function is defined as

$$\text{gain margin} = \text{GM} = 20\log_{10}\frac{1}{|L(j\omega_p)|}$$
$$= -20\log_{10}|L(j\omega_p)| \quad \text{dB} \tag{9-91}$$

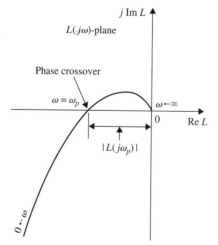

Figure 9-37   Definition of the gain margin in the polar coordinates.

On the basis of this definition, we can draw the following conclusions about the gain margin of the system shown in Fig. 9-37, depending on the properties of the Nyquist plot.

1. The $L(j\omega)$ plot does not intersect the negative real axis (no finite nonzero phase crossover).

$$|L(j\omega_p)| = 0 \qquad \text{GM} = \infty \text{ dB} \tag{9-92}$$

2. The $L(j\omega)$ plot intersects the negative real axis between (phase crossover lies between) 0 and the -1 point.

$$0 < |L(j\omega_p)| < 1 \qquad \text{GM} > 0 \text{ dB} \tag{9-93}$$

3. The $L(j\omega)$ plot passes through (phase crossover is at) the $(-1, j0)$ point.

$$|L(j\omega_p)| = 1 \qquad \text{GM} = 0 \text{ dB} \tag{9-94}$$

4. The $L(j\omega)$ plot encloses (phase crossover is to the left of) the $(-1, j0)$ point.

$$|L(j\omega_p)| > 1 \qquad \text{GM} < 0 \text{ dB} \tag{9-95}$$

• Gain margin is the amount of gain in dB that can be added to the loop before the closed-loop system becomes unstable.

Based on the foregoing discussions, the physical significance of gain margin can be summarized as:

Gain margin is the amount of gain in decibels (dB) that can be added to the loop before the closed-loop system becomes unstable.

- When the Nyquist plot does not intersect the negative real axis at any finite nonzero frequency, the gain margin is infinite in dB; this means that, theoretically, the value of the loop gain can be increased to infinity before instability occurs.

- When the Nyquist plot of $L(j\omega)$ passes through the $(-1, j0)$ point, the gain margin is 0 dB, which implies that the loop gain can no longer be increased, as the system is at the margin of instability.

- When the phase-crossover is to the left of the $(-1, j0)$ point, the phase margin is negative in dB, and the loop gain must be reduced by the gain margin to achieve stability.

Gain Margin of Nonminimum-Phase Systems

Care must be taken when attempting to extend gain margin as a measure of relative stability to systems with nonminimum-phase loop transfer functions. For such systems, a system may be stable even when the phase-crossover point is to the left of $(-1, j0)$, and thus a negative gain margin may still correspond to a stable system. Nevertheless, the closeness of the phase-crossover to the $(-1, j0)$ point still gives an indication of relative stability.

9-10-2 Phase Margin (PM)

The gain margin is only a one-dimensional representation of the relative stability of a closed-loop system. As the name implies, gain margin indicates system stability with respect to the variation in loop gain only. In principle, one would believe a system with a large gain margin should always be relatively more stable than one with a smaller gain margin. Unfortunately, gain margin alone is inadequate to indicate relative stability when system parameters other than the loop gain are subject to variation. For instance, the two systems represented by the $L(j\omega)$ plots in Fig. 9-38 apparently have the same gain margin. However, locus A actually corresponds to a more stable system than locus B, since with

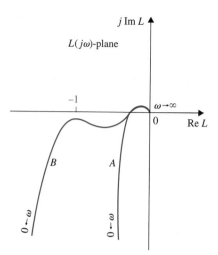

Figure 9-38 Nyquist plots showing systems with the same gain margin but different degrees of relative stability.

any change in the system parameters that affect the phase of $L(j\omega)$, locus B may easily be altered to enclose the $(-1, j0)$ point. Furthermore, we can show that the system B actually has a larger M_r than system A.

To include the effect of phase shift on stability, we introduce the **phase margin**, which requires that we first make the following definitions:

Gain Crossover: The gain-crossover is a point on the $L(j\omega)$ plot at which the magnitude of $L(j\omega)$ is equal to 1.

Gain-Crossover Frequency: The gain-crossover frequency, ω_g is the frequency of $L(j\omega)$ at the gain crossover. Or where

$$|L(j\omega_g)| = 1 \tag{9-96}$$

The definition of phase margin is stated as:
Phase margin (PM) is defined as the angle in degrees through which the L(jω) plot must be rotated about the origin so that the gain crossover passes through the $(-1, j0)$ point.

- The definition of phase margin given here is for a system with a minimum-phase loop transfer function.

Figure 9-39 shows the Nyquist plot of a typical minimum-phase $L(j\omega)$ plot, and the phase margin is shown as the angle between the line that passes through the gain crossover and the origin. In contrast to the gain margin, which is determined by loop gain, phase margin indicates the effect on system stability due to changes in system parameter, which theoretically alter the phase of $L(j\omega)$ by an equal amount at all frequencies. Phase margin is the amount of pure phase delay that can be added to the loop before the closed-loop system becomes unstable.

- Phase margin is measured at the gain crossover.

When the system is of the minimum-phase type, the analytical expression of the phase margin, as seen from Fig. 9-39, can be expressed as

$$\text{phase margin (PM)} = \angle L(j\omega_g) - 180° \tag{9-97}$$

where ω_g is the gain-crossover frequency.

- Phase margin is the amount of pure phase delay that can be added before the system becomes unstable.

Care should be taken when interpreting the phase margin from the Nyquist plot of a nonminimum-phase transfer function. When the loop transfer function if of the nonminimum-phase type, the gain crossover can occur in any quadrant of the $L(j\omega)$-plane, and the definition of phase margin given in Eq. (9-97) is no longer valid.

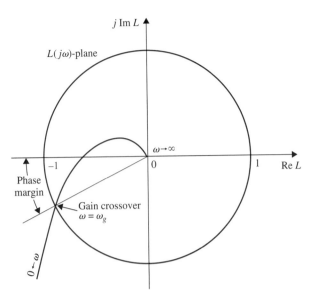

Figure 9-39 Phase margin defined in the $L(j\omega)$-plane.

▶ **EXAMPLE 9-5**   As an illustrative example on gain and phase margins, consider that the loop transfer function of a control system is

$$L(s) = \frac{2500}{s(s + 5)(s + 50)} \qquad (9\text{-}98)$$

The Nyquist plot of $L(j\omega)$ is shown in Fig. 9-40. The following results are obtained from the Nyquist plot:

Gain crossover $\omega_g = 6.22$ rad/sec

Phase crossover $\omega_p = 15.88$ rad/sec

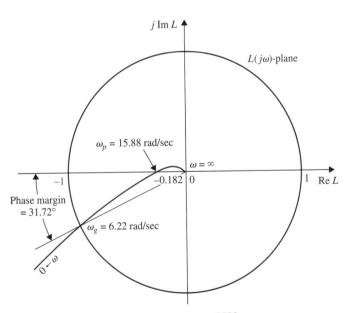

Figure 9-40   Nyquist plot of $L(s) = \dfrac{2500}{s(s + 5)(s + 50)}$.

The gain margin is measured at the phase crossover. The magnitude of $L(j\omega_p)$ is 0.182. Thus, the gain margin is obtained from Eq. (9-91),

$$\text{GM} = 20\log_{10}\frac{1}{|L(j\omega_p)|} = 20\log_{10}\frac{1}{0.182} = 14.80 \text{ dB} \qquad (9\text{-}99)$$

The phase margin is measured at the gain crossover. The phase of $L(j\omega_g)$ is 211.72°. Thus, the phase margin is obtained from Eq. (9-97),

$$\text{PM} = \angle L(j\omega_g) - 180° = 211.72° - 180° = 31.72° \qquad (9\text{-}100)$$

◀

Before embarking on the Bode plot technique of stability study, it would be beneficial to summarize advantages and disadvantages of the Nyquist plot.

Advantages of the Nyquist Plot

1. The Nyquist plot can be used for the study of stability of systems with nonminimum-phase transfer functions.

2. The stability analysis of a closed-loop system can be easily investigated by examining the Nyquist plot of the loop transfer function with reference to the $(-1, j0)$ point once the plot is made.

Disadvantage of the Nyquist Plot

1. It's not so easy to carry out the design of the controller by referring to the Nyquist plot.

▶ 9-11 STABILITY ANALYSIS WITH THE BODE PLOT

The Bode plot of a transfer function described in Appendix A is a very useful graphical tool for the analysis and design of linear control systems in the frequency domain. Before the inception of computers, Bode plots were often called the "asymptotic plots," because the magnitude and phase curves can be sketched from their asymptotic properties without detailed plotting. Modern applications of the Bode plot for control systems should be identified with the following advantages and disadvantages:

Advantages of the Bode Plot

1. In the absence of a computer, a Bode diagram can be sketched by approximating the magnitude and phase with straightline segments.

2. Gain crossover, phase crossover, gain margin, and phase margin are more easily determined on the Bode plot than from the Nyquist plot.

3. For design purposes, the effects of adding controllers and their parameters are more easily visualized on the Bode plot than on the Nyquist plot.

Disadvantage of the Bode Plot

1. Absolute and relative stability of only minimum-phase systems can be determined from the Bode plot. For instance, there is no way of telling what the stability criterion is on the Bode plot.

With reference to the definitions of gain margin and phase margin given in Figs. 9-37 and 9-39, respectively, the interpretation of these parameters from the Bode diagram is illustrated in Fig. 9-41 for a typical minimum-phase loop transfer function.

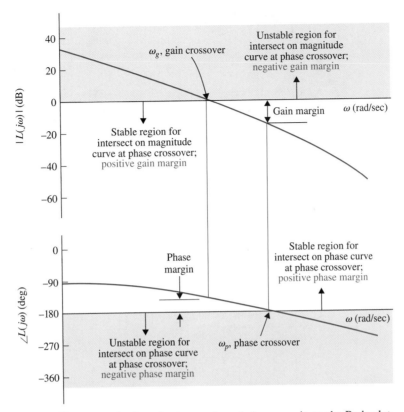

Figure 9-41 Determination of gain margin and phase margin on the Bode plot.

The following observations can be made on system stability with respect to the properties of the Bode plot:

1. The gain margin is positive and the system is stable if the magnitude of $L(j\omega)$ at the phase crossover is negative in dB. That is, the gain margin is measured below the 0-dB axis. If the gain margin is measured above the 0-dB axis, the gain margin is negative, and the system is unstable.

2. The phase margin is positive and the system is stable if the phase of $L(j\omega)$ is greater than $-180°$ at the gain crossover. That is, the phase margin is measured above the $-180°$ axis. If the phase margin is measured below the $-180°$ axis, the phase margin is negative, and the system is unstable.

• Bode plots are useful only for stability studies of systems with minimum-phase loop transfer functions.

► **EXAMPLE 9-6**

Consider the loop transfer function given in Eq. (9-98); the Bode plot of the function is drawn as shown in Fig. 9-42. The following results are observed easily from the magnitude and phase plots.

The gain crossover is the point where the magnitude curve intersects the 0-dB axis. The gain-crossover frequency ω_g is 6.22 rad/sec. The phase margin is measured at the gain crossover. The phase margin is measured from the $-180°$ axis, and is $31.72°$. Since the phase margin is measured above the $-180°$ axis, the phase margin is positive, and the system is stable.

The phase crossover is the point where the phase curve intersects the $-180°$ axis. The phase-crossover frequency is $\omega_p = 15.88$ rad/sec. The gain margin is measured at the phase crossover and is 14.8 dB. Since the gain margin is measured below the 0-dB axis, the gain margin is positive, and the system is stable.

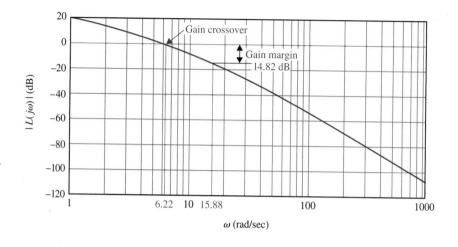

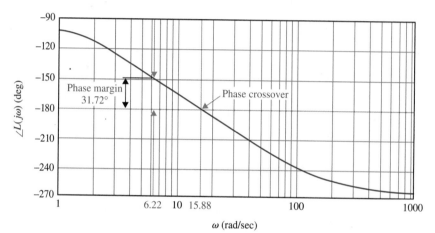

Figure 9-42 Bode plot of $L(s) = \dfrac{2500}{s(s + 5)(s + 50)}$.

The reader should compare the Nyquist plot of Fig. 9-40 with the Bode plot of Fig. 9-42, and the interpretation of ω_g, ω_p, GM, and PM on these plots. ◀

9-11-1 Bode Plots of Systems with Pure Time Delays

The stability analysis of a closed-loop system with a pure time delay in the loop can be conducted easily with the Bode plot. Example 9-7 illustrates the standard procedure.

▶ **EXAMPLE 9-7** Consider that the loop transfer function of a closed-loop system is

$$L(s) = \frac{Ke^{T_d s}}{s(s + 1)(s + 2)} \tag{9-101}$$

Figure 9-43 shows the Bode plot of $L(j\omega)$ with $K = 1$ and $T_d = 0$. The following results are obtained:

Gain-crossover frequency = 0.446 rad/sec

Phase margin = 53.4°

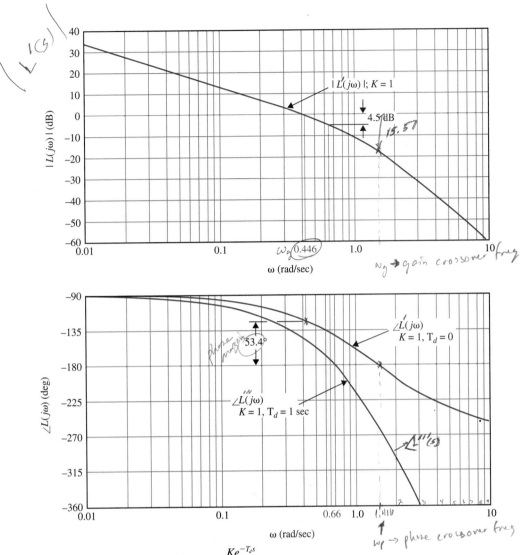

Figure 9-43 Bode plot of $L(s) = \dfrac{Ke^{-T_d s}}{s(s+1)(s+2)}$.

Phase-crossover frequency = 1.416 rad/sec

Gain margin = 15.57 dB

Thus, the system with the present parameters is stable.

The effect of the pure time delay is to add a phase of $-T_d\omega$ radians to the phase curve while not affecting the magnitude curve. The adverse effect of the time delay on stability is apparent, since the negative phase shift caused by the time delay increases rapidly with the increase in ω. To find the critical value of the time delay for stability we set

$$T_d\omega_g = 53.4°\frac{\pi}{180°} = 0.932 \quad \text{radians} \tag{9-102}$$

Solving for T_d from the last equation, we get the critical value of T_d to be 2.09 seconds.

Continuing with the example, we set T_d arbitrarily at 1 second, and find the critical value of K for stability. Figure 9-43 shows the Bode plot of $L(j\omega)$ with this new time delay. With K still equal to 1, the magnitude curve is unchanged. The phase curve droops with the increase in ω, and the following results are obtained:

$$\text{Phase-crossover frequency} = 0.66 \text{ rad/sec}$$
$$\text{Gain margin} = 4.5 \text{ dB}$$

Thus, using the definition of gain margin of Eq. (9-91), the critical value of K for stability is $10^{4.5/20} = 1.68$. ◀

▷ 9-12 RELATIVE STABILITY RELATED TO THE SLOPE OF THE MAGNITUDE CURVE OF THE BODE PLOT

In addition to GM, PM, and M_p as relative stability measures, the slope of the magnitude curve of the Bode plot of the loop transfer function at the gain crossover also gives qualitative indication on the relative stability of a closed-loop system. For example, in Fig. 9-42, if the loop gain of the system is decreased from the nominal value, the magnitude curve is shifted downward, while the phase curve is unchanged. This causes the gain-crossover frequency to be lower, and the slope of the magnitude curve at this frequency is less negative; the corresponding phase margin is increased. On the other hand, if the loop gain is increased, the gain-crossover frequency is increased, and the slope of the magnitude curve is more negative. This corresponds to a smaller phase margin, and the system is less stable. The reason behind these stability evaluations is quite simple. For a minimum-phase transfer function, the relation between its magnitude and phase is unique. Since the negative slope of the magnitude curve is a result of having more poles than zeros in the transfer function, the corresponding phase is also negative. In general, the steeper the slope of the magnitude curve, the more negative the phase. Thus, if the gain crossover is at a point where the slope of the magnitude curve is steep, it is likely that the phase margin will be small or negative.

9-12-1 Conditionally Stable System

The illustrative examples given thus far are uncomplicated in the sense that the slopes of the magnitude and phase curves are monotonically decreasing as ω increases. The following example illustrates a **conditionally stable system** that is capable of going through stable/unstable conditions as the loop gain varies.

▷ **EXAMPLE 9-8** Consider that the loop transfer function of a closed-loop system is

$$L(s) = \frac{100K(s + 5)(s + 40)}{s^3(s + 100)(s + 200)} \tag{9-103}$$

The Bode plot of $L(j\omega)$ is shown in Fig. 9-44 for $K = 1$. The following results on the system stability are obtained:

$$\text{Gain-crossover frequency} = 1 \text{ rad/sec}$$
$$\text{Phase margin} = -78°$$

There are two phase crossovers: one at 25.8 rad/sec and the other at 77.7 rad/sec. The phase characteristics between these two frequencies indicate that if the gain crossover lies in this range, the system would be stable. From the magnitude curve, the range of K for stable operation is found to be between 69 and 85.5 dB. For values of K above and below this range, the phase of $L(j\omega)$ is less than $-180°$, and the system is unstable. This example serves as a good example of the relation between relative stability and the slope of the magnitude curve at the gain crossover. As observed from Fig. 9-44, at both very low and very high frequencies, the slope of the magnitude curve is

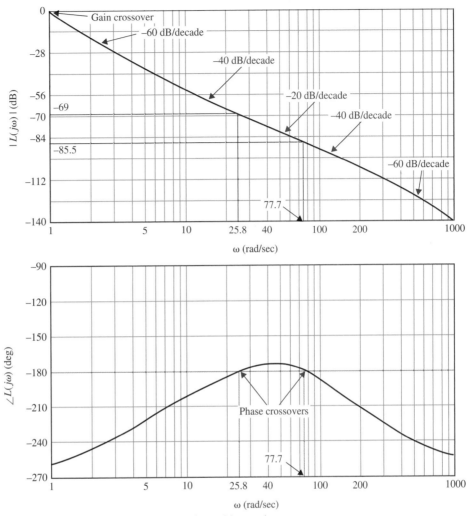

Figure 9-44   Bode plot of $L(s) = \dfrac{100K(s + 5)(s + 40)}{s^3(s + 100)(s + 200)}$, $K = 1$

-60 dB/decade; if the gain crossover falls in either one of these two regions, the phase margin is negative, and the system is unstable. In the two sections of the magnitude curve that have a slope of -40 dB/decade, the system is stable only if the gain crossover falls in about half of these regions, but even then the phase margin is small. If the gain crossover falls in the region in which the magnitude curve has a slope of -20 dB/decade, the system is stable.

Figure 9-45 shows the Nyquist plot of $L(j\omega)$. It is of interest to compare the results on stability derived from the Bode plot and the Nyquist plot. The root-locus diagram of the system is shown in Fig. 9-46. The root loci give a clear picture on the stability condition of the system with respect to K. The number of crossings of the root loci on the $j\omega$-axis of the s-plane equals the number of crossings of the phase curve of $L(j\omega)$ of the $-180°$ axis of the Bode plot, and the number of crossings of the Nyquist plot of $L(j\omega)$ with the negative real axis. The reader should check the gain margins obtained from the Bode plot and the coordinates of the crossover points on the negative real axis of the Nyquist plot with the values of K at the $j\omega$-axis crossings on the root loci.

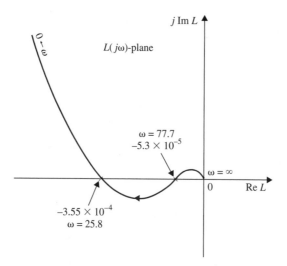

Figure 9-45 Nyquist plot of $L(s) = \dfrac{100K(s + 5)(s + 40)}{s^3(s + 100)(s + 200)}$, $K = 1$

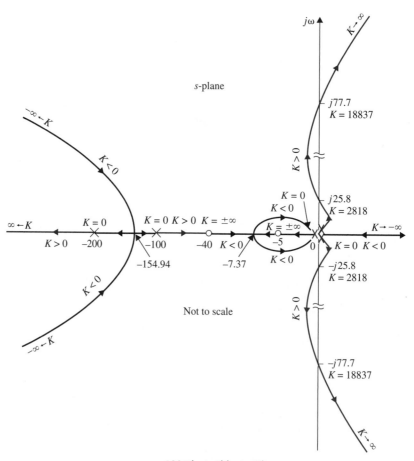

Figure 9-46 Root loci of $G(s) = \dfrac{100K(s + 5)(s + 40)}{s^3(s + 100)(s + 200)}$

▶ 9-13 STABILITY ANALYSIS WITH THE MAGNITUDE-PHASE PLOT

The magnitude-phase plot described in Section G-4 of Appendix G is another form of the frequency-domain plot that has certain advantages for analysis and design in the frequency domain. The magnitude-phase plot of a transfer function $L(j\omega)$ is done in $|L(j\omega)|$ (dB) versus $\angle L(j\omega)$ (degrees). The magnitude-phase plot of the transfer function in Eq. (9-98) is constructed in Fig. 9-47 by use of the data from the Bode plot of Fig. 9-42. The gain and phase crossovers and the gain and phase margins are clearly indicated on the magnitude-phase plot of $L(j\omega)$.

- The critical point is the intersect of the 0-dB axis and the $-180°$ axis.
- The phase crossover is where the locus intersects the $-180°$ axis.
- The gain crossover is where the locus intersects the 0-dB axis.
- The gain margin is the vertical distance in dB measured from the phase crossover to the critical point.
- The phase margin is the horizontal distance measured in degrees from the gain crossover to the critical point.

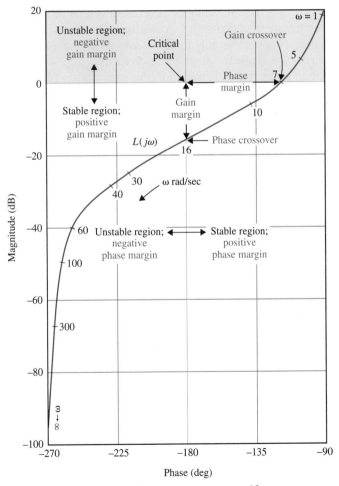

Figure 9-47 Gain-phase plot of $L(s) = \dfrac{10}{s(1 + 0.2s)(1 + 0.02s)}$.

The regions in which the gain and phase crossovers should be located for stability are also indicated. Since the vertical axis for $|L(j\omega)|$ is in dB, when the loop gain of $L(j\omega)$ changes, the locus is simply shifted up and down along the vertical axis. Similarly, when a constant phase is added to $L(j\omega)$, the locus is shifted horizontally without distortion to the curve. If $L(j\omega)$ contains a pure time delay T_d, the effect of the time delay is to add a phase equal to $-\omega T_d \times 180°/\pi$ along the curve.

Another advantage of using the magnitude-phase plot is that *for unity-feedback systems*, closed-loop system parameters such as M_r, ω_r, and BW can all be determined from the plot with the help of the constant-M loci. These closed-loop performance parameters are not represented on the Bode plot of the forward-path transfer function of a unity-feedback system.

▶ 9-14 CONSTANT-M LOCI IN THE MAGNITUDE-PHASE PLANE: THE NICHOLS CHART

It was pointed out earlier that analytically, the resonant peak M_r and bandwidth BW are difficult to obtain for high-order systems, and Bode plot provides information on the closed-loop system only in the form of gain margin and phase margin. It is necessary to develop a graphical method for the determination of M_r, ω_r, and BW using the forward-path transfer function $G(j\omega)$. As we shall see in the following development, the method is directly applicable only to unity-feedback systems, although with some modification it can also be applied to nonunity-feedback systems.

Consider that $G(s)$ is the forward-path transfer function of a unity-feedback system. The closed-loop transfer function is

$$M(s) = \frac{G(s)}{1 + G(s)} \tag{9-104}$$

For sinusoidal steady state, we replace s by $j\omega$; $G(s)$ becomes

$$\begin{aligned} G(j\omega) &= \text{Re}G(j\omega) + j\text{Im}G(j\omega) \\ &= x + jy \end{aligned} \tag{9-105}$$

where for simplicity, x denotes $\text{Re}G(j\omega)$ and y denotes $\text{Im}G(j\omega)$. The magnitude of the closed-loop transfer function is written

$$|M(j\omega)| = \left| \frac{G(j\omega)}{1 + G(j\omega)} \right| = \frac{\sqrt{x^2 + y^2}}{\sqrt{(1 + x)^2 + y^2}} \tag{9-106}$$

For simplicity of notation, let M denote $|M(j\omega)|$; then Eq. (9-106) leads to

$$M\sqrt{(1 + x)^2 + y^2} = \sqrt{x^2 + y^2} \tag{9-107}$$

Squaring both sides of Eq. (9-107) gives

$$M^2[(1 + x)^2 + y^2] = x^2 + y^2 \tag{9-108}$$

Rearranging Eq. (9-108) yields

$$(1 - M^2)x^2 + (1 - M^2)y^2 - 2M^2x = M^2 \tag{9-109}$$

This equation is conditioned by dividing through by $(1 - M^2)$ and adding the term $[M^2/(1 - M^2)]^2$ on both sides. We have

$$x^2 + y^2 - \frac{2M^2}{1 - M^2}x + \left(\frac{M^2}{1 - M^2}\right)^2 = \frac{M^2}{1 - M^2} + \left(\frac{M^2}{1 - M^2}\right)^2 \tag{9-110}$$

which is finally simplified to

$$\left(x - \frac{M^2}{1 - M^2}\right)^2 + y^2 = \left(\frac{M}{1 - M^2}\right)^2 \quad M \neq 1 \tag{9-111}$$

For a given value of M, Eq. (9-111) represents a circle with the center at

$$x = \mathrm{Re}G(j\omega) = \frac{M^2}{1 - M^2} \quad y = 0 \tag{9-112}$$

The radius of the circle is

$$r = \left|\frac{M}{1 - M^2}\right| \tag{9-113}$$

When M takes on different values, Eq. (9-111) describes in the $G(j\omega)$-plane a family of circles that are called the **constant-*M* loci**, or the **constant-*M* circles**. Figure 9-48 illustrates a typical set of constant-M circles in the $G(j\omega)$-plane. These circles are symmetrical with respect to the $M = 1$ line and the real axis. The circles to the left of the $M = 1$ locus correspond to values of M greater than 1, and those to the right of the $M = 1$ line are for M less than 1. Equations (9-111) and (9-112) show that when M becomes infinite, the circle degenerates to a point at $(-1, j0)$. Graphically, the intersection of the $G(j\omega)$ curve and the constant-M circle gives the value of M at the corresponding frequency on the $G(j\omega)$ curve. If we want to keep the value of M_r less than a certain value, the $G(j\omega)$ curve must not intersect the corresponding M circle at any point, and at the same time must not enclose the $(-1, j0)$ point. The constant-M circle with the smallest radius that is tangent to the $G(j\omega)$ curve gives the value of M_r, and the resonant frequency ω_r is read off at the tangent point on the $G(j\omega)$ curve.

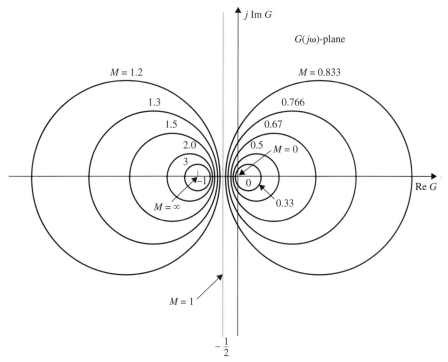

Figure 9-48 Constant-M circles in polar coordinates.

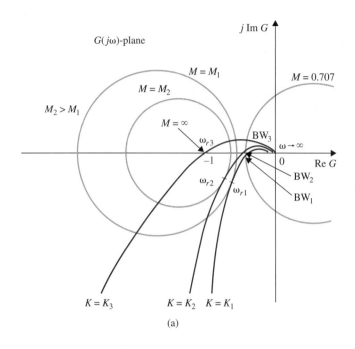

(a)

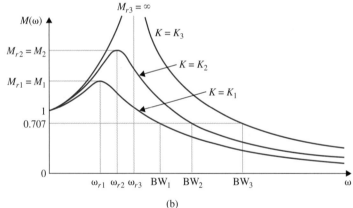

Figure 9-49 (a) Polar plots of $G(s)$ and constant-M loci. (b) Corresponding magnification curves.

(b)

Figure 9-49(a) illustrates the Nyquist plot of $G(j\omega)$ for a unity-feedback control system, together with several constant-M loci. For a given loop gain $K = K_1$, the intersects between the $G(j\omega)$ curve and the constant-M loci give the points on the $|M(j\omega)|$-versus-ω curve. The resonant peak M_r is found by locating the smallest circle that is tangent to the $G(j\omega)$ curve. The resonant frequency is found at the point of tangency, and is designated as ω_{r1}. If the loop gain is increased to K_2, and if the system is still stable, a constant-M circle with a smaller radius that corresponds to a larger M is found tangent to the $G(j\omega)$ curve, and thus the resonant peak will be larger. The resonant frequency is shown to be ω_{r2}, which is closer to the phase-crossover frequency ω_p than ω_{r1}. When K is increased to K_3, so that the $G(j\omega)$ curve now passes through the $(-1, j0)$ point, the system is marginally stable, and M_r is infinite; ω_{p3} is now the same as the resonant frequency ω_r.

When enough points of intersections between the $G(j\omega)$ curve and the constant-M loci are obtained, the magnification curves of $|M(j\omega)|$-versus-ω are plotted as shown in Fig. 9-49(b).

The bandwidth of the closed-loop system is found at the intersect of the $G(j\omega)$ curve and the $M = 0.707$ locus. For values of K beyond K_3, the system is unstable, and the constant-M loci and M_r no longer have any meaning.

- When the system is unstable, the constant-*M* loci and M_r no longer have any meaning.

A major disadvantage in working in the polar coordinates of the Nyquist plot of $G(j\omega)$ is that the curve no longer retains its original shape when a simple modification such as the change of the loop gain is made to the system. Frequently, in design situations, not only must the loop gain be altered, but a series controller may have to be added to the system. This requires the complete reconstruction of the Nyquist plot of the modified $G(j\omega)$. For design work involving M_r and BW as specifications, it is more convenient to work with the magnitude-phase plot of $G(j\omega)$, since when the loop gain is altered, the entire $G(j\omega)$ curve is shifted up or down vertically without distortion. When the phase properties of $G(j\omega)$ are changed independently, without affecting the gain, the magnitude-phase plot is affected only in the horizontal direction.

- BW is the frequency where the $G(j\omega)$ curve intersects the $M = -3dB$ locus of the Nichols chart.

For that reason, the constant-M loci in the polar coordinates are plotted in magnitude-phase coordinates, and the the resulting loci are called the **Nichols chart**. A typical Nichols chart of selected constant-M loci is shown in Fig. 9-50. Once the $G(j\omega)$ curve of the system is constructed in the Nichols chart, the intersects between the constant-M loci and the $G(j\omega)$ trajectory give the value of M at the corresponding frequencies of $G(j\omega)$. The resonant peak M_r is found by locating the smallest of the constant-M locus ($M \geq 1$) that is tangent to the $G(j\omega)$ curve from above. The resonant frequency is the frequency of $G(j\omega)$ at the point of tangency. *The bandwidth of the closed-loop system is the frequency at which the $G(j\omega)$ curve intersects the $M = 0.707$ or $M = -3dB$ locus.*

The following example illustrates the relationship among the analysis methods using the Bode plot and the Nichols chart.

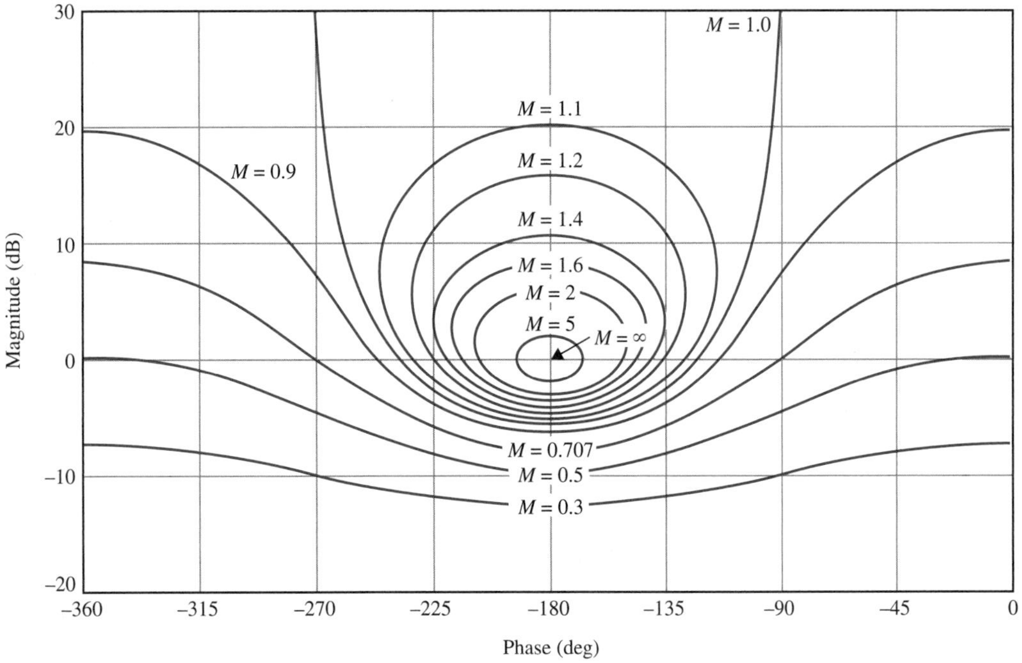

Figure 9-50   Nichols chart.

▶ **EXAMPLE 9-9** Consider the position-control system of the control surfaces of the airship analyzed in Section 7-7. The forward-path transfer function of the unity-feedback system is given by Eq. (7-147), and is repeated here.

$$G(s) = \frac{1.5 \times 10^7 K}{s(s + 400.26)(s + 3008)} \qquad (9\text{-}114)$$

The Bode plots for $G(j\omega)$ are shown in Fig. 9-51 for $K = 7.248$, 14.5, 181.2, and 273.57. The gain and phase margins of the closed-loop system for these values of K are determined and shown on the Bode plots. The magnitude-phase plots of $G(j\omega)$ corresponding to the Bode plots are shown in Fig. 9-52. These magnitude-phase plots, together with the Nichols chart, give information on the resonant peak M_r, resonant frequency ω_r, and the bandwidth BW. The gain and phase margins are also clearly marked on the magnitude-phase plots. Figure 9-53 shows the closed-loop frequency responses. Table 9-2 summarizes the results of the frequency-domain analysis for the four different values of K together with the time-domain maximum overshoots determined in Section 7-7.

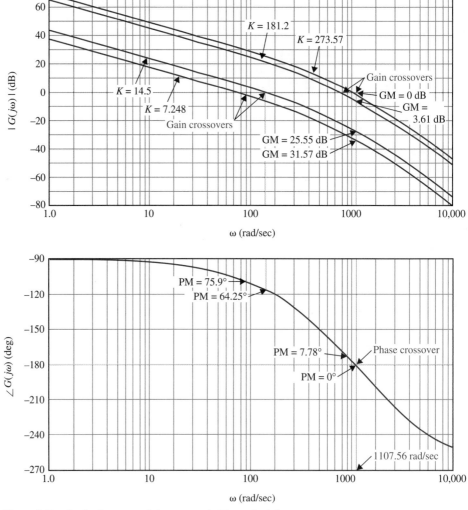

Figure 9-51 Bode diagrams of the system in Example 9-9.

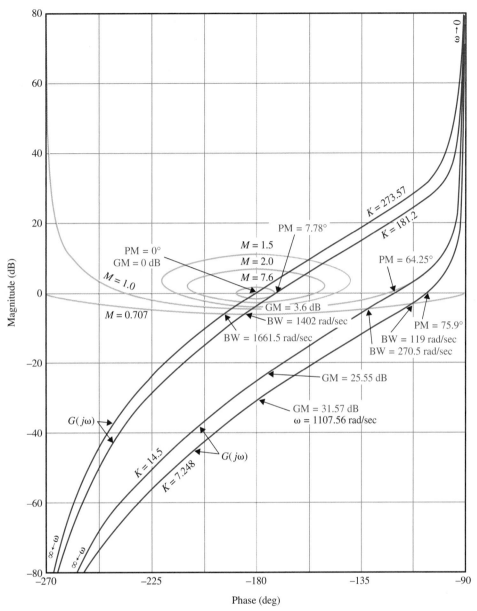

Figure 9-52 Gain-phase plots and Nichols chart of the system in Example 9-9.

TABLE 9-2 Summary of Frequency-Domain Analysis

K	Maximum Overshoot (%)	M_r	ω_r (rad/sec)	Gain Margin (dB)	Phase Margin (deg)	BW (rad/sec)
7.25	0	1.0	1.0	31.57	75.9	119.0
14.5	4.3	1.0	43.33	25.55	64.25	270.5
181.2	15.2	7.6	900.00	3.61	7.78	1402.0
273.57	100.0	∞	1000.00	0	0	1661.5

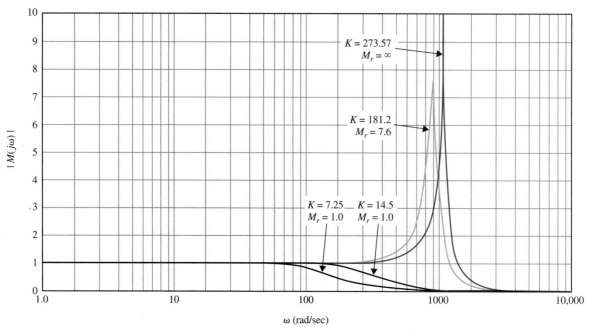

Figure 9-53   Closed-loop frequency response of the system in Example 9-9.   ◀

▶ 9-15 NICHOLS CHART APPLIED TO NONUNITY-FEEDBACK SYSTEMS

The constant-M loci and the Nichols chart presented in the preceding sections are limited to closed-loop systems with unity feedback whose transfer function is given by Eq. (9-104). When a system has nonunity feedback, the closed-loop transfer function of the system is expressed as

$$M(s) = \frac{G(s)}{1 + G(s)H(s)} \qquad (9\text{-}115)$$

where $H(s) \neq 1$. The constant-M loci and the Nichols chart cannot be applied directly to obtain the closed-loop frequency response by plotting $G(j\omega)H(j\omega)$, since the numerator of $M(s)$ does not contain $H(j\omega)$.

By proper modification, the constant-M loci and Nichols chart can still be applied to a nonunity-feedback system. Let us consider the function

$$P(s) = H(s)M(s) = \frac{G(s)H(s)}{1 + G(s)H(s)} \qquad (9\text{-}116)$$

Apparently, Eq. (9-116) is of the same form as Eq. (9-104). The frequency response of $P(j\omega)$ can be determined by plotting the function $G(j\omega)H(j\omega)$ in the amplitude-phase coordinates along with the Nichols chart. Once this is done, the frequency-response information for $M(j\omega)$ is obtained as follows.

$$|M(j\omega)| = \frac{|P(j\omega)|}{|H(j\omega)|} \qquad (9\text{-}117)$$

or in terms of dB,

$$|M(j\omega)|(dB) = |P(j\omega)|(dB) - |H(j\omega)|(dB) \tag{9-118}$$

$$\phi_m(j\omega) = \angle M(j\omega) = \angle P(j\omega) - \angle H(j\omega) \tag{9-119}$$

▶ 9-16 SENSITIVITY STUDIES IN THE FREQUENCY DOMAIN

• Sensitivity study is easily carried out in the frequency domain.

The advantage of using the frequency domain in linear control systems is that higher-order systems can be handled more easily than in the time domain. Furthermore, the sensitivity of the system with respect to parameter variations can be easily interpreted using frequency-domain plots. We shall show how the Nyquist plot and the Nichols chart can be utilized for the analysis and design of control systems based on sensitivity considerations.

Consider a linear control system with unity feedback described by the transfer function

$$M(s) = \frac{G(s)}{1 + G(s)} \tag{9-120}$$

The sensitivity of $M(s)$ with respect to the loop gain K, which is a multiplying factor in $G(s)$, is defined as

$$S_G^M(s) = \frac{\dfrac{dM(s)}{M(s)}}{\dfrac{dG(s)}{G(s)}} = \frac{dM(s)}{dG(s)} \frac{G(s)}{M(s)} \tag{9-121}$$

Taking the derivative of $M(s)$ with respect to $G(s)$, and substituting the result into Eq. (9-121) and simplifying, we have

$$S_G^M(s) = \frac{1}{1 + G(s)} = \frac{1/G(s)}{1 + 1/G(s)} \tag{9-122}$$

Clearly, the sensitivity function $S_G^M(s)$ is a function of the complex variable s. Figure 9-54 shows the magnitude plot of $S_G^M(s)$ when $G(s)$ is the transfer function given in Eq. (9-98).

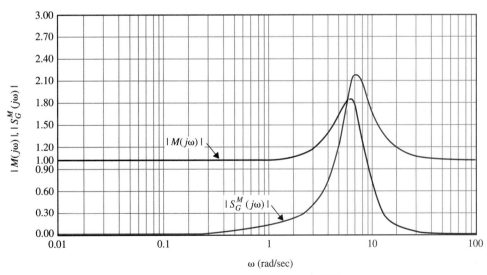

Figure 9-54 $|M(j\omega)|$ and $|S_G^M(j\omega)|$ versus ω for $G(s) = \dfrac{2500}{s(s + 5)(s + 2500)}$.

It is interesting to note that the sensitivity of the closed-loop system is inferior at frequencies greater than 4.8 rad/sec to the open-loop system whose sensitivity to the variation of K is always unity. In general, it is desirable to formulate a design criterion on sensitivity in the following manner:

$$|S_G^M(j\omega)| = \frac{1}{|1 + G(j\omega)|} = \frac{|1/G(j\omega)|}{|1 + 1/G(j\omega)|} \leq k \tag{9-123}$$

where k is a positive real number. This sensitivity criterion is in addition to the regular performance criteria on the steady-state error and the relative stability.

Equation (9-123) is analogous to the magnitude of the closed-loop transfer function, $|M(j\omega)|$, given in Eq. (9-106), with $G(j\omega)$ replaced by $1/G(j\omega)$. Thus, the sensitivity function of Eq. (9-123) can be determined by plotting $1/G(j\omega)$ in the magnitude-phase coordinates with the Nichols chart. Figure 9-55 shows the magnitude-phase plots of

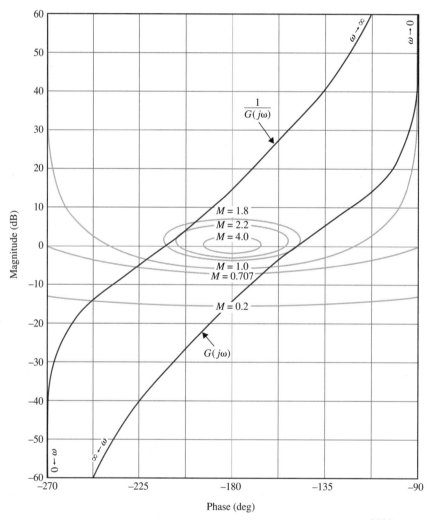

Figure 9-55 Magnitude-phase plots of $G(j\omega)$ and $1/G(j\omega)$ for $G(s) = \dfrac{2500}{s(s + 5)(s + 50)}$.

$G(j\omega)$ and $1/G(j\omega)$ of Eq. (9-98). Notice that $G(j\omega)$ is tangent to the $M = 1.8$ locus from below, which means that M_r of the closed-loop system is 1.8. The $1/G(j\omega)$ curve is tangent to the $M = 2.2$ curve from above, and according to Fig. 9-54 is the maximum value of $|S_G^M(s)|$.

Equation (9-123) shows that for low sensitivity, the loop gain of $G(j\omega)$ must be high, but it is known that in general, high gain could cause instability. Thus, the designer is again challenged by the task of designing a system with both high degree of stability and low sensitivity.

The design of robust control systems (low sensitivity) with the frequency-domain methods is discussed in Chapter 10.

▶ 9-17 MATLAB TOOLS AND CASE STUDIES

The Frequency Response Tool (freqtool) consists of a number of m-files and GUIs for the frequency analysis of simple control engineering transfer functions. The freqtool can be invoked either from the MATLAB command line by typing freqtool or from the Automatic Control Systems launch applet (ACSYS) by clicking on the appropriate button. This software enables the user to conduct the following tasks:

- Enter the transfer function values in polynomial form. (User must use the tftool discussed in Chapter 2 to convert the transfer function from the pole, zero, gain form into the polynomial form.)
- Obtain the closed loop magnitude and phase response.
- Obtain the relative stability using the transfer function loop phase and gain margin Bode plots.
- Plot the Nyquist plot of the loop transfer function.
- Understand the effect of adding zeros and poles to the closed-loop or open-loop transfer functions.
- Compare higher-order transfer functions to their approximations.
- Study position and speed-control problems.

To illustrate better how to use freqtool, let us go through the steps involved in solving some of the earlier examples.

▶ **EXAMPLE 9-1 Revisited** Consider that a single-loop feedback control system has the loop transfer function

$$L(s) = G(s)H(s) = \frac{K}{s(s + 2)(s + 10)} \tag{9-124}$$

which is of minimum-phase type. The closed-loop transfer function of Eq. (9-125) is

$$M(s) = \frac{G(s)}{1 + G(s)H(s)} = \frac{K}{s(s + 2)(s + 10) + K} = \frac{K}{s^3 + 12s^2 + 20s + K} \tag{9-125}$$

To use the Frequency Response Tool, load freqtool, as discussed earlier in this section. The window shown in Fig. 9-56 should appear. To enter the transfer function (9-124), click on the Enter Transfer Function button in the input module. The Transfer Function Input module appears next, as shown in Fig. 9-57. Press the buttons designating $G(s)$, $G\_c(s)$ or $H(s)$, in the block diagram of the

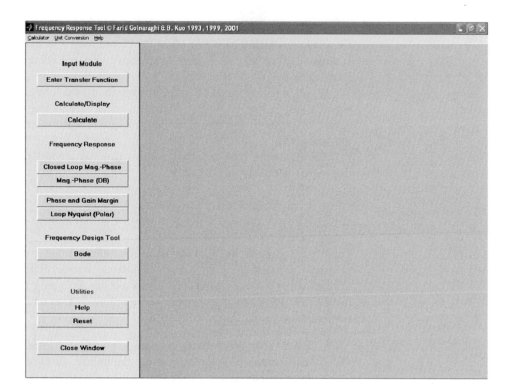

Figure 9-56 Frequency Response Tool main window.

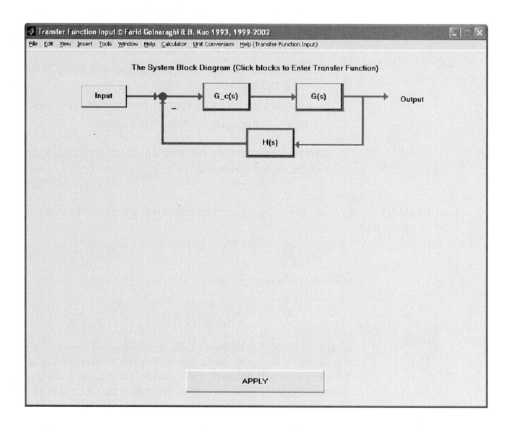

Figure 9-57 Transfer Function Input module.

single-feedback system to enter the transfer function. In this example, we set

$$G(s) = \frac{1}{s(s + 2)(s + 10)} = \frac{1}{s^3 + 12s^2 + 20s}; \quad H(s) = 1 \tag{9-126}$$

$$G\_c(s) = K = \text{Use a numerical value.}$$

As a result to enter $G(s)$ click on the "G(s)" button, and set the "Num. G(s)" to 1 and the "Den. G(s)" to 1 12 20 0, as shown in Fig. 9-58. You must be sure to enter a zero for the last coefficient in "Den. G(s)" in this case, to reflect the zero constant term in the $G(s)$ denominator. Furthermore, you must enter a numerical value for the constant K in Eq. (9-126) to be able to run the program. Let us set $G\_c(s) = K = 1$. Once you enter all values, press "APPLY" to go back to the Frequency Response main window.

Next evaluate the transfer function of the system by pressing the "Calculate" button. The closed-loop transfer function and system poles and zeros appear in the Frequency Response main window, as shown in Fig. 9-59. For more detailed results, you should refer to the MATLAB command window, as shown in Fig. 9-60. Note that evaluation of the system transfer function is a necessary step prior to any other action.

You may next perform a variety of frequency analyses such as

- Plotting closed-loop system magnitude and phase responses in Regular or log units
- Obtaining the loop transfer function $G\_c(s)G(s)H(s)$, and the gain and phase margin diagrams
- Plotting the loop transfer function Nyquist diagram
- Conducting a more thorough frequency response study by varying the value of $G\_c(s) = K$

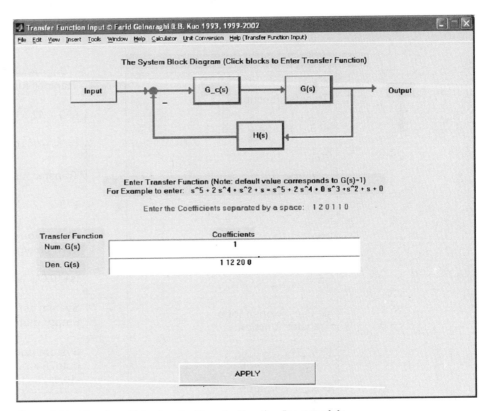

Figure 9-58 Entering $G(s)$ using the Transfer Function Input module.

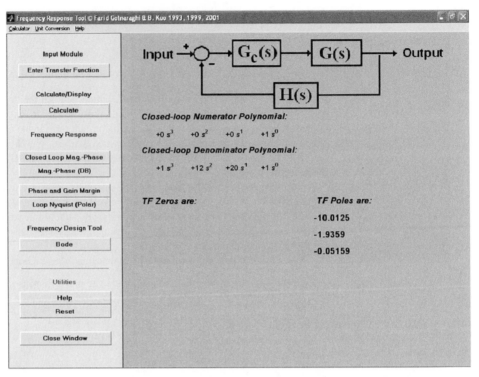

Figure 9-59 Frequency Response Tool main window, after clicking the "Calculate" button.

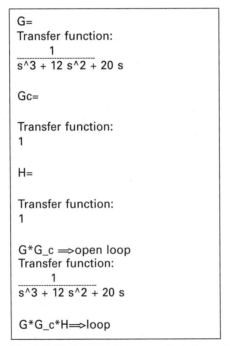

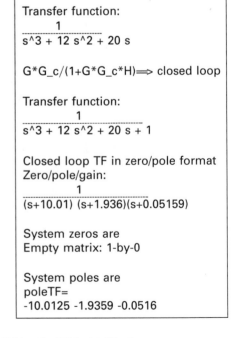

Figure 9-60 MATLAB command window, after clicking the "Calculate" button.

For example, to plot the magnitude and phase diagrams for the closed-loop system in log-DB format, click on the Closed Loop Mag. Phase (DB) button to get Fig. 9-61. The phase and gain margin diagrams of the loop transfer function, when $G\_c(s) = 1$, are shown in Fig. 9-62. This diagram may be obtained by pressing the Phase and Gain Margin button. For the value of $K = 1$, the gain margin is $GM = 47.6$ (db) at $\omega_{pc} = 4.47$ rad/sec, and the phase margin is $PM = 88.32$ (deg) at $\omega_{gc} = 0.05$ rad/sec. The Nyquist plot of the loop transfer function may be obtained by clicking on the Loop Nyquist (Polar) button.

In order to perform a more thorough frequency analysis, click on the "Bode" button. This action activates the MATLAB SISO Design GUI for the loop transfer function $G\_c(s)G(s)H(s)$. As shown in Fig. 9-63, the loop magnitude and phase Bode diagrams are plotted showing the phase and gain margin values. As in the SISO root-locus tool in timetool, discussed in Chapters 7 and 8, the frequency SISO Design GUI allows the user to vary system parameters, such as $G\_c(s) = K$, to examine their effect on time and frequency responses. For example, to change the current value of K from 1 to 10,000, simply change $C(s) = 10,000$ in the appropriate edit box in the top left corner of Fig. 9-63. The result, shown in Fig. 9-64, reflects a change in the stability of the system as the gain margin becomes $GM = -32.4$ (db) at $\omega_{pc} = 4.47$ rad/sec, and the phase margin changes to $PM = 58.8$ (deg) at $\omega_{gc} = 20.8$ rad/sec.

You may examine how the change in parameter K affects system time and frequency response by selecting the appropriate option from the "Loop Responses ..." in the Tools menu in the SISO Design window. In this case, as shown in Fig. 9-65, we have selected the closed-loop step response, the closed-loop Bode response, and the open-loop Nyquist diagram for $K = 1$. The results are shown in the diagrams in Fig. 9-66.

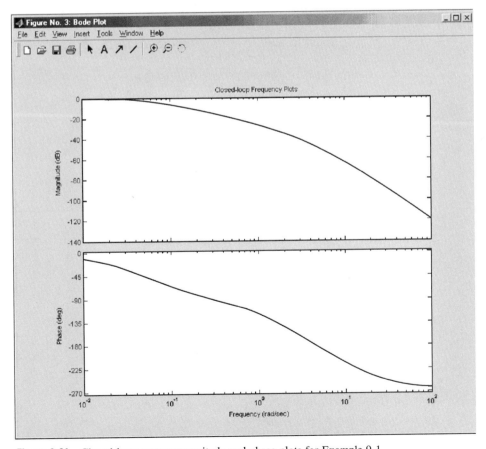

Figure 9-61 Closed-loop system magnitude and phase plots for Example 9-1.

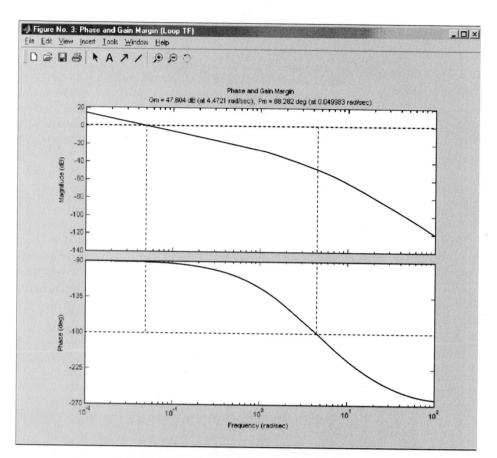

Figure 9-62 Loop transfer function phase and gain margin plots for Example 9-1.

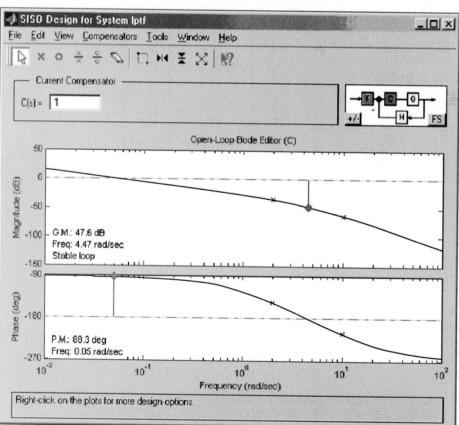

Figure 9-63 The SISO Design tool for Example 9-1, $K = 1$.

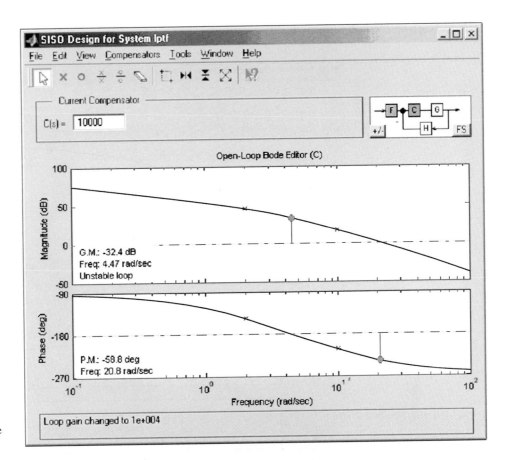

Figure 9-64 The SISO Design tool for Example 9-1, $K = 10,000$.

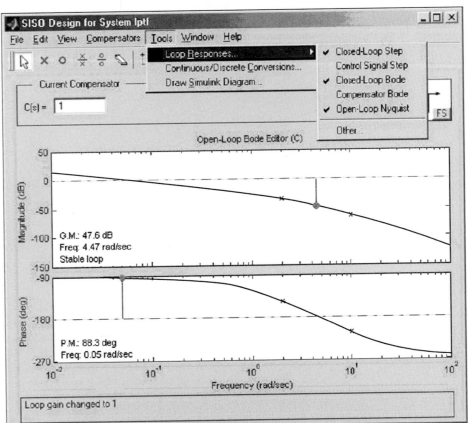

Figure 9-65 Activation of plot option within the SISO Design tool for Example 9-1.

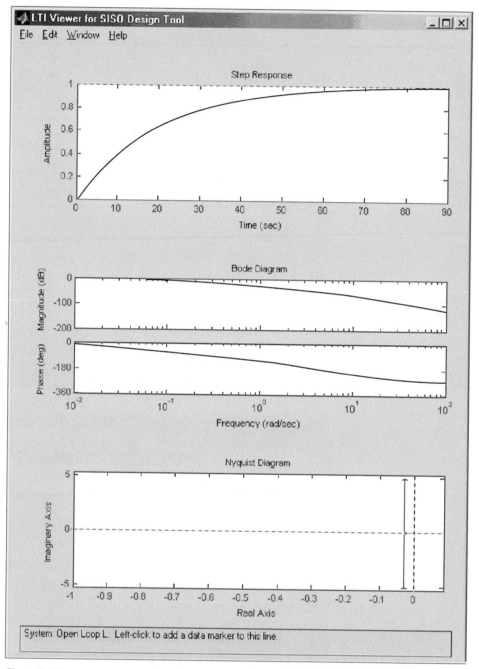

Figure 9-66 Closed-loop unit step, closed-loop Bode response, and the open-loop Nyquist diagrams for Example 9-1, $K = 1$.

As discussed earlier in Example 9-1, at $K = 240$, the system becomes marginally stable. To verify this result, set $C(s) = 240$ in Fig. 9-63. The time, frequency, and Nyquist plots of the system confirm the marginally stable response at this value of K, as shown in Fig. 9-67.

In order to examine the effect of parameter K on the poles of the system, select Root Locus from the View menu, as shown in Fig. 9-68. This allows you to get both the root-locus and frequency

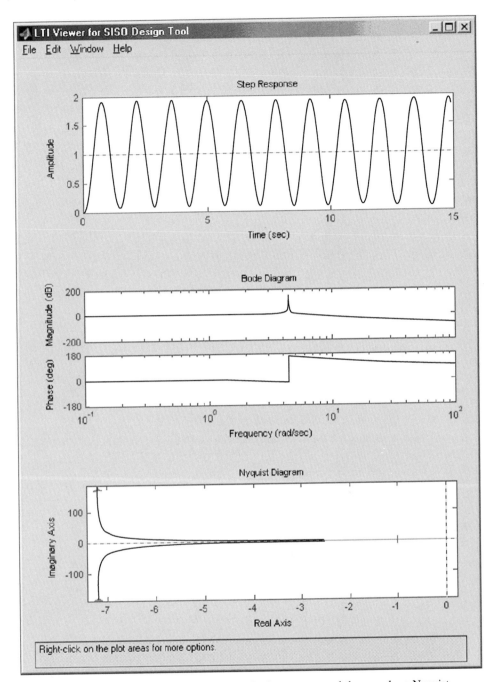

Figure 9-67 Closed-loop unit step, closed-loop Bode response, and the open-loop Nyquist diagrams for Example 9-1, $K = 240$.

diagrams side by side. If you now change the value of $C(s)$, you will notice that the poles of the closed-loop system (marked by pink squares) move along the root loci as the gain and phase margins change, as shown in Fig. 9-69.

In order to examine the effects of addition of poles and zeros on the frequency response, use the Add Pole/Zero option from the Root Locus option in the Edit menu, as shown in Fig. 9-70.

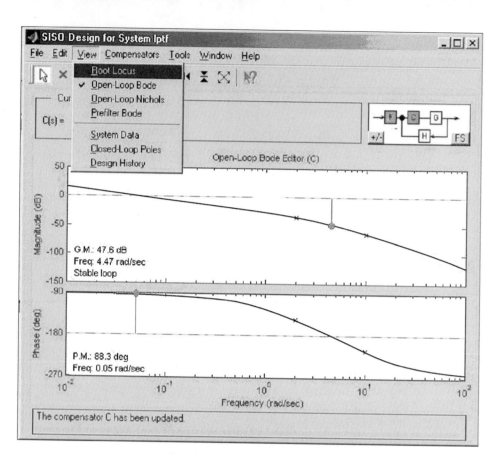

Figure 9-68 Selecting the Root Locus option in the SISO Design tool for Example 9-1.

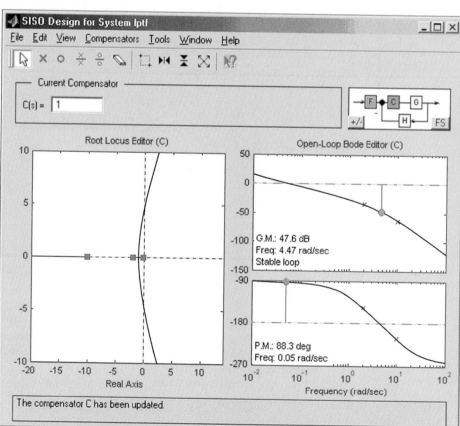

Figure 9-69 The Root Locus and Open-Loop Bode diagrams in the SISO Design tool for Example 9-1.

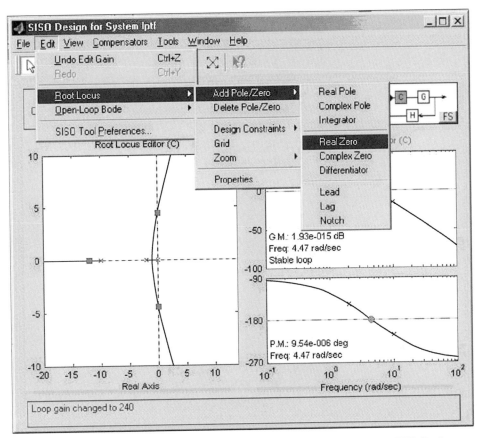

Figure 9-70 Addition of a real zero to the root-locus and Bode diagrams in the SISO Design tool for Example 9-1.

In this case a real zero has been selected at $z = -5$. The zero may be inserted on the real axis by dragging the mouse to the appropriate location on the root locus, while the left mouse button is depressed. To insert the zero, simply release the mouse button. You may vary the zero position using the mouse by bringing the mouse pointer to the zero location and pressing the left mouse button to drag the zero to a new location. Alternatively, you may click the C block in the block diagram located on the top right corner of the SISO Design window, to get the Edit Compensator C window. To insert a real zero, simply enter the desired value and click on the Add Real Zero button, as shown in Fig. 9-71. You may also enter a desired gain, for example, $K = 240$, in the Gain edit box. Figure 9-72 shows the root-locus and the phase-gain plots of the loop transfer function where

$$C(s) = 240(1 + 0.2s) \tag{9-127}$$

The root-locus and the phase and gain margin diagrams, as shown in Fig. 9-72, demonstrate that adding a real zero at $z = -5$ changes the original system from marginally stable (for $K = 240$) to stable. Notice the gain margin in this case is $GM = \infty$ and the phase margin is $PM = 39.5°$. The unit-step, Bode, and Nyquist plots also confirm this change in stability, as shown in Fig. 9-73.

In the end, the user is encouraged to apply the fretool to all examples and homework problems in this chapter.

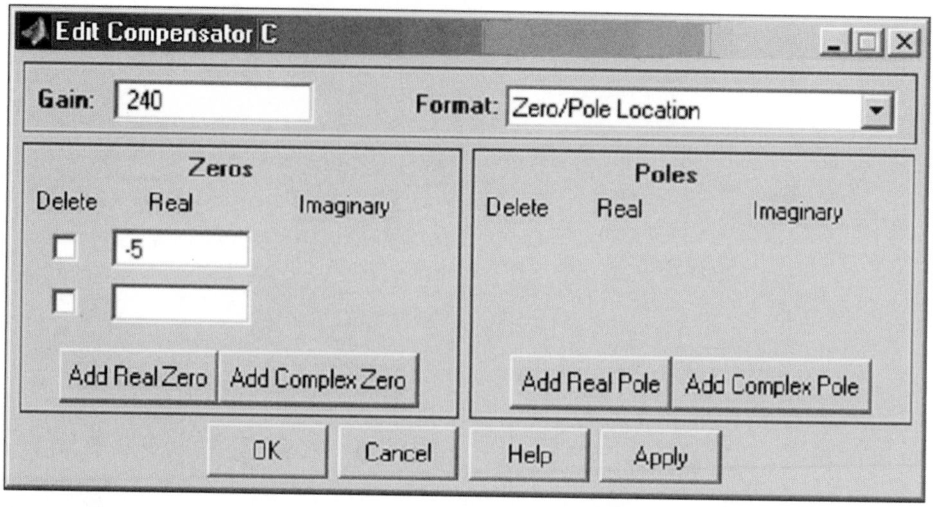

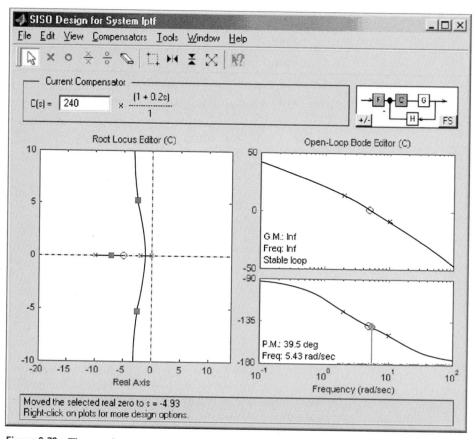

Figure 9-71   Adding a real zero and a gain to the system in Example 9-1.

Figure 9-72   The root-locus and Bode diagrams in the SISO Design tool for Example 9-1 with an added real zero at $z = -5$ and a gain $K = 240$.

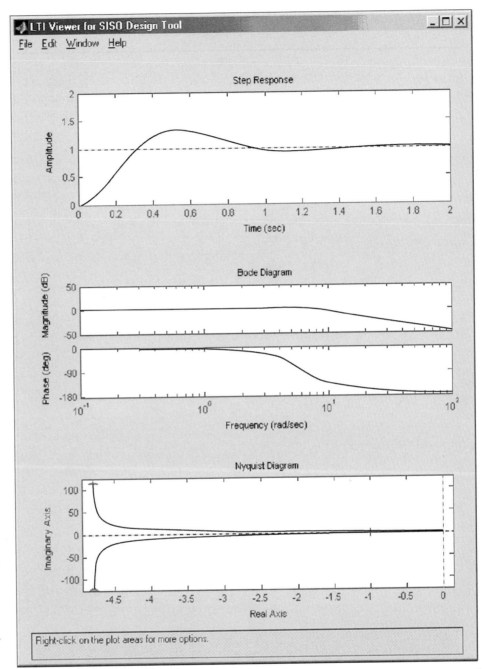

Figure 9-73 Closed-loop unit-step response, Bode diagram of the closed-loop frequency response, and open-loop Nyquist diagram for Example 9-1, when $C = G\_c(s) = 240$ $(1 + 0.2s)$.

◄

▷ 9-18 SUMMARY

The chapter began by describing typical relationships between the open-loop and closed-loop frequency responses of linear systems. Performance specifications such as the resonance peak M_r, resonant frequency ω_r, and bandwidth BW were defined in the frequency domain. The relationships among these parameters of a second-order prototype system were derived analytically. The effects of adding simple poles and zeros to the loop transfer function on M_r and BW were discussed.

The Nyquist criterion for stability analysis of linear control systems was developed. The stability of a single-loop control system can be investigated by studying the behavior of the Nyquist plot of the loop transfer function $G(s)H(s)$ for $\omega = 0$ to $\omega = \infty$ with respect to the critical point. If $G(s)H(s)$ is a minimum-phase transfer function, the condition of stability is simplified so that the Nyquist plot will not enclose the critical point.

The relationship between the root loci and the Nyquist plot was described in Section 9-7. The discussion should add more perspective to the understanding of both subjects.

Relative stability was defined in terms of gain margin and phase margin. These quantities were defined in the polar coordinates as well as on the Bode diagram. The gain-phase plot allows the Nichols chart to be constructed for closed-loop analysis. The values of M_r and BW can be easily found by plotting the $G(j\omega)$ locus on the Nichols chart.

The stability of systems with pure time delay is analyzed by use of the Bode plot.

Sensitivity function $S_G^M(j\omega)$ was defined as a measure of the variation of $M(j\omega)$ due to variations in $G(j\omega)$. It was shown that the frequency-response plots of $G(j\omega)$ and $1/G(j\omega)$ can be readily used for sensitivity studies.

Finally, using the **ACSYS** Toolbox for Frequency Response, **freqtool**, the reader may practice all the concepts discussed in this chapter.

▶ **REVIEW QUESTIONS**

1. Explain why it is important to conduct frequency-domain analyses of linear control systems.

2. Define resonance peak M_r of a closed-loop control system.

3. Define bandwidth BW of a closed-loop system.

4. List the advantages and disadvantages of studying stability with the Nyquist plot.

5. List the advantages and disadvantages of carrying out frequency-domain analysis with the Bode plot.

6. List the advantages and disadvantages of carrying out frequency-domain analysis with the magnitude-phase plot.

7. The following quantities are defined:

Z = number of zeros of $L(s)$ that are in the right-half s-plane

P = number of poles of $L(s)$ that are in the right-half s-plane

P_ω = number of poles of $L(s)$ that are on the $j\omega$-axis.

Give the conditions on these parameters for the system to be (a) open-loop stable and (b) closed-loop stable.

8. What condition must be satisfied by the function $L(j\omega)$ so that the Nyquist criterion is simplified to investigating whether the $(-1, j0)$ is enclosed by the Nyquist plot?

9. Give all the properties of a minimum-phase transfer function.

10. Give the definitions of gain margin and phase margin.

11. By applying a sinusoidal signal of frequency ω_0 to a linear system, the steady-state output of the system will also be of the same frequency. **(T)** **(F)**

12. For a prototype second-order system, the value of M_r depends solely on the damping ratio ζ. **(T)** **(F)**

13. Adding a zero to the loop transfer function will always increase the bandwidth of the closed-loop system. **(T)** **(F)**

14. The general effect of adding a pole to the loop transfer function is to make the closed-loop system less stable, while decreasing the bandwidth. **(T)** **(F)**

15. For a minimum-phase loop transfer function $L(j\omega)$, if the phase margin is negative, then the closed-loop system is always unstable. **(T)** **(F)**

16. Phase-crossover frequency is the frequency at which the phase of $L(j\omega)$ is $0°$. **(T)** **(F)**

17. Gain-crossover frequency is the frequency at which the gain of $L(j\omega)$ is 0 dB. **(T)** **(F)**

18. Gain margin is measured at the phase-crossover frequency. **(T)** **(F)**

19. Phase margin is measured at the gain-crossover frequency. **(T)** **(F)**

20. A closed-loop system with a pure time delay in the loop is usually less stable than one without a time delay. **(T)** **(F)**

21. The slope of the magnitude curve of the Bode plot of $L(j\omega)$ at the gain crossover usually gives indication on the relative stability of the closed-loop system. **(T)** **(F)**

22. Nichols chart can be used to find BW and M_r information of a closed-loop system. **(T)** **(F)**

23. Bode plot can be used for stability analysis for minimum- as well as nonminimum-phase transfer functions. **(T)** **(F)**

Answers to these true-and-false questions are found after the Problem section.

Answers to the Review Questions are found on the CD-ROM accompanying this text.

► REFERENCES

Nyquist Criterion of Continuous-Data Systems

1. H. Nyquist, "Regeneration Theory," *Bell System. Tech. J.*, Vol. 11, pp. 126–147, Jan. 1932.
2. R. W. Brockett and J. L. Willems, "Frequency Domain Stability Criteria—Part I," *IEEE Trans. Automatic Control*, Vol. AC-10, pp. 255–261, July 1965.
3. R. W. Brockett and J. L. Willems, "Frequency Domain Stability Criteria—Part II," *IEEE Trans. Automatic Control*, Vol. AC-10, pp. 407–413, Oct. 1965.
4. T. R. Natesan, "A Supplement to the Note on the Generalized Nyquist Criterion," *IEEE Trans. Automatic Control*, Vol. AC-12, pp. 215–216, April 1967.
5. K. S. Yeung, "A Reformulation of Nyquist's Criterion," *IEEE Trans. Educ.* Vol. E-28, pp. 59–60, Feb. 1985.

Sensitivity Function

6. A. Gelb, "Graphical Evaluation of the Sensitivity Function Using the Nichols Chart," *IRE Trans. Automatic Control*, Vol. AC-7, pp. 57–58, July 1962.

► PROBLEMS

Most of the following problems may also be solved by using **freqtool** computer program that has been developed for this chapter. Some of the simpler problems should be solved analytically for a better understanding of the fundamental principles.

• Analytical solutions of M_r, ω_r, and BW; analytical solutions of M_r, ω_r, and BW

9-1. The forward-path transfer function of a unity-feedback control system is

$$G(s) = \frac{K}{s(s + 6.54)}$$

Analytically, find the resonance peak M_r, resonant frequency ω_r, and bandwidth BW of the closed-loop system for the following values of K: **(a)** $K = 5$. **(b)** $K = 21.39$. **(c)** $K = 100$. Use the formulas for the second-order prototype system given in the text.

• Computer solutions of M_r, ω_r, and BW

9-2. Use **ACSYS** to solve the following problems. Do not attempt to obtain the solutions analytically. The forward-path transfer functions of unity-feedback control systems are given in the following equations. Find the resonance peak M_r, resonant frequency ω_r, and bandwidth BW of the closed-loop systems. (Reminder: Make certain that the system is stable.)

(a) $G(s) = \dfrac{5}{s(1 + 0.5s)(1 + 0.1s)}$

(b) $G(s) = \dfrac{10}{s(1 + 0.5s)(1 + 0.1s)}$

(c) $G(s) = \dfrac{500}{(s + 1.2)(s + 4)(s + 10)}$

(d) $G(s) = \dfrac{10(s + 1)}{s(s + 2)(s + 10)}$

(e) $G(s) = \dfrac{0.5}{s(s^2 + s + 1)}$

(f) $G(s) = \dfrac{100e^{-s}}{s(s^2 + 10s + 50)}$

(g) $G(s) = \dfrac{100e^{-s}}{s(s^2 + 10s + 100)}$

(h) $G(s) = \dfrac{10(s + 5)}{s(s^2 + 5s + 5)}$

9-3. The specifications on a second-order unity-feedback control system with the closed-loop transfer function

$$M(s) = \frac{Y(s)}{R(s)} = \frac{\omega_n^2}{s^2 + 2\zeta\omega_n s + \omega_n^2}$$

are that the maximum overshoot must not exceed 10 percent, and the rise time must be less than 0.1 sec. Find the corresponding limiting values of M_r and BW analytically.

- Correlation among maximum overshoot, rise time, M_r and BW

9-4. Repeat Problem 9-3 for maximum overshoot $\leq$ 20 percent, and $t_r \leq 0.2$ sec.

9-5. Repeat Problem 9-3 for maximum overshoot $\leq$ 30 percent, and $t_r \leq 0.2$ sec.

- Correlation between frequency and time responses

9-6. The closed-loop frequency response $|M(j\omega)|$-versus-frequency of a second-order prototype system is shown in Fig. 9P-6. Sketch the corresponding unit-step response of the system; indicate the values of the maximum overshoot, peak time, and the steady-state error due to a unit-step input.

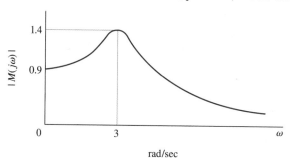

rad/sec

Figure 9P-6

- Find M_r and BW

9-7. The forward-path transfer function of a unity-feedback control system is

$$G(s) = \frac{1 + Ts}{2s(s^2 + s + 1)}$$

Find the values of BW and M_r of the closed-loop system for $T = 0.05$, 1, 2, 3, 4, and 5. Use a computer program for solutions.

- Find M_r and BW

9-8. The forward-path transfer function of a unity-feedback control system is

$$G(s) = \frac{1}{2s(s^2 + s + 1)(1 + Ts)}$$

Find the values of BW and M_r of the closed-loop system for $T = 0$, 0.5, 1, 2, 3, 4, and 5. Use a computer program for solutions.

- Stability analysis with Nyquist criterion

9-9. The loop transfer functions $L(s)$ of single-feedback-loop systems are given below. Sketch the Nyquist plot of $L(j\omega)$ for $\omega = 0$ to $\omega = \infty$. Determine the stability of the closed-loop system. If the system is unstable, find the number of poles of the closed-loop transfer function that are in the right-half s-plane. Solve for the intersect of $L(j\omega)$ on the negative real axis of the $L(j\omega)$-plane analytically. You may construct the Nyquist plot of $L(j\omega)$ using any computer program.

(a) $L(s) = \dfrac{20}{s(1 + 0.1s)(1 + 0.5s)}$ **(b)** $L(s) = \dfrac{10}{s(1 + 0.1s)(1 + 0.5s)}$

(c) $L(s) = \dfrac{100(1 + s)}{s(1 + 0.1s)(1 + 0.2s)(1 + 0.5s)}$ **(d)** $L(s) = \dfrac{10}{s^2(1 + 0.2s)(1 + 0.5s)}$

(e) $L(s) = \dfrac{3(s + 2)}{s(s^3 + 3s + 1)}$ **(f)** $L(s) = \dfrac{0.1}{s(s + 1)(s^2 + s + 1)}$

(g) $L(s) = \dfrac{100}{s(s + 1)(s^2 + 2)}$ **(h)** $L(s) = \dfrac{10(s + 10)}{s(s + 1)(s + 100)}$

• Application of Nyquist criterion

9-10. The loop transfer functions of single-feedback-loop control systems are given in the following equations. Apply the Nyquist criterion and determine the values of K for the system to be stable. Sketch the Nyquist plot of $L(j\omega)$ with $K = 1$ for $\omega = 0$ to $\omega = \infty$. You may use a computer program to plot the Nyquist plots.

(a) $L(s) = \dfrac{K}{s(s + 2)(s + 10)}$ **(b)** $L(s) = \dfrac{K(s + 1)}{s(s + 2)(s + 5)(s + 15)}$

(c) $L(s) = \dfrac{K}{s^2(s + 2)(s + 10)}$

• Application of Nyquist criterion

9-11. The forward-path transfer function of a unity-feedback control system is

$$G(s) = \frac{K}{(s + 5)^n}$$

Determine by means of the Nyquist criterion, the range of K ($-\infty < K < \infty$) for the closed-loop system to be stable. Sketch the Nyquist plot of $G(j\omega)$ for $\omega = 0$ to $\omega = \infty$.

(a) $n = 2$. **(b)** $n = 3$. **(c)** $n = 4$.

• Application of Nyquist and Routh-Hurwitz criteria

9-12. The characteristic equation of a linear control system is given in the following equation. Apply the Nyquist criterion to determine the values of K for system stability. Check the answers by means of the Routh-Hurwitz criterion.

$$s(s^3 + 2s^2 + s + 1) + K(s^2 + s + 1) = 0$$

• Stability of system with PD controller

9-13. The forward-path transfer function of a unity-feedback control system with a PD (proportional-derivative) controller is

$$G(s) = \frac{10(K_P + K_D s)}{s^2}$$

Select the value of K_P so that the parabolic-error constant K_a is 100. Find the equivalent forward-path transfer function $G_{eq}(s)$ for $\omega = 0$ to $\omega = \infty$. Determine the range of K_D for stability by the Nyquist criterion.

• Application of SyQuest and Routh-Hurwitz criterion

9-14. The block diagram of a feedback control system is shown in Fig. 9P-14.
(a) Apply the Nyquist criterion to determine the range of K for stability.
(b) Check the answer obtained in part (a) with the Routh-Hurwitz criterion.

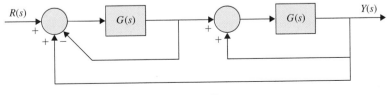

$$G(s) = \frac{K}{(s + 4)(s + 5)}$$

Figure 9P-14

9-15. The forward-path transfer function of the liquid-level control system shown in Fig. 6P-13 is

$$G(s) = \frac{K_a K_i n K_l N}{s(R_a J s + K_i K_b)(As + K_o)}$$

The following system parameters are given: $K_a = 50$, $K_i = 10$, $K_l = 50$, $J = 0.006$, $K_b = 0.0706$, $n = 0.01$, and $R_a = 10$. the values of A, N, and K_o are variable.

(a) For $A = 50$ and $K_o = 100$, sketch the Nyquist plot of $G(j\omega)$ for $\omega = 0$ to ∞ with N as a variable parameter. Find the maximum integer value of N so that the closed-loop system is stable.

(b) Let $N = 10$ and $K_o = 100$. Sketch the Nyquist plot of an equivalent transfer function $G_{eq}(j\omega)$ that has A as a multiplying factor. Find the critical value of K_o for stability.

(c) For $A = 50$ and $N = 10$, sketch the Nyquist plot of an equivalent transfer function $G_{eq}(j\omega)$ that has K_o as a multiplying factor. Find the critical value of K_o for stability.

9-16. The block diagram of a dc-motor control system is shown in Fig. 9P-16. Determine the range of K for stability using the Nyquist criterion when K_t has the following values:

(a) $K_t = 0$.   (b) $K_t = 0.01$.   (c) $K_t = 0.1$.

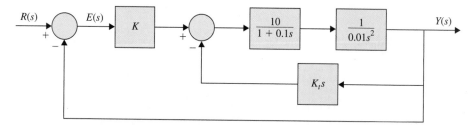

Figure 9P-16

9-17. For the system shown in Fig. 9P-16, let $K = 10$. Find the range of K_t for stability with the Nyquist criterion.

9-18. The steel-rolling control system shown in Fig. 4P-25 has the forward-path transfer function

$$G(s) = \frac{100 K e^{-T_d s}}{s(s^2 + 10s + 100)}$$

(a) When $K = 1$, determine the maximum time delay T_d in seconds for the closed-loop system to be stable.

(b) When the time delay T_d is 1 sec, find the maximum value of K for system stability.

9-19. Repeat Problem 9-18 with the following conditions.

(a) When $K = 0.1$, determine the maximum time delay T_d in seconds for the closed-loop system to be stable.

(b) When the time delay T_d is 0.1 sec, find the maximum value of K for system stability.

9-20. The system schematic shown in Fig. 9P-20 is devised to control the concentration of a chemical solution by mixing water and concentrated solution in appropriate proportions. The transfer function of the system components between the amplifier output e_a(V) and the valve position x (in.) is

$$\frac{X(s)}{E_a(s)} = \frac{K}{s^2 + 10s + 100}$$

When the sensor is viewing pure water, the amplifier output voltage e_a is zero; when it is viewing concentrated solution, $e_a = 10$ V; 0.1 in. of the valve motion changes the output concentration from zero to maximum. The valve ports can be assumed to be shaped so that the output concentration varies linearly with the valve position. The output tube has a cross-sectional area of 0.1 in.^{2}, and the rate of flow is 10^3 in./sec regardless of the valve position. To make sure that the sensor views a homogeneous solution, is it desirable to place it at some distance D in. from the valve.

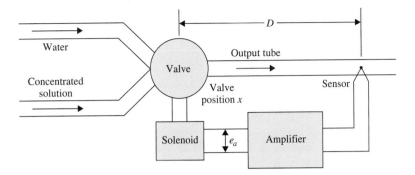

Figure 9P-20

(a) Derive the loop transfer function of the system.

(b) When $K = 10$, find the maximum distance D (in.) so that the system is stable. Use the Nyquist stability criterion.

(c) Let $D = 10$ in. Find the maximum value of K for system stability.

9-21. For the mixing system described in Problem 9-20, the following system parameters are given:

When the sensor is viewing pure water, the amplifier output voltage $e_s = 0$ V; when it is viewing concentrated solution, $e_a = 1$ V; 0.1 in. of the valve motion changes the output concentration from zero to maximum. The rest of the system characteristics are the same as in Problem 9-20. Repeat the three parts of Problem 9-20.

• Frequency-domain analysis of low-order system approximation

9-22. The approximation of high-order systems by low-order systems discussed in Section 7-10 is based on a frequency-domain criterion. The transfer functions of high-order forward-path transfer functions and the corresponding low-order approximating transfer functions determined in the illustrative examples in Chapter 7 are given in the following equations. Compute and compare the values of M_r, ω_r, and BW, of the high-order systems with those of the low-order systems, so that the accuracy of the approximations can be judged. Use a computer program for the solutions.

(a) $G(s) = \dfrac{8}{s(s^2 + 6s + 12)}$ $\quad G_L(s) = \dfrac{2.31}{s(s + 2.936)}$ $\quad$ (Example 7-9)

(b) $G(s) = \dfrac{0.909}{s(1 + 0.5455s + 0.0455s^2)}$ $\quad G_L(s) = \dfrac{0.995}{s(1 + 0.4975s)}$ $\quad$ (Example 7-10, $T = 0.1$)

(c) $G(s) = \dfrac{0.5}{s(1 + 0.75s + 0.25s^2)}$ $\quad G_L(s) = \dfrac{0.707}{s(1 + 0.3536s)}$ $\quad$ (Example 7-10, $T = 1.0$)

(d) $G(s) = \dfrac{90.3}{s(1 + 0.00283s + 8.3056 \times 10^7 s^2)}$ $\quad G_L(s) = \dfrac{92.943}{s(1 + 0.002594s)}$

(Example 7-11, $K = 7.248$)

(e) $G(s) = \dfrac{180.6}{s(1 + 0.00283s + 8.3056 \times 10^7 s^2)}$ $\quad G_L(s) = \dfrac{189.54}{s(1 + 0.002644s)}$

(Example 7-11, K=14.5)

(f) $G(s) = \dfrac{1245.52}{s(1 + 0.00283s + 8.3056 \times 10^7 s^2)}$ $\quad G_L(s) = \dfrac{2617.56}{s(1 + 0.0053s)}$

(Example 7-11, $K = 100$)

• M_r, ω_r, and BW of third-order and second-order systems

9-23. The forward-path transfer function of a unity-feedback control system is

$$G(s) = \frac{1000}{s(s^2 + 105s + 600)}$$

(a) Find the values of M_r, ω_r, and BW of the closed-loop system.

(b) Find the parameters of the second-order system with the open-loop transfer function

$$G_L(s) = \frac{\omega_n^2}{s(s + 2\zeta\omega_n)}$$

that will give the same values for M_r and ω_r as the third-order system. Compare the values of BW of the two systems.

• M_r, ω_r, and BW of third-order and second-order systems

9-24. Repeat Problem 9-23 for the transfer function given in Problem 9-22(f). (Although the approach used here may render results that are better or at least as good as those obtained by the method outlined in Section 7-10, the match here is restricted only to a prototype second-order system as the approximating low-order system.)

9-25. Sketch or plot the Bode diagrams of the forward-path transfer functions given in Problem 9-2. Find the gain margin, gain-crossover frequency, phase margin, and the phase-crossover frequency for each system.

• Gain and phase margins from Bode plots

9-26. The forward-path transfer functions of unity-feedback control systems are given in the following. Plot the Bode diagram of $G(j\omega)/K$ and do the following: **(1)** Find the value of K so that the gain margin of the system is 20 dB. **(2)** Find the value of K so that the phase margin of the system is 45°.

(a) $G(s) = \dfrac{K}{s(1 + 0.1s)(1 + 0.5s)}$ (b) $G(s) = \dfrac{K(s + 1)}{s(1 + 0.1s)(1 + 0.2s)(1 + 0.5s)}$

(c) $G(s) = \dfrac{K}{(s + 3)^3}$ (d) $G(s) = \dfrac{K}{(s + 3)^4}$

(e) $G(s) = \dfrac{Ke^{-s}}{s(1 + 0.1s + 0.01s^2)}$ (f) $G(s) = \dfrac{K(1 + 0.5s)}{s(s^2 + s + 1)}$

9-27. The forward-path transfer functions of unity-feedback control systems are given in the following equations. Plot $G(j\omega)/K$ in the gain-phase coordinates of the Nichols chart, and do the following: **(1)** Find the value of K so that the gain margin of the system is 10 dB. **(2)** Find the value of K so that the phase margin of the system is 45°. **(3)** Find the value of K so that $M_r = 1.2$.

(a) $G(s) = \dfrac{10K}{s(1 + 0.1s)(1 + 0.5s)}$ (b) $G(s) = \dfrac{5K(s + 1)}{s(1 + 0.1s)(1 + 0.2s)(1 + 0.5s)}$

(c) $G(s) = \dfrac{10K}{s(1 + 0.1s + 0.01s^2)}$ (d) $G(s) = \dfrac{10Ke^{-s}}{s(1 + 0.1s + 0.01s^2)}$

9-28. The Bode diagram of the forward-path transfer function of a unity-feedback control system is obtained experimentally, and is shown in Fig. 9P-28 when the forward gain K is set at its nominal value.

(a) Find the gain and phase margins of the system from the diagram as best you can read them. Find the gain and phase-crossover frequencies.

(b) Repeat part (a) if the gain is doubled from its nominal value.

(c) Repeat part (a) if the gain is 10 times its nominal value.

(d) Find out how much the gain must be changed from its nominal value if the gain margin is to be 40 dB.

(e) Find out how much the loop gain must be changed from its nominal value if the phase margin is to be 45°.

(f) Find the steady-state error of the system if the reference input to the system is a unit-step function.

(g) The forward path now has a pure time delay of T_d sec, so that the forward-path transfer function is multiplied by $e^{-T_d s}$. Find the gain margin and the phase margin for $T_d = 0.1$ sec. The gain is set at nominal.

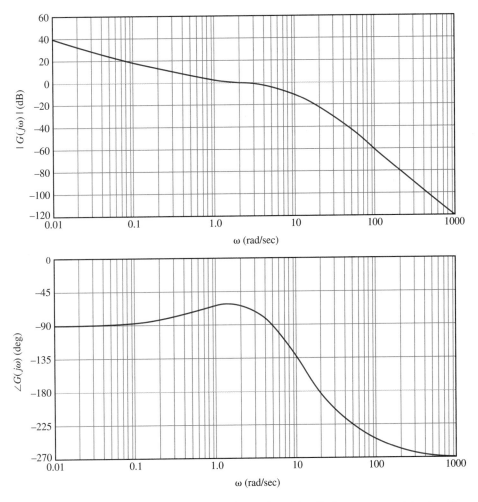

Figure 9P-28

• Applications of Bode plots

(h) With the gain set at nominal, find the maximum time delay T_d the system can tolerate without going into instability.

9-29. Repeat Problem 9-28 using Fig. 9P-28 for the following parts.

(a) Find the gain and phase margins if the gain is four times its nominal value. Find the gain- and phase-crossover frequencies.

(b) Find out how much the gain must be changed from its nominal value if the gain margin is to be 20 dB.

(c) Find the marginal value of the forward-path gain for system stability.

(d) Find out how much the gain must be changed from its nominal value if the phase margin is to be 60°.

(e) Find the steady-state error of the system if the reference input is a unit-step function and the gain is twice its nominal value.

(f) Find the steady-state error of the system if the reference input is a unit-step function and the gain is 20 times its nominal value.

(g) The system now has a pure time delay so that the forward-path transfer function is multiplied $e^{-T_d s}$. Find the gain and phase margins when $T_d = 0.1$ sec. The gain is set at its nominal value.

(h) With the gain set at 10 times its nominal, find the maximum time delay T_d the system can tolerate without going into instability.

9-30. The forward-path transfer function of a unity-feedback control system is

$$G(s) = \frac{K(1 + 0.2s)(1 + 0.1s)}{s^2(1 + s)(1 + 0.01s)^2}$$

(a) Construct the Bode and Nyquist plots of $G(j\omega)/K$ and determine the range of K for system stability.

(b) Construct the root loci of the system for $K \geq 0$. Determine the values of K and ω at the points where the root loci cross the $j\omega$-axis, using the information found from the Bode plot.

9-31. Repeat Problem 9-30 for the following transfer function.

$$G(s) = \frac{K(s + 1.5)(s + 2)}{s^2(s^2 + 2s + 2)}$$

9-32. The forward-path transfer function of the dc-motor control system described in Fig. 3P-41 is

$$G(s) = \frac{6.087 \times 10^8\,K}{s(s^3 + 423.42s^2 + 2.6667 \times 10^6 s + 4.2342 \times 10^8)}$$

Plot the Bode diagram of $G(j\omega)$ with $K = 1$, and determine the gain margin and phase margin of the system. Find the critical value of K for stability.

9-33. The transfer function between the output position $\Theta_L(s)$ and the motor current $I_a(s)$ of the robot arm modeled in Fig. 4P-20 is

$$G_p(s) = \frac{\Theta_L(s)}{I_a(s)} = \frac{K_i(Bs + K)}{\Delta_o}$$

where

$$\Delta_o(s) = s\{J_L J_m s^3 + [J_L(B_m + B) + J_m(B_L + B)]s^2$$
$$+ [B_L B_m + (B_L + B_m)B + (J_m + J_L)K]s + K(B_L + B_m)\}$$

The arm is controlled by a closed-loop system, as shown in Fig. 9P-33. The system parameters are

$$K_a = 65, K = 100, K_i = 0.4, B = 0.2, J_m = 0.2, B_L = 0.01, J_L = 0.6, \text{ and } B_m = 0.25.$$

(a) Derive the forward-path transfer function $G(s) = \Theta_L(s)/E(s)$.

(b) Draw the Bode diagram of $G(j\omega)$. Find the gain and phase margins of the system.

(c) Draw $|M(j\omega)|$ versus ω, where $M(s)$ is the closed-loop transfer function. Find M_r, ω_r, and BW.

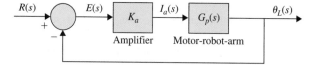

Figure 9P-33

• Relative stability analysis from gain-phase plot **9-34.** The gain-phase plot of the forward-path transfer function of $G(j\omega)/K$ of a unity-feedback control system is shown in Fig. 9P-34. Find the following performance characteristics of the system.

(a) Gain-crossover frequency (rad/sec) when $K = 1$.

(b) Phase-crossover frequency (rad/sec) when $K = 1$.

(c) Gain margin (dB) when $K = 1$.

(d) Phase margin (deg) when $K = 1$.

(e) Resonance peak M_r when $K = 1$.

(f) Resonant frequency ω_r (rad/sec) when $K = 1$.

(g) BW of the closed-loop system when $K = 1$.

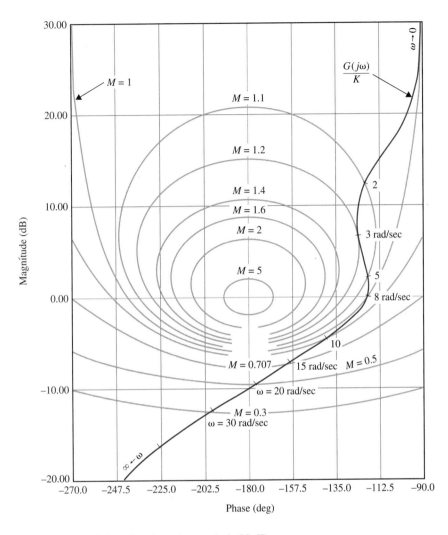

Figure 9P-34

(h) The value of K so that the gain margin is 20 dB.

(i) The value of K so that the system is marginally stable. Find the frequency of sustained oscillation in rad/sec.

(j) Steady-state error when the reference input is a unit-step function.

9-35. Repeat parts (a) through (g) of Problem 9-34 when $K = 10$. Repeat part (h) for gain margin = 40 dB.

9-36. For the gain-phase plot of $G(j\omega)/K$ shown in Fig. 9P-34, the system now has a pure time delay of T_d in the forward path, so that the forward-path transfer function becomes $G(s)e^{-T_d s}$.

(a) With $K = 1$, find T_d so that the phase margin is 40°.

(b) With $K = 1$, find the maximum value of T_d so that the system will remain stable.

9-37. Repeat Problem 9-36 with $K = 10$.

9-38. Repeat Problem 9-36 so that the gain margin is 5 dB when $K = 1$.

• Relative stability of furnace-control system

9-39. The block diagram of a furnace-control system is shown in Fig. 9P-39. The transfer function of the process is

$$G_p(s) = \frac{1}{(1 + 10s)(1 + 25s)}$$

The time delay T_d is 2 sec.

(a) Plot the Bode diagram of $G(s) = Y(s)/E(s)$ and find the gain-crossover and phase-crossover frequencies. Find the gain margin and the phase margin.

(b) Approximate the time delay by [Eq. (4-145)]

$$e^{-T_d s} \cong \frac{1}{1 + T_d s + T_d s^2/2}$$

and repeat part (a). Comment on the accuracy of the approximation. What is the maximum frequency below which the polynomial approximation is accurate?

(c) Repeat part (b) for approximating the time delay term by [Eq. (4-146)]

$$e^{-T_d s} \cong \frac{1 - T_d s/2}{1 + T_d s/2}$$

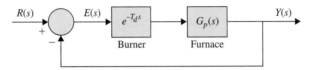

Figure 9P-39

9-40. Repeat Problem 9-39 with $T_d = 1$ sec.

9-41. Plot the $|S_G^M(j\omega)|$-versus-ω plot for the system described in Problem 9-33 for $K = 1$. Find the frequency at which the sensitivity is maximum and the value of the maximum sensitivity.

► ADDITIONAL COMPUTER PROBLEMS

9-42. Use **ACSYS** any computer program to analyze the frequency response of the following unity-feedback control systems. Plot the Bode diagrams, polar plots, and gain-phase plots, and compute the phase margin, gain margin, M_r, and BW.

(a) $G(s) = \dfrac{1 + 0.1s}{s(s + 1)(1 + 0.01s)}$.

(b) $G(s) = \dfrac{0.5(s + 1)}{s(1 + 0.2s)(1 + s + 0.5s^2)}$

(c) $G(s) = \dfrac{(s + 1)}{s(1 + 0.2s)(1 + 0.5s)}$

(d) $G(s) = \dfrac{1}{s(1 + s)(1 + 0.5s)}$

(e) $G(s) = \dfrac{50}{s(s + 1)(1 + 0.5s^2)}$

(f) $G(s) = \dfrac{(1 + 0.1s)e^{-0.1s}}{s(s + 1)(1 + 0.01s)}$

(g) $G(s) = \dfrac{10e^{-0.1s}}{s^2 + 2s + 2}$

Answers to True-and-False Review Questions
11. (T)   **12.** (T)   **13.** (T)   **14.** (T)   **15.** (T)   **16.** (F)   **17.** (T)   **18.** (T)   **19.** (T)
20. (T)   **21.** (T)   **22.** (T)   **23.** (F)

CHAPTER 10

Design of Control Systems

10-1 INTRODUCTION

All the foundations of analysis that we have laid in the preceding chapters led to the ultimate goal of design of control systems. Starting with the controlled process such as that shown by the block diagram in Fig. 10-1, control system design involves the following three steps:

1. Determine what the system should do and how to do it (design specifications).

2. Determine the controller or compensator configuration, relative to how it is connected to the controlled process.

3. Determine the parameter values of the controller to achieve the design goals.

These design tasks are explored further in the following sections.

10-1-1 Design Specifications

We often use design specifications to describe what the system should do and how it is done. These specifications are unique to each individual application, and often include specifications about **relative stability**, **steady-state accuracy (error)**, **transient-response**, and **frequency-response characteristics**. In some applications there may be additional specifications on **sensitivity to parameter variations**, that is, **robustness**, or **disturbance rejection**.

The design of linear control systems can be carried out in either the time domain or the frequency domain. For instance, **steady-state accuracy** is often specified with respect to a step input, a ramp input, or a parabolic input, and the design to meet a certain requirement is more conveniently carried out in the time domain. Other specifications such as **maximum overshoot**, **rise time**, and **settling time**, are all defined for a unit-step input, and therefore are used specifically for time-domain design. We have learned that relative stability is also measured in terms of **gain margin**, **phase margin**, and M_r. These are typical frequency-domain specifications, which should be used in conjunction with such tools as Bode plot, polar plot, gain-phase plot, and Nichols chart.

We have shown that for a second-order prototype system, there are simple analytical relationships between some of these time-domain and frequency-domain specifications. However, for higher-order systems, correlation between time-domain and frequency-domain specifications are difficult to establish. As pointed out earlier, the analysis and design of control systems is pretty much an exercise of selecting from several alternative methods

$\mathbf{u}(t)$ Control vector → CONTROLLED PROCESS G_P → $\mathbf{y}(t)$ Controlled variables (output vector)

Figure 10-1 Controlled process.

for solving the same problem. Thus, the choice of whether the design should be conducted in the time domain or the frequency domain depends often on the preference of the designer. We should be quick to point out, however, that in most cases, time-domain specifications such as maximum overshoot, rise time, and settling time are usually used as the final measure of system performance. To an inexperienced designer, it is difficult to comprehend the physical connection between frequency-domain specifications such as gain and phase margins, and resonance peak, to actual system performance. For instance, does a gain margin of 20 dB guarantee a maximum overshoot of less than 10 percent? To a designer it makes more sense to specify, for example, that the maximum overshoot should be less than 5 percent, and a settling time less than 0.01 sec. It is less obvious what, for example, a phase margin of 60° and an M_r of less than 1.1 may bring in system performance. The following outline will hopefully clarify and explain the choices and reasons for using time-domain versus frequency-domain specifications.

1. Historically, the design of linear control systems was developed with a wealth of graphical tools such as Bode plot, Nyquist plot, gain-phase plot, and Nichols chart, which are all carried out in the frequency domain. The advantage of these tools is that they can all be sketched by following approximation methods without detailed plotting. Therefore, the designer can carry out designs using frequency-domain specifications such as **gain margin**, **phase margin**, M_r, and the like. High-order systems do not generally pose any particular problem. For certain types of controllers, design procedures in the frequency domain are available to reduce the trial-and-error effort to a minimum.

2. Design in the time domain using such performance specifications as **rise time**, **delay time**, **settling time**, **maximum overshoot**, and the like, is possible *analytically* only for second-order systems, or for systems that can be approximated by second-order systems. General design procedures using time-domain specifications are difficult to establish for system with order higher than the second.

The development and availability of high-powered and user-friendly computer software is rapidly changing the practice of control system design, which until recently had been dictated by historical development. With modern computer software tools, the designer can go through a large number of design runs using the time-domain specifications within a matter of minutes. This diminishes considerably the historical edge of the frequency-domain design, which is based on the convenience of performing graphical design manually.

At the end of the chapter we present the MATLAB tools (controls and statetool) to solve the design problems associated with this chapter. As most tools developed for **A**utomatic **C**ontrol **Sys**tems software package (**ACSYS**), the **controls** tool is an easy-to-use and fully graphics-based to eliminate the users need to write codes. The **statetools** is used for the state space problems, and has already been discussed in Chapter 5. These programs have been developed using MATLAB R12 and prerelease R13 and should also work with the student version of MATLAB (optimized for MATLAB 6.1[1]).

Finally, it is generally difficult (except for an experienced designer) to select a meaningful set of frequency-domain specifications that will correspond to the desired time-domain performance requirements. For example, specifying a phase margin of 60° would be meaningless unless we know that it corresponds to a certain maximum overshoot. As it turns out, to control maximum overshoot, usually one has to specify at least phase margin and M_r. Eventually, establishing an intelligent set of frequency-domain specifications becomes a trial-and-error process that precedes the actual design, which often is also a trial-and-error effort.

[1]visit www.mathworks.com

However, frequency-domain methods are still valuable in interpreting noise rejection and sensitivity properties of the system, and, most important, they offer another perspective to the design process. Therefore, in this chapter the design techniques in the time domain and the frequency domain are treated side by side, so that the methods can be easily compared.

10-1-2 Controller Configurations

In general, the dynamics of a linear controlled process can be represented by the block diagram shown in Fig. 10-1. The design objective is to have the controlled variables, represented by the output vector $\mathbf{y}(t)$, behave in certain desirable ways. The problem essentially involves the determination of the control signal $\mathbf{u}(t)$ over the prescribed time interval so that the design objectives are all satisfied.

Most of the conventional design methods in control systems rely on the so-called **fixed-configuration design** in that the designer at the outset decides the basic configuration of the overall designed system, and the place where the controller is to be positioned relative to the controlled process. The problem then involves the design of the elements of the controller. Because most control efforts involve the modification or compensation of the system-performance characteristics, the general design using fixed configuration is also called **compensation**.

Figure 10-2 illustrates several commonly used system configurations with controller compensation. These are described briefly as follows.

- *Series (cascade) compensation.* Figure 10-2(a) shows the most commonly used system configuration with the controller placed in series with the controlled process, and the configuration is referred to as **series** or **cascade compensation**.

- *Feedback compensation.* In Fig. 10-2(b), the controller is placed in the minor feedback path, and the scheme is called **feedback compensation**.

- *State-feedback compensation.* Figure 10-2(c) shows a system that generates the control signal by feeding back the state variables through constant real gains, and the scheme is known as **state feedback**. The problem with state-feedback control is that for high-order systems, the large number of state variables involved would require a large number of transducers to sense the state variables for feedback. Thus, the actual implementation of the state-feedback control scheme may be costly or impractical. Even for low-order systems, often not all the state variables are directly accessible, and an **observer** or **estimator** may be necessary to create the estimated state variables from measurements of the output variables.

The compensation schemes shown in Fig. 10-2(a), (b), and (c) all have one degree of freedom in that there is only one controller in each system, even though the controller may have more than one parameter that can be varied. The disadvantage with a one-degree-of-freedom controller is that the performance criteria that can be realized are limited. For example, if a system is to be designed to achieve a certain amount of relative stability, it may have poor sensitivity to parameter variations. Or if the roots of the characteristic equation are selected to provide a certain amount of relative damping, the maximum overshoot of the step response may still be excessive, because of the zeros of the closed-loop transfer function. The compensation schemes shown in Fig. 10-2(d), (e), and (f) all have two degrees of freedom.

- *Series-feedback compensation.* Figure 10-2(d) shows the series-feedback compensation for which a series controller and a feedback controller are used.

- *Feedforward compensation.* Figures 10-2(e) and (f) show the so-called **feedforward compensation**. In Fig. 10-2(e), the feedforward controller $G_{cf}(s)$ is placed in series

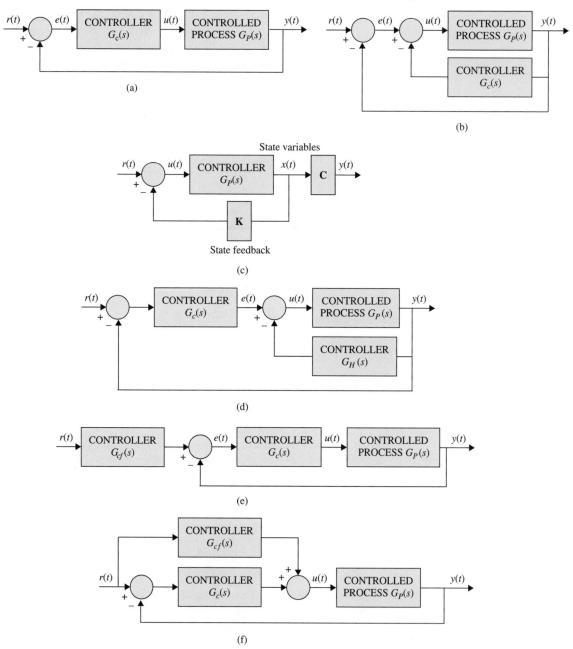

Figure 10-2 Various controller configurations in control-system compensation. (a) Series or cascade compensation. (b) Feedback compensation. (c) State-feedback control. (d) Series-feedback compensation (two degrees of freedom). (e) Forward compensation with series compensation (two degrees of freedom). (f) Feedforward compensation (two degrees of freedom).

with the closed-loop system, which has a controller $G_c(s)$ in the forward path. In Fig. 10-2(f), the feedforward controller $G_{cf}(s)$ is placed in parallel with the forward path. The key to the feedforward compensation is that the controller $G_{cf}(s)$ is not in the loop of the system, so it does not affect the roots of the characteristic equation of the original system. The poles and zeros of $G_{cf}(s)$ may be selected to add or cancel the poles and zeros of the closed-loop system transfer function.

One of the commonly used controllers in the compensation schemes just described is a PID controller, which applies a signal to the process that is proportional to the actuating signal in addition to adding integral and derivative of the actuating signal. Since these signal components are easily realized and visualized in the time domain, PID controllers are commonly designed using time-domain methods. In addition to the PID-type controllers, lead, lag, lead-lag, and notch controllers are also frequently used. The names of these controllers come from properties of their respective frequency-domain characteristics. As a result, these controllers are often designed using frequency-domain concepts. Despite these design tendencies, however, all control system designs will benefit by viewing the resulting design from both time- and frequency-domain viewpoints. Thus, both methods will be used extensively in this chapter.

It should be pointed out that these compensation schemes are by no means exhaustive. The details of these compensation schemes will be discussed in later sections of this chapter. Although the systems illustrated in Fig. 10-2 are all for continuous-data control, the same configurations can be applied to discrete-data control, in which case the controllers are all digital, with the necessary interfacings and signal converters.

10-1-3 Fundamental Principles of Design

After a controller configuration is chosen, the designer must choose a controller type that, with proper selection of its element values, will satisfy all the design specifications. The types of controllers available for control-system design are bounded only by one's imagination. Engineering practice usually dictates that one choose the simplest controller that meets all the design specifications. In most cases, the more complex a controller is, the more it costs, the less reliable it is, and more difficult it is to design. Choosing a specific controller for a specific application is often based on the designer's past experience and sometimes intuition, and entails as much *art* as it does *science*. As a novice, you may find it difficult to make intelligent choices of controllers initially with confidence. By understanding that confidence comes only through experience, this chapter provides guided experiences that illustrate the basic elements of control system designs.

After a controller is chosen, the next task is to choose controller parameter values. These parameter values are typically the coefficients of one or more transfer functions making up the controller. The basic design approach is to use the analysis tools discussed in the previous chapters to determine how individual parameter values influence the design specifications, and, finally, system performance. Based on this information, controller parameters are selected so that all design specifications are met. While this process is sometimes straightforward, more often than not it involves many design iterations since controller parameters usually interact with each other and influence design specifications in conflicting ways. For example, a particular parameter value may be chosen so that the maximum overshoot is satisfied, but in the process of varying another parameter value in an attempt to meet the rise time requirement, the maximum overshoot specification may no longer be met! Clearly, the more design specifications there are and the more controller parameters there are, the more complicated the design process becomes.

In carrying out the design either in the time domain or the frequency domain, it is important to establish some basic guidelines or design rules. Keep in mind that time-domain design usually relies heavily on the *s*-plane and the root loci. Frequency-domain design is based on manipulating the gain and phase of the loop transfer function so that the specifications are met.

In general, it is useful to summarize the time-domain and frequency-domain characteristics so that they can be used as guidelines for design purposes.

1. Complex conjugate poles of the closed-loop transfer function leads to a step response that is underdamped. If all system poles are real, the step response is overdamped. However, zeros of the closed-loop transfer function may cause overshoot even if the system is overdamped.

2. The response of a system is dominated by those poles closest to the origin in the s-plane. Transients due to those poles farther to the left decay faster.

3. The farther to the left in the s-plane the system's dominant poles are, the faster the system will respond and the greater its bandwidth will be.

4. The farther to the left in the s-plane the system's dominant poles are, the more expensive it will be and the larger its internal signals will be. While this can be justified analytically, it is obvious that striking a nail harder with a hammer drives the nail in faster but requires more energy per strike. Similarly, a sports car can accelerate faster, but it uses more fuel than an average car.

5. When a pole and zero of a system transfer function nearly cancel each other, the portion of the system response associated with the pole will have a small magnitude.

6. Time-domain and frequency-domain specifications are loosely associated with each other. Rise time and bandwidth are inversely proportional. Larger phase margin, gain margin, and lower M_r, will improve damping.

▶ 10-2 DESIGN WITH THE PD CONTROLLER

In all the examples of control systems we have discussed thus far, the controller has been typically a simple amplifier with a constant gain K. This type of control action is formally known as **proportional control**, since the control signal at the output of the controller is simply related to the input of the controller by a proportional constant.

Intuitively, one should also be able to use the derivative or integral of the input signal, in addition to the proportional operation. Therefore, we can consider a more general continuous-data controller to be one that contains such components as adders (addition or subtraction), amplifiers, attenuators, differentiators, and integrators. The designer's task is to determine which of these components should be used, in what proportion, and how they are connected. For example, one of the best-known controllers used in practice is the PID controller, described in Appendix E, where the letters stand for **proportional**, **integral**, and **derivative**. The integral and derivative components of the PID controller have individual performance implications, and their applications require an understanding of the basics of these elements. To gain an understanding of this controller, we consider just the PD portion of the controller first.

Figure 10-3 shows the block diagram of a feedback control system that arbitrarily has a second-order prototype process with the transfer function

$$G_p(s) = \frac{\omega_n^2}{s(s + 2\zeta\omega_n)} \tag{10-1}$$

The series controller is a proportional-derivative (PD) type with the transfer function

$$G_c(s) = K_P + K_D s \tag{10-2}$$

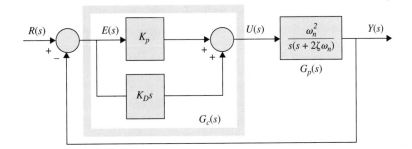

Figure 10-3 Control system with PD controller.

Thus, the control signal applied to the process is

$$u(t) = K_P e(t) + K_D \frac{de(t)}{dt} \tag{10-3}$$

where K_P and K_D are the proportional and derivative constants, respectively. Using the components given in Table E-1 in Appendix E, two electronic-circuit realizations of the PD controller are shown in Fig. 10-4. The transfer function of the circuit in Fig. 10-4(a) is

$$\frac{E_o(s)}{E_{in}(s)} = \frac{R_2}{R_1} + R_2 C_1 s \tag{10-4}$$

Comparing Eq. (10-2) with Eq. (10-4), we have

$$K_P = R_2/R_1 \qquad K_D = R_2 C_1 \tag{10-5}$$

The transfer function of the circuit in Fig. 10-4(b) is

$$\frac{E_o(s)}{E_{in}(s)} = \frac{R_2}{R_1} + R_d C_d s \tag{10-6}$$

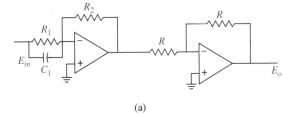

(a)

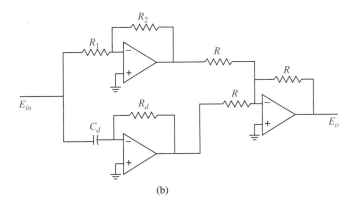

(b)

Figure 10-4 Op-amp circuit realization of the PD controller.

Comparing Eq. (10-2) with Eq. (10-6), we have

$$K_P = R_2/R_1 \qquad K_D = R_d C_d \tag{10-7}$$

The advantage with the circuit in Fig. 10-4(a) is that only two op-amps are used. However, the circuit does not allow the independent selection of K_P and K_D because they are commonly dependent on R_2. An important concern of the PD controller is that if the value of K_D is large, a large capacitor C_1 would be required. The circuit in Fig. 10-4(b) allows K_P and K_D to be independently controlled. A large K_D can be compensated by choosing a large value for R_d, thus resulting in a realistic value for C_d. Although the scope of this text does not include all the practical issues involved in controller transfer function implementation, these issues are of the utmost importance in practice.

The forward-path transfer function of the compensated system is

• PD control adds a simple zero at $s = -K_P/K_D$ to the forward-path transfer function.

$$G(s) = \frac{Y(s)}{E(s)} = G_c(s)G_p(s) = \frac{\omega_n^2(K_P + K_D s)}{s(s + 2\zeta\omega_n)} \tag{10-8}$$

which shows that the PD control is equivalent to adding a simple zero at $s = -K_P/K_D$ to the forward-path transfer function.

10-2-1 Time-Domain Interpretation of PD Control

The effect of the PD control on the transient response of a control system can be investigated by referring to the time responses shown in Fig. 10-5. Let us assume that the unit-step response of a stable system with only proportional control is as shown in Fig. 10-5(a),

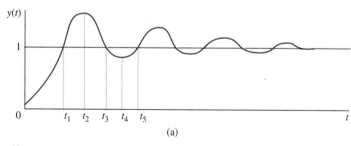

(a)

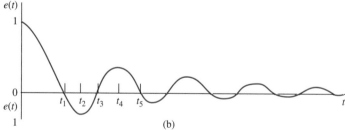

(b)

Figure 10-5 Waveforms of $y(t)$, $e(t)$, and $de(t)/dt$, showing the effect of derivative control.
(a) Unit-step response.
(b) Error signal. (c) Time rate of change of the error signal.

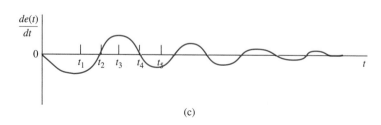

(c)

which has a relatively high maximum overshoot and is rather oscillatory. The corresponding error signal, which is the difference between the unit-step input and the output $y(t)$, and its time derivative $de(t)/dt$ are shown in Figs. 10-5(b) and (c), respectively. The overshoot and oscillation characteristics are also reflected in $e(t)$ and $de(t)/dt$. For the sake of illustration, we assume that the system contains a motor of some kind with its torque proportional to $e(t)$. The performance of the system with proportional control is analyzed as follows.

1. During the time interval, $0 < t < t_1$: The error signal $e(t)$ is positive. The motor torque is positive and rising rapidly. The large overshoot and subsequent oscillations in the output $y(t)$ are due to the excessive amount of torque developed by the motor and the lack of damping during this time interval.

2. During the time interval, $t_1 < t < t_3$: The error signal $e(t)$ is negative, and the corresponding motor torque is negative. This negative torque tends to slow down the output acceleration and eventually causes the direction of the output $y(t)$ to reverse and undershoot.

3. During the time interval, $t_3 < t < t_5$: The motor torque is again positive, thus tending to reduce the undershoot in the response caused by the negative torque in the previous time interval. Since the system is assumed to be stable, the error amplitude is reduced with each oscillation, and the output eventually settles to its final value.

Considering the above analysis of the system time response, we can say that the contributing factors to the high overshoot are

1. The positive correcting torque in the interval $0 < t < t_1$ is too large.
2. The retarding torque in the time interval $t_1 < t < t_2$ is inadequate.

Therefore, in order to reduce the overshoot in the step response, without significantly increasing the rise time, a logical approach would be to

1. Decrease the amount of positive correcting torque during $0 < t < t_1$.
2. Increase the retarding torque during $t_1 < t < t_2$.

Similarly, during the time interval, $t_2 < t < t_4$, the negative corrective torque in $t_2 < t < t_3$ should be reduced, and the retarding torque during $t_3 < t < t_4$, which is now in the positive direction, should be increased in order to improve the undershoot of $y(t)$.

The PD control described by Eq. (10-2) gives precisely the compensation effect required. Since the control signal of the PD control is given by Eq. (10-3), Fig. 10-5(c) shows the following effects provided by the PD controller:

1. For $0 < t < t_1$, $de(t)/dt$ is negative; this will reduce the original torque developed due to $e(t)$ alone.

2. For $t_1 < t < t_2$, both $e(t)$ and $de(t)/dt$ are negative, which means that the negative retarding torque developed will be greater than that with only proportional control.

3. For $t_2 < t < t_3$, $e(t)$ and $de(t)/dt$ have opposite signs. Thus, the negative torque that originally contributes to the undershoot is reduced also.

Therefore, all these effects will result in smaller overshoots and undershoots in $y(t)$.

Another way of looking at the derivative control is that since $de(t)/dt$ represents the slope of $e(t)$, the PD control is essentially an *anticipatory* control. That is, by knowing the slope, the controller can anticipate direction of the error and use it to better control the process. Normally, in linear systems, if the slope of $e(t)$ or $y(t)$ due to a step input is

• PD is essentially an anticipatory control.

large, a high overshoot will subsequently occur. The derivative control measures the instantaneous slope of $e(t)$, predicts the large overshoot ahead of time, and makes a proper corrective effort before the excessive overshoot actually occurs.

• Derivative or PD control will have an effect on a steady-state error only if the error varies with time.

Intuitively, derivative control affects the steady-state error of a system only if the steady-state error varies with time. If the steady-state error of a system is constant with respect to time, the time derivative of this error is zero, and the derivative portion of the controller provides no input to the process. But if the steady-state error increases with time, a torque is again developed in proportion to $de(t)/dt$, which reduces the magnitude of the error. Equation (10-8) also clearly shows that the PD control does not alter the system type that governs the steady-state error of a unity-feedback system.

10-2-2 Frequency-Domain Interpretation of PD Control

For frequency-domain design, the transfer function of the PD controller is written

$$G_c(s) = K_P + K_D s = K_P\left(1 + \frac{K_D}{K_P}s\right) \tag{10-9}$$

• The PD controller is a high-pass filter.

so that it is more easily interpreted on the Bode plot. The Bode plot of Eq. (10-9) is shown in Fig. 10-6 with $K_P = 1$. In general, the proportional-control gain K_P can be combined with a series gain of the system, so that the zero-frequency gain of the PD controller can be regarded as unity. The high-pass filter characteristics of the PD controller are clearly shown by the Bode plot in Fig. 10-6. The phase-lead property may be utilized to improve the phase margin of a control system. Unfortunately, the magnitude characteristics of the PD controller push the gain-crossover frequency to a higher value. Thus, *the design principle of the PD controller involves the placing of the corner frequency of the controller, $\omega = K_P/K_D$, such that an effective improvement of the phase margin is realized at the new gain-crossover frequency.* For a given system, there is a range of values of K_P/K_D that is optimal for improving the damping of the system. Another practical consideration in selecting the values of K_P and K_D is in the physical implementation of the PD controller. Other apparent effects of the PD control in the frequency domain are that, due to its high-pass characteristics, in most cases it will increase the BW of the system and reduce the rise time of the step response. The practical disadvantage of the PD controller is that the differentiator portion is a high-pass filter, which usually accentuates any high-frequency noise that enters at the input.

• The PD controller has the disadvantage that it accentuates high-frequency noise.

• The PD controller will generally increase the BW and reduce the rise time of the step response.

10-2-3 Summary of Effects of PD Control

Though it is not effective with lightly damped or initially unstable systems, a properly designed PD controller can affect the performance of a control system in the following ways:

1. Improving damping and reducing maximum overshoot.
2. Reducing rise time and settling time.
3. Increasing BW.
4. Improving GM, PM, and M_r.
5. Possibly accentuating noise at higher frequencies.
6. Possibly requiring a relatively large capacitor in circuit implementation.

The following example illustrates the effects of the PD controller on the time-domain and frequency-domain responses of a second-order system.

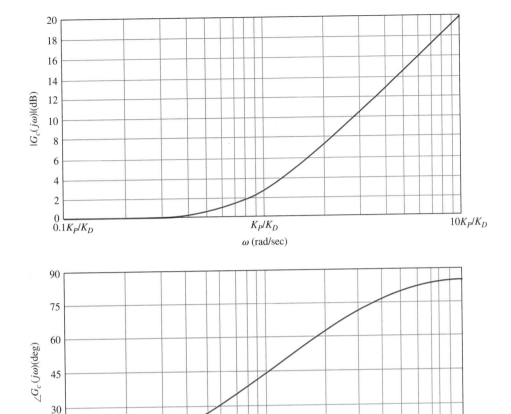

Figure 10-6 Bode diagram of $1 + \dfrac{K_D s}{K_P}, K_P = 1.$

▶ **EXAMPLE 10-1** Consider the second-order model of the aircraft attitude control system shown in Fig. 7-25. The forward-path transfer function of the system is given in Eq. (7-126) and is repeated here:

$$G(s) = \frac{4500K}{s(s + 361.2)} \qquad (10\text{-}10)$$

Let us set the performance specifications as follows:

> Steady-state error due to unit-ramp input ≤ 0.000443
> Maximum overshoot ≤ 5 percent
> Rise time $t_r \leq 0.005$ sec
> Settling time $t_s \leq 0.005$ sec

To satisfy the maximum value of the specified steady-state error requirement, K should be set at 181.17. However, with this value of K, the damping ratio of the system is 0.2, and the maximum overshoot is 52.7 percent, as shown by the unit-step response in Fig. 7-27 and again in Fig. 10-9. Let us consider inserting a PD controller in the forward path of the system so that the damping and the maximum overshoot of the system are improved while maintaining the steady-state error due to the unit-ramp input at 0.000443.

Time-Domain Design

With the PD controller of Eq. (10-9) and $K = 181.17$, the forward-path transfer function of the system becomes

$$G(s) = \frac{\Theta_y(s)}{\Theta_e(s)} = \frac{815,265(K_P + K_D s)}{s(s + 361.2)} \tag{10-11}$$

The closed-loop transfer function is

$$\frac{\Theta_y(s)}{\Theta_r(s)} = \frac{815,265(K_P + K_D s)}{s^2 + (361.2 + 815,265 K_D)s + 815,265 K_P} \tag{10-12}$$

The ramp-error constant is

$$K_v = \lim_{s \to 0} sG(s) = \frac{815,265 K_P}{361.2} = 2257.1 K_P \tag{10-13}$$

The steady-state error due to a unit-ramp input is $e_{ss} = 1/K_v = 0.000443/K_P$.

Equation (10-12) shows that the effects of the PD controller are

1. adding a zero at $s = -K_P/K_D$ to the closed-loop transfer function
2. increasing the *damping term*, which is the coefficient of the s term in the denominator, from 361.2 to $361.2 + 815,265 K_D$.

The characteristic equation is written

$$s^2 + (361.2 + 815,265 K_D)s + 815,265 K_P = 0 \tag{10-14}$$

We can set $K_P = 1$ which is acceptable from the steady-state error requirement. The damping ratio of the system is

$$\zeta = \frac{361.2 + 815,265 K_D}{1805.84} = 0.2 + 451.46 K_D \tag{10-15}$$

which clearly shows the positive effect of K_D on damping. If we wish to have critical damping, $\zeta = 1$, Eq. (10-15) gives $K_D = 0.001772$. We should quickly point out that Eq. (10-11) no longer represents a prototype second-order system, since the transient response is also affected by the zero of the transfer function at $s = -K_P/K_D$. It turns out that for this second-order system, as the value of K_D increases, the zero will move very close to the origin and effectively cancel the pole of $G(s)$ at $s = 0$. Thus, as K_D increases, the transfer function in Eq. (10-11) approaches that of a first-order system with the pole at $s = -361.2$, and the closed-loop system will not have any overshoot. In general, for higher-order systems, however, the zero at $s = -K_P/K_D$ may increase the overshoot when K_D becomes very large.

We can apply the root-contour method to the characteristic equation in Eq. (10-14) to examine the effect of varying K_P and K_D. First, by setting K_D to zero, Eq. (10-14) becomes

$$s^2 + 361.2s + 815,265 K_P = 0 \tag{10-16}$$

The root loci of the last equation as K_P varies between 0 and ∞ are shown in Fig. 10-7. When $K_D \neq 0$, the characteristic equation in Eq. (10-14) is conditioned as

$$1 + G_{eq}(s) = 1 + \frac{815,265 K_D s}{s^2 + 361.2s + 815,265 K_P} = 0 \tag{10-17}$$

The root contours of Eq. (10-14) with K_P = constant and K_D varying are constructed based on the pole-zero configuration of $G_{eq}(s)$, and are shown in Fig. 10-8 for $K_P = 0.25$ and $K_P = 1$. We see that when $K_P = 1$ and $K_D = 0$, the characteristic equation roots are at $-180.6 + j884.67$ and $-180.6 - j884.67$, and the damping ratio of the closed-loop system is 0.2. When the value of K_D is increased, the two characteristic equation roots move toward the real axis along a circular arc. When K_D is increased to 0.00177, the roots are real and equal at -902.92, and the damping is critical. When K_D is increased beyond 0.00177, the two roots become real and unequal, and the system is overdamped.

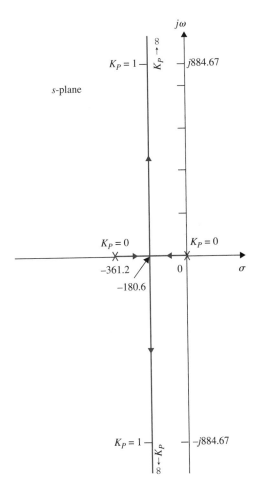

Figure 10-7 Root loci of Eq. (10-16).

When K_P is 0.25 and $K_D = 0$, the two characteristic equation roots are at $-180.6 + j413.76$ and $-180.6 - j413.76$. As K_D increases in value, the root contours again show the improved damping due to the PD controller. Figure 10-9 shows the unit-step responses of the closed-loop system without PD control, and with $K_P = 1$ and $K_D = 0.00177$. With the PD control, the maximum overshoot is 4.2 percent. In the present case, although K_D is chosen for critical damping, the overshoot is due to the zero at $s = -K_P/K_D$ of the closed-loop transfer function. Table 10-1 gives the results on maximum overshoot, rise time, and settling time for $K_P = 1$, $K_D = 0$, 0.0005, 0.00177, and 0.0025. The results in Table 10-1 show that the performance requirements are all satisfied with $K_D \geq 0.00177$. It should be kept in mind that K_D should only be large enough to satisfy the performance requirements. Large K_D corresponds to large BW, which may cause high-frequency noise problems, and there is also the concern of the capacitor value in the op-amp-circuit implementation.

The general conclusion is that the PD controller decreases the maximum overshoot, the rise time, and the settling time.

Another analytic way of studying the effects of the parameters K_P and K_D is to evaluate the performance characteristics in the parameter plane of K_P and K_D. From the characteristic equation of Eq. (10-14), we have

$$\zeta = \frac{0.2 + 451.46K_D}{\sqrt{K_P}} \tag{10-18}$$

Applying the stability requirement to Eq. (10-14), we find that for system stability,

$$K_P > 0 \quad \text{and} \quad K_D > -0.000443$$

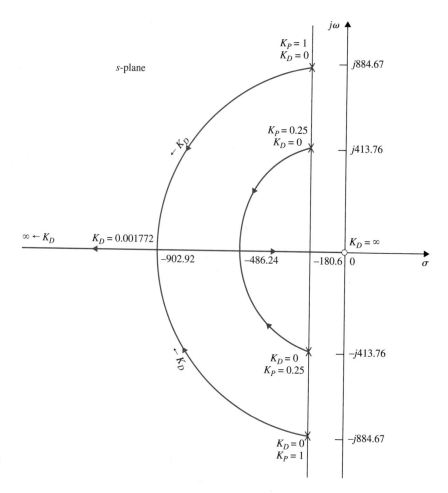

Figure 10-8 Root contours of Eq. (10-14) when $K_P = 0.25$ and 1.0; K_D varies.

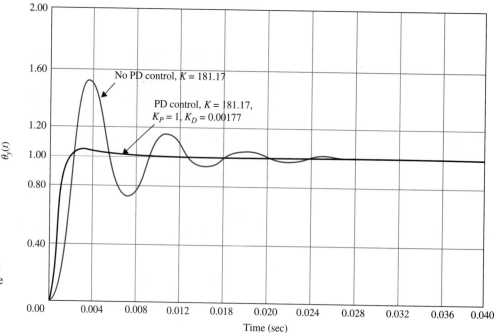

Figure 10-9 Unit-step response of the attitude control system in Fig. 7-25 with and without PD controller.

446

TABLE 10-1 **Attributes of the Unit-Step Responses of the System in Example 10-1 with PD Controller**

K_D	t_r (sec)	t_s (sec)	Maximum Overshoot (%)
0	0.00125	0.0151	52.2
0.0005	0.0076	0.0076	25.7
0.00177	0.00119	0.0049	4.2
0.0025	0.00103	0.0013	0.7

The boundaries of stability in the K_P-versus-K_D parameter plane are shown in Fig. 10-10. The constant-damping-ratio trajectory is described by Eq. (10-18), and is a parabola. Figure 10-10 illustrates the constant-ζ trajectories for $\zeta = 0.5$, 0.707, and 1.0. The ramp-error constant K_v is given by Eq. (10-13), which describes a horizontal line in the parameter plane, as shown in Fig. 10-10. The figure gives a clear picture as to how the values of K_P and K_D affect the various performance criteria of the system. For instance, if K_v is set at 2257.1, which corresponds to $K_P = 1$, the constant-ζ loci show that the damping is increased monotonically with the increase in K_D. The intersection between the constant-K_v locus and the constant-ζ locus gives the value of K_D for the desired K_v and ζ.

Frequency-Domain Design

Now let us carry out the design of the PD controller in the frequency domain. Figure 10-11 shows the Bode plot of $G(s)$ in Eq. (10-11) with $K_P = 1$ and $K_D = 0$. The phase margin of the uncompensated system is 22.68°, and the resonant peak M_r is 2.522. These values correspond to a lightly

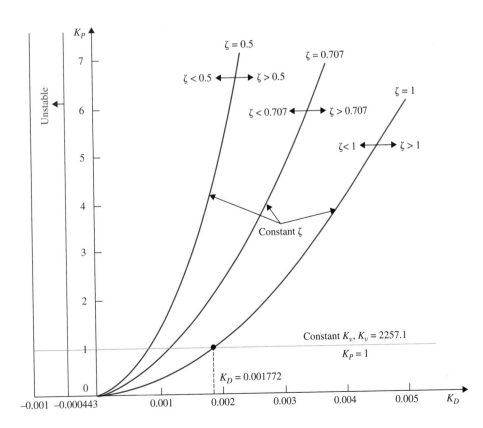

Figure 10-10 K_P-versus-K_D parameter plane for the attitude control system with a PD controller.

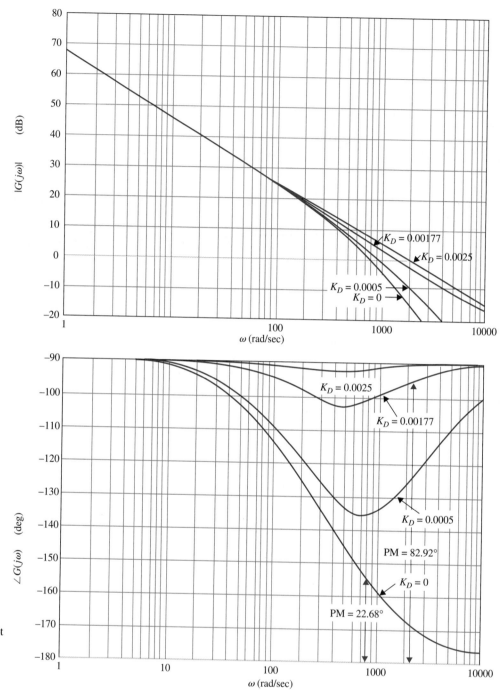

Figure 10-11 Bode plot
of $G(s) =$
$\dfrac{815, 265(1 + K_D s)}{s(s + 361.2)}$.

damped system. Let us give the following performance criteria:

Steady-state error due to a unit-ramp input ≤ 0.00443

Phase margin $\geq 80°$

Resonant peak $M_r \leq 1.05$

BW ≤ 2000 rad/sec

TABLE 10-2 Frequency-Domain Characteristics of the System in Example 10-1 with PD Controller

K_D	GM (dB)	PM (deg)	Gain CO (rad/sec)	BW (rad/sec)	M_r	t_r (sec)	t_s (sec)	Maximum Overshoot (%)
0	∞	22.68	868	1370	2.522	0.00125	0.0151	52.2
0.0005	∞	46.2	913.5	1326	1.381	0.0076	0.0076	25.7
0.00177	∞	82.92	1502	1669	1.025	0.00119	0.0049	4.2
0.0025	∞	88.95	2046	2083	1.000	0.00103	0.0013	0.7

The Bode plots of $G(s)$ for $K_P = 1$, $K_D = 0$, 0.005, 0.00177, and 0.0025 are shown in Fig. 10-11. The performance measures in the frequency domain for the compensated system with these controller parameters are tabulated in Table 10-2, along with the time-domain attributes for comparison. The Bode plots as well as the performance data were easily generated by using MATLAB tools. Use ACSYS component controls to reproduce the results in Table 10-2.

The results in Table 10-2 show that the gain margin is always infinite, and thus the relative stability is measured by the phase margin. This is one example where the gain margin is not an effective measure of the relative stability of the system. When $K_D = 0.00177$, which corresponds to critical damping, the phase margin is 82.92°, the resonant peak M_r is 1.025, and BW is 1669 rad/sec. The performance requirements in the frequency domain are all satisfied. Other effects of the PD control are that the BW and the gain crossover frequency are increased. The phase crossover frequency is always infinite in this case. ◀

▶ **EXAMPLE 10-2** Consider the third-order aircraft attitude control system discussed in Chapter 7 with the forward-path transfer function given in Eq. (7-147),

$$G(s) = \frac{1.5 \times 10^7 K}{s(s^2 + 3408.3s + 1,204,000)} \tag{10-19}$$

The same set of time-domain specifications given in Example 10-1 is to be used. It was shown in Chapter 7 that when $K = 181.17$, the maximum overshoot of the system is 78.88%.

Let us attempt to meet the transient performance requirements by use of a PD controller with the transfer function given in Eq. (10-2). The forward-path transfer function of the system with the PD controller and $K = 181.17$ is

$$G(s) = \frac{2.718 \times 10^9 (K_P + K_D s)}{s(s^2 + 3408.3s + 1,204,000)} \tag{10-20}$$

You may use ACSYS to solve this problem. See section 10-15.

Time-Domain Design
Setting $K_P = 1$ arbitrarily, the characteristic equation of the closed-loop system is written

$$s^3 + 3408.3s^2 + (1,204,000 + 2.718 \times 10^9 K_D)s + 2.718 \times 10^9 = 0 \tag{10-21}$$

To apply the root-contour method, we condition Eq. (10-21) as

$$1 + G_{eq}(s) = 1 + \frac{2.718 \times 10^9 K_D s}{s^3 + 3408.2s^2 + 1,204,000s + 2.718 \times 10^9} = 0 \tag{10-22}$$

where

$$G_{eq}(s) = \frac{2.718 \times 10^9 K_D s}{(s + 3293.3)(s + 57.49 + j906.6)(s + 57.49 - j906.6)} \tag{10-23}$$

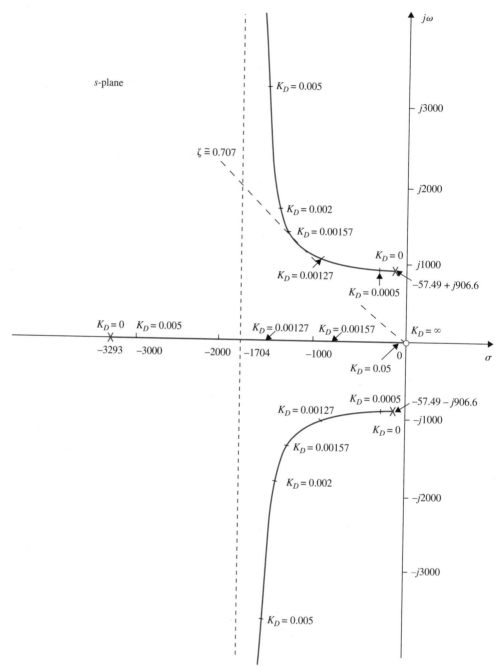

Figure 10-12 Root contours of $s^3 + 3408.3s^2 + (1,204,000 + 2.718 \times 10^9 K_D)s + 2.718 \times 10^9 = 0$.

The root contours of Eq. (10-21) are plotted as shown in Fig. 10-12, based on the pole-zero configuration of $G_{eq}(s)$. The root contours of Fig. 10-12 reveal the effectiveness of the PD controller for the improvement on the relative stability of the system. Notice that as the value of K_D increases, one root of the characteristic equation moves from -3293.3 toward the origin, while the two complex roots start out toward the left and eventually approach the vertical asymptotes that intersect at $s = -1704$. The immediate assessment of the situation is that if the value of K_D is too large, *the*

TABLE 10-3 Time-Domain Attributes of the Third-Order System in Example 10-2 with PD Controller

K_D	% Maximum Overshoot	t_r (sec)	t_s (sec)	Characteristic Equation Roots	
0	78.88	0.00125	0.0495	-3293.3,	$-57.49 \pm j906.6$
0.0005	41.31	0.00120	0.0106	-2843.07,	$-282.62 \pm j936.02$
0.00127	17.97	0.00100	0.00398	-1523.11,	$-942.60 \pm j946.58$
0.00157	14.05	0.00091	0.00337	-805.33,	$-1301.48 \pm j1296.59$
0.00200	11.37	0.00080	0.00255	-531.89,	$-1438.20 \pm j1744.00$
0.00500	17.97	0.00042	0.00130	-191.71,	$-1608.29 \pm j3404.52$
0.01000	31.14	0.00026	0.00093	-96.85,	$-1655.72 \pm j5032$
0.05000	61.80	0.00010	0.00144	-19.83,	$-1694.30 \pm j11583$

• If a system has very low damping or is unstable, the PD control may not be effective in improving the stability of the system.

two complex roots will actually have reduced damping, while increasing the natural frequency of the system. It appears that the ideal location for the two complex characteristic equation roots, from the standpoint of relative stability, is near the bend of the root contour, where the relative damping ratio is approximately 0.707. The root contours of Fig. 10-12 clearly show that if the original system has low damping or is unstable, the zero introduced by the PD controller may not be able to add sufficient damping or even stabilize the system.

Table 10-3 gives the results of maximum overshoot, rise time, settling time, and the roots of the characteristic equation as functions of the parameter K_D. The following conclusions are drawn on the effects of the PD controller on the third-order system.

1. The minimum value of the maximum overshoot, 11.37%, occurs when K_D is approximately 0.002.

2. Rise time is improved (reduced) with the increase of K_D.

3. Too high a value of K_D will actually increase the maximum overshoot and the settling time substantially. The latter is because the damping is reduced as K_D is increased indefinitely.

Figure 10-13 shows the unit-step responses of the system with the PD controller for several values of K_D. The conclusion is that while the PD control does improve the damping of the system, it does not meet the maximum-overshoot requirement.

Frequency-Domain Design

The Bode plot of Eq. (10-20) is used to conduct the frequency-domain design of the PD controller. Figure 10-14 shows the Bode plot for $K_P = 1$ and $K_D = 0$. The following performance data are obtained for the uncompensated system:

$$\text{Gain margin} = 3.6 \text{ dB}$$
$$\text{Phase margin} = 7.77°$$
$$\text{Resonant Peak } M_r = 7.62$$
$$\text{Bandwidth BW} = 1408.83 \text{ rad/sec}$$
$$\text{Gain crossover (GCO)} = 888.94 \text{ rad/sec}$$
$$\text{Phase crossover (PCO)} = 1103.69 \text{ rad/sec}$$

• For PD control, we should first examine the amount of phase needed to realize the desired PM.

Let us use the same set of frequency-domain performance requirements listed in Example 10-1. The logical way to approach this problem is to first examine how much additional phase is needed to realize a phase margin of 80°. Since the uncompensated system with the gain set to meet the steady-state requirement is only 7.77°, the PD controller must provide an additional phase of 72.23°. This additional phase must be placed at the gain crossover of the compensated system in order to realize a PM of 80°. Referring to the Bode plot of the PD controller in Fig. 10-6, we see that the

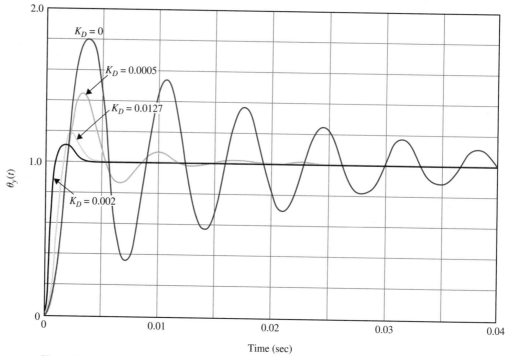

Figure 10-13   Unit-step responses of the system in Example 10-2 with PD controller.

additional phase is always accompanied by a gain in the magnitude curve. As a result, the gain crossover of the compensated system will be pushed to a higher frequency at which the phase of the uncompensated system would correspond to an even smaller PM. Thus, we may run into the problem of diminishing returns. This symptom is parallel to the situation illustrated by the root-contour plot in Fig. 10-12, in which case the larger K_D would simply push the roots to a higher frequency, and the damping would actually be decreased. The frequency-domain performance data of the compensated system with the values of K_D used in Table 10-3 are obtained from the Bode plots for each case, and the results are shown in Table 10-4. The Bode plots of some of these cases are shown in Fig. 10-14. Notice that the gain margin becomes infinite when the PD controller is added, and the phase margin becomes the dominant measure of relative stability. This is because the phase curve of the PD-compensated system stays above the $-180°$ axis, and the phase crossover is at infinity.

TABLE 10-4   Frequency-Domain Characteristics of the Third-Order System in Example 10-2 with PD Controller

K_D	GM (dB)	PM (deg)	M_r	BW (rad/sec)	Gain CO (rad/sec)	Phase CO (rad/sec)
0	3.6	7.77	7.62	1408.83	888.94	1103.69
0.0005	∞	30.94	1.89	1485.98	935.91	∞
0.00127	∞	53.32	1.19	1939.21	1210.74	∞
0.00157	∞	56.83	1.12	2198.83	1372.30	∞
0.00200	∞	58.42	1.07	2604.99	1620.75	∞
0.00500	∞	47.62	1.24	4980.34	3118.83	∞
0.01000	∞	35.71	1.63	7565.89	4789.42	∞
0.0500	∞	16.69	3.34	17989.03	11521.00	∞

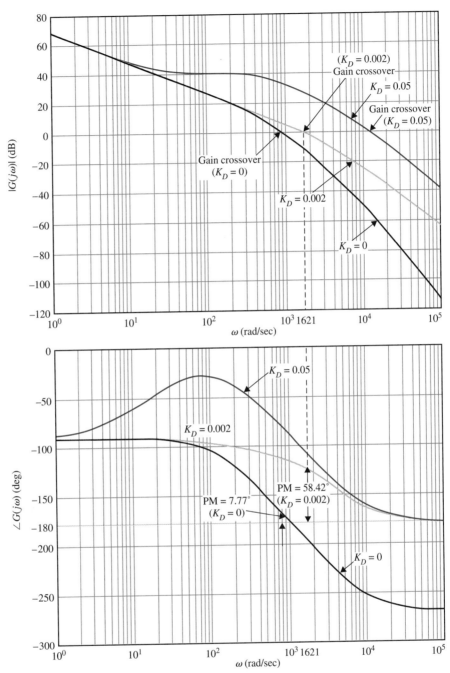

Figure 10-14 Bode diagram of $G(s)$ of system in Example 10-2 with PD controller.

When $K_D = 0.002$, the phase margin is at a maximum of 58.42°, and M_r is also minimum at 1.07, which happens to agree with the optimal value obtained in the time-domain design summarized in Table 10-3. When the value of K_D is increased beyond 0.002, the phase margin decreases, which agrees with the findings from the time-domain design that large values of K_D actually decreases damping. However, the BW and the gain crossover increase continuously with the increase in K_D. The frequency-domain design again shows that the PD control falls short in meeting the performance

• If the slope of the phase curve near the gain crossover is very steep, PD control may not be effective.

requirements imposed on the system. Just as in the time-domain design, we have demonstrated that if the original system has very low damping, or is unstable, PD control may not be effective in improving the stability of the system. Another situation under which PD control may be ineffective is if the slope of the phase curve near the gain-crossover frequency is steep, in which case the rapid decrease of the phase margin due to the increase of the gain crossover from the added gain of the PD controller may render the additional phase ineffective. ◀

▶ 10-3 DESIGN WITH THE PI CONTROLLER

We see from the Section 10-2 that the PD controller can improve the damping and rise time of a control system at the expense of higher bandwidth and resonant frequency, and the steady-state error is not affected unless it varies with time, which is typically not the case for step-function inputs. Thus, PD controller may not fulfill the compensation objectives in many situations.

The integral part of the PID controller produces a signal that is proportional to the time integral of the input of the controller. Figure 10-15 illustrates the block diagram of a the prototype second-order system with a series PI controller. The transfer function of the PI controller is

• Given K_P, the value of K_I should not be too small, or the capacitor C_2 of the op-amp-circuit implemention would be too large.

$$G_c(s) = K_P + \frac{K_I}{s} \tag{10-24}$$

Using the circuit elements given in Table E-1 in Appendix E, two op-amp-circuit realizations of Eq. (10-24) are shown in Fig. 10-16. The transfer function of the two-op-amp circuit in Fig. 10-16(a) is

$$G_c(s) = \frac{E_o(s)}{E_{in}(s)} = \frac{R_2}{R_1} + \frac{R_2}{R_1 C_2 s} \tag{10-25}$$

Comparing Eq. (10-24) with Eq. (10-25), we have

$$K_P = \frac{R_2}{R_1} \qquad K_I = \frac{R_2}{R_1 C_2} \tag{10-26}$$

The transfer function of the three-op-amp circuit in Fig. 10-16(b) is

$$G_c(s) = \frac{E_o(s)}{E_{in}(s)} = \frac{R_2}{R_1} + \frac{1}{R_i C_i s} \tag{10-27}$$

Thus, the parameters of the PI controller are related to the circuit parameters as

$$K_P = \frac{R_2}{R_1} \qquad K_I = \frac{1}{R_i C_i} \tag{10-28}$$

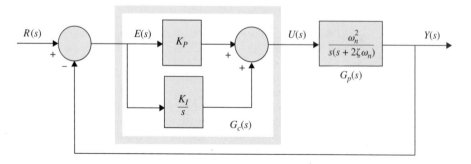

Figure 10-15 Control system with PI controller.

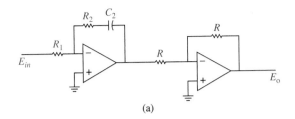

(a)

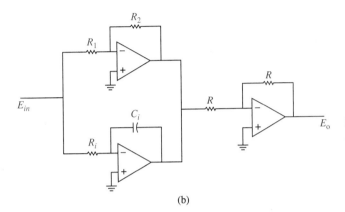

(b)

Figure 10-16 Op-amp-circuit realization of the PI controller, $G_c(s) = K_P + \dfrac{K_I}{s}$.

(a) Two-op-amp circuit.
(b) Three-op-amp circuit.

The advantage with the circuit in Fig. 10-16(b) is that the values of K_P and K_I are independently related to the circuit parameters. However, in either circuit, K_I is inversely proportional to the value of the capacitor. Unfortunately, effective PI-control designs usually result in small values of K_I, and thus we must again watch out for unrealistically large capacitor values.

The forward-path transfer function of the compensated system is

$$G(s) = G_c(s)G_p(s) = \frac{\omega_n^2(K_Ps + K_I)}{s^2(s + 2\zeta\omega_n)} \tag{10-29}$$

Clearly, the immediate effects of the PI controller are

- The PI control increases the system type by 1, thus improving the steady-state error by one order.

1. Adding a zero at $s = -K_I/K_P$ to the forward-path transfer function.

2. Adding a pole at $s = 0$ to the forward-path transfer function. This means that the system type is increased by 1 to a type 2 system. Thus, the steady-state error of the original system is improved by one order; that is, if the steady-state error to a given input is constant, the PI control reduces it to zero (provided that the compensated system remains stable).

The system in Fig. 10-15, with the forward-path transfer function in Eq. (10-29), will now have a zero steady-state error when the reference input is a ramp function. However, because the system is now of the third order, *it may be less stable* than the original second-order system or even become *unstable* if the parameters K_P and K_I are not properly chosen.

In the case of a type 1 system with a PD control, the value of K_P is important because the ramp-error constant K_v is directly proportional to K_P, and thus the magnitude of the steady-state error is inversely proportional to K_P when the input is a ramp. On the other hand, if K_P is too large, the system may become unstable. Similarly, for a type 0 system, the steady-state error due to a step input will be inversely proportional to K_P.

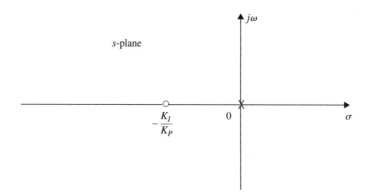

Figure 10-17 Pole-zero configuration of a PI controller.

• The PI controller is essentially a low-pass filter.

When a type 1 system is converted to type 2 by the PI controller, K_P no longer affects the steady-state error, and the latter is always zero for a stable system with a ramp input. The problem is then to choose the proper combination of K_P and K_I so that the transient response is satisfactory.

10-3-1 Time-Domain Interpretation and Design of PI Control

The pole-zero configuration of the PI controller in Eq. (10-24) is shown in Fig. 10-17. At first glance, it may seem that PI control will improve the steady-state error at the expense of stability. However, we shall show that if the location of the zero of $G_c(s)$ is selected properly, both the damping and the steady-state error can be improved. Since the PI controller is essentially a low-pass filter, the compensated system usually will have a slower rise time and longer settling time. *A viable method of designing the PI control is to select the zero at $s = -K_I/K_P$ so that it is relatively close to the origin and away from the most significant poles of the process; the values of K_P and K_I should be relatively small.*

10-3-2 Frequency-Domain Interpretation and Design of PI Control

For frequency-domain design, the transfer function of the PI controller is written

$$G_c(s) = K_P + \frac{K_I}{s} = \frac{K_I\left(1 + \dfrac{K_P}{K_I}s\right)}{s} \tag{10-30}$$

The Bode plot of $G_c(j\omega)$ is shown in Fig. 10-18. Notice that the magnitude of $G_c(j\omega)$ at $\omega = \infty$ is $20\log_{10}K_P$ dB, which represents an attenuation if the value of K_P is less than 1. This attenuation may be utilized to improve the stability of the system. The phase of $G_c(j\omega)$ is always negative, which is detrimental to stability. Thus, we should place the corner frequency of the controller, $\omega = K_I/K_P$, as far to the left as the bandwidth requirement allows, so the phase-lag properties of $G_c(j\omega)$ do not degrade the achieved phase margin of the system.

The frequency-domain design procedure for the PI control to realize a given phase margin is outlined as follows:

1. The Bode plot of the forward-path transfer function $G_p(s)$ of the uncompensated system is made with the loop gain set according to the steady-state performance requirement.

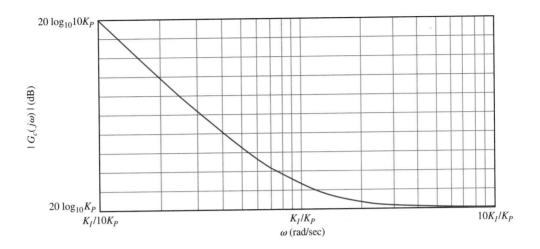

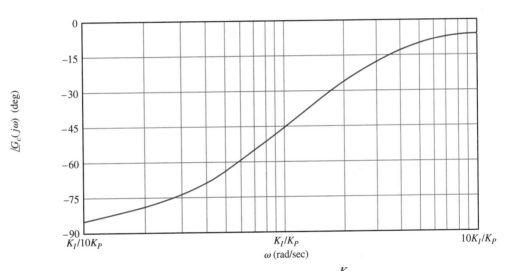

Figure 10-18 Bode diagram of the PI controller. $G_c(s) = K_P + \dfrac{K_I}{s}$.

2. The phase margin and the gain margin of the uncompensated system are determined from the Bode plot. For a specified phase margin requirement, the new gain-crossover frequency ω'_g corresponding to this phase margin is found on the Bode plot. The magnitude plot of the compensated transfer function must pass through the 0-dB axis at this new gain-crossover frequency in order to realize the desired phase margin.

3. To bring the magnitude curve of the uncompensated transfer function down to 0 dB at the new gain-crossover frequency ω'_g, the PI controller must provide the amount of attenuation equal to the gain of the magnitude curve at the new gain-crossover frequency. In other words, set

$$|G_P(j\omega'_g)|_{dB} = -20\log_{10}K_P \text{ dB} \qquad K_P < 1 \qquad (10\text{-}31)$$

from which we have

$$K_P = 10^{-|G_P(j\omega_g')|_{dB}/20} \qquad K_P < 1 \qquad (10\text{-}32)$$

Once the value of K_P is determined, it is necessary only to select the proper value of K_I to complete the design. Up to this point, we have assumed that although the gain-crossover frequency is altered by attenuating the magnitude of $G_c(j\omega)$ at ω_g', the original phase is not affected by the PI controller. This is not possible, however, since, as shown in Fig. 10-18, the attenuation property of the PI controller is accompanied with a phase lag that is detrimental to the phase margin. It is apparent that if the corner frequency $\omega = K_I/K_P$ is placed far below ω_g', the phase lag of the PI controller will have a negligible effect on the phase of the compensated system near ω_g'. On the other hand, the value of K_I/K_P should not be too small or the bandwidth of the system will be too low, causing the rise time and settling time to be too long. As a general guideline, K_I/K_P should correspond to a frequency that is at least one decade, sometimes as much as two decades, below ω_g'. That is, we set

$$\frac{K_I}{K_P} = \frac{\omega_g'}{10} \quad \text{rad/sec} \qquad (10\text{-}33)$$

Within the general guideline, the selection of the value of K_I/K_P is pretty much at the discretion of the designer, who should be mindful of its effect on BW and its practical implementation by an op-amp circuit.

4. The Bode plot of the compensated system is investigated to see if the performance specifications are all met.

5. The values of K_I and K_P are substituted in Eq. (10-30) to give the desired transfer function of the PI controller.

If the controlled process $G_P(s)$ is type 0, the value of K_I may be selected based on the ramp-error-constant requirement, and then there would only be one parameter, K_P, to determine. By computing the phase margin, gain margin, M_r, and BW of the closed-loop system with a range of values of K_P, the best value for K_P can be easily selected.

Based on the preceding discussions, we can summarize the advantages and disadvantages of a properly designed PI controller as

1. Improving damping and reducing maximum overshoot.
2. Increasing rise time.
3. Decreasing BW.
4. Improving gain margin, phase margin, and M_r.
5. Filtering out high-frequency noise.
6. Selecting a proper combination of K_I and K_P so that the capacitor in the circuit implementation of the controller is not excessively large is more difficult than in the case of the PD controller.

The following examples will illustrate how the PI control is designed and what its effects are.

▶ **EXAMPLE 10-3** Consider the second-order attitude-control system discussed in Example 10-1. Applying the PI controller of Eq. (10-24), the forward-path transfer function of the system becomes

$$G(s) = G_c(s)G_P(s) = \frac{4500KK_P(s + K_I/K_P)}{s^2(s + 361.2)} \qquad (10\text{-}34)$$

You may use **ACSYS** to solve this problem.

Time-Domain Design

Let the time-domain performance requirements be

Steady-state error due to parabolic input $t^2 u_s(t)/2 \leq 0.2$

Maximum overshoot ≤ 5 percent

Rise time $t_r \leq 0.01$ sec

Settling time $t_s \leq 0.02$ sec

We have to relax the rise time and settling time requirements from those in Example 10-1 so that we will have a meaningful design for this system. The significance of the requirement on the steady-state error due to a parabolic input is that it indirectly places a minimum requirement on the speed of the transient response.

The parabolic-error constant is

$$K_a = \lim_{s \to 0} s^2 G(s) = \lim_{s \to 0} s^2 \frac{4500 K K_P (s + K_I/K_P)}{s^2(s + 361.2)}$$

$$= \frac{4500 K K_I}{361.2} = 12.46 K K_I \qquad (10\text{-}35)$$

The steady-state error due to the parabolic input $t^2 u_s(t)/2$ is

$$e_{ss} = \frac{1}{K_a} = \frac{0.08026}{K K_I} (\leq 0.2) \qquad (10\text{-}36)$$

Let us set $K = 181.17$, simply because this was the value used in Example 10-1. Apparently, to satisfy a given steady-state error requirement for a parabolic input, the larger the K, the smaller K_I can be. Substituting $K = 181.17$ in Eq. (10-36) and solving K_I for the minimum steady-state error requirement of 0.2, we get the minimum value of K_I to be 0.002215. If necessary, the value of K can be adjusted later.

With $K = 181.17$, the characteristic equation of the closed-loop system is

$$s^3 + 361.2 s^2 + 815,265 K_P s + 815,265 K_I = 0 \qquad (10\text{-}37)$$

Applying Routh's test to Eq. (10-37) yields the result that the system is stable for $0 < K_I/K_P < 361.2$. This means that the zero of $G(s)$ at $s = -K_I/K_P$ cannot be placed too far to the left in the left-half s-plane, or the system will be unstable. Let us place the zero at $-K_I/K_P$ relatively close to the origin. For the present case, the most significant pole of $G_P(s)$, besides the pole at $s = 0$, is at -361.2. Thus, K_I/K_P should be chosen so that the following condition is satisfied:

$$\frac{K_I}{K_P} \ll 361.2 \qquad (10\text{-}38)$$

The root loci of Eq. (10-37) with $K_I/K_P = 10$ are shown in Fig. 10-19. Notice that other than the small loop around the zero at $s = -10$, these root loci for the most part are very similar to those shown in Fig. 10-7, which are for Eq. (10-16). With the condition in Eq. (10-38) satisfied, Eq. (10-34) can be approximated by

$$G(s) \cong \frac{815,265 K_P}{s(s + 361.2)} \qquad (10\text{-}39)$$

where the term K_I/K_P in the numerator is neglected when compared with the magnitude of s, which takes on values along the operating points on the complex portion of the root loci that correspond to, say, a relative damping ratio in the range of $0.7 < \zeta < 1.0$. Let us assume that we wish to have a relative damping ratio of 0.707. From Eq. (10-39), the required value of K_P for this damping ratio is 0.08. This should also be true for the third-order system with the PI controller if the value of K_I/K_P satisfies Eq. (10-38). Thus, with $K_P = 0.08$, $K_I = 0.8$; the root loci in Fig. 10-19 show that the relative damping ratio of the two complex roots is approximately 0.707. In fact, the three characteristic equation roots are at

$$s = -10.605 \qquad -175.3 + j175.4 \qquad \text{and} \qquad -175.3 - j175.4$$

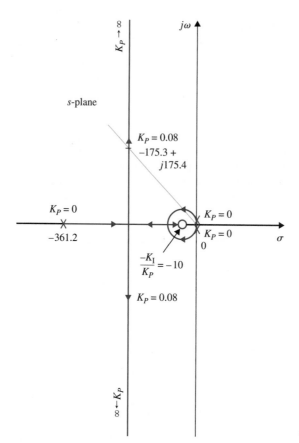

Figure 10-19 Root loci of Eq. (10-37) with $K_I/K_P = 10$; K_P varies.

The reason for this is that when we "stand" at the root at $-175.3 + j175.4$ and "look" toward the neighborhood near the origin, we see that the zero at $s = -10$ is relatively close to the origin, and, thus, practically cancels one of the poles at $s = 0$. In fact, we can show that as long as $K_P = 0.08$, and the value of K_I is chosen such that Eq. (10-38) is satisfied, the relative damping ratio of the complex roots will be very close to 0.707. For example, let us select $K_I/K_P = 5$; the three characteristic equation roots are at

$$s = -5.145 \qquad -178.03 + j178.03 \qquad \text{and} \qquad -178.03 - j178.03$$

and the relative damping ratio is still 0.707. Although the real pole of the closed-loop transfer function is moved, it is very close to the zero at $s = -K_I/K_P$ so that the transient due to the real pole is negligible. For example, when $K_P = 0.08$ and $K_I = 0.4$, the closed-loop transfer function of the compensated system is

$$\frac{\Theta_y(s)}{\Theta_r(s)} = \frac{65,221.2(s + 5)}{(s + 5.145)(s + 178.03 + j178.03)(s + 178.03 - j178.03)} \tag{10-40}$$

Since the pole at $s = 5.145$ is very close to the zero at $s = -5$, the transient response due to this pole is negligible, and the system dynamics are essentially dominated by the two complex poles.

Table 10-5 gives the attributes of the unit-step responses of the system with PI control for various values of K_I/K_P, with $K_P = 0.08$ which corresponds to a relative damping ratio of 0.707.

The results in Table 10-5 verify the fact that PI control reduces the overshoot but at the expense of longer rise time. For $K_I \le 1$, the settling times in Table 10-5 actually show a sharp reduction, which is misleading. This is because the settling times for these cases are measured at the points

TABLE 10-5 Attributes of the Unit-Step Responses of the
System in Example 10-3 with PI Controller

K_I/K_P	K_I	K_P	Maximum Overshoot (%)	t_r (sec)	t_s (sec)
0	0	1.00	52.7	0.00135	0.015
20	1.60	0.08	15.16	0.0074	0.049
10	0.80	0.08	9.93	0.0078	0.0294
5	0.40	0.08	7.17	0.0080	0.023
2	0.16	0.08	5.47	0.0083	0.0194
1	0.08	0.08	4.89	0.0084	0.0114
0.5	0.04	0.08	4.61	0.0084	0.0114
0.1	0.008	0.08	4.38	0.0084	0.0115

where the response enters the band between 0.95 and 1.00, since the maximum overshoots are less than 5 percent.

The maximum overshoot of the system can still be reduced further than those shown in Table 10-5 by using smaller values of K_P than 0.08. However, the rise time and settling time will be excessive. For example, with $K_P = 0.04$ and $K_I = 0.04$, the maximum overshoot is 1.1 percent, but the rise time is increased to 0.0182 seconds, and the settling time is 0.024 seconds.

For the system considered, improvement on the maximum overshoot slows down for K_I less than 0.08, unless K_P is also reduced. As mentioned earlier, the value of the capacitor C_2 is inversely proportional to K_I. Thus, for practical reasons, there is a lower limit on the value of K_I.

Figure 10-20 shows the unit-step responses of the attitude-control system with PI control, with $K_P = 0.08$ and several values of K_P. The unit-step response of the same system with the PD controller designed in Example 10-1, with $K_P = 1$ and $K_D = 0.00177$, is also plotted in the same figure as a comparison.

Frequency-Domain Design

The forward-path transfer function of the uncompensated system is obtained by setting $K_P = 1$ and $K_I = 0$ in the $G(s)$ in Eq. (10-34), and the Bode plot is shown in Fig. 10-21. The phase margin is 22.68°, and the gain crossover frequency is 868 rad/sec.

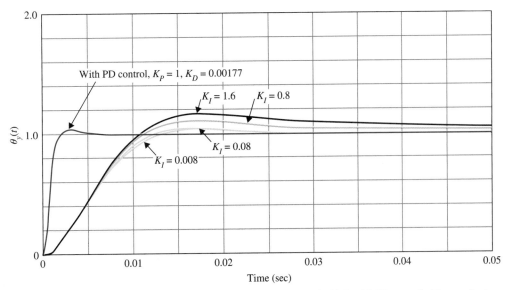

Figure 10-20 Unit-step responses of the system in Example 10-3 with PI control. Also, unit-step response of the system in Example 10-1 with PD controller.

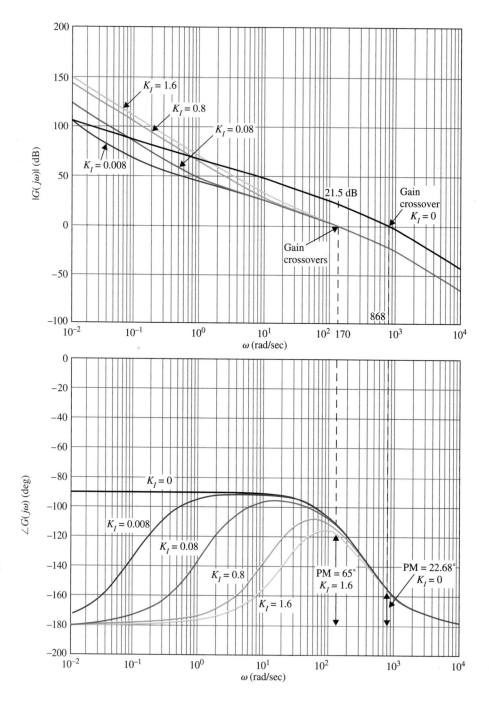

Figure 10-21 Bode plots of control system in Example 10-3 with PI controller.

$$G(s) = \frac{815,265K_P(s + K_I/K_P)}{s^2(s + 361.2)}$$

$K_P = 0.08$

Let us specify that the required phase margin should be at least 65°, and this is to be achieved with the PI controller of Eq. (10-30). Following the procedure outlined earlier in Eqs. (10-31) through (10-33) on the design of the PI controller, we conduct the following steps:

1. Look for the new gain crossover frequency ω_g' at which the phase margin of 65° is realized. From Fig. 10-21, ω_g' is found to be 170 rad/sec. The magnitude of $G(j\omega)$ at this frequency is 21.5 dB. Thus, the PI controller should provide an attenuation of −21.5 dB at

TABLE 10-6 Frequency-Domain Performance Data of the System in Example 10-3 with PI Controller

K_I/K_P	K_I	K_P	GM (dB)	PM (deg)	M_r	BW (rad/sec)	Gain CO (rad/sec)	Phase CO (rad/sec)
0	0	1.00	∞	22.6	2.55	1390.87	868	∞
20	1.6	0.08	∞	58.45	1.12	268.92	165.73	∞
10	0.8	0.08	∞	61.98	1.06	262.38	164.96	∞
5	0.4	0.08	∞	63.75	1.03	258.95	164.77	∞
1	0.08	0.08	∞	65.15	1.01	256.13	164.71	∞
0.1	0.008	0.08	∞	65.47	1.00	255.49	164.70	∞

$\omega'_g = 170$ rad/sec. Substituting $|G(j\omega'_g)| = 21.5$ dB into Eq. (10-32), and solving for K_P, we get

$$K_P = 10^{-|G(j\omega'_g)|_{dB}/20} = 10^{-21.5/20} = 0.084 \qquad (10\text{-}41)$$

Notice that in the time-domain design conducted earlier, K_P was selected to be 0.08 so that the relative damping ratio of the complex characteristic equation roots will be approximately 0.707. (Perhaps we have cheated a little by selecting the desired phase margin to be 65°. This could not be just a coincidence. Can you believe that we have had no prior knowledge that in this case, $\zeta = 0.707$ corresponds to PM = 65°?)

2. Let us choose $K_P = 0.08$, so that we can compare the design results of the frequency domain with those of the time-domain design obtained earlier. Equation (10-33) gives the general guideline of finding K_I once K_P is determined. Thus,

$$K_I = \frac{\omega'_g K_P}{10} = \frac{170 \times 0.08}{10} = 1.36 \qquad (10\text{-}42)$$

As pointed out earlier, the value of K_I is not rigid, as long as the ratio K_I/K_P is sufficiently smaller than the magnitude of the pole of $G(s)$ at -361.2. As it turns out, the value of K_I given by Eq. (10-42) is not sufficiently small for this system.

The Bode plots of the forward-path transfer function with $K_P = 0.08$ and $K_I = 0, 0.008, 0.08, 0.8,$ and 1.6 are shown in Fig. 10-21. Table 10-6 shows the frequency-domain properties of the uncompensated system and the compensated system with various values of K_I. Notice that for values of K_I/K_P that are sufficiently small, the phase margin, M_r, BW, and gain crossover frequency all vary little.

It should be noted that the phase margin of the system can be improved further by reducing the value of K_P below 0.08. However, the bandwidth of the system will be further reduced. For example, for $K_P = 0.04$ and $K_I = 0.04$, the phase margin is increased to 75.7°, and $M_r = 1.01$, but BW is reduced to 117.3 rad/sec. ◄

▶ **EXAMPLE 10-4** Now let us consider using the PI control for the third-order attitude control system described by Eq. (10-19). First, the time-domain design is carried out as follows. You may use **ACSYS** to solve this problem.

Time-Domain Design
Let the time-domain specifications be

Steady-state error due to the parabolic input $t^2 u_s(t)/2 \le 0.2$

Maximum overshoot ≤ 5 percent

Rise time $t_r \le 0.01$ sec

Settling time $t_s \le 0.02$ sec

These are identical to the specifications given for the second-order system in Example 10-3.

Applying the PI controller of Eq. (10-24), the forward-path transfer function of the system becomes

$$G(s) = G_c(s)G_p(s) = \frac{1.5 \times 10^9 \, KK_P(s + K_I/K_P)}{s^2(s^2 + 3408.3s + 1,204,000)}$$

$$= \frac{1.5 \times 10^9 \, KK_P(s + K_I/K_P)}{s^2(s + 400.26)(s + 3008)} \tag{10-43}$$

We can show that the steady-state error of the system due to the parabolic input is again given by Eq. (10-36), and arbitrarily setting $K = 181.17$, the minimum value of K_I is 0.002215.

The characteristic equation of the closed-loop system with $K = 181.17$ is

$$s^4 + 3408.3s^3 + 1,204,000s^2 + 2.718 \times 10^9 \, K_P s + 2.718 \times 10^9 \, K_I = 0 \tag{10-44}$$

The Routh's tabulation of the last equation is performed as follows:

s^4	1	1,204,000	$2.718 \times 10^9 K_I$
s^3	3408.3	$2.718 \times 10^9 K_P$	0
s^2	$1,204,000 - 797465K_P$	$2.718 \times 10^9 K_I$	0
s^1	$\dfrac{1,204,000K_P - 797465K_P^2 - 3408.3K_I}{1,204,000 - 797465K_P}$	0	
s^0	$2.718 \times 10^9 K_I$	0	

The stability requirements are:

$$K_I > 0$$
$$K_P < 1.5098 \tag{10-45}$$

and
$$K_I < 353.255K_P - 233.98K_P^2$$

The design of the PI controller calls for the selection of a small value for K_I/K_P, relative to the nearest pole of $G(s)$ to the origin, which is at -400.26. The root loci of Eq. (10-44) are plotted using the pole-zero configuration of Eq. (10-43). Figure 10-22(a) shows the root loci as K_P varies for $K_I/K_P = 2$. The root loci near the origin due to the pole and zero of the PI controller again form a small loop, and the root loci at a distance away from the origin will be very similar to those of the uncompensated system, which are shown in Fig. 7-30. By selecting the value of K_P properly along the root loci, it may be possible to satisfy the performance specifications given above. To minimize the rise time and settling time, we should select K_P so that the dominant roots are complex conjugate. Table 10-7 gives the performance attributes of several combinations of K_I/K_P and K_P. Notice that although several combinations of these parameters correspond to systems that satisfy the performance specifications, the one with $K_P = 0.075$ and $K_I = 0.15$ gives the best rise and settling times among those shown.

Frequency-Domain Design

The Bode plot of Eq. (10-43) for $K = 181.17$, $K_P = 1$, and $K_I = 0$ is shown in Fig. 10-22(b). The performance data of the uncompensated system are

Gain margin = 3.578 dB

Phase margin = 7.788°

$M_r = 6.572$

BW = 1378 rad/sec

Let us require that the compensated system has a phase margin of at least 65°, and this is to be achieved with the PI controller of Eq. (10-30). Following the procedure outlined in Eqs. (10-31) through (10-33) on the design of the PI controller, we carry out the following steps.

1. Look for the new gain crossover frequency ω_g' at which the phase margin of 65° is realized. From Fig. 10-20, ω_g' is found to be 163 rad/sec, and the magnitude of $G(j\omega)$ at this frequency is 22.5 dB. Thus, the PI controller should provide an attenuation of -22.5 dB at $\omega_g' = 163$ rad/sec. Substituting $|G(j\omega_g')| = 22.5$ dB into Eq. (10-32), and solving for K_P, we get

$$K_P = 10^{-|G(j\omega_g')|_{dB}/20} = 10^{-22.5/20} = 0.075 \tag{10-46}$$

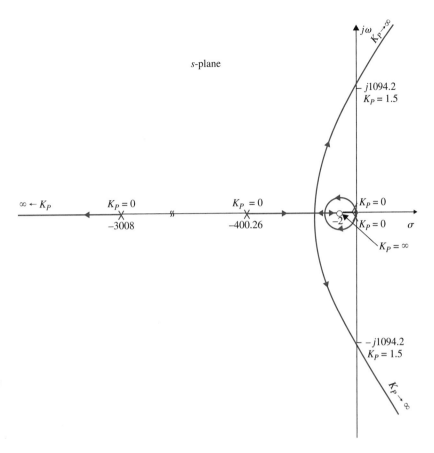

Figure 10-22(a) (a) Root loci of control system in Example 10-4 with PI controller $K_I/K_P = 2$, $0 \leq K_P < \infty$.

This is exactly the same result that was selected for the time-domain design that resulted in a system with a maximum overshoot of 4.9 percent when $K_I = 0.15$, or $K_I/K_P = 2$.

2. The suggested value of K_I is found from Eq. (10-33),

$$K_I = \frac{\omega_g' K_P}{10} = \frac{163 \times 0.075}{10} = 1.222 \qquad (10\text{-}47)$$

TABLE 10-7 Attributes of the Unit-Step Responses of the System in Example 10-4 with PI Controller

K_I/K_P	K_I	K_P	Maximum Overshoot (%)	t_r (sec)	t_s (sec)	Roots of Characteristic Equation			
0	0	1	76.2	0.00158	0.0487	−3293.3	−57.5	$\pm j$ 906.6	
20	1.6	0.08	15.6	0.0077	0.0471	−3035	−22.7	− 175.3	$\pm j$ 180.3
20	0.8	0.04	15.7	0.0134	0.0881	−3021.6	−259	− 99	−28
5	0.4	0.08	6.3	0.00883	0.0202	−3035	−5.1	−184	$\pm j$ 189.2
2	0.08	0.04	2.1	0.02202	0.01515	−3021.7	−234.6	−149.9	−2
5	0.2	0.04	4.8	0.01796	0.0202	−3021.7	−240	−141.2	−5.3
2	0.16	0.08	5.8	0.00787	0.01818	−3035.2	−185.5	$\pm j$ 190.8	−2
1	0.08	0.08	5.2	0.00792	0.01616	−3035.2	−186	$\pm j$ 191.4	−1
2	0.15	0.075	4.9	0.0085	0.0101	−3033.5	−187.2	$\pm j$ 178	−1
2	0.14	0.070	4.0	0.00917	0.01212	−3031.8	−187.2	$\pm j$ 164	−1

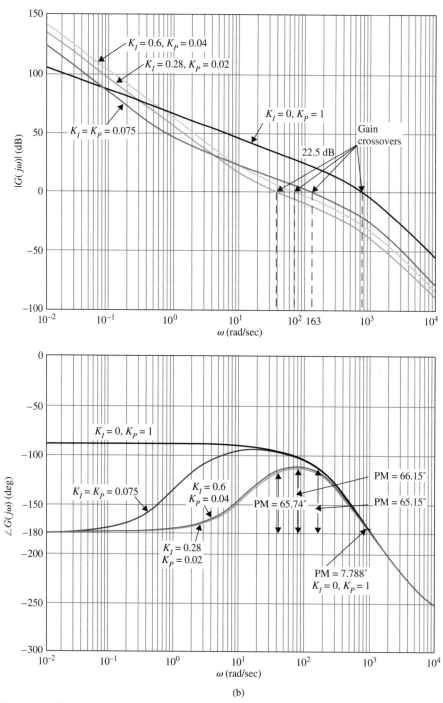

Figure 10-22(b) (b) Bode plots of control system in Example 10-4 with PI control.

Thus, $K_I/K_P = 16.3$. However, the phase margin of the system with these design parameters is only 59.52.

To realize the desired PM of 65°, we can reduce the value of K_P or K_I. Table 10-8 gives the results of several designs with various combinations of K_P and K_I. Notice that the last three designs

TABLE 10-8 Performance Summary of System in Example 10-4 with PI Controller

K_I/K_P	K_I	K_P	GM (dB)	PM (deg)	M_r	BW (rad/sec)	Maximum Overshoot (%)	t_r (sec)	t_s (sec)
0	0	1	3.578	7.788	6.572	1378	77.2	0.0015	0.0490
16.3	1.222	0.075	25.67	59.52	1.098	264.4	13.1	0.0086	0.0478
1	0.075	0.075	26.06	65.15	1.006	253.4	4.3	0.0085	0.0116
15	0.600	0.040	31.16	66.15	1.133	134.6	12.4	0.0142	0.0970
14	0.280	0.020	37.20	65.74	1.209	66.34	17.4	0.0268	0.1616

in the table all satisfy the PM requirements. However, the design ramifications show that

Reducing K_P would reduce BW and increase M_r.

Reducing K_I would increase the capacitor value in the implementing circuit.

In fact, only the $K_I = K_P = 0.075$ case gives the best all-around performance in both the frequency domain and the time domain. In attempting to increase K_I, the maximum overshoot becomes excessive. This is one example that shows the inadequacy of specifying phase margin only. The purpose of this example is to bring out the properties of the PI controller and the important considerations in its design. No details are explored further.

Figure 10-23 shows the unit-step responses of the uncompensated system and several systems with PI control.

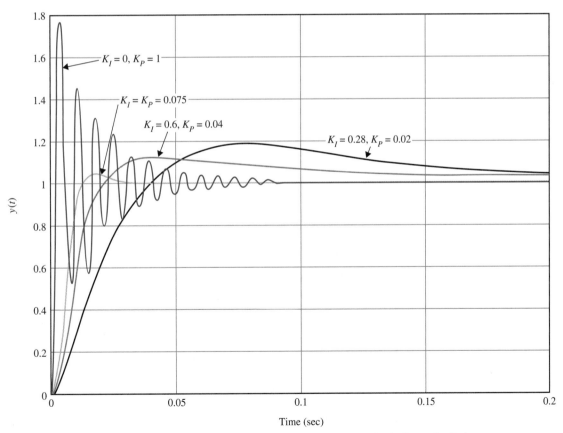

Figure 10-23 Unit-step response of system with PI controller in Example 10-4.

$$G(s) = \frac{2.718 \times 10^9 K_P(s + K_I/K_P)}{s^2(s + 400.26)(s + 3008)}.$$

▶ 10-4 DESIGN WITH THE PID CONTROLLER

From the preceding discussions, we see that the PD controller could add damping to a system, but the steady-state response is not affected. The PI controller could improve the relative stability and improve the steady-state error at the same time, but the rise time is increased. This leads to the motivation of using a PID controller so that the best features of each of the PI and PD controllers are utilized. We can outline the following procedure for the design of the PID controller.

1. Consider that the PID controller consists of a PI portion connected in cascade with a PD portion. The transfer function of the PID controller is written as

$$G_c(s) = K_P + K_D s + \frac{K_I}{s} = (1 + K_{D1}s)\left(K_{P2} + \frac{K_{I2}}{s}\right) \qquad (10\text{-}48)$$

The proportional constant of the PD portion is set to unity, since we need only three parameters in the PID controller. Equating both sides of Eq. (10-48), we have

$$K_P = K_{P2} + K_{D1}K_{I2} \qquad (10\text{-}49)$$

$$K_D = K_{D1}K_{P2} \qquad (10\text{-}50)$$

$$K_I = K_{I2} \qquad (10\text{-}51)$$

2. Consider that the PD portion only is in effect. Select the value of K_{D1} so that a portion of the desired relative stability is achieved. In the time domain, this relative stability may be measured by the maximum overshoot, and in the frequency domain it is the phase margin.

3. Select the parameters K_{I2} and K_{P2} so that the total requirement on relative stability is satisfied.

As an alternative, the PI portion of the controller can be designed first for a portion of the requirement on relative stability, and finally, the PD portion is designed.

The following example illustrates how the PID controller is designed in the time domain and the frequency domain.

▶ **EXAMPLE 10-5** Consider the third-order attitude control system represented by the forward-path transfer function given in Eq. (10-19). With $K = 181.17$, the transfer function is

$$G_p(s) = \frac{2.718 \times 10^9}{s(s + 400.26)(s + 3008)} \qquad (10\text{-}52)$$

You may use **ACSYS** to solve this problem.

Time-Domain Design

Let the time-domain performance specifications be

Steady-state error due to a ramp input $t^2 u_s(t)/2 \leq 0.2$

Maximum overshoot ≤ 5 percent

Rise time $t_r \leq 0.005$ sec

Settling time $t_s \leq 0.005$ sec

We realize from the previous examples that these requirements cannot be fulfilled by either the PI or PD control acting alone. Let us apply the PD control with the transfer function $(1 + K_{D1}s)$. The forward-path transfer function becomes

$$G(s) = \frac{2.718 \times 10^9(1 + K_{D1}s)}{s(s + 400.26)(s + 3008)} \qquad (10\text{-}53)$$

TABLE 10-9 Time-Domain Performance Characteristics of Third-Order Attitude
Control System with PID Controller Designed in Example 10-5

K_{P2}	Maximum Overshoot (%)	t_r (sec)	t_s (sec)	Roots of Characteristic Equation		
1.0	11.1	0.00088	0.0025	−15.1	−533.2	−1430 ± j 1717.5
0.9	10.8	0.00111	0.00202	−15.1	−538.7	−1427 ± j 1571.8
0.8	9.3	0.00127	0.00303	−15.1	−546.5	−1423 ± j 1385.6
0.7	8.2	0.00130	0.00303	−15.1	−558.4	−1417 ± j 1168.7
0.6	6.9	0.00155	0.00303	−15.2	−579.3	−1406 ± j 897.1
0.5	5.6	0.00172	0.00404	−15.2	−629	−1382 ± j 470.9
0.4	5.1	0.00214	0.00505	−15.3	−1993	−700 ± j 215.4
0.3	4.8	0.00271	0.00303	−15.3	−2355	−519 ± j 263.1
0.2	4.5	0.00400	0.00404	−15.5	−2613	−390 ± j 221.3
0.1	5.6	0.00747	0.00747	−16.1	−284	−284 ± j 94.2
0.08	6.5	0.00895	0.04545	−16.5	−286.3	−266 ± j 4.1

Table 10-3 shows that the best PD controller that can be obtained from the maximum over-shoot standpoint is with $K_{D1} = 0.002$, and the maximum overshoot is 11.37 percent. The rise time and settling time are well within the required values. Next, we add the PI controller, and the forward-path transfer function becomes

$$G(s) = \frac{5.436 \times 10^6 \, K_{P2}(s + 500)(s + K_{I2}/K_{P2})}{s^2(s + 400.26)(s + 3008)} \tag{10-54}$$

Following the guideline of choosing a relatively small value for K_{I2}/K_{P2}, we let $K_{I2}/K_{P2} = 15$. Equation (10-54) becomes

$$G(s) = \frac{5.436 \times 10^6 \, K_{P2}(s + 500)(s + 15)}{s^2(s + 400.26)(s + 3008)} \tag{10-55}$$

Table 10-9 gives the time-domain performance characteristics along with the roots of the characteristic equation for various values of K_{P2}. Apparently, the optimal value of K_{P2} is in the neighborhood of between 0.2 and 0.4.

Selecting $K_{P2} = 0.3$, and with $K_{D1} = 0.002$ and $K_{I2} = 15K_{P2} = 4.5$, the following results are obtained for the parameters of the PID controller using Eqs. (10-49) through (10-51):

$$\begin{aligned}
K_I &= K_{I2} = 4.5 \\
K_P &= K_{P2} + K_{D1}K_{I2} = 0.3 + 0.002 \times 4.5 = 0.309 \\
K_D &= K_{D1}K_{P2} = 0.002 \times 0.3 = 0.0006
\end{aligned} \tag{10-56}$$

Notice that the PID design resulted in a smaller K_D and a larger K_I which correspond to smaller capacitors in the implementing circuit.

Figure 10-24 shows the unit-step responses of the system with the PID controller, as well as those with PD and PI controls designed in Examples 10-2 and 10-4, respectively. Notice that the PID control, when designed properly, captures the advantages of both the PD and the PI controls.

Frequency-Domain Design
The PD control of the third-order attitude control systems was already carried out in Example 10-2, and the results were tabulated in Table 10-3. When $K_P = 1$ and $K_D = 0.002$, the maximum overshoot is 11.37 percent, but this is the best that the PD control could offer. Using this PD controller, the forward-path transfer function of the system is

$$G(s) = \frac{2.718 \times 10^9(1 + 0.002s)}{s(s + 400.26)(s + 3008)} \tag{10-57}$$

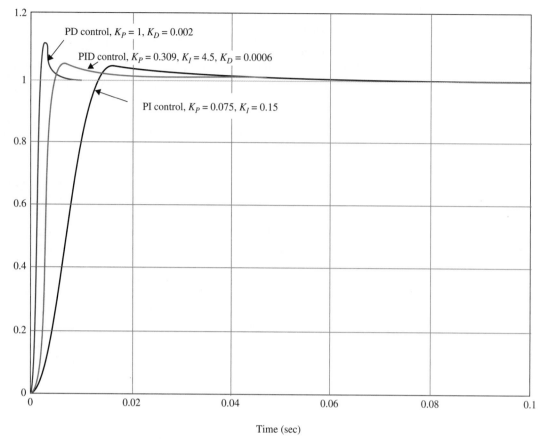

Figure 10-24 Step responses of system in Example 10-5 with PD, PI, and PID controllers.

and its Bode plot is shown in Fig. 10-25. Let us estimate that the following set of frequency-domain criteria corresponds to the time-domain specifications given in this problem.

$$\text{Phase margin} \geq 70°$$
$$M_r \leq 1.1$$
$$\text{BW} \geq 1.000 \text{ rad/sec}$$

TABLE 10-10   Frequency-Domain Performance of System in Example 10-5 with PID Controller

K_{P2}	K_{I2}	GM (dB)	PM (deg)	M_r	BW (rad/sec)	t_r (sec)	t_s (sec)	Maximum Overshoot (%)
1.00	0	∞	58.45	1.07	2607	0.0008	0.00255	11.37
0.45	6.75	∞	68.5	1.03	1180	0.0019	0.0040	5.6
0.40	6.00	∞	69.3	1.027	1061	0.0021	0.0050	5.0
0.30	4.50	∞	71.45	1.024	1024	0.0027	0.00303	4.8
0.20	3.00	∞	73.88	1.031	528.8	0.0040	0.00404	4.5
0.10	1.5	∞	76.91	1.054	269.5	0.0076	0.0303	5.6
0.08	1.2	∞	77.44	1.065	216.9	0.0092	0.00469	6.5

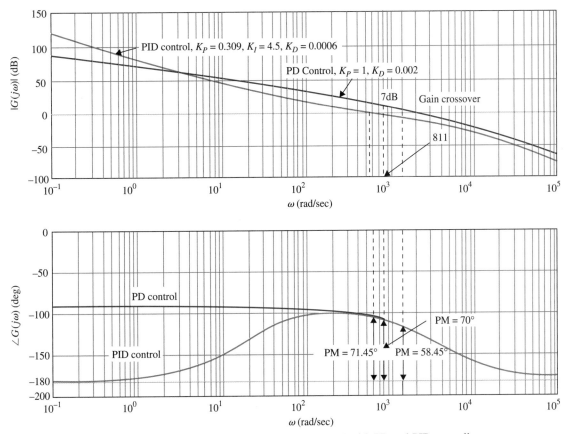

Figure 10-25 Bode plot of system in Example 10-5 with PD and PID controllers.

From the Bode diagram in Fig. 10-25 we see that to achieve a phase margin of 70°, the new phase crossover frequency should be $\omega'_g = 811$ rad/sec, at which the magnitude of $G(j\omega)$ is 7 dB. Thus, using Eq. (10-32), the value of K_{P2} is calculated to be

$$K_{P2} = 10^{-7/20} = 0.45 \qquad (10\text{-}58)$$

Notice that the desirable range of K_{P2} found from the time-domain design with $K_{I2}/K_{P2} = 15$ is from 0.2 to 0.4. The result given in Eq. (10-58) is slightly out of the range. Table 10-10 shows the frequency-domain performance results with $K_D = 0.002$, $K_{I2}/K_{P2} = 15$, and several values of K_{P2} starting with 0.45. It is interesting to note that as K_{P2} continues to decrease, the phase margin increases monotonically, but below $K_{P2} = 0.2$, the maximum overshoot actually increases. In this case, the phase margin results are misleading, but the resonant peak M_r is a more accurate indication of this. ◄

► 10-5 DESIGN WITH PHASE-LEAD CONTROLLER

The PID controller and its components in the form of PD and PI controls represent simple forms of controllers that utilize derivative and integration operations in the compensation of control systems. In general, we can regard the design of controllers of control systems as a filter design problem; then there are a large number of possible schemes. From the filtering standpoint, the PD controller is a high-pass filler, the PI controller is a low-pass filter, and the PID controller is a band-pass or band-attenuate filter, depending on the values of the controller parameters. The high-pass filter is often referred to as a

phase-lead controller since positive phase is introduced to the system over some frequency range. The low-pass filter is also known as **phase-lag controller**, since the corresponding phase introduced is negative. These ideas related to filtering and phase shifts are useful if designs are carried out in the frequency domain.

The transfer function of a simple lead or lag controller is expressed as

$$G_c(s) = K_c \frac{s + z_1}{s + p_1} \tag{10-59}$$

where the controller is high-pass or phase-lead if $p_1 > z_1$, and low-pass or phase-lag if $p_1 < z_1$.

The op-amp circuit implementation of Eq. (10-59) is given in Table E-1(g) of Appendix E, and is repeated in Fig. 10-26 with an inverting amplifier. The transfer function of the circuit is

$$G_c(s) = \frac{E_o(s)}{E_{in}(s)} = \frac{C_1}{C_2} \frac{s + \dfrac{1}{R_1 C_1}}{s + \dfrac{1}{R_2 C_2}} \tag{10-60}$$

Comparing the last two equations, we have

$$K_c = C_1/C_2$$
$$z_1 = 1/R_1 C_1 \tag{10-61}$$
$$p_1 = 1/R_2 C_2$$

We can reduce the number of design parameters from four to three by setting $C = C_1 = C_2$. Then Eq. (10-60) is written as

$$\begin{aligned} G_c(s) &= \frac{R_2}{R_1}\left(\frac{1 + R_1 Cs}{1 + R_2 Cs}\right) \\ &= \frac{1}{a}\left(\frac{1 + aTs}{1 + Ts}\right) \end{aligned} \tag{10-62}$$

where

$$a = \frac{R_1}{R_2} \tag{10-63}$$

$$T = R_2 C \tag{10-64}$$

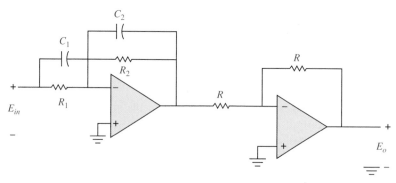

Figure 10-26 Op-amp circuit implementation of $G(s) = K_c \dfrac{s + z_1}{s + P_1}$.

10-5-1 Time-Domain Interpretation and Design of Phase-Lead Control

In this section, we shall first consider that Eqs. (10-60) and (10-62) represent a phase-lead controller ($z_1 < p_1$ or $a > 1$). In order that the phase-lead controller will not degrade the steady-state error, the factor a in Eq. (10-62) should be absorbed by the forward-path gain K. Then, for design purposes, $G_c(s)$ can be written as

$$G_c(s) = \frac{1 + aTs}{1 + Ts} \qquad (a > 1) \qquad (10\text{-}65)$$

The pole-zero configuration of Eq. (10-65) is shown in Fig. 10-27. Based on the discussions given in Chapter 8 on the effects of adding a pole-zero pair (with the zero closer to the origin) to the forward-path transfer function, the phase-lead controller can improve the stability of the closed-loop system if its parameters are chosen properly. The design of phase-lead control is essentially that of placing the pole and zero of $G_c(s)$ so that the design specifications are satisfied. The root-contour method can be used to indicate the proper ranges of the parameters. The **ACSYS** MATLAB tool can be used to speed up the cut-and-try procedure considerably. The following guidelines can be made with regard to the selection of the parameters a and T.

1. Moving the zero $-1/aT$ toward the origin should improve rise time and settling time. If the zero is moved too close to the origin, the maximum overshoot may again increase, since $-1/aT$ also appears as a zero of the closed-loop transfer function.

2. Moving the pole at $-1/T$ farther away from the zero and the origin should reduce the maximum overshoot, but if the value of T is too small, rise time and settling time will again increase.

We can make the following general statements with respect to the effects of phase-lead control on the time-domain performance of a control system:

1. When used properly, it can increase damping of the system.

2. It improves rise time and settling time.

3. In the form of Eq. (10-65), phase-lead control does not affect the steady-state error, since $G_c(0) = 1$.

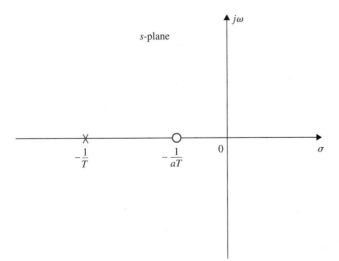

Figure 10-27 Pole-zero configuration of the phase-lead controller.

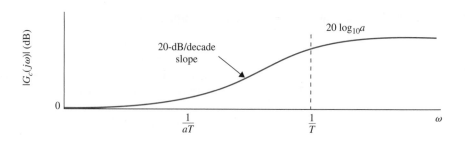

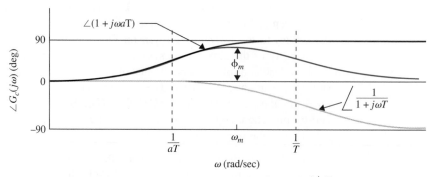

Figure 10-28 Bode plot of phase-lead controller $G_c(s) = a\dfrac{s + 1/aT}{s + 1/T}$ $a > 1$.

10-5-2 Frequency-Domain Interpretation and Design of Phase-Lead Control

The Bode plot of the phase-lead controller of Eq. (10-65) is shown in Fig. 10-28. The two corner frequencies are at $\omega = 1/aT$ and $\omega = 1/T$. The maximum value of the phase, ϕ_m, and the frequency at which it occurs, ω_m, are derived as follows. Since ω_m is the geometric mean of the two corner frequencies, we write

$$\log_{10}\omega_m = \frac{1}{2}\left(\log_{10}\frac{1}{aT} + \log_{10}\frac{1}{T}\right) \tag{10-66}$$

Thus,

$$\omega_m = \frac{1}{\sqrt{a}\,T} \tag{10-67}$$

To determine the maximum phase ϕ_m, the phase of $G_c(j\omega)$ is written

$$\angle G_c(j\omega) = \phi(j\omega) = \tan^{-1}\omega aT - \tan^{-1}\omega T \tag{10-68}$$

from which we get

$$\tan\phi(j\omega) = \frac{\omega aT - \omega T}{1 + (\omega aT)(\omega T)} \tag{10-69}$$

Substituting Eq. (10-67) into Eq. (10-69), we have

$$\tan\phi_m = \frac{a - 1}{2\sqrt{a}} \tag{10-70}$$

or

$$\sin\phi_m = \frac{a - 1}{a + 1} \tag{10-71}$$

Thus, by knowing ϕ_m, the value of a is determined from

$$a = \frac{1 + \sin\phi_m}{1 - \sin\phi_m} \tag{10-72}$$

The relationship between the phase ϕ_m and a and the general properties of the Bode plot of the phase-lead controller provide an advantage of designing in the frequency domain. The difficulty is, of course, in the correlation between the time-domain and frequency-domain specifications. The general outline of phase-lead controller design in the frequency domain is given as follows. It is assumed that the design specifications simply include steady-state error and phase-margin requirements.

1. The Bode diagram of the uncompensated process $G_p(j\omega)$ is constructed with the gain constant K set according to the steady-state error requirement. The value of K has to be adjusted upward once the value of a is determined.

2. The phase margin and the gain margin of the uncompensated system are determined, and the additional amount of phase lead needed to realize the phase margin is determined. From the additional phase lead required, the desired value of ϕ_m is estimated accordingly, and the value of a is calculated from Eq. (10-72).

3. Once a is determined, it is necessary only to determine the value of T, and the design is in principle completed. This is accomplished by placing the corner frequencies of the phase-lead controller, $1/aT$ and $1/T$, such that ϕ_m is located at the new gain-crossover frequency ω_g', so the phase margin of the compensated system is benefited by ϕ_m. It is known that the high-frequency gain of the phase-lead controller is $20\log_{10}a$ dB. Thus, to have the new gain crossover at ω_m, which is the geometric mean of $1/aT$ and $1/T$, we need to place ω_m at the frequency where the magnitude of the uncompensated $G_p(j\omega)$ is $-10\log_{10}a$ dB so that adding the controller gain of $10\log_{10}a$ dB to this makes the magnitude curve go through 0 dB at ω_m.

4. The Bode diagram of the forward-path transfer function of the compensated system is investigated to check that all performance specifications are met; if not, a new value of ϕ_m must be chosen and the steps repeated.

5. If the design specifications are all satisfied, the transfer function of the phase-lead controller is established from the values of a and T.

If the design specifications also include M_r and/or BW, then these must be checked using either the Nichols chart or the output data from a computer program.

We use the following example to illustrate the design of the phase-lead controller in the time domain and frequency domain.

▶ **EXAMPLE 10-6** The block diagram of the sun-seeker control system described in Section 4-9 is again shown in Fig. 10-29. The system may be mounted on a space vehicle so that it will track the sun with high accuracy. The variable θ_r represents the reference angle of the solar ray, and θ_o denotes the vehicle axis. The objective of the sun-seeker system is to maintain the error α between θ_r and θ_o near zero. The parameters of the system are as follows:

$R_F = 10{,}000\ \Omega$	$K_b = 0.0125$ V/rad/sec
$K_i = 0.0125$ N-m/A	$R_a = 6.25\ \Omega$
$J = 10^{-6}$ kg-m^2	$K_s = 0.1$ A/rad
$K =$ to be determined	$B = 0$
$n = 800$	

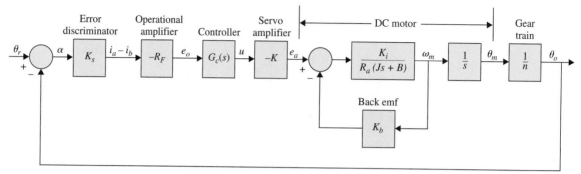

Figure 10-29   Block diagram of sun-seeker control system.

The forward-path transfer function of the uncompensated system is

$$G_p(s) = \frac{\Theta_o(s)}{A(s)} = \frac{K_s R_F K K_i/n}{R_a J s^2 + K_i K_b s} \tag{10-73}$$

where $\Theta_o(s)$ and $A(s)$ are the Laplace transforms of $\theta_o(t)$ and $\alpha(t)$, respectively.

Substituting the numerical values of the system parameters in Eq. (10-73), we get

$$G_p(s) = \frac{\Theta_o(s)}{A(s)} = \frac{2500K}{s(s + 25)} \tag{10-74}$$

You may use **ACSYS** to solve this problem after reducing the block diagram in Fig. 10-29 to a standard form. See section 10-15.

Time-Domain Design

The time-domain specifications of the system are as follows:

1. The steady-state error of $\alpha(t)$ due to a unit-ramp function input for $\theta_r(t)$ should be ≤ 0.01 rad per rad/sec of the final steady-state output velocity. In other words, the steady-state error due to a ramp input should be ≤ 1 percent.

2. The maximum overshoot of the step response should be less than 5%, or as small as possible.

3. Rise time $t_r \leq 0.02$ sec.

4. Settling time $t_s \leq 0.02$ sec.

The minimum value of the amplifier gain, K, is determined initially from the steady-state requirement. Applying the final-value theorem to $\alpha(t)$, we have

$$\lim_{t \to \infty} \alpha(t) = \lim_{s \to 0} sA(s) = \lim_{s \to 0} \frac{s\Theta_r(s)}{1 + G_p(s)} \tag{10-75}$$

For a unit-ramp input, $\Theta_r(s) = 1/s^2$. By using Eq. (10-74), Eq. (10-75) leads to

$$\lim_{t \to \infty} \alpha(t) = \frac{0.01}{K} \tag{10-76}$$

Thus, for the steady-state value of $\alpha(t)$ to be ≤ 0.01, K must be ≥ 1. Let us set $K = 1$, the worst case from the steady-state error standpoint, the characteristic equation of the uncompensated system is

$$s^2 + 25s + 2500 = 0 \tag{10-77}$$

We can show that the damping ratio of the uncompensated system with $K = 1$ is only 0.25, which corresponds to a maximum overshoot of 44.4%. Figure 10-30 shows the unit-step response of the system with $K = 1$.

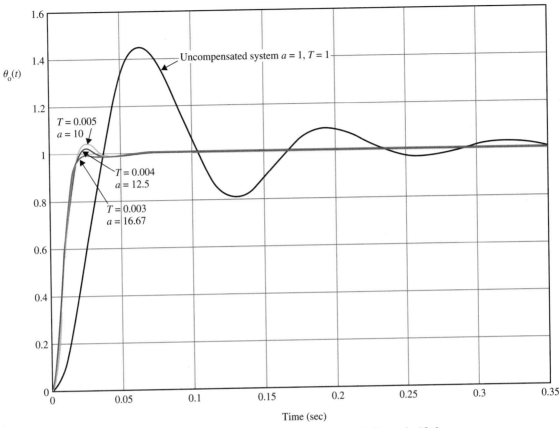

Figure 10-30 Unit-step response of sun-seeker system in Example 10-6.

A space has been reserved in the forward path of the block diagram of Fig. 10-29 for a controller with transfer function $G_c(s)$. Let us consider using the phase-lead controller of Eq. (10-62), although in the present case, a PD controller or one of the other types of phase-lead controllers may also be effective in satisfying the performance criteria given.

The forward-path transfer function of the compensated system is written

$$G(s) = \frac{2500K(1 + aTs)}{as(s + 25)(1 + Ts)} \tag{10-78}$$

For the compensated system to satisfy the steady-state error requirement, K must satisfy

$$K \geq a \tag{10-79}$$

Let us set $K = a$. The characteristic equation of the system is

$$(s^2 + 25s + 2500) + Ts^2(s + 25) + 2500aTs = 0 \tag{10-80}$$

We can use the root-contour method to show the effects of varying a and T of the phase-lead controller. Let us first set $a = 0$. The characteristic equation of Eq. (10-80) becomes

$$s^2 + 25s + 2500 + Ts^2(s + 25) = 0 \tag{10-81}$$

Dividing both sides of the last equation by the terms that do not contain T, we get

$$1 + G_{eq1}(s) = 1 + \frac{Ts^2(s + 25)}{s^2 + 25s + 2500} = 0 \tag{10-82}$$

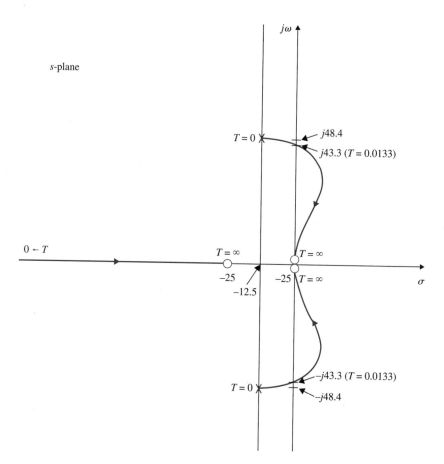

Figure 10-31 Root contours of the sun-seeker system with $a = 0$, and T varies from 0 to ∞.

Thus, the root contours of Eq. (10-81) when T varies are determined using the pole-zero configuration of $G_{eq1}(s)$ in Eq. (10-82). These root contours are drawn as shown in Fig. 10-31. Notice that the poles of $G_{eq1}(s)$ are the roots of the characteristic equation when $a = 0$ and $T = 0$. The root contours in Fig. 10-31 clearly show that adding the factor $(1 + Ts)$ to the denominator of Eq. (10-74) alone would not improve the system performance, since the characteristic equation roots are pushed toward the right-half plane. In fact, the system becomes unstable when T is greater than 0.0133. To achieve the full effect of the phase-lead controller, we must restore the value of a in Eq. (10-80). To prepare for the root contours with a as the variable parameter, we divide both sides of Eq. (10-80) by the terms that do not contain a, and the following equation results:

$$1 + aG_{eq2}(s) = 1 + \frac{2500\,aTs}{s^2 + 25s + 2500 + Ts^2(s + 25)} = 0 \tag{10-83}$$

For a given T, the root contours of Eq. (10-80) when a varies are obtained based on the poles and zeros of $G_{eq2}(s)$. Notice that the poles of $G_{eq2}(s)$ are the same as the roots of Eq. (10-81). Thus, for a given T, the root contours of Eq. (10-80) when a varies must start $(a = 0)$ at points on the root contours of Fig. 10-31. These root contours end $(a = \infty)$ at $s = 0, \infty, \infty$, which are the zeros of $G_{eq2}(s)$. The complete root contours of Eq. (10-80) are now shown in Fig. 10-32 for several values of T, and a varies from 0 to ∞.

• For effective phase-lead control, the value of T should be small.

From the root contours of Fig. 10-32, we see that for effective phase-lead control, the value of T should be small. For large values of T, the natural frequency of the system increases rapidly as a increases, and very little improvement is made on the damping of the system.

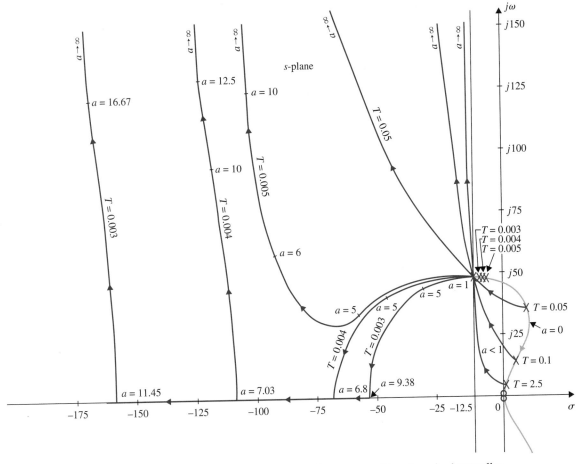

Figure 10-32 Root contours of the sun-seeker system with a phase-lead controller.

Let us choose $T = 0.01$ arbitrarily. Table 10-11 shows the attributes of the unit-step response when the value of aT is varied from 0.02 to 0.1. The **ACSYS** MATLAB tool was used for the calculations of the time responses. The results show that the smallest maximum overshoot is obtained when $aT = 0.05$, although the rise and settling times decrease continuously as aT increases. However, the smallest value of the maximum overshoot is 16.2%, which exceeds the design specification.

TABLE 10-11 **Attributes of Unit-Step Response of System with Phase-Lead Controller in Example 10-6: $T = 0.01$**

aT	a	Maximum Overshoot (%)	t_r (sec)	t_s (sec)
0.02	2	26.6	0.0222	0.0830
0.03	3	18.9	0.0191	0.0665
0.04	4	16.3	0.0164	0.0520
0.05	5	16.2	0.0146	0.0415
0.06	6	17.3	0.0129	0.0606
0.08	8	20.5	0.0112	0.0566
0.10	10	23.9	0.0097	0.0485

TABLE 10-12 **Attributes of Unit-Step Responses of System with Phase-Lead Controller in Example 10-6: $aT = 0.05$**

T	a	Maximum Overshoot (%)	t_r (sec)	t_s (sec)
0.01	5.0	16.2	0.0146	0.0415
0.005	10.0	4.1	0.0133	0.0174
0.004	12.5	1.1	0.0135	0.0174
0.003	16.67	0	0.0141	0.0174
0.002	25.0	0	0.0154	0.0209
0.001	50.0	0	0.0179	0.0244

Next, we set $aT = 0.05$, and vary T from 0.01 to 0.001, as shown in Table 10-12. Table 10-12 shows the attributes of the unit-step responses. As the value of T decreases, the maximum overshoot decreases, but the rise time and settling time increase. The cases that satisfy the design requirements are indicated in Table 10-12 for a $T = 0.05$. Figure 10-30 shows the unit-step responses of the phase-lead-compensated system with three sets of controller parameters.

Choosing $T = 0.004$, $a = 12.5$, the transfer function of the phase-lead controller is

$$G_c(s) = a\frac{s + 1/aT}{s + 1/T} = 12.5\frac{s + 20}{s + 250} \tag{10-84}$$

The transfer function of the compensated system is

$$G(s) = G_c(s)G_p(s) = \frac{31250(s + 20)}{s(s + 25)(s + 250)} \tag{10-85}$$

To find the op-amp-circuit realization of the phase-lead controller, we arbitrarily set $C = 0.1$ μf, and the resistors of the circuit are found using Eqs. (10-63) and (10-64) as $R_1 = 500,000$ Ω and $R_2 = 40,000$ Ω.

Frequency-Domain Design

Let us specify that the steady-state error requirement is the same as that given earlier. For frequency-domain design, the phase margin is to be greater than 45°. The following design steps are taken:

1. The Bode diagram of Eq. (10-74) with $K = 1$ is plotted as shown in Fig. 10-33.
2. The phase margin of the uncompensated system, read at the gain-crossover frequency, $\omega_c = 47$ rad/sec, is 28°. Since the minimum desired phase margin is 45°, at least 17° more phase lead should be added to the loop at the gain-crossover frequency.
3. The phase-lead controller of Eq. (10-65) must provide the additional 17° at the gain-crossover frequency of the compensated system. However, by applying the phase-lead controller, the magnitude curve of the Bode plot is also affected in such a way that the gain-crossover frequency is shifted to a higher frequency. Although it is a simple matter to adjust the corner frequencies, $1/aT$ and $1/T$, of the controller so that the maximum phase of the controller ϕ_m falls exactly at the new gain-crossover frequency, the original phase curve at this point is no longer 28° (and could be considerably less) because the phase of most control processes decreases with the increase in frequency. In fact, if the phase of the uncompensated process decreases rapidly with increasing frequency near the gain-crossover frequency, the single-stage phase-lead controller will no longer be effective.
   In view of the difficulty estimating the necessary amount of phase lead, it is essential to include some safety margin to account for the inevitable phase drop-off. Therefore, in the present case, instead of selecting a ϕ_m of a mere 17°, let ϕ_m be 25°. Using Eq. (10-72), we have

$$a = \frac{1 + \sin 25°}{1 - \sin 25°} = 2.46 \tag{10-86}$$

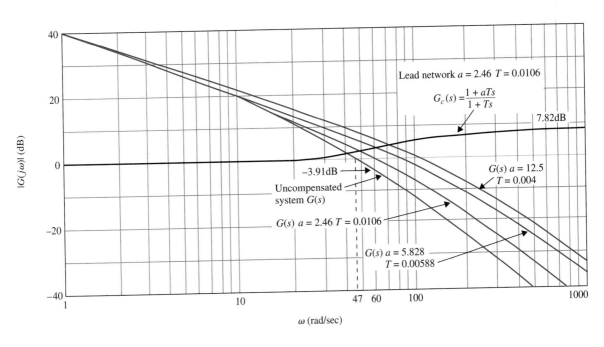

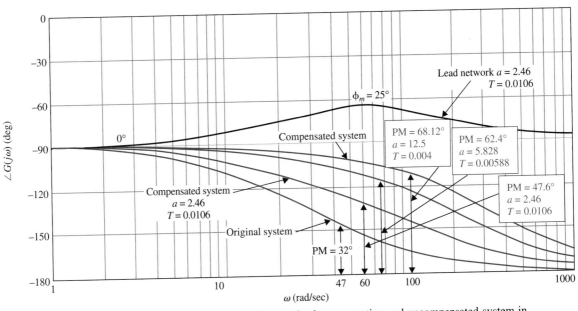

Figure 10-33 Bode diagram of phase-lead compensation and uncompensated system in Example 10-6, $G(s) = \dfrac{2500(1 + aTs)}{s(s + 25)(s + Ts)}$.

4. To determine the proper location of the two corner frequencies, $1/aT$ and $1/T$, of the controller, it is known from Eq. (10-67) that the maximum phase lead ϕ_m occurs at the geometric mean of the two corner frequencies. To achieve the maximum phase margin with the value of a determined, ϕ_m should occur at the new gain-crossover frequency ω_g', which is not known. The following steps are taken to ensure that ϕ_m occurs at ω_g'.

(a) The high-frequency gain of the phase-lead controller of Eq. (10-65) is

$$20 \log_{10} a = 20 \log_{10} 2.46 = 7.82 \text{ dB} \qquad (10\text{-}87)$$

(b) The geometric mean ω_m of the two corner frequencies $1/aT$ and $1/T$ should be located at the frequency at which the magnitude of the uncompensated process transfer function $G_p(j\omega)$ in dB is equal to the negative value in dB of one-half of this gain. This way, the magnitude curve of the compensated transfer function will pass through the 0-dB axis at $\omega = \omega_m$. Thus, ω_m should be located at the frequency where

$$|G_p(j\omega)|_{dB} = -10\log_{10} 2.46 = -3.91 \text{ dB} \tag{10-88}$$

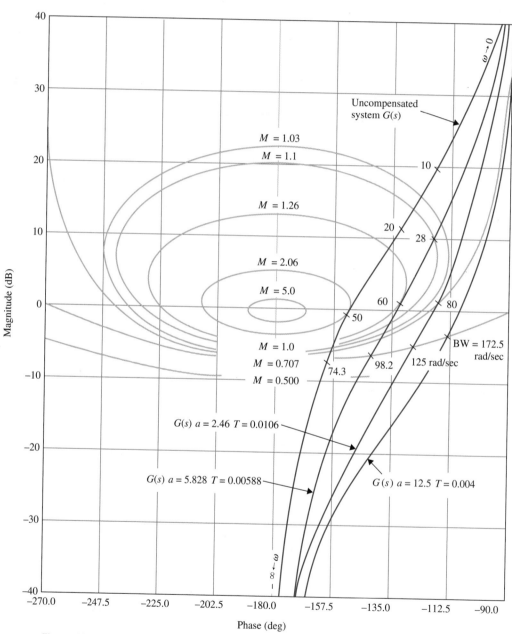

Figure 10-34 Plots of $G(s)$ in the Nichols chart for the system in Example 10-6.

$$G(s) = \frac{2500(1 + aTs)}{s(s + 25)(1 + Ts)}.$$

TABLE 10-13 Attributes of System with Phase-Lead Controller in Example 10-6

a	T	PM (deg)	M_r	Gain CO (rad/sec)	BW (rad/sec)	Maximum Overshoot (%)	t_r (sec)	t_s (sec)
1	1	28.03	2.06	47.0	74.3	44.4	0.0255	0.2133
2.46	0.0106	47.53	1.26	60.2	98.2	22.3	0.0204	0.0744
5.828	0.00588	62.36	1.03	79.1	124.7	7.7	0.0169	0.0474
12.5	0.0040	68.12	1.00	113.1	172.5	1.1	0.0135	0.0174

From Fig. 10-33, this frequency is found to be $\omega_m = 60$ rad/sec. Now using Eq. (10-67), we have

$$\frac{1}{T} = \sqrt{a}\,\omega_m = \sqrt{2.46} \times 60 = 94.1 \text{ rad/sec} \tag{10-89}$$

Then, $1/aT = 94.1/2.46 = 38.21$ rad/sec. The transfer function of the phase-lead controller is

$$G_c(s) = a\frac{s + 1/aT}{s + 1/T} = 2.46\frac{s + 38.21}{s + 94.1} \tag{10-90}$$

The forward-path transfer function of the compensated system is

$$G(s) = G_c(s)G_p(s) = \frac{6150(s + 38.21)}{s(s + 25)(s + 94.1)} \tag{10-91}$$

Figure 10-33 shows that the phase margin of the compensated system is actually 47.6°.

In Fig. 10-34, the magnitude and phase of the original and the compensated systems are plotted on the Nichols chart for display only. These plots can be made by taking the data directly from the Bode plots of Fig. 10-33. The values of M_r, ω_r, and BW can all be determined from the Nichols chart. However, the performance data are more easily obtained with **ACSYS**.

Checking the time-domain performance of the compensated system, we have the following results:

$$\text{Maximum overshoot} = 22.3\% \qquad t_r = 0.02045 \text{ sec} \qquad t_s = 0.07439 \text{ sec}$$

which fall short of the time-domain specifications listed earlier. Figure 10-33 also shows the Bode plot of the system compensated with a phase-lead controller with $a = 5.828$ and $T = 0.00588$. The phase margin is improved to 62.4°. Using Eq. (10-71), we can show that the result of $a = 12.5$ obtained in the time-domain design actually corresponds to $\phi_m = 58.41$. Adding this to the original phase of 28°, the corresponding phase margin would be 86.41°. The time-domain and frequency-domain attributes of the system with the three phase-lead controllers are summarized in Table 10-13. The results show that with $a = 12.5$ and $T = 0.004$, even the projected phase margin is 86.41°; the actual value is 68.12 due to the fall-off of the phase curve at the new gain crossover. ◄

► **EXAMPLE 10-7** In this example we illustrate the application of a phase-lead controller to a third-order system with relatively high loop gain.

Let us consider that the inductance of the dc motor of the sun-seeker system described in Fig. 10-29 is not zero. The following set of system parameters is given:

$$R_F = 10,000 \ \Omega \qquad K_b = 0.0125 \text{ V/rad/sec}$$
$$K_i = 0.0125 \text{ N-m/A} \qquad R_a = 6.25 \ \Omega$$
$$J = 10^{-6} \text{ kg-m}^2 \qquad K_s = 0.3 \text{ A/rad}$$
$$K = \text{to be determined} \qquad B = 0$$
$$n = 800 \qquad L_a = 10^{-3} \text{ H}$$

The transfer function of the dc motor is written

$$\frac{\Omega_m(s)}{E_a(s)} = \frac{K_i}{s(L_aJs^2 + JR_as + K_iK_b)} \tag{10-92}$$

The forward-path transfer function of the system is

$$G_p(s) = \frac{\Theta_o(s)}{A(s)} = \frac{K_sR_FKK_i}{s(L_aJs^2 + JR_as + K_iK_b)} \tag{10-93}$$

Substituting the values of the system parameters in Eq. (10-92), we get

$$G_p(s) = \frac{\Theta_o(s)}{A(s)} = \frac{4.6875 \times 10^7 K}{s(s^2 + 625s + 156,250)} \tag{10-94}$$

You may use **ACSYS** to solve this problem.

Time-Domain Design

The time-domain specifications of the system are given as follows:

1. The steady-state error of $\alpha(t)$ due to a unit-ramp function input for $\theta_r(t)$ should be $\leq 1/300$ rad/rad/sec of the final steady-state output velocity.

2. The maximum overshoot of the step response should be less than 5 percent, or as small as possible.

3. Rise time $t_r \leq 0.004$ sec.

4. Settling time $t_s \leq 0.02$ sec.

The minimum value of the amplifier gain K is determined initially from the steady-state requirement. Applying the final-value theorem to $\alpha(t)$, we get

$$\lim_{t \to \infty} \alpha(t) = \lim_{s \to 0} sA(s) = \lim_{s \to 0} \frac{s\Theta_r(s)}{1 + G_p(s)} \tag{10-95}$$

Substituting Eq. (10-94) into Eq. 10-95, and $\Theta_r(s) = 1/s^2$, we have

$$\lim_{t \to \infty} \alpha(t) = \frac{1}{300 \, K} \tag{10-96}$$

Thus, for the steady-state value of $\alpha(t)$ to be $\leq 1/300$, K must be ≥ 1. Let us set $K = 1$; the forward-path transfer function in Eq. (10-94) becomes

$$G_p(s) = \frac{4.6875 \times 10^7}{s(s^2 + 625s + 156,250)} \tag{10-97}$$

We can show that the closed-loop sun-seeker system with $K = 1$ has the following attributes for the unit-step response.

Maximum overshoot = 43 percent   Rise time $t_r = 0.004797$ sec   Settling time $t_s = 0.04587$ sec

To improve the system response, let us select the phase-lead controller described by Eq. (10-62). The forward-path transfer function of the compensated system is

$$G(s) = G_c(s)G_p(s) = \frac{4.6875 \times 10^7 K(1 + aTs)}{as(s^2 + 625s + 156,250)(1 + Ts)} \tag{10-98}$$

Now to satisfy the steady-state requirement, K must be readjusted so that $K \geq a$. Let us set $K = a$. The characteristic equation of the phase-lead compensated system becomes

$$(s^3 + 625s^2 + 156,250s + 4.6875 \times 10^7) + Ts^2(s^2 + 625s + 156,250) + 4.6875 \times 10^7 aTs = 0 \tag{10-99}$$

We can use the root-contour method to examine the effects of varying a and T of the phase-lead controller. Let us first set a to zero. The characteristic equation of Eq. (10-99) becomes

$$(s^3 + 625s^2 + 156,250s + 4.6875 \times 10^7) + Ts^2(s^2 + 625s + 156,250) = 0 \tag{10-100}$$

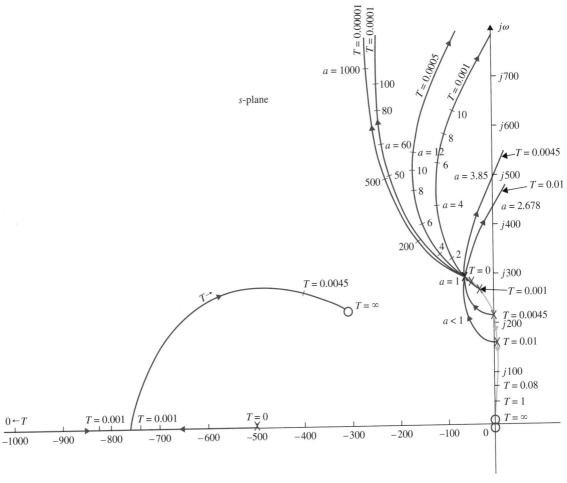

Figure 10-35 Root contours of sun-seeker system in Example 10-7 with phase-lead controller.
$$G_c(s) = \frac{1 + aTs}{1 + Ts}.$$

Dividing both sides of the last equation by the terms that do not contain T, we get

$$1 + G_{eq1}(s) = 1 + \frac{Ts^2(s^2 + 625s + 156{,}250)}{s^3 + 625s^2 + 156{,}250s + 4.6875 \times 10^7} = 0 \qquad (10\text{-}101)$$

The root contours of Eq. (10-100) when T varies are determined from the pole-zero configuration of $G_{eq1}(s)$ in Eq. (10-101), and are drawn as shown in Fig. 10-35. When a varies from 0 to ∞, we divide both sides of Eq. (10-99) by the terms that do not contain a, and we have

$$1 + G_{eq2}(s) = 1 + \frac{4.6875 \times 10^7 \, aTs}{s^3 + 625s^2 + 156{,}250s + 4.6875 \times 10^7 + Ts^2(s^2 + 625s + 156{,}250)} = 0$$
$$(10\text{-}102)$$

For a given T, the root contours of Eq. (10-99) when a varies are obtained based on the poles and zeros of $G_{eq2}(s)$. The poles of $G_{eq2}(s)$ are the same as the roots of Eq. (10-100). Thus, the root contours when a varies start ($a = 0$) at the root contours for variable T. Figure 10-34 shows the dominant portions of the root contours when a varies for $T = 0.01$, 0.0045, 0.001, 0.0005, 0.0001, and 0.00001. Notice that due to the fact that the uncompensated system is lightly damped, for the

TABLE 10-14 Roots of Characteristic Equation and Time Response Attributes of System with Phase-Lead Controller in Example 10-7

T	a	Roots of Characteristic Equation			Maximum Overshoot (%)	t_r (sec)	t_s (sec)
0.001	4	-189.6	-1181.6	$-126.9 \pm j439.5$	21.7	0.0037	0.0184
0.0005	9	-164.6	-2114.2	$-173.1 \pm j489.3$	13.2	0.00345	0.0162
0.0001	50	-147	-10024	$-227 \pm j517$	5.4	0.00348	0.0150
0.00005	100	-147	-20012	$-233 \pm j515$	4.5	0.00353	0.0150
0.00001	500	-146.3	-10^5	$-238 \pm j\,513.55$	3.8	0.00357	0.0146

phase-lead controller to be effective, the value of T should be very small. Even for very small values of T, there is only a small range of a that could bring increased damping, but the natural frequency of the system increases with the increase in a. The root contours in Fig. 10-35 show the approximate locations of the dominant characteristic equation roots where maximum damping occur. Table 10-14 gives the roots of the characteristic equation and the unit-step-response attributes for the cases that correspond to near-smallest maximum overshoot for the T selected. Figure 10-36 shows the unit-step response when $a = 500$ and $T = 0.00001$. Although the maximum overshoot is only 3.8%, the undershoot in this case is greater than the overshoot.

Frequency-Domain Design

The Bode plot of $G_p(s)$ in Eq. (10-97) is shown in Fig. 10-37. The performance attributes of the uncompensated system are

$$PM = 29.74°$$
$$M_r = 2.156$$
$$BW = 426.5 \text{ rad/sec}$$

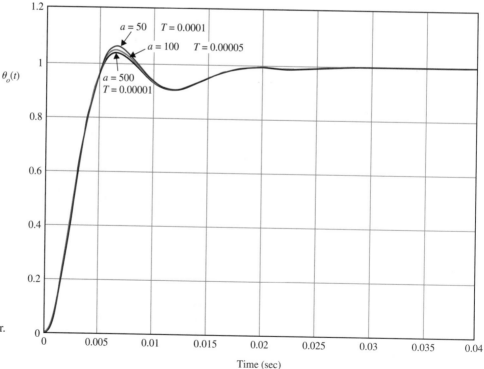

Figure 10-36 Unit-step responses of sun-seeker system in Example 10-7 with phase-lead controller.
$$G_c(s) = \frac{1 + aTs}{1 + Ts}.$$

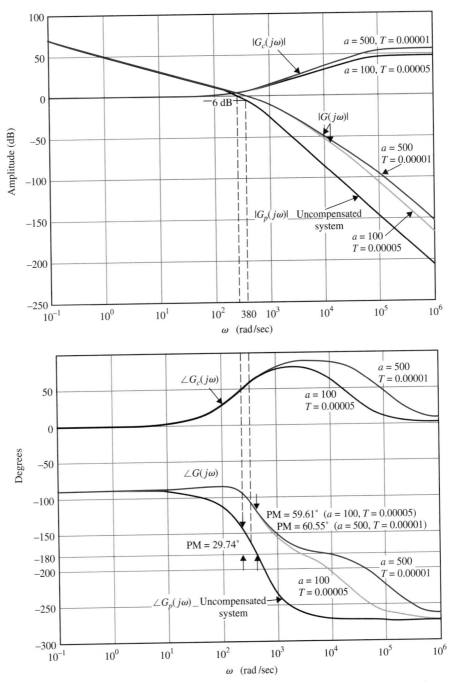

Figure 10-37 Bode plots of phase-lead controller and forward-path transfer function of sun-seeker system in Example 10-7. $G_c(s) = \dfrac{1 + aTs}{1 + Ts}$.

We would like to show that the frequency-domain design procedure outlined earlier does not work effectively here, since the phase curve of $G_p(j\omega)$ shown in Fig. 10-37 has a very steep slope near the gain crossover. For example, if we wish to realize a phase margin of 65°, we need at least $65 - 29.74 = 35.26°$ of phase lead. Or, $\phi_m = 35.26°$. Using Eq. (10-72), the value of

TABLE 10-15 Attributes of System with Phase-Lead Controller in Example 10-7

T	a	PM (deg)	GM (dB)	M_r	BW (rad/sec)	Maximum Overshoot (%)	t_r (sec)	t_r (sec)
1	1	29.74	6.39	2.16	430.4	43.0	0.00478	0.0459
0.00005	100	59.61	31.41	1.009	670.6	4.5	0.00353	0.015
0.00001	500	60.55	45.21	1.000	664.2	3.8	0.00357	0.0146

a is calculated to be

$$a = \frac{1 + \sin \phi_m}{1 - \sin \phi_m} = \frac{1 + \sin 35.26°}{1 - \sin 35.26°} = 3.732 \tag{10-103}$$

Let us choose $a = 4$. Theoretically, to maximize the utilization of ϕ_m, ω_m should be placed at the new gain crossover which is located at the frequency where the magnitude of $G_p(j\omega)$ is $-10 \log_{10} a$ dB $= -10\log_{10}4 = -6$ dB. From the Bode plot in Fig. 10-37 this frequency is found to be 380 rad/sec. Thus, we let $\omega_m = 380$ rad/sec. The value of T is found by using Eq. (10-67).

$$T = \frac{1}{\omega_m \sqrt{a}} = \frac{1}{380\sqrt{4}} = 0.0013 \tag{10-104}$$

However, checking the frequency response of the phase-lead compensated system with $a = 4$ and $T = 0.0013$, we found that the phase margin is only improved to 38.27°, and $M_r = 1.69$. The reason is due to the steep negative slope of the phase curve of $G_p(j\omega)$. The fact is that at the new gain crossover frequency of 380 rad/sec, the phase of $G_p(j\omega)$ is $-170°$, as against $-150.26°$ at the original gain crossover—a drop of almost 20° ! From the time-domain design, the first line in Table 10-14 shows that when $a = 4$ and $T = 0.001$, the maximum overshoot is 21.7 percent.

Checking the frequency response of the phase-lead compensated system with $a = 500$ and $T = 0.00001$, the following performance data are obtained:

$$\text{PM} = 60.55 \text{ degrees} \qquad M_r = 1 \qquad \text{BW} = 664.2 \text{ rad/sec}$$

This shows that the value of a has to be increased substantially just to overcome the steep drop of the phase characteristics when the gain crossover is moved upward.

Figure 10-37 shows the Bode plots of the phase-lead controller and the forward-path transfer functions of the compensated system with $a = 100$, $T = 0.0005$, and $a = 500$, $T = 0.00001$. A summary of performance data is given in Table 10-15.

Selecting $a = 100$ and $T = 0.00005$, the phase-lead controller is described by the transfer function

$$G_c(s) = \frac{1}{a} \frac{1 + aTs}{1 + Ts} = \frac{1}{100} \frac{1 + 0.005s}{1 + 0.00005s} \tag{10-105}$$

Using Eqs. (10-63) and (10-64), and letting $C = 0.01 \ \mu\text{F}$, the circuit parameters of the phase-lead controller are found to be

$$R_2 = \frac{T}{C} = \frac{5 \times 10^{-5}}{10^{-8}} = 5000 \ \Omega \tag{10-106}$$

$$R_1 = aR_2 = 500,000 \ \Omega \tag{10-107}$$

The forward-path transfer function of the compensated system is

$$\frac{\Theta_o(s)}{A(s)} = \frac{4.6875 \times 10^7 (1 + 0.005s)}{s(s^2 + 625s + 156,250)(1 + 0.00005s)} \tag{10-108}$$

where the amplifier gain K has been set to 100, to satisfy the steady-state requirement. ◀

From the results of the last two illustrative examples, we can summarize the effects and limitations of the single-stage phase-lead controller as follows.

10-5-3 Effects of Phase-Lead Compensation

1. The phase-lead controller adds a zero and a pole, with the zero to the right of the pole, to the forward-path transfer function. The general effect is to add more damping to the closed-loop system. The rise time and settling time are reduced in general.

2. The phase of the forward-path transfer function in the vicinity of the gain-crossover frequency is increased. This improves the phase margin of the closed-loop system.

3. The slope of the magnitude curve of the Bode plot of the forward-path transfer function is reduced at the gain-crossover frequency. This usually corresponds to an improvement in the relative stability of the system in the form of improved gain and phase margins.

4. The bandwidth of the closed-loop system is increased. This corresponds to faster time response.

5. The steady-state error of the system is not affected.

10-5-4 Limitations of Single-Stage Phase-Lead Control

In general, phase-lead control is not suitable for all systems. Successful application of single-stage phase-lead compensation to improve the stability of a control system is hinged on the following conditions:

1. Bandwidth Considerations: If the original system is unstable or with a low stability margin, the additional phase lead required to realize a certain desired phase margin may be excessive. This may require a relatively large value of a for the controller, which, as a result, will give rise to a large bandwidth for the compensated system, and the transmission of high-frequency noise entering the system at the input may become objectionable. However, if the noise enters the system near the output, as shown in Fig. 10-2(g), then the increased bandwidth may be beneficial to noise rejection. The larger bandwidth also has the advantage of robustness; that is, the system is insensitive to parameter variations and noise rejection as described before.

2. If the original system is unstable, or with low stability margin, the phase curve of the Bode plot of the forward-path transfer function has a steep negative slope near the gain-crossover frequency. Under this condition, the single-stage phase-lead controller may not be effective because the additional phase lead at the new gain crossover is added to a much smaller phase angle than that at the old gain crossover. The desired phase margin can be realized only by using a very large value of a for the controller. The amplifier gain K must be set to compensate a, so a large value for a requires a high-gain amplifier, which could be costly.

   As shown in Example 10-7, the compensated system may have a larger undershoot than overshoot. Often, a portion of the phase curve may still dip below the 180°-axis, resulting in a **conditionally stable** system, even though the desired phase margin is satisfied.

3. The maximum phase lead available from a single-stage phase-lead controller is less than 90°. Thus, if a phase lead of more than 90° is required, a multistage controller should be used.

10-5-5 Multistage Phase-Lead Controller

When the design with a phase-lead controller requires an additional phase of more than 90°, a multistage controller should be used. Figure 10-38 shows an op-amp-circuit realization of

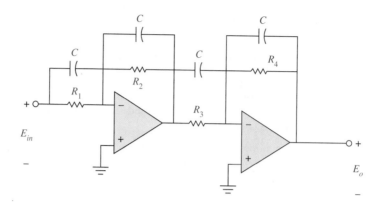

Figure 10-38 Two-stage phase-lead (phase-lag) controller.

a two-stage phase-lead controller. The input-output transfer function of the circuit is

$$G_c(s) = \frac{E_o(s)}{E_{in}(s)} = \left(\frac{s + \dfrac{1}{R_1 C}}{s + \dfrac{1}{R_2 C}}\right)\left(\frac{s + \dfrac{1}{R_3 C}}{s + \dfrac{1}{R_4 C}}\right) \tag{10-109}$$

$$= \frac{R_2 R_4}{R_1 R_3}\left(\frac{1 + R_1 C s}{1 + R_2 C s}\right)\left(\frac{1 + R_3 C s}{1 + R_4 C s}\right)$$

or

$$G_c(s) = \frac{1}{a_1 a_2}\left(\frac{1 + a_1 T_1 s}{1 + T_1 s}\right)\left(\frac{1 + a_2 T_2 s}{1 + T_2 s}\right) \tag{10-110}$$

where $a_1 = R_1/R_2$, $a_2 = R_3/R_4$, $T_1 = R_2 C$, and $T_2 = R_4 C$.

The design of a multistage phase-lead controller in the time domain becomes more cumbersome, since now there are more poles and zeros to be placed. The root-contour method also becomes impractical, since there are more variable parameters. The frequency-domain design in this case does represent a better choice of the design method. For example, for a two-stage controller, we can choose the parameters of the first stage of a two-stage controller so that a portion of the phase margin requirement is satisfied, and then the second stage fulfills the remaining requirement. In general, there is no reason why the two stages cannot be identical. The following example illustrates the design of a system with a two-stage phase-lead controller.

► **EXAMPLE 10-8** For the sun-seeker system designed in Example 10-7, let us alter the rise time and settling time requirements to be

Rise time $t_r \le 0.001$ sec

Settling time $t_s \le 0.005$ sec

The other requirements are not altered. One way to meet faster rise time and settling time requirements is to increase the forward-path gain of the system. Let us consider that the forward-path transfer function is

$$G_p(s) = \frac{\Theta_o(s)}{A(s)} = \frac{156{,}250{,}000}{s(s^2 + 625s + 156{,}250)} \tag{10-111}$$

Another way of interpreting the change in the forward-path gain is that the ramp-error constant is increased to 1000 (up from 300 in Example 10-6). The Bode plot of $G_p(s)$ is shown in Fig. 10-39. The closed-loop system is unstable, with a phase margin of $-15.43°$.

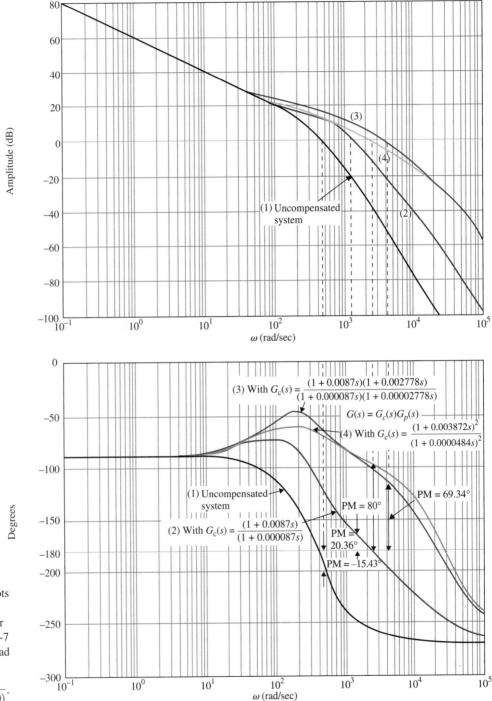

Figure 10-39 Bode plots of uncompensated and compensated sun-seeker systems in Example 10-7 with two-state phase-lead controller. $G_p(s) = \dfrac{156,250,000}{s(s^2 + 625s + 156,250)}$.

Since the compensated system in Example 10-7 had a phase margin of 60.55°, we would expect that to satisfy the more stringent time response requirements in this example, the corresponding phase margin would have to be greater. Apparently, this increased phase margin cannot be realized with a single-stage phase-lead controller. It appears that a two-stage controller would be adequate.

The design involves some trial-and-error steps in arriving at a satisfied controller. Since we have two stages of controllers at our disposal, the design has a multitude of flexibility. We can set out by arbitrarily setting $a_1 = 100$ for the first stage of the phase-lead controller. The phase lead provided by the controller is obtained from Eq. (10-71),

$$\phi_m = \sin^{-1}\left(\frac{a_1 - 1}{a_1 + 1}\right) = \sin^{-1}\left(\frac{99}{101}\right) = 78.58° \tag{10-112}$$

To maximize the effect of ϕ_m, the new gain crossover should be at

$$-10\log_{10} a_1 = -10\log_{10} 100 = -20 \text{ dB} \tag{10-113}$$

From Fig. 10-39 the frequency that corresponds to this gain on the amplitude curve is approximately 1150 rad/sec. Substituting $\omega_{m1} = 1150$ rad/sec and $a_1 = 100$ in Eq. (10-67), we get

$$T_1 = \frac{1}{\omega_{m1}\sqrt{a_1}} = \frac{1}{1150\sqrt{100}} = 0.000087 \tag{10-114}$$

The forward-path transfer function with the one-stage phase-lead controller is

$$G(s) = \frac{156,250,000(1 + 0.0087s)}{s(s^2 + 625s + 156,250)(1 + 0.000087s)} \tag{10-115}$$

The Bode plot of the last equation is drawn as curve (2) in Fig. 10-39. We see that the phase margin of the interim design is only 20.36°. Next, we arbitrarily set the value of a_2 of the second stage at 100. From the Bode plot of the transfer function of Eq. (10-115) in Fig. 10-39, we find that the frequency at which the magnitude of $G(j\omega)$ is -20 dB is approximately 3600 rad/sec. Thus,

$$T_2 = \frac{1}{\omega_{m2}\sqrt{a_2}} = \frac{1}{3600\sqrt{100}} = 0.00002778 \tag{10-116}$$

The forward-path transfer function of the sun-seeker system with the two-stage phase-lead controller is ($a_1 = a_2 = 100$)

$$G(s) = \frac{156,250,000(1 + 0.0087s)(1 + 0.002778s)}{s(s^2 + 625s + 156,250)(1 + 0.000087s)(1 + 0.00002778s)} \tag{10-117}$$

Figure 10-39 shows the Bode plot of the sun-seeker system with the two-stage phase-lead controller designed above (Curve 3). As seen from Fig. 10-39, the phase margin of the system with $G(s)$ given in Eq. (10-117) is 69.34°. As shown by the system attributes tabulated in Table 10-16, the system satisfies all the time-domain specifications. In fact, the selection of $a_1 = a_2 = 100$ appears to be overly stringent. To show that the design is not critical, we can select $a_1 = a_2 = 80$, and then 70 and the time-domain specifications are still satisfied. Following similar design steps, we arrived at $T_1 = 0.0001117$ and $T_2 = 0.000039$ for $a_1 = a_2 = 70$, and $T_1 = T_2 = 0.0000484$ for $a_1 = a_2 = 80$. Curve (4) of Fig. 10-39 shows the Bode plot of the compensated system with $a_1 = a_2 = 80$. Table 10-16 summarizes all the attributes of the system performance with these three controllers.

The unit-step responses of the system with the two-stage phase-lead controller for $a_1 = a_2 = 80$ and 100 are shown in Fig. 10-40.

TABLE 10-16 Attributes of Sun-Seeker System in Example 10-8 with Two-Stage Phase-Lead Controller

$a_1 = a_2$	T_1	T_2	PM (deg)	M_r	BW (rad/sec)	Maximum Overshoot (%)	t_r (sec)	t_s (sec)
80	0.0000484	0.0000484	80	1	5686	0	0.00095	0.00475
100	0.000087	0.0000278	69.34	1	5686	0	0.000597	0.00404
70	0.0001117	0.000039	66.13	1	5198	0	0.00063	0.00404

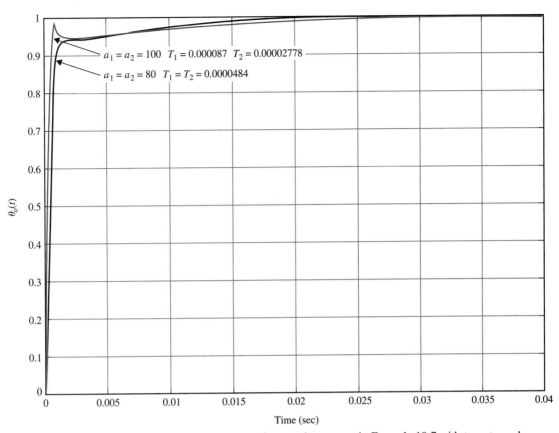

Figure 10-40 Unit-step responses of sun-seeker system in Example 10-7 with two-stage phase-lead controller. $G_c(s) = \left(\dfrac{1 + a_1 T_1 s}{1 + T_1 s}\right)\left(\dfrac{1 + a_2 T_2 s}{1 + T_2 s}\right)$ $\quad G_p(s) = \dfrac{156{,}250{,}000}{s(s^2 + 625s + 156{,}250)}$ ◀

10-5-6 Sensitivity Considerations

The sensitivity function defined in Section 9-16, Eq. (9-122) can be used as a design specification to indicate the robustness of the system. In Eq. (9-122), the sensitivity of the closed-loop transfer function with respect to the variations of the forward-path transfer function is defined as

$$S_G^M(s) = \frac{\partial M(s)/M(s)}{\partial G(s)/G(s)} = \frac{G^{-1}(s)}{1 + G^{-1}(s)} = \frac{1}{1 + G(s)} \tag{10-118}$$

The plot of $|S_G^M(j\omega)|$ versus frequency gives an indication of the sensitivity of the system as a function of frequency. The ideal robust situation is for $|S_G^M(j\omega)|$ to assume a small value ($\ll 1$) over a wide range of frequency. As an example, the sensitivity function of the sun-seeker system designed in Example 10-7 with the one-stage phase-lead controller with $a = 100$ and $T = 0.00005$ is plotted as shown in Fig. 10-41. Note that the sensitivity function is low at low frequencies, and is less than unity for $\omega < 400$ rad/sec. Although the sun-seeker system in Example 10-7 does not need a multistage phase-lead controller, we shall show that if a two-stage phase-lead controller is used, not only the value of a will be substantially reduced, resulting in lower gains for the op-amps, but the system will be more robust. Following the design procedure outlined in Example 10-8, a two-stage phase-lead controller is designed for the sun-seeker system with the process transfer function

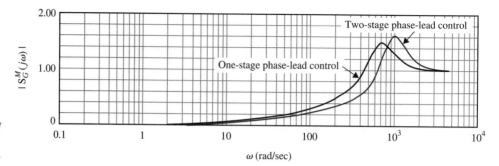

Figure 10-41 Sensitivity functions of sun-seeker system in Example 10-7.

described by Eq. (10-96). The parameters of the controller are: $a_1 = a_2 = 5.83$ and $T_1 = T_2 = 0.000673$. The forward-path transfer function of the compensated system is

• Systems with phase-lead control will generally be more robust due to the increase in bandwidth.

$$G(s) = \frac{4.6875 \times 10^7 (1 + 0.0039236s)^2}{s(s^2 + 625s + 156,250)(1 + 0.000673s)^2} \qquad (10\text{-}119)$$

Figure 10-41 shows that the sensitivity function of the system with the two-stage phase-lead controller is less than unity for $\omega < 600$ rad/sec. Thus, the system with the two-stage phase-lead controller is more robust than the system with the single-stage controller. The reason for this is that the more robust system has a higher bandwidth. In general, systems with phase-lead control will be more robust due to the higher bandwidth. However, Fig. 10-41 shows that the system with the two-stage phase-lead controller has a higher sensitivity at high frequencies.

▶ 10-6 DESIGN WITH PHASE-LAG CONTROLLER

The transfer function in Eq. (10-62) represents a phase-lag controller or low-pass filter when $a < 1$. The transfer function is repeated as follows.

$$G_c(s) = \frac{1}{a}\left(\frac{1 + aTs}{1 + Ts}\right) \qquad a < 1 \qquad (10\text{-}120)$$

10-6-1 Time-Domain Interpretation and Design of Phase-Lag Control

The pole-zero configuration of $G_c(s)$ is shown in Fig. 10-42. Unlike the PI controller, which provides a pole at $s = 0$, the phase-lag controller affects the steady-state error only in the sense that the zero-frequency gain of $G_c(s)$ is greater than unity. Thus, any error constant that is finite and nonzero will be increased by the factor $1/a$ from the phase-lag controller.

Since the pole at $s = -1/T$ is to the right of the zero at $-1/aT$, effective use of the phase-lag controller to improve damping would have to follow the same design principle of the PI control presented in Section 10-3. Thus, *the proper way of applying the phase-lag control is to place the pole and zero close together. For type 0 and type 1 systems the combination should be located near the origin in the s-plane.* Figure 10-43 illustrates the design strategies in the s-plane for type 0 and type 1 systems. Phase-lag control should not be applied to a type 2 system.

The design principle described above can be explained by considering that the controlled process of a type 0 control system is

$$G_p(s) = \frac{K}{(s + p_1)(s + \overline{p}_1)(s + p_3)} \qquad (10\text{-}121)$$

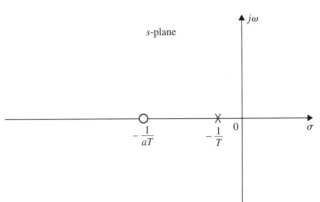

Figure 10-42 Pole-zero configuration of phase-lag controller.

$$G_c(s) = \frac{1}{a} \frac{1 + aTs}{1 + Ts} \quad a < 1.$$

where p_1 and $\bar{p}_1$ are complex conjugate poles, such as the situation shown in Fig. 10-43.

Just as in the case of the phase-lead controller, we can drop the gain factor $1/a$ in Eq. (10-120), since whatever the value of a is, the value of K can be adjusted to compensate for it. Applying the phase-lag controller of Eq. (10-120), without the factor $1/a$, to the system, the forward-path transfer function becomes

$$G(s) = G_c(s)G_p(s) = \frac{K(1 + aTs)}{(s + p_1)(s + \bar{p}_1)(s + p_3)(1 + Ts)} \quad (a < 1) \quad (10\text{-}122)$$

Let us assume that the value of K is set to meet the steady-state-error requirement. Also, assume that with the selected value of K, the system damping is low or is even unstable. Now let $1/T \cong 1/aT$, and place the pole-zero pair near the pole at $-1/p_3$, as shown in Fig. 10-43. Figure 10-44 shows the root loci of the system with and without the phase-lag controller. Since the pole-zero combination of the controller is very close to the pole at $-1/p_3$, the shape of the loci of the dominant roots with and without the phase-lag control will be very similar. This is easily explained by writing Eq. (10-122) as

$$G(s) = \frac{Ka(s + 1/aT)}{(s + p_1)(s + \bar{p}_1)(s + p_3)(s + 1/T)}$$
$$\cong \frac{Ka}{(s + p_1)(s + \bar{p}_1)(s + p_3)} \quad (10\text{-}123)$$

Since a is less than 1, the application of phase-lag control is equivalent to reducing the forward-path gain from K to Ka, *while not affecting the steady-state performance of the system.* Figure 10-44 shows that the value of a can be chosen so that the damping of the compensated system is satisfactory. Apparently, the amount of damping that can be added is limited if the poles $-p_1$ and $-\bar{p}_1$ are very close to the imaginary axis. Thus, we can select a using the following equation:

• Phase-lag control will generally increase rise time and settling time.

$$a = \frac{K \text{ to realize the desired damping}}{K \text{ to realize the steady-state performance}} \quad (10\text{-}124)$$

The value of T should be so chosen that the pole and zero of the controller are very close together and close to $-1/p_3$.

In the time domain, phase-lag control generally has the effect of increasing the rise time and settling time.

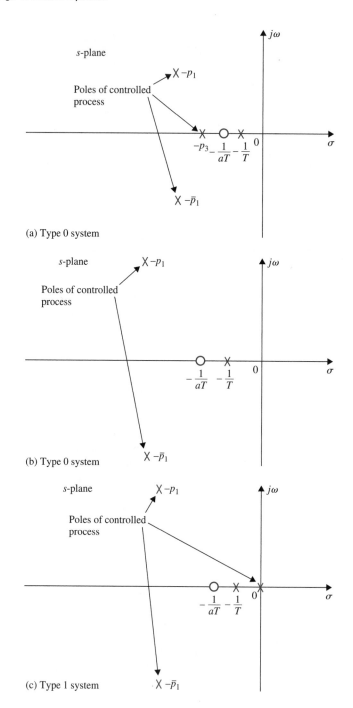

Figure 10-43 Design strategies for phase-lag control for type 0 and type 1 systems.

10-6-2   Frequency-Domain Interpretation and Design of Phase-Lag Control

The transfer function of the phase-lag controller can again be written as

$$G_c(s) = \frac{1 + aTs}{1 + Ts} \qquad (a < 1) \tag{10-125}$$

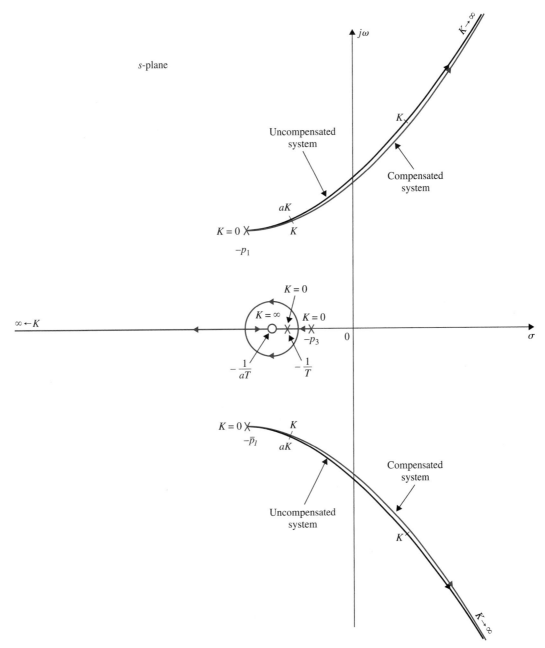

Figure 10-44 Root loci of uncompensated and phase-lag-compensated systems.

by assuming that the gain factor $-1/a$ is eventually absorbed by the forward gain K. The Bode diagram of Eq. (10-125) is shown in Fig. 10-45. The magnitude curve has corner frequencies at $\omega = 1/aT$ and $1/T$. Since the transfer functions of the phase-lead and phase-lag controllers are identical in form, except for the value of a, the maximum phase lag ϕ_m of the phase curve of Fig. 10-45 is given by

$$\phi_m = \sin^{-1}\left(\frac{a-1}{a+1}\right) \quad (a < 1) \quad\quad (10\text{-}126)$$

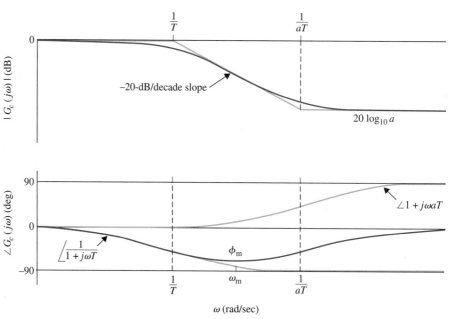

Figure 10-45 Bode diagram of the phase-lag controller. $G_c(s) = \dfrac{(1 + aTs)}{(1 + Ts)}$ $a < 1$.

• The objective of phase-lag control in the frequency domain is to move the gain crossover to the frequency where the desired phase margin is realized, while keeping the phase curve relatively unchanged at the new gain crossover.

Figure 10-45 shows that the phase-lag controller essentially provides an attenuation of $20 \log_{10} a$ at high frequencies. Thus, unlike the phase lead control that utilizes the maximum phase-lead of the controller, phase-lag control utilizes the attenuation of the controller at high frequencies. This is parallel to the situation of introducing an attenuation of a to the forward-path gain in the root-locus design. For phase-lead control, the objective of the controller is to increase the phase of the open-loop system in the vicinity of the gain crossover while attempting to locate the maximum phase lead at the new gain crossover. In phase-lag control, the objective is to move the gain crossover to a lower frequency where the desired phase margin is realized, while keeping the phase curve of the Bode plot relatively unchanged at the new gain crossover.

The design procedure for phase-lag control using the Bode plot is outlined as follows:

1. The Bode plot of the forward-path transfer function of the uncompensated system is drawn. The forward-path gain K is set according to the steady-state performance requirement.

2. The phase and gain margins of the uncompensated system are determined from the Bode plot.

3. Assuming that the phase margin is to be increased, the frequency at which the desired phase margin is obtained is located on the Bode plot. This frequency is also the new gain crossover frequency ω_g', where the compensated magnitude curve crosses the 0-dB axis.

4. To bring the magnitude curve down to 0 dB at the new gain-crossover frequency ω_g', the phase-lag controller must provide the amount of attenuation equal to the value of the magnitude curve at ω_g'. In other words,

$$|G_p(j\omega_g')| = -20\log_{10} a \text{ dB} \qquad (a < 1) \qquad (10\text{-}127)$$

Solving for a from the last equation, we get

$$a = 10^{-|G_p(j\omega'_g)|/20} \qquad (a < 1) \qquad \text{(10-128)}$$

Once the value of a is determined, it is necessary only to select the proper value of T to complete the design. Using the phase characteristics shown in Fig. 10-45, if the corner frequency $1/aT$ is placed far below the new gain-crossover frequency ω'_g, the phase lag of the controller will not appreciably affect the phase of the compensated system near ω'_g. On the other hand, the value of $1/aT$ should not be too small because the bandwidth of the system will be too low, causing the system to be too sluggish and less robust. Usually, as a general guideline, the frequency $1/aT$ should be approximately one decade below ω'_g; that is,

$$\frac{1}{aT} = \frac{\omega'_g}{10} \text{ rad/sec} \qquad \text{(10-129)}$$

Then,

$$\frac{1}{T} = \frac{a\omega'_g}{10} \text{ rad/sec} \qquad \text{(10-130)}$$

5. The Bode plot of the compensated system is investigated to see if the phase margin requirement is met; if not, the values of a and T are readjusted, and the procedure is repeated. If design specifications involve gain margin, M_r, or BW, then these values should be checked and satisfied.

Since the phase-lag control brings in more attenuation to a system, then if the design is proper, the stability margins will be improved but at the expense of lower bandwidh. The only benefit of lower bandwidth is reduced sensitivity to high-frequency noise and disturbances.

The following example illustrates the design of the phase-lag controller and all its ramifications.

▶ **EXAMPLE 10-9** In this example we shall use the second-order sun-seeker system described in Example 10-6 to illustrate the principle of design of phase-lag control. The forward-path transfer function of the uncompensated system is

$$G_p(s) = \frac{\Theta_o(s)}{A(s)} = \frac{2500K}{s(s + 25)} \qquad \text{(10-131)}$$

You may use **ACSYS** to solve this problem.

Time-Domain Design

The time-domain specifications of the system are

1. The steady-state error of $\alpha(t)$ due to a unit-ramp function input for $\theta_r(t)$ should be ≤ 1 percent.

2. The maximum overshoot of the step response should be less than 5 percent, or as small as possible.

3. Rise time $t_r \leq 0.5$ sec.

4. Settling time $t_s \leq 0.5$ sec.

5. Due to noise problems, the bandwidth of the system must be < 50 rad/sec.

Notice that the rise-time and settling-time requirements have been relaxed considerably from the phase-lead design in Example 10-6. The root loci of the uncompensated system are shown in Fig. 10-46(a).

As in Example 10-6, we set $K = 1$ initially. The damping ratio of the uncompensated system is 0.25, and the maximum overshoot is 44.4 percent. Figure 10-47 shows the unit-step response of the system with $K = 1$.

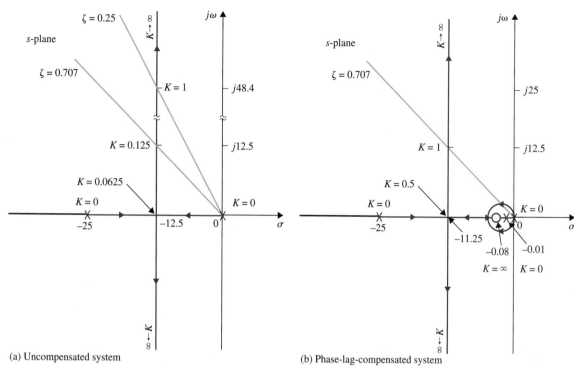

(a) Uncompensated system   (b) Phase-lag-compensated system

Figure 10-46 Root loci of sun-seeker system in Example 10-9.

$$G_p(s) = \frac{2500\,K}{s(s+25)} \quad G_c(s) = \frac{1+aTs}{1+Ts} \quad a = 0.125 \quad T = 100.$$

Let us select the phase-lag controller with the transfer function given in Eq. (10-120). The forward-path transfer function of the compensated system is

$$G(s) = G_c(s)\,G_p(s) = \frac{2500K(s + 1/aT)}{s(s+25)(s+1/T)} \tag{10-132}$$

If the value of K is maintained at 1, the steady-state error will be a percent, which is better than that of the uncompensated system, since $a < 1$. *For effective phase-lag control, the pole and zero of the controller transfer function should be placed close together, and then for the type 1 system, the combination should be located relatively close to the origin of the s-plane.* From the root loci of the uncompensated system in Fig. 10-46(a), we see that if K could be set to 0.125, the damping ratio would be 0.707, and the maximum overshoot of the system would be 4.32%. By setting the pole and zero of the controller close to $s = 0$, the shape of the loci of the dominant roots of the compensated system will be very similar to those of the uncompensated system. We can find the value of a using Eq. (10-124); that is,

$$a = \frac{K \text{ to realize the desired damping}}{K \text{ to realize the steady-state performance}} = \frac{0.125}{1} = 0.125 \tag{10-133}$$

Thus, if the value of T is sufficiently large, when $K = 1$, the dominant roots of the characteristic equation will correspond to a damping ratio of approximately 0.707. Let us arbitrarily select $T = 100$. The root loci of the compensated system are shown in Fig. 10-46(b). The roots of the characteristic equation when $K = 1$, $a = 0.125$, and $T = 100$ are

$$s = -0.0805, \quad -12.465 + j12.465 \quad \text{and} \quad -12.465 - j12.465$$

which corresponds to a damping ratio of exactly 0.707. If we had chosen a smaller value for T, then the damping ratio would be slightly off 0.707. From a practical standpoint, the value of T cannot be

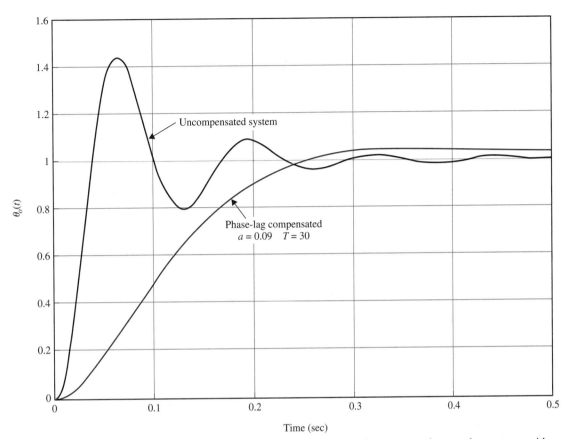

Figure 10-47 Unit-step responses of uncompensated and compensated sun-seeker systems with phase-lag controller in Example 10-9. $G_p(s) = \dfrac{2500K}{s(s+25)}$ $G_c(s) = \dfrac{1 + aTs}{1 + Ts}$ $a = 0.09$ $T = 30$.

too large, since from Eq. (10-64), $T = R_2C$, a large T would correspond to either a large capacitor or an unrealistically large resistor. To reduce the value of T, and simultaneously satisfy the maximum overshoot requirement, a should also be reduced. However, a cannot be reduced indefinitely, or the zero of the controller at $-1/aT$ would be too far to the left on the real axis. Table 10-17 gives

TABLE 10-17 Attributes of Performance of Sun-Seeker System in Example 10-9 with Phase-Lag Controller

a	T	Maximum Overshoot (%)	t_r (sec)	t_s (sec)	BW (rad/sec)	Roots of Characteristic Equation	
1.000	1	44.4	0.0255	0.2133	75.00		$-12.500 \pm j48.412$
0.125	100	4.9	0.1302	0.1515	17.67	-0.0805	$-12.465 \pm j12.465$
0.100	100	2.5	0.1517	0.2020	13.97	-0.1009	$-12.455 \pm j9.624$
0.100	50	3.4	0.1618	0.2020	14.06	-0.2037	$-12.408 \pm j9.565$
0.100	30	4.5	0.1594	0.1515	14.19	-0.3439	$-12.345 \pm j9.484$
0.100	20	5.9	0.1565	0.4040	14.33	-0.5244	$-12.263 \pm j9.382$
0.090	50	3.0	0.1746	0.2020	12.53	-0.2274	$-12.396 \pm j8.136$
0.090	30	4.4	0.1719	0.2020	12.68	-0.3852	$-12.324 \pm j8.029$
0.090	20	6.1	0.1686	0.5560	12.84	-0.5901	$-12.230 \pm j7.890$

the attributes of the time-domain performance of the phase-lag compensated sun-seeker system with various a and T. The ramifications of the various design parameters are clearly displayed.

Thus, a suitable set of controller parameters would be $a = 0.09$ and $T = 30$. With $T = 30$, selecting $C = 1\ \mu F$ would require R_2 to be 30 MΩ. A smaller value for T can be realized by using a two-stage phase-lag controller. The unit-step response of the compensated system with $a = 0.09$ and $T = 30$ is shown in Fig. 10-47. Notice that the maximum overshoot is reduced at the expense of rise time and settling time. Although the settling time of the compensated system is shorter than

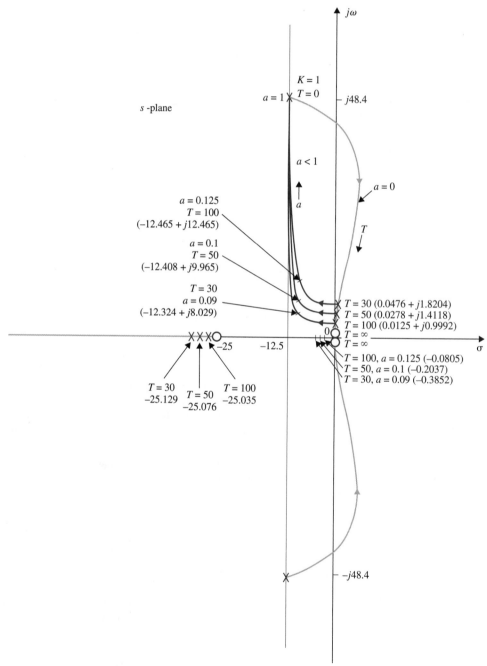

Figure 10-48 Root contours of sun-seeker system in Example 10-9 with phase-lag controller.

that of the uncompensated system, actually, it takes much longer time for the phase-lag-compensated system to reach steady state.

It would be enlightening to explain the design of the phase-lag controller by means of the root contours. The root-contour design conducted earlier in Example 10-6 using Eqs. (10-80) through (10-83) for phase-lead control and Figs. 10-31 and 10-32 is still valid for phase-lag control, except that in the present case, $a < 1$. Thus, in Fig. 10-32 only the portions of the root contours that correspond to $a < 1$ are applicable for phase-lag control. These root contours clearly show that for effective phase-lag control, the value of T should be relatively large. In Fig. 10-48 we illustrate further that the complex poles of the closed-loop transfer function are rather insensitive to the value of T when the latter is relatively large.

Frequency-Domain Design

The Bode plot of $G_p(j\omega)$ of Eq. (10-131) is shown in Fig. 10-49 for $K = 1$. The Bode plot shows that the phase margin of the uncompensated system is only 28°. Not knowing what phase margin will correspond to a maximum overshoot of less than 5 percent, we conduct the following series of designs using the Bode plot in Fig. 10-49. Starting with a phase margin of 45°, we observe that this phase margin can be realized if the gain-crossover frequency ω_g' is at 25 rad/sec. This means that the phase-lag controller must reduce the magnitude curve of $G_p(j\omega)$ to 0 dB at $\omega = 25$ rad/sec while it does not appreciably affect the phase curve near this frequency. Since the phase-lag controller still contributes a small negative phase when the corner frequency $1/aT$ is placed at $1/10$ of the value of ω_g', it is a safe measure to choose ω_g' at somewhat less than 25 rad/sec, say, 20 rad/sec.

From the Bode plot, the value of $|G_p(j\omega_g')|_{dB}$ at $\omega_g' = 20$ rad/sec is 11.7 dB. Thus, using Eq. (10-128), we have

$$a = 10^{-|G_p(j\omega_g')|/20} = 10^{-11.7/20} = 0.26 \qquad (10\text{-}134)$$

The value of $1/aT$ is chosen to be at $1/10$ the value of $\omega_g' = 20$ rad/sec. Thus,

$$\frac{1}{aT} = \frac{\omega_g'}{10} = \frac{20}{10} = 2 \text{ rad/sec} \qquad (10\text{-}135)$$

and

$$T = \frac{1}{2a} = \frac{1}{0.52} = 1.923 \qquad (10\text{-}136)$$

Checking out the unit-step response of the system with the designed phase-lag control, we found that the maximum overshoot is 24.5 percent. The next step is to try aiming at a higher phase margin. Table 10-18 gives the various design results by using various desired phase margin up to 80°.

Examining the results in Table 10-18, we see that none of the cases satisfies the maximum overshoot requirement of ≤ 5 percent. The $a = 0.044$ and $T = 52.5$ case yields the best maximum overshoot, but the value of T is too large to be practical. Thus, we single out the case with $a = 0.1$ and $T = 10$ and refine the design by increasing the value of T. As shown in Table 10-17, when $a = 0.1$ and $T = 30$, the maximum overshoot is reduced to 4.5 percent. The Bode plot of the compensated system is shown in Fig. 10-49. The phase margin is 67.61°.

TABLE 10-18 Performance Attributes of Sun-Seeker System in Example 10-9 with Phase-Lag Controller

Desired PM (deg)	a	T	Actual PM (deg)	M_r	BW (rad/sec)	Maximum Overshoot (%)	t_r (sec)	t_s (sec)
45	0.26	1.923	46.78	1.27	33.37	24.5	0.0605	0.2222
60	0.178	3.75	54.0	1.19	25.07	17.5	0.0823	0.303
70	0.1	10	63.87	1.08	14.72	10.0	0.1369	0.7778
80	0.044	52.5	74.68	1.07	5.7	7.1	0.3635	1.933

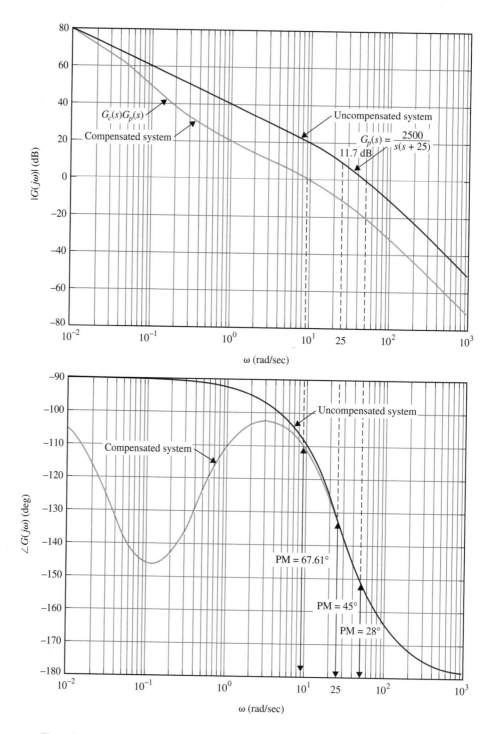

Figure 10-49 Bode plot of uncompensated and compensated systems with phase-lag controller in Example 10-9.

$$G_c(s) = \frac{1 + 3s}{1 + 30s}$$

$$G_p(s) = \frac{2500}{s(s + 25)}$$

The unit-step response of the phase-lag-compensated system shown in Fig. 10-47 points out a major disadvantage of the phase-lag control. Since the phase-lag controller is essentially a low-pass filter, the rise time and settling time of the compensated system are usually increased. However, we shall show by the following example that phase-lag control can be more versatile and has a wider range of effectiveness in improving stability than the single-stage phase-lead controller, especially if the system has low or negative damping. ◀

▶ **EXAMPLE 10-10** Consider the sun-seeker system designed in Example 10-8, with the forward-path transfer function given in Eq. (10-111). Let us restore the gain K, so that a root-locus plot can be made for the system. Then, Eq. (10-111) is written

$$G_p(s) = \frac{156{,}250{,}000\,K}{s(s^2 + 625s + 156{,}250)} \tag{10-137}$$

The root loci of the closed-loop system are shown in Fig. 10-50. When $K = 1$, the system is unstable, and the characteristic equation roots are at -713.14, $44.07 + j466.01$ and $44.07 - j466.01$.

Example 10-8 shows that the performance specification on stability cannot be achieved with a single-stage phase-lead controller. Let the performance criteria be

$$\text{Maximum overshoot} \leq 5 \text{ percent}$$
$$\text{Rise time } t_r \leq 0.02 \text{ sec}$$
$$\text{Settling time } t_s \leq 0.02 \text{ sec}$$

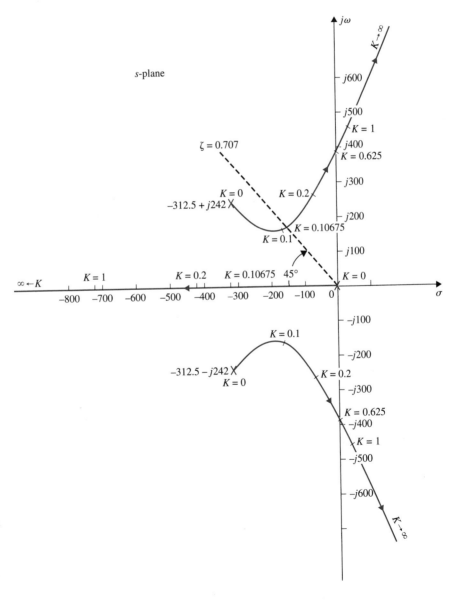

Figure 10-50 Root loci of uncompensated system in Example 10-10.
$G_p(s) =$
$\dfrac{156{,}250{,}000\,K}{s(s^2 + 625s + 156{,}250)}.$

TABLE 10-19 Performance Attributes of Sun-Seeker System in Example 10-10 with Phase-Lag Controller

a	T	BW (rad/sec)	PM (deg)	% Max Overshoot	t_r (sec)	t_s (sec)
0.1	20	173.5	66.94	1.2	0.01273	0.01616
0.1	10	174	66.68	1.6	0.01262	0.01616
0.1	5	174.8	66.15	2.5	0.01241	0.01616
0.1	2	177.2	64.56	4.9	0.01601	0.0101

Let us assume that the desired relative damping ratio is 0.707. Figure 10-50 shows that when $K = 0.10675$, the dominant characteristic equation roots of the uncompensated system are at $-172.77 \pm j172.73$, which correspond to a damping ratio of 0.707. Thus, the value of a is determined from Eq. (10-124),

$$a = \frac{K \text{ to realize the desired damping}}{K \text{ to realize the steady-state performance}} = \frac{0.10675}{1} = 0.10675 \qquad (10\text{-}138)$$

Let $a = 0.1$. Since the loci of the dominant roots are far away from the origin in the s-plane, the value of T has a wide range of flexibility. Table 10-19 shows the performance results when $a = 0.1$ and various values of T.

Therefore, the conclusion is that only one stage of the phase-lag controller is needed to satisfy the stability requirement, whereas two stages of the phase-lead controller are needed, as shown in Example 10-8.

Sensitivity Function

The sensitivity function $|S_G^M(j\omega)|$ of the phase-lag compensated system with $a = 0.1$ and $T = 20$ is shown in Fig. 10-51. Notice that the sensitivity function is less than unity for frequencies up to only 102 rad/sec. This is due to the low bandwidth of the system as a result of phase-lag control.

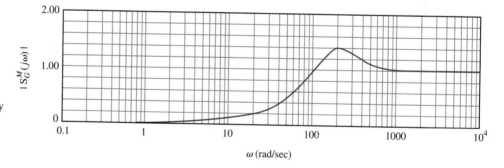

Figure 10-51 Sensitivity function of phase-lag-compensated system in Example 10-10.

10-6-3 Effects and Limitations of Phase-Lag Control

From the results of the preceding illustrative examples, the effects and limitations of phase-lag control on the performance of linear control systems can be summarized as follows.

1. For a given forward-path gain K, the magnitude of the forward-path transfer function is attenuated near the above the gain-crossover frequency, thus improving the relative stability of the system.

2. The gain-crossover frequency is decreased, and thus the bandwidth of the system is reduced.

3. The rise time and settling time of the system are usually longer, because the bandwidth is usually decreased.

4. The system is more sensitive to parameter variations because the sensitivity function is greater than unity for all frequencies approximately greater than the bandwidth of the system.

▷ 10-7 DESIGN WITH LEAD–LAG CONTROLLER

We have learned from preceding sections that phase-lead control generally improves rise time and damping, but increases the natural frequency of the closed-loop system. However, phase-lag control when applied properly improves damping, but usually results in a longer rise time and settling time. Therefore, each of these control schemes has its advantages, disadvantages, and limitations, and there are many systems that cannot be satisfactorily compensated by either scheme acting alone. It is natural, therefore, whenever necessary, to consider using a combination of the lead and lag controllers, so that the advantages of both schemes are utilized.

The transfer function of a simple lag–lead (or lead–lag) controller can be written

$$G_c(s) = G_{c1}(s)G_{c2}(s) = \left(\frac{1 + a_1T_1s}{1 + T_1s}\right)\left(\frac{1 + a_2T_2s}{1 + T_2s}\right) \qquad (a_1 > 1, a_2 < 1) \qquad (10\text{-}139)$$
$$|\leftarrow \text{lead} \rightarrow||\leftarrow \text{lag} \rightarrow|$$

The gain factors of the lead and lag controllers are not included because, as shown previously, these gain and attenuation are compensated eventually by the adjustment of the forward gain K.

Since the lead–lag controller transfer function in Eq. (10-139) now has four unknown parameters, its design is not as straightforward as the single-stage phase-lead or phase-lag controller. *In general, the phase-lead portion of the controller is used mainly to achieve a shorter rise time and higher bandwidth, and the phase-lag portion is brought in to provide major damping of the system.* Either the phase-lead or the phase-lag control can be designed first. We shall use Example 10-11 to illustrate the design steps.

▶ **EXAMPLE 10-11** As an illustrative example of designing a lead–lag controller, let us consider the sun-seeker system of Example 10-8. The uncompensated system with $K = 1$ was shown to be unstable. A two-stage phase-lead controller was designed in Example 10-9, and a single-stage phase-lag controller was designed in Example 10-10.

Based on the design in Example 10-8, we can first select a phase-lead control with $a = 70$ and $T_1 = 0.00004$. The remaining phase-lag control can be designed using either the root-locus method or the Bode plot method. Table 10-20 gives the results by letting $T_2 = 2$, which is an insensitive parameter, and various values of a. The results in Table 10-20 show that the optimal value of a_2, from the standpoint of minimizing the maximum overshoot, for $a_1 = 70$ and

TABLE 10-20 Performance Attributes of Sun-Seeker System in Example 10-11 with Lead–Lag Controller: $a_1 = 70$, $T_1 = 0.00004$

a_2	T_2	PM (deg)	M_r	BW (rad/sec)	Maximum Overshoot (%)	t_r (sec)	t_s (sec)
0.1	20	81.81	1.004	122.2	0.4	0.01843	0.02626
0.15	20	76.62	1.002	225.5	0.2	0.00985	0.01515
0.20	20	70.39	1.001	351.4	0.1	0.00668	0.00909
0.25	20	63.87	1.001	443.0	4.9	0.00530	0.00707

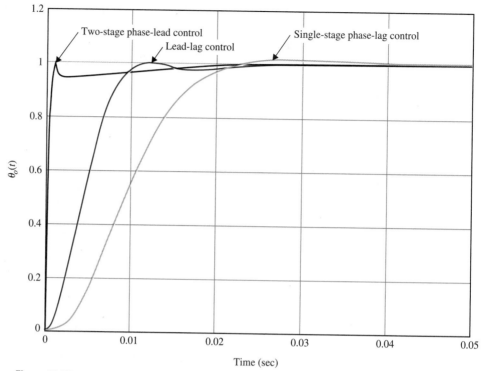

Figure 10-52 Sun-seeker system in Example 10-11 with single-stage phase-lag controller, lead–lag controller, and two-stage phase-lead controller.

$T_2 = 0.00004$, is approximately 0.2. Compared with the single-stage phase-lag control designed in Example 10-9, the BW is increased to 351.4 rad/sec from 66.94 rad/sec, and the rise time is reduced to 0.00668 sec from 0.01273 sec. The system with the lead–lag controller should be more robust, since the magnitude of the sensitivity function should not increase to unity until near the BW of 351.4 rad/sec. As a comparison, the unit-step responses of the system with the two-stage phase-lead control, the single-stage phase-lag control, and the lead–lag control are shown in Fig. 10-52.

It should be noted that the bandwidth and rise time of the sun-seeker system can be further increased and reduced, respectively, by using a larger value of a_1 for the phase-lead portion of the controller. However, the resulting step response will have a large undershoot, although the maximum overshoot can be kept small. ◀

▶ 10-8 POLE-ZERO CANCELLATION DESIGN: NOTCH FILTER

The transfer functions of many controlled processes contain one or more pairs of complex-conjugate poles that are very close to the imaginary axis of the s-plane. These complex poles usually cause the closed-loop system to be lightly damped or unstable. One immediate solution is to use a controller that has a transfer function with zeros selected, which would cancel the undesirable poles of the controlled process, and to place the poles of the controller at more desirable locations in the s-plane to achieve the desired dynamic performance. For example, if the transfer function of a process is

$$G_p(s) = \frac{K}{s(s^2 + s + 10)} \tag{10-140}$$

in which the complex-conjugate poles may cause stability problems in the closed-loop system when the value of K is large, the suggested series controller may be of the form

$$G_c(s) = \frac{s^2 + s + 10}{s^2 + as + b} \tag{10-141}$$

The constants a and b may be selected according to the performance specifications of the closed-loop system.

There are practical difficulties with the pole-zero-cancellation design scheme that should prevent the method from being used indiscriminately. The problem is that in practice *exact* cancellation of poles and zeros of transfer functions is rarely possible. In practice, the transfer function of the process, $G_p(s)$, is usually determined through testing and physical modeling; linearization of a nonlinear process and approximation of a complex process are unavoidable. Thus, the *true* poles and zeros of the transfer function of the process may not be accurately modeled. In fact, the true order of the system may even be higher than that represented by the transfer function used for modeling purposes. Another difficulty is that the dynamic properties of the process may vary, even very slowly, due to aging of the system components or changes in the operating environment, so the poles and zeros of the transfer function may move during the operation of the system. The parameters of the controller are constrained by the actual physical components available, and cannot be assigned arbitrarily. For these and other reasons, even if we could precisely design the poles and zeros of the transfer function of the controller, exact pole-zero cancellation is almost never possible in practice. We will now show that in most cases, exact cancellation is *not* really necessary to effectively negate the influence of the undesirable poles using pole-zero-cancellation compensation schemes.

• In the real world, exact cancellation of poles and zeros is almost never possible.

Let us assume that a controlled process is represented by

$$G_p(s) = \frac{K}{s(s + p_1)(s + \bar{p}_1)} \tag{10-142}$$

where p_1 and $\bar{p}_1$ are the two complex-conjugate poles that are to be canceled. Let the transfer function of the series controller be

$$G_c(s) = \frac{(s + p_1 + \varepsilon_1)(s + \bar{p}_1 + \bar{\varepsilon}_1)}{s^2 + as + b} \tag{10-143}$$

where ε_1 is a complex number whose magnitude is very small, and $\bar{\varepsilon}_1$ is its complex conjugate. The open-loop transfer function of the compensated system is

$$G(s) = G_c(s)G_p(s) = \frac{K(s + p_1 + \varepsilon_1)(s + \bar{p}_1 + \bar{\varepsilon}_1)}{s(s + p_1)(s + \bar{p}_1)(s^2 + as + b)} \tag{10-144}$$

Because of inexact cancellation, we cannot discard the terms $(s + p_1)(s + \bar{p}_1)$ in the denominator of Eq. (10-144). The closed-loop transfer function is

$$\frac{Y(s)}{R(s)} = \frac{K(s + p_1 + \varepsilon_1)(s + \bar{p}_1 + \bar{\varepsilon}_1)}{s(s + p_1)(s + \bar{p}_1)(s^2 + as + b) + K(s + p_1 + \varepsilon_1)(s + \bar{p}_1 + \varepsilon_1)} \tag{10-145}$$

The root-locus diagram in Fig. 10-53 explains the effect of inexact pole-zero cancellation. Notice that the two closed-loop poles as a result of inexact cancellation lie between the pairs of poles and zeros at $s = -p_1, -\bar{p}_1$, and $-p_1 - \varepsilon_1, -\bar{p}_1 - \bar{\varepsilon}_1$, respectively. Thus, these closed-loop poles are very close to the open-loop poles and zeros that are meant to be canceled. Eq. (10-145) can be approximated as

$$\frac{Y(s)}{R(s)} \cong \frac{K(s + p_1 + \varepsilon_1)(s + \bar{p}_1 + \bar{\varepsilon}_2)}{(s + p_1 + \delta_1)(s + \bar{p}_1 + \bar{\delta}_1)(s^3 + as + b + K)} \tag{10-146}$$

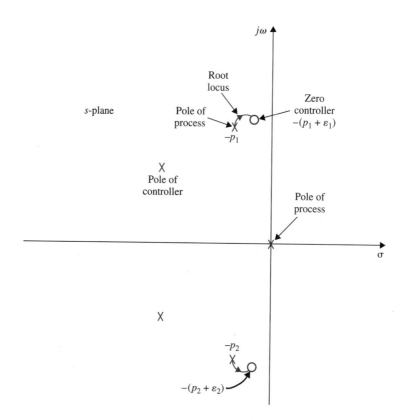

Figure 10-53 Pole-zero configuration and root loci of inexact cancellation.

where δ_1 and $\bar{\delta}_1$ are a pair of very small complex-conjugate numbers that depend on ε_1, $\bar{\varepsilon}_1$, and all the other parameters. The partial-fraction expansion of Eq. (10-146) is

$$\frac{Y(s)}{R(s)} \cong \frac{K_1}{s + p_1 + \delta_1} + \frac{K_2}{s + \bar{p}_1 + \bar{\delta}_1} + \text{terms due to the remaining poles} \quad (10\text{-}147)$$

We can show that K_1 is proportional to $\varepsilon_1 - \delta_1$, which is a very small number. Similarly, K_2 is also very small. *This exercise simply shows that although the poles at $-p_1$ and $-p_2$ cannot be canceled precisely, the resulting transient-response terms due to inexact cancellation will have insignificant amplitudes, so unless the controller zeros earmarked for cancellation are too far off target, the effect can be neglected for all practical purposes.* Another way of viewing this problem is that the zeros of $G(s)$ are retained as the zeros of closed-loop transfer function $Y(s)/R(s)$, so from Eq. (10-146), we see that the two pairs of poles and zeros are close enough to be canceled from the transient-response standpoint.

Keep in mind that we should never attempt to cancel poles that are in the right-half s-plane, since any inexact cancellation will result in an unstable system. Inexact cancellation of poles could cause difficulties if the unwanted poles of the process transfer function are very close to or right on the imaginary axis of the s-plane. In this case, inexact cancellation may also result in an unstable system. Figure 10-54(a) illustrates a situation in which the relative positions of the poles and zeros intended for cancellation result in a stable system, whereas in Fig. 10-54(b), the inexact cancellation is unacceptable. Although the relative distance between the poles and zeros intended for cancellation is small, which results in terms in the time response that have very small amplitudes, these responses will grow without bound as time increases.

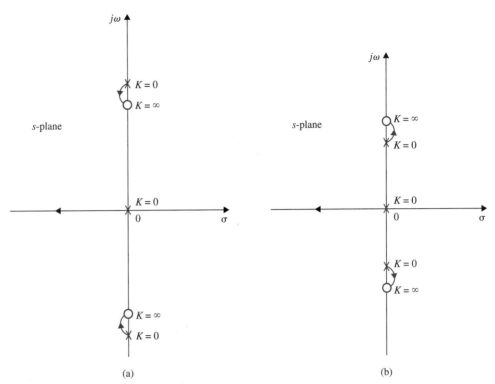

Figure 10-54 Root loci showing the effects of inexact pole-zero cancellations.

10-8-1 Second-Order Active Filter

Transfer functions with complex poles and/or zeros can be realized by electric circuits with op-amps. Consider the transfer function

$$G_c(s) = \frac{E_2(s)}{E_1(s)} = K\frac{s^2 + b_1 s + b_2}{s^2 + a_1 s + a_2} \tag{10-148}$$

where a_1, a_2, b_1, and b_2 are real constants. The active-filter realization of Eq. (10-148) can be accomplished by using the direct decomposition scheme of state variables discussed in Section 5-9. A typical op-amp circuit is shown in Fig. 10-55. The parameters of the transfer function in Eq. (10-148) are related to the circuit parameters as follows:

$$K = -\frac{R_6}{R_7} \tag{10-149}$$

$$a_1 = \frac{1}{R_1 C_1} \tag{10-150}$$

$$a_2 = \frac{1}{R_2 R_4 C_1 C_2} \tag{10-151}$$

$$b_1 = \left(1 - \frac{R_1 R_7}{R_3 R_8}\right)a_1 \qquad (b_1 < a_1) \tag{10-152}$$

$$b_2 = \left(1 - \frac{R_2 R_7}{R_3 R_9}\right)a_2 \qquad (b_2 < a_2) \tag{10-153}$$

Since $b_1 < a_1$, the zeros of $G_c(s)$ in Eq. (10-148) are less damped and are closer to the origin in the s-plane than the poles. By setting various combinations of R_7 and R_8, and R_9

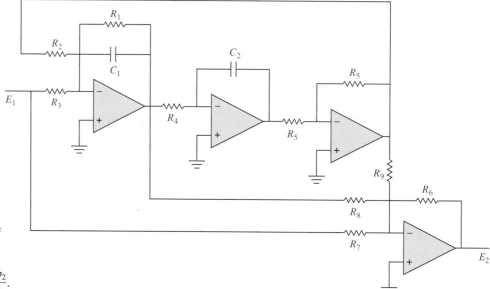

Figure 10-55 Op-amp circuit realization of the second-order transfer function.
$$\frac{E_2(s)}{E_1(s)} = K\frac{s^2 + b_1 s + b_2}{s^2 + a_1 s + a_2}.$$

to infinity, a variety of second-order transfer functions can be realized. Note that all the parameters can be adjusted independently of one another. For example, R_1 can be adjusted to set a_1; R_4 can be adjusted to set a_2; and b_1 and b_2 are set by adjusting R_8 and R_9, respectively. The gain factor K is controlled independently by R_6.

10-8-2 Frequency-Domain Interpretation and Design

While it is simple to grasp the idea of pole-zero-cancellation design in the s-domain, the frequency-domain provides added perspective to the design principles. Figure 10-56 illustrates the Bode plot of the transfer function of a typical second-order controller with complex zeros. The magnitude plot of the controller typically has a "notch" at the resonant frequency ω_n. The phase plot is negative below and positive above the resonant frequency, while passing through zero degrees at the resonant frequency. The attenuation of the magnitude curve and the positive-phase characteristics can be used effectively to improve the stability of a linear system. Because of the "notch" characteristic in the magnitude curve, the controller is also referred to in the industry as a **notch filter** or **notch controller**.

From the frequency-domain standpoint, the notch controller has advantages over the phase-lead and phase-lag controllers in certain design conditions, since the magnitude and phase characteristics do not affect the high- and low-frequency properties of the system. Without using the pole-zero-cancellation principle, the design of the notch controller for compensation in the frequency domain involves the determination of the amount of attenuation required and the resonant frequency of the controller.

Let us express the transfer function of the notch controller in Eq. (10-148) as

$$G_c(s) = \frac{s^2 + 2\zeta_z \omega_n s + \omega_n^2}{s^2 + 2\zeta_p \omega_n s + \omega_n^2} \tag{10-154}$$

where we have made the simplification by assuming that $a_2 = b_2$.

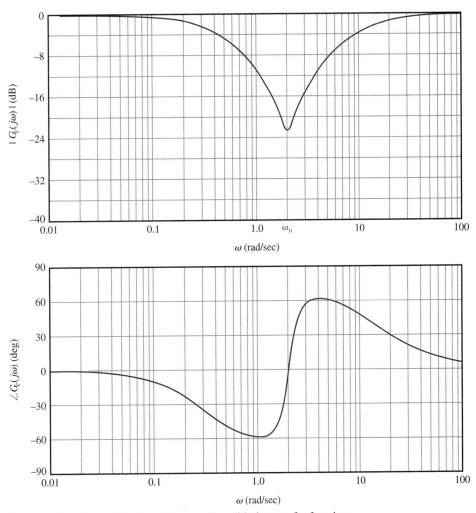

Figure 10-56 Bode plot of a notch controller with the transfer function.

$$G(s) = \frac{(s^2 + 0.8s + 4)}{(s + 0.384)(s + 10.42)}.$$

The attenuation provided by the magnitude of $G_c(j\omega)$ at the resonant frequency ω_n is

$$|G_c(j\omega_n)| = \frac{\zeta_z}{\zeta_p} \tag{10-155}$$

Thus, knowing the maximum attenuation required at ω_n, the ratio of ζ_z/ζ_p is known.

The following example illustrates the design of the notch controller based on pole-zero cancellation and required attenuation at the resonant frequency.

► **EXAMPLE 10-12** Complex-conjugate poles in system transfer functions are often due to compliances in the coupling between mechanical elements. For instance, if the shaft between the motor and load is nonrigid, the shaft is modeled as a torsional spring, which could lead to complex-conjugate poles in the process transfer function. Figure 10-57 shows a speed-control system in which the coupling between the motor and the load is modeled as a torsional spring. The system equations are

$$T_m(t) = J_m \frac{d\omega_m(t)}{dt} + B_m\omega_m(t) + J_L \frac{d\omega_L(t)}{dt} \tag{10-156}$$

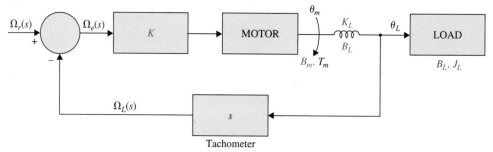

Figure 10-57 Block diagram of speed-control system in Example 10-12.

$$K_L[\theta_m(t) - \theta_L(t)] + B_L[\omega_m(t) - \omega_L(t)] = J_L \frac{d\omega_L(t)}{dt} \tag{10-157}$$

$$T_m(t) = K\omega_e(t) \tag{10-158}$$
$$\omega_e(t) = \omega_r(t) - \omega_L(t) \tag{10-159}$$

$T_m(t)$ = motor torque
$\omega_m(t)$ = motor angular velocity
$\omega_L(t)$ = load angular velocity
$\theta_L(t)$ = load angular displacement
$\theta_m(t)$ = motor angular displacement
J_m = motor inertia = 0.0001 oz-in.-sec^2
J_L = load inertia = 0.0005 oz-in.-sec^2
B_m = viscous-friction coefficient of motor = 0.01 oz-in.-sec
B_L = viscous-friction coefficient of shaft = 0.001 oz-in.-sec
K_L = spring constant of shaft = 100 on-in./rad
K = amplifier gain = 1

The loop transfer function of the system is

$$G_p(s) = \frac{\Omega_L(s)}{\Omega_e(s)} = \frac{B_L s + K_L}{J_m J_L s^3 + (B_m J_L + B_L J_m + B_L J_L)s^2 + (K_L J_L + B_m B_L + K_L J_m)s + B_m K_L} \tag{10-160}$$

By substituting the system parameters in the last equation, $G_p(s)$ becomes

$$G_p(s) = \frac{20,000(s + 100,000)}{s^3 + 112s^2 + 1,200,200s + 20,000,000}$$
$$= \frac{20,000(s + 100,000)}{(s + 16.69)(s + 47.66 + j1094)(s + 47.66 - j1094)} \tag{10-161}$$

Thus, the shaft compliance between the motor and the load creates two complex-conjugate poles in $G_p(s)$ that are lightly damped. The resonant frequency is approximately 1095 rad/sec, and the closed-loop system is unstable. The complex poles of $G_p(s)$ would cause the speed response to oscillate even if the system were stable.

Pole-Zero Cancellation Design with Notch Controller
The following are the performance specifications of the system

1. The steady-state speed of the load due to a unit-step input should have an error of not more than 1 percent.
2. Maximum overshoot of output speed ≤ 5 percent.
3. Rise time $t_r < 0.5$ sec.
4. Settling time $t_s < 0.5$ sec.

To compensate the system, we need to get rid, or, perhaps more realistically, minimize the effect, of the complex poles of $G_p(s)$ at $s = -47.66 + j1094$ and $-47.66 - j1094$. Let us select a notch controller with the transfer function given in Eq. (10-154) to improve the performance of the system. The complex-conjugate zeros of the controller should be so placed that they will cancel the undesirable poles of the process. Therefore, the transfer function of the notch controller should be

$$G_c(s) = \frac{s^2 + 95.3s + 1,198,606.6}{s^2 + 2\zeta_p\omega_n s + \omega_n^2} \tag{10-162}$$

The forward-path transfer function of the compensated system is

$$G(s) = G_c(s)G_p(s) = \frac{20,000(s + 100,000)}{(s + 16.69)(s^2 + 2\zeta_p\omega_n s + \omega_n^2)} \tag{10-163}$$

Since the system is type 0, the step-error constant is

$$K_P = \lim_{s \to 0} G(s) = \frac{2 \times 10^9}{16.69 \times \omega_n^2} = \frac{1.198 \times 10^8}{\omega_n^2} \tag{10-164}$$

For a unit-step input, the steady-state error of the system is written

$$e_{ss} = \lim_{t \to \infty} \omega_e(t) = \lim_{s \to 0} s\,\Omega_e(s) = \frac{1}{1 + K_P} \tag{10-165}$$

Thus, for the steady-state error to be less than or equal to 1 percent, $K_P \geq 99$. The corresponding requirement on ω_n is found from Eq. (10-164),

$$\omega_n \leq 1210 \tag{10-166}$$

We can show that from the stability standpoint, it is better to select a large value for ω_n. Thus, let $\omega_n = 1200$ rad/sec, which is at the high end of the allowable range from the steady-state error standpoint. However, the design specifications given above can only be achieved by using a very large value for ζ_p. For example, when $\zeta_p = 15,000$, the time response has the following performance attributes:

> Maximum overshoot = 3.7 percent
>
> Rise time $t_r = 0.1897$ sec
>
> Settling time $t_s = 0.256$ sec

Although the performance requirements are satisfied, the solution is unrealistic, since the extremely large value for ζ_p cannot be realized by physically available controller components.

Let us choose $\zeta_p = 10$ and $\omega_n = 1000$ rad/sec. The forward-path transfer function of the system with the notch controller is

$$G(s) = G_c(s)G_p(s) = \frac{20,000(s + 100,000)}{(s + 16.69)(s + 50)(s + 19,950)} \tag{10-167}$$

We can show that the system is stable, but the maximum overshoot is 71.6%. Now we can regard the transfer function in Eq. (10-167) as a new design problem. There are a number of possible solutions to the problem of meeting the design specifications given. We can introduce a phase-lag controller or a PI controller, among other possibilities.

Second-Stage Phase-Lag Controller Design

Let us design a phase-lag controller as the second-stage controller for the system. The roots of the characteristic equation of the system with the notch controller are at $s = -19954$, $-31.328 + j316.36$, and $-31.328 - j316.36$. The transfer function of the phase-lag controller is

$$G_{c1}(s) = \frac{1 + aTs}{1 + Ts} \qquad (a < 1) \tag{10-168}$$

where for design purposes we have omitted the gain factor $1/a$ in Eq. (10-168).

TABLE 10-21 Time-Domain Performance Attributes of System in Example 10-12 with Notch-Phase-Lag Controller

a	T	aT	Maximum Overshoot (%)	t_r (sec)	t_s (sec)
0.001	10	0.01	14.8	0.1244	0.3836
0.002	10	0.02	10.0	0.1290	0.3655
0.004	10	0.04	3.2	0.1348	0.1785
0.005	10	0.05	1.0	0.1375	0.1818
0.0055	10	0.055	0.3	0.1386	0.1889
0.006	10	0.06	0	0.1400	0.1948

Let us select $T = 10$ for the phase-lag controller. Table 10-21 gives time-domain performance attributes for various values of a. The best value of a from the overall performance standpoint appears to be 0.005. Thus, the transfer function of the phase-lag controller is

$$G_{c1}(s) = \frac{1 + aTs}{1 + Ts} = \frac{1 + 0.05s}{1 + 10s} \qquad (10\text{-}169)$$

The forward-path transfer function of the compensated system with the notch-phase-lag controller is

$$G(s) = G_c(s)G_{c1}(s)G_p(s) = \frac{20,000(s + 100,000)(1 + 0.05s)}{(s + 16.69)(s + 50)(s + 19,950)(1 + 10s)} \qquad (10\text{-}170)$$

The unit-step response of the system is shown in Fig. 10-58. Since the step-error constant is 120.13, the steady-state speed error due to a step input is 1/120.13, or 0.83%.

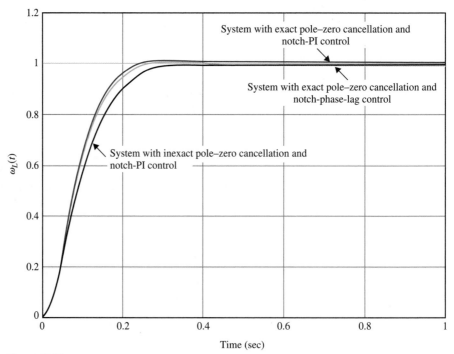

Figure 10-58 Unit-step responses of speed-control system in Example 10-12.

Second-Stage PI Controller Design

A PI controller can be applied to the system to improve the steady-state error and the stability simultaneously. The transfer function of the PI control is written

$$G_{c2}(s) = K_P + \frac{K_I}{s} = K_P\left(\frac{s + K_I/K_P}{s}\right) \tag{10-171}$$

We can design the PI controller based on the phase-lag controller by writing Eq. (10-169) as

$$G_{c1}(s) = 0.005\left(\frac{s + 20}{s + 0.1}\right) \tag{10-172}$$

Thus, we can set $K_P = 0.005$, and $K_I/K_P = 20$. Then, $K_I = 0.1$. Figure 10-58 shows the unit-step response of the system with the notch-PI controller. The attributes of the step response are

$$\% \text{ Maximum overshoot} = 1\%$$

$$\text{Rise time } t_r = 0.1380 \text{ sec.}$$

$$\text{Settling time } t_s = 0.1818 \text{ sec.}$$

which are extremely close to those with the notch-phase-lag controller, except that in the notch-PI case the steady-state velocity error is zero when the input is a step function.

Sensitivity Due to Imperfect Pole-Zero Cancellation

As mentioned earlier, exact cancellation of poles and zeros is almost never possible in real life. Let us consider that the numerator polynomial of the notch controller in Eq. (10-162) cannot be exactly realized by physical resistor and capacitor components. Rather, the transfer function of the notch controller is more realistically chosen as

$$G_c(s) = \frac{s^2 + 100s + 1,000,000}{s^2 + 20,000s + 1,000,000} \tag{10-173}$$

Figure 10-58 shows the unit-step response of the system with the notch controller in Eq. (10-173). The attributes of the unit-step response are

$$\% \text{ Maximum overshoot} = 0.4\%$$

$$\text{Rise time } t_r = 0.17 \text{ sec.}$$

$$\text{Settling time } t_s = 0.2323 \text{ sec.}$$

Frequency-Domain Design

To carry out the design of the notch controller, we refer to the Bode plot of Eq. (10-161) shown in Fig. 10-59. Due to the complex-conjugate poles of $G_p(s)$, the magnitude plot has a peak of 24.86 dB at 1095 rad/sec. From the Bode plot in Fig. 10-59 we see that we may want to bring the magnitude plot down to -20 dB at the resonant frequency of 1095 rad/sec so that the resonance is smoothed out. This requires an attenuation of -44.86 dB. Thus, from Eq. (10-155),

$$|G_c(j\omega_c)| = -44.86 \text{ dB} = \frac{\zeta_z}{\zeta_p} = \frac{0.0435}{\zeta_p} \tag{10-174}$$

where ζ_z is found from Eq. (10-162). Solving for ζ_p from the last equation, we get $\zeta_p = 7.612$. The attenuation should be placed at the resonant frequency of 1095 rad/sec; thus, $\omega_n = 1095$ rad/sec. The notch controller of Eq. (10-162) becomes

$$G_c(s) = \frac{s^2 + 95.3s + 1,198,606.6}{s^2 + 16,670.28s + 1,199,025} \tag{10-175}$$

The Bode plot of the system with the notch controller in Eq. (10-175) is shown in Fig. 10-59. We can see that the system with the notch controller has a phase margin of only 13.7°, and M_r is 3.92.

To complete the design, we can use a PI controller as a second-stage controller. Following the guideline given in Section 10-3 on the design of a PI controller, we assume that the desired phase margin is 80°. From the Bode plot in Fig. 10-59, we see that to realize a phase margin of 80°, the

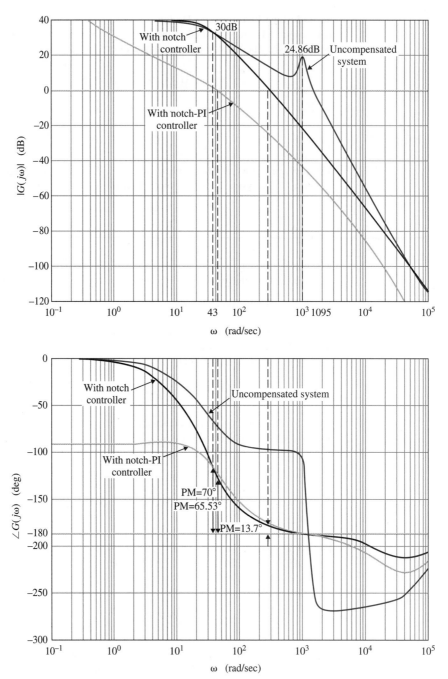

Figure 10-59 Bode plots of uncompensated speed-control system in Example 10-12, with notch controller and with notch-PI controller.

$$G_c(s) = \frac{s^2 + 95.3s + 1,198,606.6}{s^2 + 16,670.28s + 1,199,025} \qquad G_{c2}(s) = 0.0316 + \frac{0.35}{s}.$$

new gain crossover frequency should be $\omega'_g = 43$ rad/sec, and the magnitude of $G(j\omega'_g)$ is 30 dB. Thus, from Eq. (10-32),

$$K_P = 10^{-|G(j\omega'_g)|_{dB}/20} = 10^{-30/20} = 0.0316 \qquad (10\text{-}176)$$

TABLE 10-22 Performance Attributes of System in Example 10-12 with Notch-PI Controller Designed in Frequency Domain.

K_P	K_I	PM (deg)	M_r	Maximum Overshoot (%)	t_r (sec)	t_s (sec)
0.0316	0.1	76.71	1.00	0	0.2986	0.5758
0.0316	0.135	75.15	1.00	0	0.2036	0.4061
0.0316	0.200	72.22	1.00	0	0.0430	0.2403
0.0316	0.300	67.74	1.00	0	0.0350	0.1361
0.0316	0.350	65.53	1.00	1.6	0.0337	0.0401
0.0316	0.400	63.36	1.00	4.3	0.0323	0.0398

The value of K_I is determined using the guideline given by Eq. (10-25),

$$K_I = \frac{\omega_g' K_P}{10} = \frac{43 \times 0.0316}{10} = 0.135 \tag{10-177}$$

Since the original system is type 0, the final design needs to be refined by adjusting the value of K_I. Table 10-22 gives the performance attributes when $K_P = 0.0316$ and K_I is varied from 0.135. From the best maximum overshoot, rise time, and settling time measures, the best value of K_I appears to be 0.35. The forward-path transfer function of the compensated system with the notch-PI controller is

$$G(s) = \frac{20,000(s + 100,000)(0.0316s + 0.35)}{s(s + 16.69)(s^2 + 16,670.28s + 1,199,025)} \tag{10-178}$$

Figure 10-59 shows the Bode plot of the system with the notch-PI controller, with $K_P = 0.0316$ and $K_I = 0.35$. The unit-step responses of the compensated system with $K_P = 0.0316$ and $K_I = 0.135$, 0.35, and 0.40, are shown in Fig. 10-60.

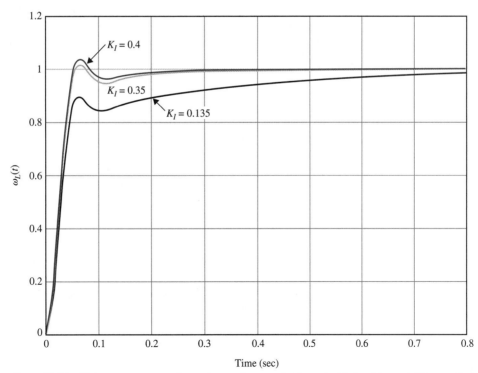

Figure 10-60 Unit-step responses of speed-control system in Example 10-12 with notch-PI controller.

$$G_c(s) = \frac{s^2 + 95.3s + 1,198,606.6}{s^2 + 16,670.28s + 1,199,025} \qquad G_{c2}(s) = 0.0316 + \frac{0.35}{s}.$$

◀

▷ 10-9 FORWARD AND FEEDFORWARD CONTROLLERS

The compensation schemes discussed in the preceding sections all have one degree of freedom in that there is essentially one controller in the system, although the controller can contain several stages connected in series or in parallel. The limitations of a one-degree-of-freedom controller were discussed in Section 10-1. The two-degree-of-freedom compensation scheme shown in Fig. 10-2(d) through Fig. 10-2(f) offer design flexibility when a multiple number of design criteria have to be satisfied simultaneously.

From Fig. 10-1(e), the closed-loop transfer function of the system is

$$\frac{Y(s)}{R(s)} = \frac{G_{cf}(s)G_c(s)G_p(s)}{1 + G_c(s)G_p(s)} \tag{10-179}$$

and the error transfer function is

$$\frac{E(s)}{R(s)} = \frac{1}{1 + G_c(s)G_p(s)} \tag{10-180}$$

Thus, the controller $G_c(s)$ can be designed so that the error transfer function will have certain desirable characteristics, and the controller $G_{cf}(s)$ can be selected to satisfy performance requirements with reference to the input-output relationship. Another way of describing the flexibility of a two-degree-of-freedom design is that the controller $G_c(s)$ is usually designed to provide a certain degree of system stability and performance, but since the zeros of $G_c(s)$ always become the zeros of the closed-loop transfer function, unless some of the zeros are canceled by the poles of the process transfer function, $G_p(s)$, these zeros may cause a large overshoot in the system output even when the relative damping as determined by the characteristic equation is satisfactory. In this case and for other reasons, the transfer function $G_{cf}(s)$ may be used for the control or cancellation of the undesirable zeros of the closed-loop transfer function, while keeping the characteristic equation intact. Of course, we can also introduce zeros in $G_{cf}(s)$ to cancel some of the undesirable poles of the closed-loop transfer function that could not be otherwise affected by the controller $G_c(s)$. The feedforward compensation scheme shown in Fig. 10-2(f) serves the same purpose as the forward compensation, and the difference between the two configurations depends on system and hardware implementation considerations.

It should be kept in mind that while the forward and feedforward compensations may seem powerful because they can be used directly for the addition or deletion of poles and zeros of the closed-loop transfer function, there is a fundamental question involving the basic characteristics of feedback. If the forward or feedforward controller is so powerful, then why do we need feedback at all? Since $G_{cf}(s)$ in the systems of Figs. 10-2(e) and 10-2(f) are outside the feedback loop, the system is susceptible to parameter variations in $G_{cf}(s)$. Therefore, in reality, these types of compensation cannot be satisfactorily applied to all situations.

▷ **EXAMPLE 10-13** As an illustration of the design of the forward and feedforward compensators, consider the second-order sun-seeker system with phase-lag control designed in Example 10-9. One of the disadvantages of phase-lag compensation is that the rise time is usually quite long. Let us consider that the phase-lag-compensated sun-seeker system has the forward-path transfer function,

$$G(s) = G_c(s)G_p(s) = \frac{2500(1 + 10s)}{s(s + 25)(1 + 100s)} \tag{10-181}$$

The time-response attributes are

$$\text{Maximum overshoot} = 2.5\% \qquad t_r = 0.1637 \text{ sec} \qquad t_s = 0.2020 \text{ sec}$$

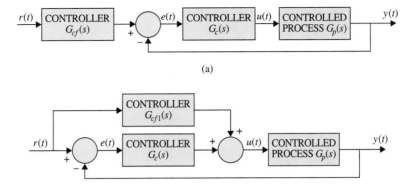

Figure 10-61 (a) Forward compensation with series compensation. (b) Feedforward compensation with series compensations.

We can improve the rise time and the settling time while not appreciably increasing the overshoot by adding a PD controller $G_{cf}(s)$ to the system, as shown in Fig. 10-61(a). This effectively adds a zero to the closed-loop transfer function, while not affecting the characteristic equation. Selecting the PD controller as

$$G_{cf}(s) = 1 + 0.05s \qquad (10\text{-}182)$$

the time-domain performance attributes are

$$\text{Maximum overshoot} = 4.3\% \quad t_r = 0.1069 \text{ sec} \quad t_s = 0.1313 \text{ sec}$$

If instead, the feedforward configuration of Fig. 10-61(b) is chosen, the transfer function of $G_{cf1}(s)$ is directly related to $G_{cf}(s)$; that is, equating the closed-loop transfer functions of the two systems in Figs. 10-62(a) and 10-62(b), we have

$$\frac{[G_{cf1}(s) + G_c(s)]G_p(s)}{1 + G_c(s)G_p(s)} = \frac{G_{cf}G_c(s)G_p(s)}{1 + G_c(s)G_p(s)} \qquad (10\text{-}183)$$

Solving for $G_{cf1}(s)$ from Eq. (10-183) yields

$$G_{cf1}(s) = [G_{cf}(s) - 1]G_c(s) \qquad (10\text{-}184)$$

Thus, with $G_{cf}(s)$ as given in Eq. (10-179), we have the transfer function of the feedforward controller,

$$G_{cf1}(s) = 0.05s\left(\frac{1 + 10s}{1 + 100s}\right) \qquad (10\text{-}185)$$

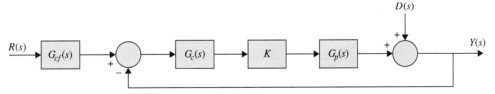

Figure 10-62 Control system with disturbance.

◄

► 10-10 DESIGN OF ROBUST CONTROL SYSTEMS

In many control-system applications, not only must the system satisfy the damping and accuracy specifications, but the control must also yield performance that is **robust** (insensitive) to external disturbance and parameter variations. We have shown that feedback in conventional control systems has the inherent ability to reduce the effects of external disturbance and parameter variations. Unfortunately, robustness with the conventional feedback configuration is achieved only with a high loop gain, which is normally detrimental to stability. Let us consider the control system shown in Fig. 10-62. The external disturbance is denoted by the signal $d(t)$, and we assume that the amplifier gain K is subject to variation during operation. The input-output transfer function of the system when $d(t) = 0$ is

$$M(s) = \frac{Y(s)}{R(s)} = \frac{KG_{cf}(s)G_c(s)G_p(s)}{1 + KG_c(s)G_p(s)} \tag{10-186}$$

and the disturbance-output transfer function when $r(t) = 0$ is

$$T(s) = \frac{Y(s)}{D(s)} = \frac{1}{1 + KG_c(s)G_p(s)} \tag{10-187}$$

In general, the design strategy is to select the controller $G_c(s)$ so that the output $y(t)$ is insensitive to the disturbance over the frequency range in which the latter is dominant, and to design the feedforward controller $G_{cf}(s)$ to achieve the desired transfer function between the input $r(t)$ and the output $y(t)$.

Let us define the sensitivity of $M(s)$ due to the variation of K as

$$S_K^M = \frac{\text{percent change in } M(s)}{\text{percent change in } K} = \frac{dM(s)/M(s)}{dK/K} \tag{10-188}$$

Then, for the system in Fig. 10-62,

$$S_K^M = \frac{1}{1 + KG_c(s)G_p(s)} \tag{10-189}$$

which is identical to Eq. (10-187). Thus, the sensitivity function and the disturbance-output transfer function are identical, which means that disturbance suppression and robustness with respect to variations of K can be designed with the same control schemes.

The following example shows how the two-degree-of-freedom control system of Fig. 10-62 can be used to achieve a high-gain system that will satisfy the performance and robustness requirements, as well as noise rejection.

▶ **EXAMPLE 10-14** Let us consider the second-order sun-seeker system in Example 10-9, which is compensated with phase-lag control. The forward-path transfer function is

$$G_p(s) = \frac{2500K}{s(s + 25)} \tag{10-190}$$

where $K = 1$. The forward-path transfer function of the phase-lag-compensated system with $a = 0.1$ and $T = 100$ is

$$G(s) = G_c(s)G_p(s) = \frac{2500K(1 + 10s)}{s(s + 25)(1 + 100s)} \quad (K = 1) \tag{10-191}$$

Since the phase-lag controller is a low-pass filter, the sensitivity of the closed-loop transfer function $M(s)$ with respect to K is poor. The bandwidth of the system is only 13.97 rad/sec, but it is expected that $|S_K^M(j\omega)|$ will be greater than unity at frequencies beyond 13.97 rad/sec. Figure 10-63 shows the unit-step responses of the system when $K = 1$, the nominal value, and $K = 0.5$ and 2.0. Notice that if for some reason, the forward gain K is changed from its nominal value, the system response of the phase-lag-compensated system would vary substantially. The attributes of the step responses and the characteristic equation roots are shown in Table 10-23 for the three values of K.

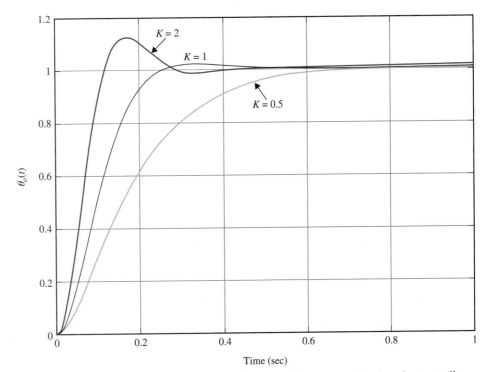

Figure 10-63 Unit-step responses of second-order sun-seeker system with phase-lag controller.

$$G(s) = \frac{2500(1 + 10s)}{s(s + 25)(1 + 100s)}.$$

Figure 10-64 shows the root loci of the system with the phase-lag controller. The two complex roots of the characteristic equation vary substantially as K varies from 0.5 to 2.0.

The design strategy of the robust controller is to place two zeros of the controller near the desired closed-loop poles, which according to the phase-lag-compensated system are at $s = -12.455 \pm j9.624$. Thus, we let the controller transfer function be

$$G_c(s) = \frac{(s + 13 + j10)(s + 13 - j10)}{269} = \frac{(s^2 + 26s + 269)}{269} \qquad (10\text{-}192)$$

The forward-path transfer function of the system with the robust controller is

$$G(s) = \frac{9.2937K(s^2 + 26s + 269)}{s(s + 25)} \qquad (10\text{-}193)$$

Figure 10-65 shows the root loci of the system with the robust controller. By placing the two zeros of $G_c(s)$ near the desired characteristic equation roots, the sensitivity of the system is greatly

TABLE 10-23 Attributes of Unit-Step Response of Second-Order Sun-Seeker System with Phase-Lag Controller in Example 10-14.

K	Maximum Overshoot (%)	t_r (sec)	t_s (sec)	Roots of Characteristic Equation		
2.0	12.6	0.07854	0.2323	-0.1005	$-12.4548 \pm j18.51$	
1.0	2.6	0.1519	0.2020	-0.1009	$-12.4545 \pm j9.624$	
0.5	1.5	0.3383	0.4646	-0.1019	-6.7628	-18.1454

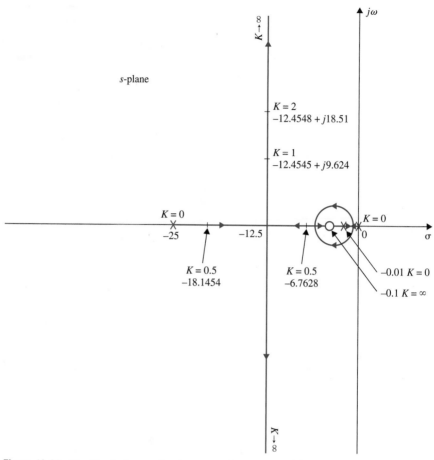

Figure 10-64 Root loci of second-order sun-seeker system with phase-lag controller.

$$G(s) = \frac{2500(1 + 10s)}{s(s + 25)(1 + 100s)}.$$

improved. In fact, the root sensitivity near the two complex zeros at which the root loci terminate is very low. Figure 10-65 shows that when K approaches infinity, the two characteristic equation roots approach $-13 \pm j10$.

Since the zeros of the forward-path transfer function are identical to the zeros of the closed-loop transfer function, the design is not complete by using only the series controller $G_c(s)$, since the closed-loop zeros will essentially cancel the closed-loop poles. This means that we must add the forward controller, as shown in Fig. 10-62, where $G_{cf}(s)$ should contain poles to cancel the zeros of $s^2 + 26s + 269$ of the closed-loop transfer function. Thus, the transfer function of the forward controller is

$$G_{cf}(s) = \frac{269}{s^2 + 26s + 269} \tag{10-194}$$

The block diagram of the overall system is shown in Fig. 10-66. The closed-loop transfer function of the compensated system with $K = 1$ is

$$\frac{\Theta_o(s)}{\Theta_r(s)} = \frac{242.88}{s^2 + 25.903s + 242.88} \tag{10-195}$$

The unit-step responses of the system for $K = 0.5$, 1.0, and 2.0 are shown in Fig. 10-67, and their attributes are given in Table 10-24. As shown, the system is now very insensitive to the variation of K.

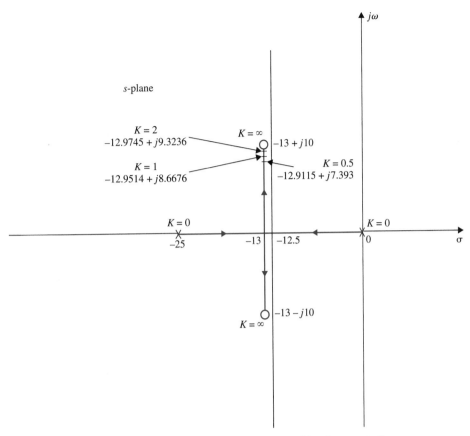

Figure 10-65 Root loci of second-order sun-seeker system with robust controller.

$$G(s) = \frac{9.2937K(s^2 + 26s + 269)}{s(s + 25)}.$$

Since the system in Fig. 10-66 is now more robust, it is expected that the disturbance effect will be reduced. However, we cannot evaluate the effect of the controllers in the system of Fig. 10-66 by applying a unit-step function as $d(t)$. The true improvement on the noise rejection properties is more appropriately analyzed by investigating the frequency response of $\Theta_o(s)/D(s)$. The noise-to-output transfer function is, written from Fig. 10-66,

$$\frac{\Theta_o(s)}{D(s)} = \frac{1}{1 + G_c(s)G_p(s)} = \frac{s(s + 25)}{10.2937s^2 + 266.636s + 2500} \quad (10\text{-}196)$$

The amplitude Bode plot of Eq. (10-196) is shown in Fig. 10-68, along with those of the uncompensated system and the system with phase-lag control. Notice that the magnitude of the frequency response between $D(s)$ and $\Theta_o(s)$ is much smaller than those of the system without compensation

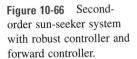

Figure 10-66 Second-order sun-seeker system with robust controller and forward controller.

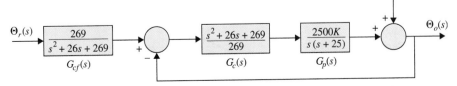

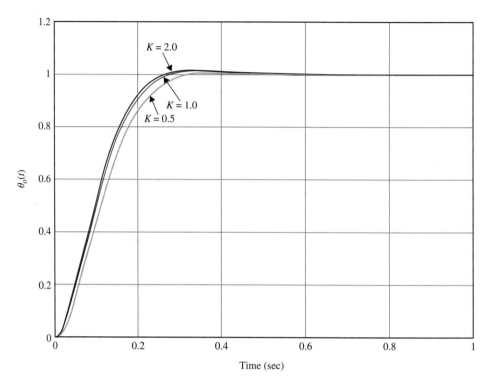

Figure 10-67 Unit-step responses of second-order sun-seeker system with robust controller and forward controller.

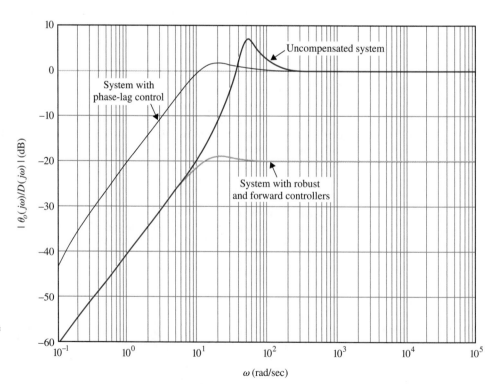

Figure 10-68 Amplitude Bode plot of response due to noise of second-order sun-seeker system.

TABLE 10-24 Attributes of Unit-Step Response of Second-Order Sun-Seeker System with Robust Controller in Example 10-14.

K	Maximum Overshoot (%)	t_r (sec)	t_s (sec)	Roots of Characteristic Equation
2.0	1.3	0.1576	0.2121	$-12.9745 \pm j9.3236$
1.0	0.9	0.1664	0.2222	$-12.9514 \pm j8.6676$
0.5	0.5	0.1846	0.2525	$-12.9115 \pm j7.3930$

and with phase-lag control. The phase-lag control also accentuates the noise for frequencies up to approximately 40 rad/sec, adding more stability to the system. ◄

► **EXAMPLE 10-15** In this example, a robust controller with forward compensation is designed for the third-order sun-seeker system in Example 10-10 with phase-lag control. The forward-path transfer function of the uncompensated system is

$$G_p(s) = \frac{156,250,000K}{s(s^2 + 625s + 156,250)} \tag{10-197}$$

where $K = 1$. The root loci of the closed-loop system are shown in Fig. 10-50, which lead to the phase-lag control with results shown in Table 10-19. Let us select the parameters of the phase-lag controller as $a = 0.1$ and $T = 20$. The dominant roots of the characteristic equation are $s = -187.73 \pm j164.93$.

Let us place the two zeros of the second-order robust controller at $-180 \pm j166.13$, so that the controller transfer function is

$$G_c(s) = \frac{s^2 + 360s + 60,000}{60,000} \tag{10-198}$$

To ease the high-frequency realization problem of the controller, we may add two nondominant poles to $G_c(s)$. The following analysis is carried out with $G_c(s)$ given in Eq. (10-198), however. The root loci of the compensated system are shown in Fig. 10-69. Thus, by placing the zeros of the controller very close to the desired dominant roots, the system is very insensitive to changing value of K near and beyond the nominal value of K. The forward controller has the transfer function

$$G_{cf}(s) = \frac{60,000}{s^2 + 360s + 60,000} \tag{10-199}$$

The attributes of the unit-step response for $K = 0.5$, 1.0, 2.0, and 10.0 and the corresponding characteristic equation roots are given in Table 10-25.

TABLE 10-25 Attributes of Unit-Step Response and Characteristic Equation Roots of Third-Order Sun-Seeker System with Robust and Forward Controllers in Example 10-15.

K	Maximum Overshoot (%)	t_r (sec)	t_s (sec)	Characteristic Equation Roots	
0.5	1.0	0.01115	0.01616	-1558.1	$-184.5 \pm j126.9$
1.0	2.1	0.01023	0.01414	-2866.6	$-181.3 \pm j147.1$
2.0	2.7	0.00966	0.01313	-5472.6	$-180.4 \pm j156.8$
10.0	3.2	0.00924	0.01263	-26307	$-180.0 \pm j164.0$

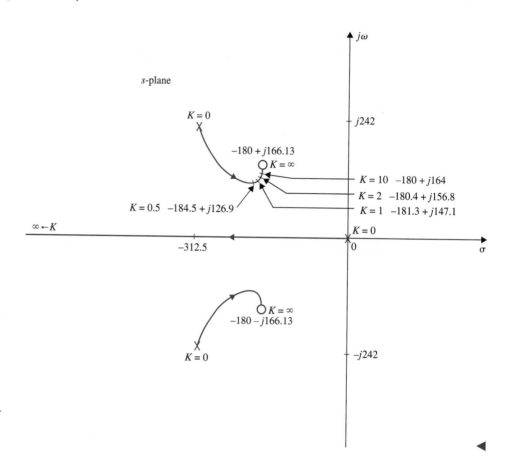

Figure 10-69 Root loci of third-order sun-seeker system with robust and forward controllers.

▶ **EXAMPLE 10-16** In this example, we consider the design of a position-control system that has a variable load inertia. This type of situation is quite common in control systems. For example, the load inertia seen by the motor in an electronic printer will change when different printwheels are used. The system should have satisfactory performance for all the printwheels intended to be used with the system.

To illustrate the design of a robust system that is insensitive to the variation of load inertia, consider that the forward-path transfer function of a unity-feedback control system is

$$G_p(s) = \frac{KK_i}{s[(Js + B)(Ls + B) + K_iK_b]} \tag{10-200}$$

The system parameters are

K_i = motor torque constant = 1 N-m/A

K_b = motor back-emf constant = 1 V/rad/sec

R = motor resistance = 1 Ω

L = motor inductance = 0.01 H

B = motor and load viscous-friction coefficient ≅ 0

J = motor and load inertia, varies between 0.01 and 0.02 N-m/rad/sec²

K = amplifier gain

Substituting these system parameters into Eq. (10-200), we get

For $J = 0.01$
$$G_p(s) = \frac{10,000K}{s(s^2 + 100s + 10,000)} \tag{10-201}$$

For $J = 0.02$ $$G_p(s) = \frac{5000K}{s(s^2 + 100s + 5000)}$$ (10-202)

The performance specifications are

Ramp error constant $K_v \geq 200$

Maximum overshoot $\leq 5\%$ or as small as possible

Rise time $t_r \leq 0.05$ sec

Settling time $t_s \leq 0.05$ sec

These specifications are to be maintained for $0.01 \leq J \leq 0.02$.

To satisfy the ramp-error constant requirement, the value of K must be at least 200. Figure 10-70 shows the root loci of the uncompensated system for $J = 0.01$ and $J = 0.02$. We see that regardless of the value of J, the uncompensated system is unstable for $K > 100$.

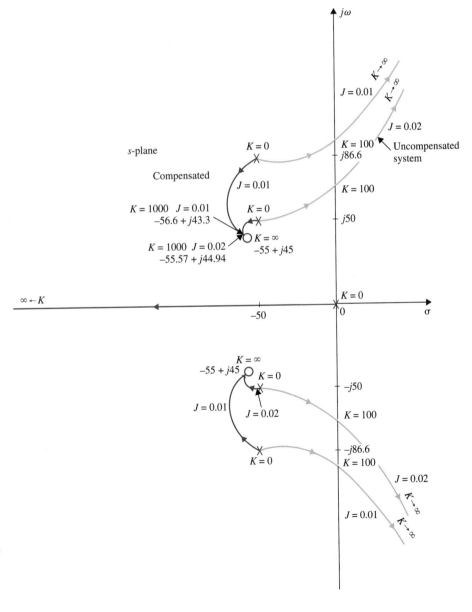

Figure 10-70 Root loci of position-control system in Example 10-16 with robust and forward controllers.

TABLE 10-26 **Attributes of Unit-Step Response and Characteristic Equation Roots of System with Robust and Forward Controllers in Example 10-16.**

J N-m/rad/sec^2	Maximum Overshoot (%)	t_r (sec)	t_s (sec)	Roots of Characteristic Equation	
0.01	1.6	0.03453	0.04444	-1967	$-56.60 \pm j43.3$
0.02	2.0	0.03357	0.04444	-978.96	$-55.57 \pm j44.94$

To achieve robust control, let us choose the system configuration of Fig. 10-61(a). We introduce a second-order series controller with the zeros placed near the desired dominant characteristic equation of the compensated system. The zeros should be so placed that the dominant characteristic equation roots would be insensitive to the variation in J. This is done by placing the two zeros at $-55 \pm j45$, although the exact location is unimportant. By choosing the two controller zeros as designated, the root loci of the compensated system show that the two complex roots of the characteristic equation will be very close to these zeros for various J, especially when the value of K is large. The transfer function of the robust controller is

$$G_c(s) = \frac{(s^2 + 110s + 5050)}{5050} \tag{10-203}$$

As in the last example, we may add two nondominant poles to $G_c(s)$ to ease the high-frequency realization problem of the controller. The analysis is carried out with $G_c(s)$ given in Eq. (10-203).

Let $K = 1000$, although 200 would have been adequate to satisfy the K_v requirement. Then, for $J = 0.01$, the forward-path transfer function of the compensated system is

$$G(s) = G_c(s)G_p(s) = \frac{1980.198(s^2 + 110s + 5050)}{s(s^2 + 100s + 10{,}000)} \tag{10-204}$$

and for $J = 0.02$,

$$G(s) = \frac{990.99(s^2 + 110s + 5050)}{s(s^2 + 100s + 5000)} \tag{10-205}$$

To cancel the two zeros of the closed-loop transfer function, the transfer function of the forward controller is

$$G_{cf}(s) = \frac{5050}{s^2 + 110s + 5050} \tag{10-206}$$

The attributes of the unit-step response and the characteristic equation roots of the compensated system with $K = 1000$, $J = 0.01$, and $J = 0.02$ are given in Table 10-26. ◀

▶ 10-11 MINOR-LOOP FEEDBACK CONTROL

The control schemes discussed in the preceding sections have all utilized series controllers in the forward path of the main loop or feedforward path of the control system. Although series controllers are the most common because of their simplicity in implementation, depending on the nature of the system, sometimes there are advantages in placing the controller in a minor feedback loop, as shown in Fig. 10-2(b). For example, a tachometer may be coupled directly to a dc motor not only for the purpose of speed indication, but more often for improving the stability of the closed-loop system by feeding back the output signal of the tachometer. The motor speed can also be generated by processing the back emf of the motor electronically. In principle, the PID controller or phase-lead and phase-lag controllers can all, with varying degree of effectiveness, be applied as minor-loop feedback controllers. Under certain conditions, minor-loop control can yield systems that are more robust, that is, less sensitive to external disturbance or internal parameter variations.

10-11-1 Rate-Feedback or Tachometer-Feedback Control

The principle of using the derivative of the actuating signal to improve the damping of a closed-loop system can be applied to the output signal to achieve a similar effect. In other words, the derivative of the input signal is fed back and added algebraically to the actuating signal of the system. In practice, if the output variable is mechanical displacement, a tachometer may be used to convert mechanical displacement into an electrical signal that is proportional to the derivative of the displacement. Figure 10-71 shows the block diagram of a control system with a secondary path that feeds back the derivative of the output. The transfer function of the tachometer is denoted by $K_t s$, where K_t is the tachometer constant, usually expressed in volts per radian per second for analytical purposes. Commercially, K_t is given in the data sheet of the tachometer, typically in volts per 1000 rpm. The effects of rate or tachometer feedback can be illustrated by applying it to a second-order prototype system. Consider that the controlled process of the system shown in Fig. 10-71 has the transfer function

$$G_p(s) = \frac{\omega_n^2}{s(s + 2\zeta\omega_n)} \tag{10-207}$$

The closed-loop transfer function of the system is

$$\frac{Y(s)}{R(s)} = \frac{\omega_n^2}{s^2 + (2\zeta\omega_n + K_t\omega_n^2)s + \omega_n^2} \tag{10-208}$$

and the characteristic equation is

$$s^2 + (2\zeta\omega_n + K_t\omega_n^2)s + \omega_n^2 = 0 \tag{10-209}$$

From the characteristic equation, *it is apparent that the effect of the tachometer feedback is the increase of the damping of the system, since K_t appears in the same term as the damping ratio ζ.*

In this respect, tachometer-feedback control has exactly the same effect as the PD control. However, the closed-loop transfer function of the system with PD control in Fig. 10-3 is

$$\frac{Y(s)}{R(s)} = \frac{\omega_n^2(K_P + K_D s)}{s^2 + (2\zeta\omega + K_D\omega_n^2)s + \omega_n^2 K_P} \tag{10-210}$$

Comparing the two transfer functions in Eqs. (10-208) and (10-210), we see that the two characteristic equations are identical if $K_P = 1$ and $K_D = K_t$. However, Eq. (10-210) has a zero at $s = -K_P/K_D$, whereas Eq. (10-208) does not. Thus, the response of the system with tachometer feedback is uniquely defined by the characteristic equation, whereas the response of the system with the PD control also depends on the zero at $s = -K_P/K_D$, which could have a significant effect on the overshoot of the step response.

With reference to the steady-state analysis, the forward-path transfer function of the system with tachometer feedback is

$$\frac{Y(s)}{E(s)} = \frac{\omega_n^2}{s(s + 2\zeta\omega_n + K_t\omega_n^2)} \tag{10-211}$$

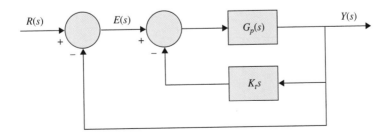

Figure 10-71 Control system with tachometer feedback.

• For a type 1 system, tachometer feedback decreases the ramp-error constant, but does not affect the step-error constant.

Since the system is still type 1, the basic characteristics of the steady-state error are not altered by the tachometer feedback; that is, when the input is a step function, the steady-state error is zero. For a unit-ramp function input, the steady-state error of the system is $(2\zeta + K_t\omega_n)/\omega_n$, whereas that of the system with PD control in Fig. 10-3 is $2\zeta/\omega_n$. Thus, *for a type 1 system, tachometer feedback decreases the ramp-error constant K_v but does not affect the step-error constant K_P.*

10-11-2 Minor-Loop Feedback Control with Active Filter

Instead of using a tachometer, an active filter with RC elements and op-amps can be used to reduce cost, and save space in the minor feedback loop for compensation. We illustrate this approach with the following example.

► **EXAMPLE 10-17** Consider that for the second-order sun-seeker system in Example 10-9, instead of using a series controller in the forward path, we adopt the minor-loop feedback control, as shown in Fig. 10-72(a), with

$$G_p(s) = \frac{2500}{s(s + 25)} \tag{10-212}$$

and

$$H(s) = \frac{K_t s}{1 + Ts} \tag{10-213}$$

To maintain the system as type 1, it is necessary that $H(s)$ contain a zero at $s = 0$. Equation (10-213) can be realized by the op-amp circuit shown in Fig. 10-72(b). This circuit cannot be applied as a series controller in the forward path, since it acts as an open circuit in the steady state when the frequency is zero. As a minor-loop controller, the zero-transmission property to dc signals does not pose any problem.

The forward-path transfer function of the system in Fig. 10-72(a) is

$$
\begin{aligned}
\frac{\Theta_o(s)}{\Theta_e(s)} = G(s) &= \frac{G_p(s)}{1 + G_p(s)H(s)} \\
&= \frac{2500(1 + Ts)}{s[(s + 25)(1 + Ts) + 2500K_t]}
\end{aligned}
\tag{10-214}
$$

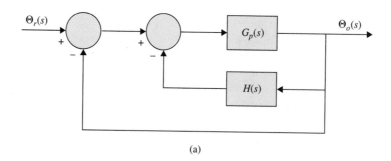

(a)

Figure 10-72 (a) Sun-seeker control system with minor-loop control. (b) Op-amp circuit realization of $\dfrac{K_t s}{1 + Ts}$.

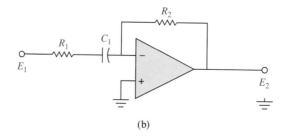

(b)

The characteristic equation of the system is

$$Ts^3 + (25T + 1)s^2 + (25 + 2500T + 2500K_t)s + 2500 = 0 \qquad (10\text{-}215)$$

To show the effects of the parameters K_t and T, we construct the root contours of Eq. (10-215) by first considering that K_t is fixed and T is variable. Dividing both sides of Eq. (10-215) by the terms that do not contain T, we get

$$1 + \frac{Ts(s^2 + 25s + 2500)}{s^2 + (25 + 2500K_t)s + 5000} = 0 \qquad (10\text{-}216)$$

When the value of K_t is relatively large, the two poles of the last equation are real with one very close to the origin. It is more effective to choose K_t so that the poles of Eq. (10-216) are complex.

Figure 10-73 shows the root contours of Eq. (10-215) with $K_t = 0.02$, and T varies from 0 to ∞.

When $T = 0.006$, the characteristic equation roots are at -56.72, $-67.47 + j52.85$, and $-67.47 - j52.85$. The attributes of the unit-step response are

$$\text{Maximum overshoot} = 0 \qquad t_r = 0.04485 \text{ sec} \qquad t_s = 0.06061 \text{ sec} \qquad t_{max} = 0.4 \text{ sec}$$

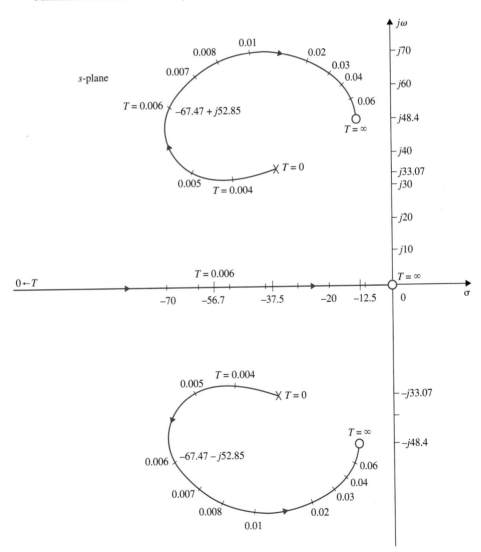

Figure 10-73 Root contours of $Ts^3 + (25T + 1)s^2 + (25 + 2500K_t + 2500T)s + 2500 = 0$, $K_t = 0.02$.

The ramp-error constant of the system is

$$K_v = \lim_{s \to 0} sG(s) = \frac{100}{1 + 100K_t} \tag{10-217}$$

Thus, just as with tachometer feedback, the minor-loop feedback controller of Eq. (10-213) reduces the ramp-error constant K_v, although the system is still type 1. ◄

► 10-12 STATE-FEEDBACK CONTROL

A majority of the design techniques in modern control theory is based on the state-feedback configuration. That is, instead of using controllers with fixed configurations in the forward or feedback path, control is achieved by feeding back the state variables through real constant gains. The block diagram of a system with state-feedback control is shown in Fig. 10-2(c). A more detailed block diagram is shown in Fig. 10-74.

We can show that the PID control and the tachometer-feedback control discussed earlier are all special cases of the state-feedback control scheme. In the case of tachometer-feedback control, let us consider the second-order prototype system described in Eq. (10-207). The process is decomposed by direct decomposition and is represented by the state diagram of Fig. 10-75(a). If the states $x_1(t)$ and $x_2(t)$ are physically accessible, these variables may be fed back through constant real gains $-k_1$ and $-k_2$, respectively, to form the control $u(t)$, as shown in Fig. 10-75(b). The transfer function of the system with state feedback is

$$\frac{Y(s)}{R(s)} = \frac{\omega_n^2}{s^2 + (2\zeta\omega_n + K_2)s + K_1} \tag{10-218}$$

For comparison purposes, we display the transfer functions of the systems with tachometer feedback, Eq. (10-208), and with PD control in Eq. (10-210) as follows:

Tachometer feedback: $\quad \dfrac{Y(s)}{R(s)} = \dfrac{\omega_n^2}{s^2 + (2\zeta\omega_n + K_t\omega_n^2)s + \omega_n^2} \tag{10-219}$

PD control: $\quad \dfrac{Y(s)}{R(s)} = \dfrac{\omega_n^2(K_P + K_D s)}{s^2 + (2\zeta\omega + K_D\omega_n^2)s + \omega_n^2 K_P} \tag{10-220}$

Thus, tachometer feedback is equivalent to state feedback if $k_1 = \omega_n^2$ and $k_2 = K_t\omega_n^2$. Comparing Eq. (10-218) with Eq. (10-220), we see that the characteristic equation of the system

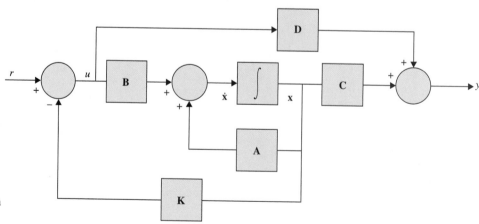

Figure 10-74 Block diagram of control system with state feedback.

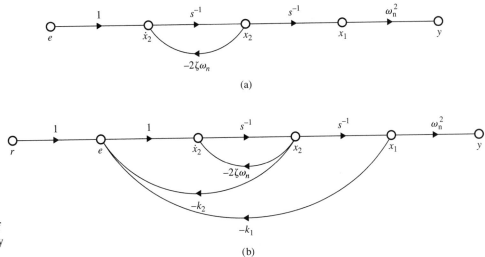

Figure 10-75 Control of a second-order system by state feedback.

with state feedback would be identical to that of the system with PD control if $k_1 = \omega_n^2 K_P$ and $k_2 = \omega_n^2 K_D$. However, the numerators of the two transfer functions are different.

The systems with zero reference input, $r(t) = 0$, are commonly known as **regulators**. When $r(t) = 0$, the control objective is to drive any arbitrary initial conditions of the system to zero in some prescribed manner, for example, "as quickly as possible." Then a second-order system with PD control is the same as state-feedback control.

It should be emphasized that the comparisons just made are all for second-order systems. For higher-order systems, the PD control and tachometer-feedback control are equivalent to feeding back only the state variables x_1 and x_2, while state-feedback control feeds back all the state variables.

Since PI control increases the order of the system by one, it cannot be made equivalent to state feedback through constant gains. We show in Section 10-13 that if we combine state feedback with integral control we can again realize PI control in the sense of state-feedback control.

▶ 10-13 POLE-PLACEMENT DESIGN THROUGH STATE FEEDBACK

When root loci are utilized for the design of control systems, the general approach may be described as that of **pole placement**; the poles here refer to that of the closed-loop transfer function, which are also the roots of the characteristic equation. Knowing the relation between the closed-loop poles and the system performance, we can effectively carry out the design by specifying the location of these poles.

The design methods discussed in the preceding sections are all characterized by the property that the poles are selected based on what can be achieved with the fixed-controller configuration and the physical range of the controller parameters. A natural question would be: *Under what condition can the poles be placed arbitrarily?* This is an entirely new design philosophy and freedom that apparently can be achieved only under certain conditions.

When we have a controlled process of the third order or higher, the PD, PI, the single-stage phase-lead or phase-lag controller would not be able to control independently all the poles of the system, since there are only two free parameters in each of these controllers.

To investigate the condition required for arbitrary pole placement in an nth-order system, let us consider that the process is described by the following state equation:

$$\frac{d\mathbf{x}(t)}{dt} = \mathbf{A}\mathbf{x}(t) + \mathbf{B}u(t) \tag{10-221}$$

where $\mathbf{x}(t)$ is an $n \times 1$ state vector, and $u(t)$ is the scalar control. The state-feedback control is

$$u(t) = -\mathbf{K}\mathbf{x}(t) + r(t) \tag{10-222}$$

where $\mathbf{K}$ is the $1 \times n$ feedback matrix with constant-gain elements. By substituting Eq. (10-222) into Eq. (10-221), the closed-loop system is represented by the state equation

$$\frac{d\mathbf{x}(t)}{dt} = (\mathbf{A} - \mathbf{B}\mathbf{K})\mathbf{x}(t) + \mathbf{B}r(t) \tag{10-223}$$

It will be shown in the following that if the pair $[\mathbf{A}, \mathbf{B}]$ is completely controllable, then a matrix $\mathbf{K}$ exists that can give an arbitrary set of eigenvalues of $(\mathbf{A} - \mathbf{B}\mathbf{K})$; that is, the n roots of the characteristic equation

$$|s\mathbf{I} - \mathbf{A} + \mathbf{B}\mathbf{K}| = 0 \tag{10-224}$$

can be arbitrarily placed. To show that this is true, we refer to the findings in Chapter 5, Section 5-8, that if a system is completely controllable, it can always be represented in the controllable canonical form (CCF); that is, in Eq. (10-221),

$$\mathbf{A} = \begin{bmatrix} 0 & 1 & 0 & \cdots & 0 \\ 0 & 0 & 1 & \cdots & 0 \\ \vdots & \vdots & \vdots & \ddots & \vdots \\ 0 & 0 & 0 & \cdots & 1 \\ -a_0 & -a_1 & -a_2 & \cdots & -a_{n-1} \end{bmatrix} \qquad \mathbf{B} = \begin{bmatrix} 0 \\ 0 \\ \vdots \\ 0 \\ 1 \end{bmatrix} \tag{10-225}$$

The feedback gain matrix $\mathbf{K}$ is expressed as

$$\mathbf{K} = \begin{bmatrix} k_1 & k_2 & \cdots & k_n \end{bmatrix} \tag{10-226}$$

where $k_1, k_2, \ldots, k_n$ are real constants. Then,

$$\mathbf{A} - \mathbf{B}\mathbf{K} = \begin{bmatrix} 0 & 1 & 0 & \cdots & 0 \\ 0 & 0 & 1 & \cdots & 0 \\ \vdots & \vdots & \vdots & \ddots & \vdots \\ 0 & 0 & 0 & \cdots & 1 \\ -a_0 - k_1 & -a_1 - k_2 & -a_2 - k_3 & \cdots & -a_{n-1} - k_n \end{bmatrix} \tag{10-227}$$

The eigenvalues of $\mathbf{A} - \mathbf{B}\mathbf{K}$ are then found from the characteristic equation

$$|s\mathbf{I} - (\mathbf{A} - \mathbf{B}\mathbf{K})| = s^n + (a_{n-1} + k_n)s^{n-1} + (a_{n-2} + k_{n-1})s^{n-2} + \cdots + (a_0 + k_1) = 0 \tag{10-228}$$

Clearly, the eigenvalues can be arbitrarily assigned, since the feedback gains $k_1, k_2, \ldots k_n$ are isolated in each coefficient of the characteristic equation. Intuitively, it makes sense that a system must be controllable for the poles to be placed arbitrarily. If one or more state variables are uncontrollable, then the poles associated with these state variables are also uncontrollable, and cannot be moved as desired. The following example illustrates the design of a control system with state feedback.

► **EXAMPLE 10-18** Consider the magnetic-ball suspension system analyzed in Section 5-14. This is a typical regulator system for which the control problem is to maintain the ball at its equilibrium position. It is shown in Section 5-14 that the system without control is unstable.

The linearized state model of the magnetic-ball system is represented by the state equation,

$$\frac{d\Delta\mathbf{x}(t)}{dt} = \mathbf{A}^*\Delta\mathbf{x}(t) + \mathbf{B}^*\Delta v(t) \tag{10-229}$$

where $\Delta\mathbf{x}(t)$ denotes the linearized state vector, and $\Delta v(t)$ the linearized input voltage. The coefficient matrices are

$$\mathbf{A}^* = \begin{bmatrix} 0 & 1 & 0 \\ 64.4 & 0 & -16 \\ 0 & 0 & -100 \end{bmatrix} \quad \mathbf{B}^* = \begin{bmatrix} 0 \\ 0 \\ 100 \end{bmatrix} \tag{10-230}$$

The eigenvalues of $\mathbf{A}^*$ are $s = -100, -8.025$, and 8.025. Thus, the system without feedback control is unstable.

Let us give the following design specifications:

1. The system must be stable.

2. For any initial disturbance on the position of the ball from its equilibrium position, the ball must return to the equilibrium position with zero steady-state error.

3. The time response should settle to within 5 percent of the initial disturbance in not more than 0.5 sec.

4. The control is to be realized by state feedback

$$\Delta v(t) = -\mathbf{K}\Delta\mathbf{x}(t) = -[k_1 \quad k_2 \quad k_3]\Delta\mathbf{x}(t) \tag{10-231}$$

where k_1, k_2, and k_3 are real constants.

A state diagram of the "open-loop" ball-suspension system is shown in Fig. 10-76(a), and the same of the "closed-loop" system with state feedback is shown in Fig. 10-76(b).

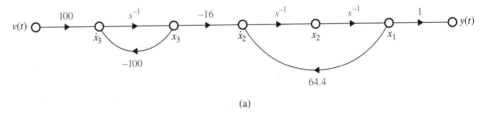

(a)

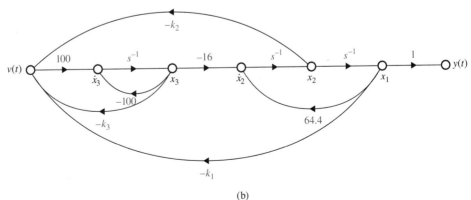

(b)

Figure 10-76 (a) State diagram of magnetic-ball suspension system. (b) State diagram of magnetic-ball-suspension system with state feedback.

We must select the desired location of the eigenvalues of $s\mathbf{I} - \mathbf{A}^* + \mathbf{B}^*\mathbf{K}$ so that requirement 3 in the preceding list on the time response is satisfied. Without entirely resorting to trial and error, we can start with the following decisions:

1. The system dynamics should be controlled by two dominant roots.
2. To achieve a relatively fast response, the two dominant roots should be complex.
3. The damping that is controlled by the real parts of the complex roots should be adequate, and the imaginary parts should be high enough for the transient to die out sufficiently fast.

After a few trial-and-error runs, using the **ACSYS/MATLAB** tool, we found that the following characteristic equation roots should satisfy the design requirements:

$$s = -20 \qquad s = -6 + j4.9 \qquad s = -6 - j4.9$$

The corresponding characteristic equation is

$$s^3 + 32s^2 + 300s + 1200 = 0 \qquad (10\text{-}232)$$

The characteristic equation of the closed-loop system with state feedback is written

$$|s\mathbf{I} - \mathbf{A}^* + \mathbf{B}^*\mathbf{K}| = \begin{vmatrix} s & -1 & 0 \\ -64.4 & s & 16 \\ 100k_1 & 100k_2 & s + 100 + 100k_3 \end{vmatrix} \qquad (10\text{-}233)$$

$$= s^3 + 100(k_3 + 1)s^2 - (64.4 + 1600k_2)s - 1600k_1 - 6440(k_3 + 1) = 0$$

which can also be obtained directly from Fig. 10-76(b) using the SFG gain formula. Equating like coefficients of Eqs. (10-232) and (10-233), we get the following simultaneous equations:

$$\begin{aligned} 100(k_3 + 1) &= 32 \\ -64.4 - 1600k_2 &= 300 \\ -1600k_1 - 6440(k_3 + 1) &= 1200 \end{aligned} \qquad (10\text{-}234)$$

Solving the last three equations, and being assured that the solutions exist and are unique, we get the feedback-gain matrix,

$$\mathbf{K} = \begin{bmatrix} k_1 & k_2 & k_3 \end{bmatrix} = \begin{bmatrix} -2.038 & -0.22775 & -0.68 \end{bmatrix} \qquad (10\text{-}235)$$

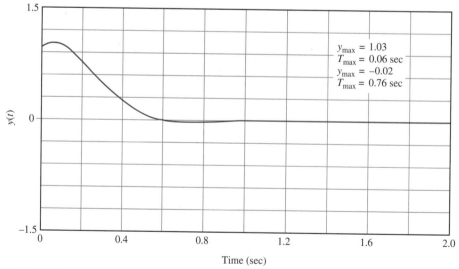

$y_{max} = 1.03$
$T_{max} = 0.06$ sec
$y_{max} = -0.02$
$T_{max} = 0.76$ sec

Figure 10-77 Output response of magnetic-ball suspension system with state feedback, subject to initial condition $y(0) = x_1(0) = 1$.

Figure 10-77 shows the output response $y(t)$ when the system is subject to the initial condition

$$\mathbf{x}(0) = \begin{bmatrix} 1 \\ 0 \\ 0 \end{bmatrix} \tag{10-236}$$

◄

▶ **EXAMPLE 10-19** In this example, we shall design a state-feedback control for the second-order sun-seeker system treated in Example 10-9. The transfer function of the controlled process is given in Eq. (10-131) with $K = 1$. The CCF state diagram of the process is shown in Fig. 10-78(a). The problem involves the design of a state-feedback control with

$$\theta_e(t) = -\mathbf{Kx}(t) = -[k_1 \quad k_2]\mathbf{x}(t) \tag{10-237}$$

The state equations are represented in vector-matrix form as

$$\frac{d\mathbf{x}(t)}{dt} = \mathbf{Ax}(t) + \mathbf{B}\theta_e(t) \tag{10-238}$$

where

$$\mathbf{A} = \begin{bmatrix} 0 & 1 \\ 0 & -25 \end{bmatrix} \quad \mathbf{B} = \begin{bmatrix} 0 \\ 1 \end{bmatrix} \tag{10-239}$$

The output equation is

$$\theta_o(t) = \mathbf{Cx}(t) \tag{10-240}$$

where

$$\mathbf{C} = [1 \quad 0] \tag{10-241}$$

The design objectives are

1. The steady-state error due to a step function input should equal 0.
2. With the state-feedback control, the unit-step response should have minimum overshoot, rise time, and settling time.

The transfer function of the system with state feedback is written

$$\frac{\Theta_o(s)}{\Theta_r(s)} = \frac{2500}{s^2 + (25 + k_2)s + k_1} \tag{10-242}$$

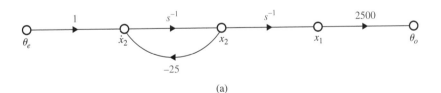

(a)

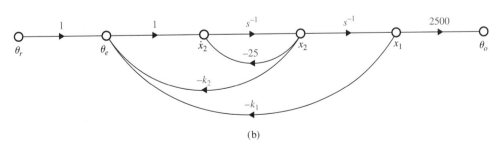

(b)

Figure 10-78 (a) State diagram of second-order sun-seeker system. (b) State diagram of second-order sun-seeker system with state feedback.

Thus, for zero steady-state error to a step input, the constant terms in the numerator and denominator must equal, or $k_1 = 2500$. This means that while the system is completely controllable, we cannot arbitrarily assign the two roots of the characteristic equation, which is now

$$s^2 + (25 + k_2)s + 2500 = 0 \qquad (10\text{-}243)$$

In other words, only one of the roots of Eq. (10-243) can be arbitrarily assigned. The problem is solved using **ACSYS**. After a few trial-and-error runs, we found out that the maximum overshoot, rise time, and settling time are all at a minimum when $k_2 = 75$. The two roots are $s = -50$ and -50. The attributes of the unit-step response are

$$\text{maximum overshoot} = 0\% \quad t_r = 0.06717 \text{ sec} \qquad t_s = 0.09467 \text{ sec}$$

The state-feedback gain matrix is

$$\mathbf{K} = \begin{bmatrix} 2500 & 75 \end{bmatrix} \qquad (10\text{-}244)$$

The lesson that we learned from this illustrative example is that state-feedback control generally produces a system that is type 0. In order for the system to track a step input without steady-state error, which requires a type 1 or higher-type system, the feedback gain k_1 of the system in the CCF state diagram cannot be assigned arbitrarily. This means that for an nth-order system, only $n - 1$ roots of the characteristic equation can be placed arbitrarily. ◀

▶ 10-14 STATE FEEDBACK WITH INTEGRAL CONTROL

The state-feedback control structured in the preceding section has one deficiency in that it does not improve the type of the system. As a result, the state-feedback control with constant-gain feedback is generally useful only for regulator systems for which the system does not track inputs, if all the roots of the characteristic equation are to be placed at will.

In general, most control systems must track inputs. One solution to this problem is to introduce integral control, just as with PI controller, together with the constant-gain state feedback. The block diagram of a system with constant-gain state feedback and integral control feedback of the output is shown in Fig. 10-79. The system is also subject to a noise input $n(t)$. For a SISO system, the integral control adds one integrator to the system. As shown in Fig. 10-79, the output of the $(n + 1)$st integrator is designated as x_{n+1}. The dynamic equations of the system in Fig. 10-79 are written as

$$\frac{d\mathbf{x}(t)}{dt} = \mathbf{A}\mathbf{x}(t) + \mathbf{B}u(t) + \mathbf{E}n(t) \qquad (10\text{-}245)$$

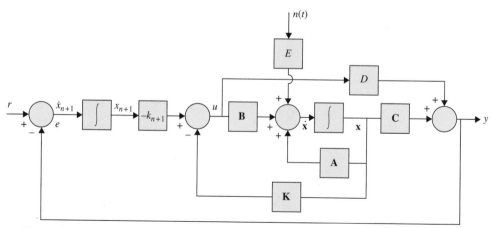

Figure 10-79 Block diagram of control system with state feedback and integral output feedback.

$$\frac{dx_{n+1}(t)}{dt} = r(t) - y(t) \tag{10-246}$$

$$y(t) = \mathbf{C}\mathbf{x}(t) + Du(t) \tag{10-247}$$

where $\mathbf{x}(t)$ is the $n \times 1$ state vector, $u(t)$ and $y(t)$ are the scalar actuating signal and output, respectively; $r(t)$ is the scalar reference input, and $n(t)$ is the scalar disturbance input. The coefficient matrices are represented by $\mathbf{A}$, $\mathbf{B}$, $\mathbf{C}$, D and $\mathbf{E}$, with appropriate dimensions. The actuating signal $u(t)$ is related to the state variables through constant-state and integral feedback,

$$u(t) = -\mathbf{K}\mathbf{x}(t) - k_{n+1}x_{n+1}(t) \tag{10-248}$$

where

$$\mathbf{K} = \begin{bmatrix} k_1 & k_2 & k_3 & \cdots & k_n \end{bmatrix} \tag{10-249}$$

with constant real gain elements, and k_{n+1} is the scalar integral-feedback gain.

Substituting Eq. (10-248) into Eq. (10-245), and combining with Eq. (10-246), the $n + 1$ state equations of the overall system with constant-gain and integral feedback are written

$$\frac{d\overline{\mathbf{x}}(t)}{dt} = (\overline{\mathbf{A}} - \overline{\mathbf{B}}\,\overline{\mathbf{K}})\overline{\mathbf{x}}(t) + \begin{bmatrix} \mathbf{0} \\ 1 \end{bmatrix} r(t) + \overline{\mathbf{E}}n(t) \tag{10-250}$$

where

$$\overline{\mathbf{x}}(t) = \begin{bmatrix} \dfrac{d\mathbf{x}(t)}{dt} \\ \dfrac{dx_{n+1}(t)}{dt} \end{bmatrix} \qquad (n+1) \times 1 \tag{10-251}$$

$$\overline{\mathbf{A}} = \begin{bmatrix} \mathbf{A} & \mathbf{0} \\ -\mathbf{C} & 0 \end{bmatrix} \quad (n+1) \times (n+1) \qquad \overline{\mathbf{B}} = \begin{bmatrix} \mathbf{B} \\ D \end{bmatrix} \quad (n+1) \times 1 \tag{10-252}$$

$$\overline{\mathbf{K}} = \begin{bmatrix} \mathbf{K} & k_{n+1} \end{bmatrix} = \begin{bmatrix} k_1 & K_2 & \cdots & k_n & k_{n+1} \end{bmatrix} \quad 1 \times (n+1) \tag{10-253}$$

$$\overline{\mathbf{E}} = \begin{bmatrix} \mathbf{E} \\ 0 \end{bmatrix} \quad [(n+1) \times 1] \tag{10-254}$$

Substituting Eq. (10-248) into Eq. (10-247), the output equation of the overall system is written

$$y(t) = \overline{\mathbf{C}}\overline{\mathbf{x}}(t) \tag{10-255}$$

where

$$\overline{\mathbf{C}} = \begin{bmatrix} \mathbf{C} - D\mathbf{K} & D\mathbf{K} \end{bmatrix} \qquad [1 \times (n+1)] \tag{10-256}$$

The design objectives are

(1) The steady-state value of the output $y(t)$ follows a step-function input with zero error; that is,

$$e_{ss} = \lim_{t \to \infty} e(t) = 0 \tag{10-257}$$

(2) The $n + 1$ eigenvalues of $(\overline{\mathbf{A}} - \overline{\mathbf{B}}\mathbf{G})$ are placed at desirable locations. For the last condition to be possible, the pair $[\overline{\mathbf{A}}, \overline{\mathbf{B}}]$ must be completely controllable.

The following example illustrates the applications of state-feedback with integral control.

▶ **EXAMPLE 10-20** We have shown in Example 10-19 that with constant-gain state-feedback control, the second-order sun-seeker system can have only one of its two roots placed at will, for the system to track a step input without steady-state error. Now let us consider the same second-order sun-seeker system in Example 10-19, except that an integral control is added to the forward path. The state diagram of the overall system is shown in Fig. 10-80. The coefficient matrices are

$$\mathbf{A} = \begin{bmatrix} 0 & 1 \\ 0 & 25 \end{bmatrix} \quad \mathbf{B} = \begin{bmatrix} 0 \\ 1 \end{bmatrix} \quad \mathbf{C} = [2500 \quad 0] \quad D = 0 \tag{10-258}$$

From Eq. (10-252),

$$\overline{\mathbf{A}} = \begin{bmatrix} \mathbf{A} & \mathbf{0} \\ -\mathbf{C} & 0 \end{bmatrix} = \begin{bmatrix} 0 & 1 & 0 \\ 0 & -25 & 0 \\ -2500 & 0 & 0 \end{bmatrix} \quad \overline{\mathbf{B}} = \begin{bmatrix} \mathbf{B} \\ D \end{bmatrix} = \begin{bmatrix} 0 \\ 1 \\ 0 \end{bmatrix} \tag{10-259}$$

We can show that the pair $[\overline{\mathbf{A}}, \overline{\mathbf{B}}]$ is completely controllable. Thus, the eigenvalues of $(s\mathbf{I} - \overline{\mathbf{A}} + \overline{\mathbf{B}}\,\overline{\mathbf{K}})$ can be arbitrarily placed. Substituting of $\overline{\mathbf{A}}, \overline{\mathbf{B}}$, and $\overline{\mathbf{K}}$ in the characteristic equation of the closed-loop system with state and integral feedback, we have

$$|s\mathbf{I} - \overline{\mathbf{A}} + \overline{\mathbf{B}}\,\overline{\mathbf{K}}| = \begin{vmatrix} s & -1 & 0 \\ k_1 & s + 25 + k_2 & k_3 \\ -2500 & 0 & s \end{vmatrix} \tag{10-260}$$
$$= s^3 + (25 + k_2)s^2 + k_1 s + 2500k_3 = 0$$

which can also be found from Fig. 10-80 using the SFG gain formula.

The design objectives are

(1) The steady-state output must follow a step function input with zero error.
(2) The rise time and settling time must be less than 0.05 sec.
(3) The maximum overshoot of the unit-step response must be less than 5 percent.

Since all three roots of the characteristic equation can be placed arbitrarily, it is not realistic to require minimum rise and settling times, as in Example 10-19.

Again, to realize a fast rise time and settling time, the roots of the characteristic equation should be placed far to the left in the s-plane, and the natural frequency should be high. Keep in mind that *roots with large magnitudes will lead to high gains for the state-feedback matrix.*

The **ACSYS**/MATLAB software was used to carry out the design. After a few trial-and-error runs, the design specifications can be satisfied by placing the roots at

$$s = -200 \quad -50 + j50 \quad \text{and} \quad -50 - j50$$

The desired characteristic equation is

$$s^3 + 300s^2 + 25,000s + 1,000,000 = 0 \tag{10-261}$$

Equating like coefficients of Eq. (10-260) and (10-261), we get

$$k_1 = 25,000 \quad k_2 = 275 \quad \text{and} \quad k_3 = 400$$

The attributes of the unit-step response are

$$\text{Maximum overshoot} = 4\% \quad t_r = 0.03247 \text{ sec} \quad t_s = 0.04667 \text{ sec}$$

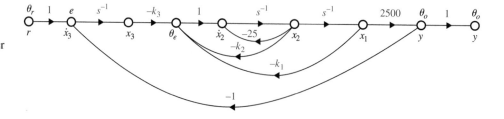

Figure 10-80 Sun-seeker system with state feedback and integral control in Example 10-20.

Notice that the high feedback gain of k_1, which is due to the large values of the roots selected, may pose physical problems; if so, the design specifications may have to be revised. ◀

▶ **EXAMPLE 10-21** In this example we illustrate the application of state-feedback with integral control to a system with a disturbance input.

Consider a dc-motor control system that is described by the following state equations:

$$\frac{d\omega(t)}{dt} = \frac{-B}{J}\omega(t) + \frac{K_i}{J}i_a(t) - \frac{1}{J}T_L \tag{10-262}$$

$$\frac{di_a(t)}{dt} = \frac{-K_b}{L}\omega(t) - \frac{R}{L}i_a(t) + \frac{1}{L}e_a(t) \tag{10-263}$$

where

$i_a(t)$ = armature current, A
$e_a(t)$ = armature applied voltage, V
$\omega(t)$ = motor velocity, rad/sec
B = viscous-friction coefficient of motor and load = 0
J = moment of inertia of motor and load = 0.02 N-m/rad/sec^2
K_i = motor torque constant = 1 N-m/A
K_b = motor back-emf constant = 1 V/rad/sec
T_L = constant load torque (magnitude not known), N-m
L = armature inductance = 0.005 H
R = armature resistance = 1 Ω

The output equation is

$$y(t) = \mathbf{C}\mathbf{x}(t) = \begin{bmatrix} 1 & 0 \end{bmatrix}\mathbf{x}(t) \tag{10-264}$$

The design problem is to find the control $u(t) = e_a(t)$ through state feedback and integral control such that

(1) $\lim\limits_{t\to\infty} i_a(t) = 0$ and $\lim\limits_{t\to\infty}\dfrac{d\omega(t)}{dt} = 0$ $\tag{10-265}$

(2) $\lim\limits_{t\to\infty}\omega(t)$ = step input $r(t) = u_s(t)$ $\tag{10-266}$

(3) The eigenvalues of the closed-loop system with state feedback and integral control are at $s = -300, -10 + j10,$ and $-10 - j10$.

Let the state variables be defined as $x_1(t) = \omega(t)$ and $x_2(t) = i_a(t)$. The state equations in Eqs. (10-262) and (10-263) are written in vector-matrix form.

$$\frac{d\mathbf{x}(t)}{dt} = \mathbf{A}\mathbf{x}(t) + \mathbf{B}u(t) + \mathbf{E}n(t) \tag{10-267}$$

where $n(t) = T_L u_s(t)$.

$$\mathbf{A} = \begin{bmatrix} -\dfrac{B}{J} & \dfrac{K_i}{J} \\ -\dfrac{K_b}{L} & -\dfrac{R}{L} \end{bmatrix} = \begin{bmatrix} 0 & 50 \\ -200 & -200 \end{bmatrix} \tag{10-268}$$

$$\mathbf{B} = \begin{bmatrix} 0 \\ \dfrac{1}{L} \end{bmatrix} = \begin{bmatrix} 0 \\ 200 \end{bmatrix} \tag{10-269}$$

$$\mathbf{E} = \begin{bmatrix} -\dfrac{1}{J} \\ 0 \end{bmatrix} = \begin{bmatrix} -50 \\ 0 \end{bmatrix} \tag{10-270}$$

From Eq. (10-252),

$$\overline{\mathbf{A}} = \begin{bmatrix} \mathbf{A} & \mathbf{0} \\ -\mathbf{C} & 0 \end{bmatrix} = \begin{bmatrix} 0 & 50 & 0 \\ -200 & -200 & 0 \\ -1 & 0 & 0 \end{bmatrix} \qquad \overline{\mathbf{B}} = \begin{bmatrix} \mathbf{B} \\ 0 \end{bmatrix} = \begin{bmatrix} 0 \\ 200 \\ 0 \end{bmatrix} \qquad (10\text{-}271)$$

$$\overline{\mathbf{C}} = [\mathbf{C} \quad 0] = [1 \quad 0 \quad 0] \qquad \overline{\mathbf{E}} = \begin{bmatrix} \mathbf{E} \\ 0 \end{bmatrix} = \begin{bmatrix} -50 \\ 0 \\ 0 \end{bmatrix} \qquad (10\text{-}272)$$

The control is given by

$$u(t) = -\mathbf{Kx}(t) - k_{n+1}x_{n+1}(t) = \overline{\mathbf{K}}\overline{\mathbf{x}}(t) \qquad (10\text{-}273)$$

where

$$\overline{\mathbf{K}} = [k_1 \quad k_2 \quad k_3] \qquad (10\text{-}274)$$

Figure 10-81 shows the state diagram of the overall designed system.

The coefficient matrix of the closed-loop system is

$$\overline{\mathbf{A}} - \overline{\mathbf{B}}\overline{\mathbf{K}} = \begin{bmatrix} 0 & 50 & 0 \\ -200 - 200k_1 & -200 - 200k_2 & -200k_3 \\ -1 & 0 & 0 \end{bmatrix} \qquad (10\text{-}275)$$

The characteristic equation is

$$|s\mathbf{I} - \overline{\mathbf{A}} + \overline{\mathbf{B}}\overline{\mathbf{K}}| = s^3 + 200(1 + k_2)s^2 + 10{,}000(1 + k_1)s - 10{,}000k_3 = 0 \quad (10\text{-}276)$$

which is more easily determined by applying the gain formula of SFG to Fig. 10-81.

For the three roots assigned, the last equation must equal

$$s^3 + 320s^2 + 6{,}200s + 60{,}000 = 0 \qquad (10\text{-}277)$$

Equating the like coefficients of Eqs. (10-276) and (10-277), we get

$$k_1 = -0.38 \qquad k_2 = 0.6 \qquad k_3 = -6.0$$

Applying the SFG gain formula to Fig. 10-81 between the inputs $r(t)$ and $n(t)$ and the states $\omega(t)$ and $i_a(t)$, we have

$$\begin{bmatrix} \Omega(s) \\ I_a(s) \end{bmatrix} = \frac{1}{\Delta_c(s)} \begin{bmatrix} -\dfrac{1}{J}\left(s^2 + \dfrac{R}{L}s + \dfrac{k_2}{L}s\right) & -\dfrac{k_3 K_i}{JL} \\ -\dfrac{1}{J}\left(-\dfrac{K_b}{L}s - \dfrac{k_1}{L}s + \dfrac{k_3}{L}\right) & -\dfrac{k_3}{L}\left(s + \dfrac{B}{J}\right) \end{bmatrix} \begin{bmatrix} T_L \\ s \\ \dfrac{1}{s} \end{bmatrix} \qquad (10\text{-}278)$$

where $\Delta_c(s)$ is the characteristic polynomial given in Eq. (10-277).

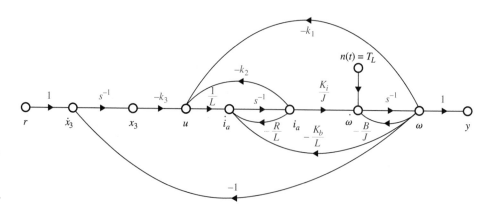

Figure 10-81 DC motor control system with state feedback and integral control and disturbance torque in Example 10-21.

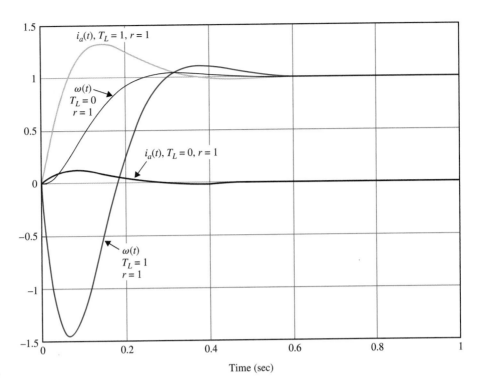

Figure 10-82 Time responses of dc-motor control system with state feedback and integral control and disturbance torque in Example 10-21.

Applying the final-value theorem to the last equation, the steady-state values of the state variables are found to be

$$\lim_{t \to \infty}\begin{bmatrix} \omega(t) \\ i_a(t) \end{bmatrix} = \lim_{s \to 0} s \begin{bmatrix} \Omega(s) \\ I_a(s) \end{bmatrix} = \begin{bmatrix} 0 & K_i \\ 1 & B \end{bmatrix} \begin{bmatrix} T_L \\ 1 \end{bmatrix} = \begin{bmatrix} 1 \\ T_L \end{bmatrix} \tag{10-279}$$

Thus, the motor velocity $\omega(t)$ will approach the constant reference input step function $r(t) = u_s(t)$ as t approaches infinity, independent of the disturbance torque T_L. Substituting the system parameters into Eq. (10-278), we get

$$\begin{bmatrix} \Omega(s) \\ I_a(s) \end{bmatrix} = \frac{1}{\Delta_c(s)} \begin{bmatrix} -50(s + 320)s & 60{,}000 \\ 6200s + 60{,}000 & 1{,}200s \end{bmatrix} \begin{bmatrix} \dfrac{T_L}{s} \\ \dfrac{1}{s} \end{bmatrix} \tag{10-280}$$

Figure 10-82 shows the time responses of $\omega(t)$ and $i_a(t)$ when $T_L = 1$ and $T_L = 0$. The reference input is a unit-step function. ◂

▷ 10-15 MATLAB TOOLS AND CASE STUDIES

The MATLAB tools used for solving the problem in this chapter are **controls** and **statetool** (see Chapter 5 for more detail on the statetool). The Controller Design Tool (controls) consists of a number of m-files and GUIs for time and frequency analyses and the design of simple control systems using transfer functions. The controls tool can be invoked either from the MATLAB command line by simply typing controls or from the Automatic Control Systems launch applet (**ACSYS**) by clicking on the appropriate button. The controls tool is the **main component** of ACSYS, which contains most of the software you have used throughout this book. This comprehensive tool essentially incorporates the functionality of **tftime** (see Chapter 7) and **freqtool** (see Chapter 9) to create a controller design tool. Virtually all the problems discussed in this chapter (and most of this text) may be solved by the controls tool. This software allows the user to be able to conduct the following tasks:

- Enter the transfer function values in polynomial form. (User must use the tftool as discussed in Chapter 2 to convert the transfer function from pole, zero, gain form into polynomial form.)
- Obtain the step, impulse, parabolic, ramp, or other type input time responses.
- Obtain the closed-loop frequency plots.
- Obtain the phase and gain margin Bode plots and the polar plot of the loop transfer functions (in a single feedback loop configuration).
- Understand the effect of adding zeros and poles to the closed-loop or open-loop transfer functions.
- Design and compare various controllers including PID, lead, and lag compensators.

To better illustrate how to use the controls tool, let us go through the steps involved in solving the earlier example involving position control in Sections 7-6 and 7-11.

▶ **EXAMPLE 10-22:** Recall a plant with the following transfer function
EXAMPLE 10-1
REVISITED

$$G(s) = \frac{4500}{s(s + 361.2)} \qquad (10\text{-}281)$$

Time-Domain Design: Design a controller so that the closed-loop system shown in Fig. 10-83 meets the following criteria:

> Steady-state error due to unit-ramp input ≤ 0.000443
>
> Maximum overshoot ≤ 5 percent
>
> Rise time t_r ≤ 0.005 sec
>
> Settling time t_s ≤ 0.005 sec

In order to examine the system performance, we start by using a proportional controller, also discussed in Section 7-11. To enter the transfer function, invoke the Controller Design tool by typing controls at the MATLAB prompt in MATLAB command window, or by pressing the appropriate button within ACSYS. Figure 10-84 will appear on the computer screen. Click the "Enter Transfer Function" button to move to the input module. To enter the plant equation press the G(s) button, and enter the plant numerator and denominator coefficients as directed. In this case Eq. (10-281) is entered as

> Num G(s) is **4500**
> Den G(s) is **1 361.2 0**   (Numbers are separated by space)

Next enter the values for G_c(s) and H(s). Note that the value of H(s) is unity. We also keep the default value of one for G_c(s) for the time being. Use the APPLY button to move back to the main window, and press "Calculate" to evaluate the closed loop transfer function. **It is recommended that you always refer to the MATLAB command window for an accurate representation of transfer functions.** See Section 7-11 for a more thorough discussion of the time response of this system.

To examine the performance of the proportional controller, we need to find the system root locus. The root locus of the system may be obtained by clicking the corresponding button in the controls main window (Fig. 10-85), which will activate the MATLAB SISO design tool. Figure 10-86a shows the root locus of system. In order to see the poles and zeros of G and H, go to the View menu and

Figure 10-83 The control system with unity feedback for Example 10-22.

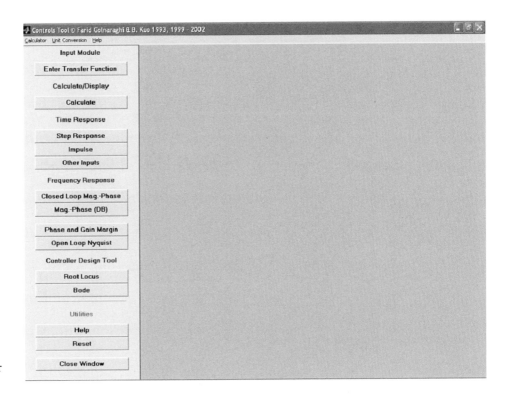

Figure 10-84 Controller Design main window.

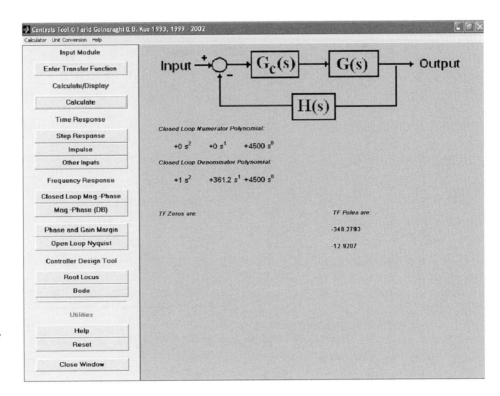

Figure 10-85 Controller Design main window showing the closed-loop transfer function.

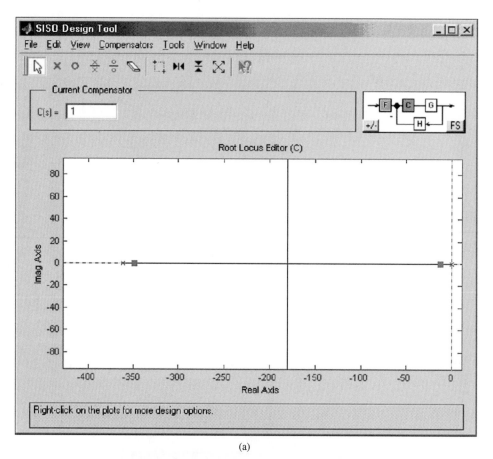

(a)

(b)

Figure 10-86   (a) Root locus of Eq. (7-201); (b) G and H poles and zeros.

select System Data, or alternatively, double click the blocks G or H in the top right corner of the block diagram in the top right corner of the figure. The System Data window is shown in Fig. 10-86b.

The red squares in Fig. 10-86a correspond to the closed-loop system poles for $K = 1$. In order to see the closed-loop system time response to a unit-step input, select the Closed-Loop Step from the Loop Responses category within the Tools menu option, which is located at the top of the screen shown in Fig. 10-86a. Figure 10-87a shows the unit-step response of the closed-loop system

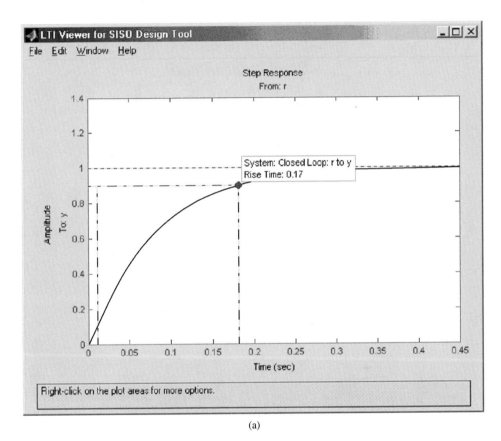

(a)

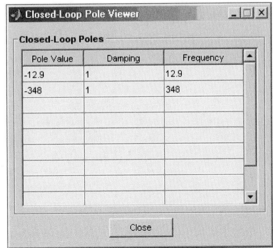

(b)

Figure 10-87 (a) Unit-step response of Eq. (7-201) for $K=1$; (b) poles and zeros of (7-201) for $K = 1$.

for $K = 1$. You may also obtain the closed-loop system poles by selecting Closed-Loop Poles from the View menu in the SISO Design Tool window. Recall the poles of the closed-loop system are

$$s_{1,2} = -180.6 \pm \sqrt{32616 - 4500K} \qquad (10\text{-}282)$$

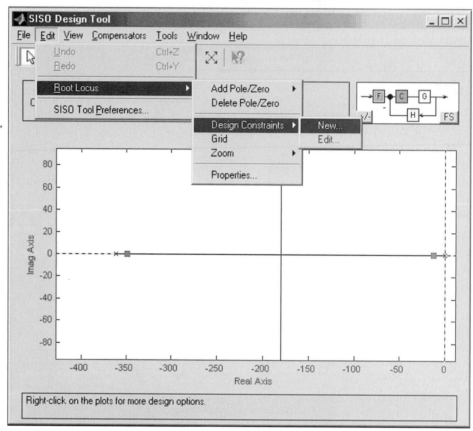

(a)

(b)

Figure 10-88 (a) Using the Design Constraints option for Example 10-22. (b) Settling time < 0.005 and percent overshoot < 4.33.

Changing K, therefore, affects the pole locations. In the Root Locus windows, "C(s)" represents the controller transfer function, in this case, $C(s) = K = 1$. Hence, if $C(s)$ is increased, the effective value of K increases, forcing the closed-loop poles to move together on the real axis, and then ultimately to move apart to become complex. See Fig. 10-87b.

Incorporation of the Design Criteria: As a first step to design a controller, we use the built-in design criteria option within the SISO Design Tool to establish the desired pole regions on the root locus. To add the design constraints, use the Edit menu and choose the Root Locus option. "New" must then be selected within the Design Constrains option, as shown in Fig. 10-88a. The Design Constraints option allows the user to investigate the effect of the following:

- Settling time
- Percent overshoot
- Damping ratio
- Natural frequency

In this case, as shown in Fig. 10-88b, we have included settling time and percent overshoot as design constrains. In order also to enter the rise time as a constraint, the user must first establish a relation between the damping ratio and the natural frequency using an equation for rise time. Recall that the approximate equations for rise time for a second-order prototype system were provided in Chapter 7. Since the settling time and the percent overshoot are more important criteria in this example, we will use them as primary constraints. After designing a controller based on these constraints, we will determine whether the system complies with the rise time constraint.

Figure 10-89 shows the desired closed-loop system pole locations on the root locus after inclusion of the design constraints. Obviously, the poles of the system for $K = 1$ are not in the desired area.

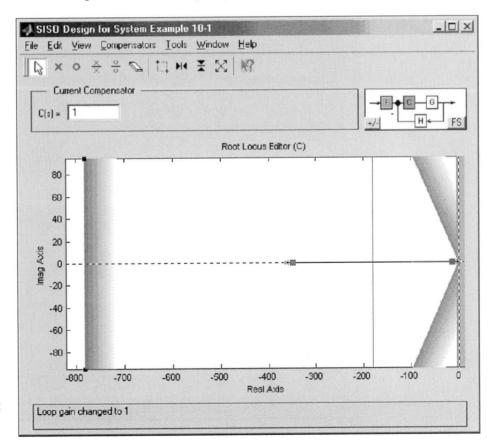

Figure 10-89 The root-locus diagram for Example 10-22, after incorporating the percent overshoot and the settling time as design constraints.

The desired poles of the system must lie to the left of the boundary imposed by the settling time between -700 and -800 markers in Fig. 10-89. Obviously, it is impossible to use the proportional controller (for any value of K) to move the poles of the closed-loop system farther to the left-hand plane. However, a PD controller may be used to accomplish this task. As proposed earlier in the solution to Example 10-1, a PD controller with a zero at $z = -1/0.001772$ may be used for this purpose. Recall from the solution of Example 10-1 that this number was obtained from the steady-state error criterion by examining Eq. (10-13). In this case, if the proportional component of a PD controller is set to $K_P = 1$, the steady-state error due to a unit-ramp input is $e_{ss} = 0.000443$ for $K_D = 0.001772$.

To enter a zero to the controller, click the C(s) block in the block diagram in the top right corner of Fig. 10-89, or simply follow the instructions on the bottom of the screen shown in Fig. 10-89 and right-click the mouse to edit the controller transfer function. Figure 10-90 shows the Edit Compensator window and how the PD controller is added.

The new root-locus diagram for the system appears in Fig. 10-91. It is now very easy to establish the required gain to push the poles to the desired region. Using the value of $K = 181.2$, which was proposed in the original solution of Example 10-1, would force the closed-loop poles to the right region, as shown in Fig. 10-92a. The system poles are also shown in Fig. 10-92b. The step response of the controlled system in Fig. 10-93 shows the system has now complied with all design criteria. The 2 percent settling time is now 0.0488 sec, while the percent overshoot is 4.17. It is interesting that although the closed-loop poles are both real, the system has a nonoscillatory response with an overshoot. This is because of the effect of zero on the response. Review the effect of adding a zero to a closed-loop transfer function, which was discussed in Section 7-8, to further appreciate this behavior.

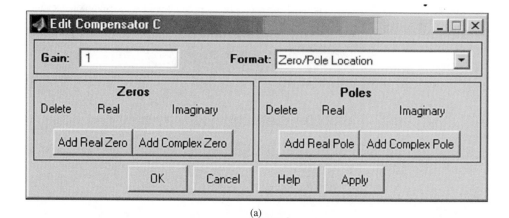

(a)

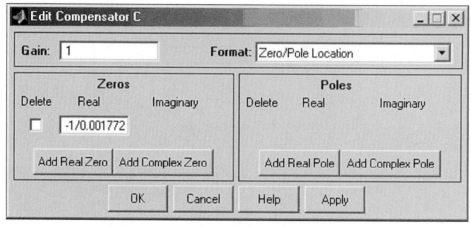

Figure 10-90 Addition of a zero to the controller $C(s)$ to create a PD controller.

(b)

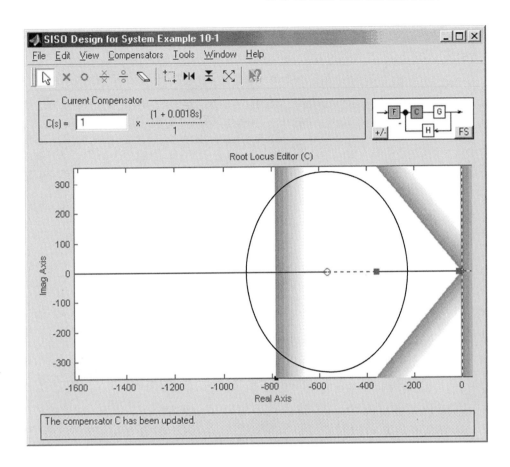

Figure 10-91 The root-locus diagram for Example 10-22, after incorporating a zero in the PD controller at $-1/0.001772$.

• Verify your actuator torque and power limitations.

Finally, in practice, always verify that the actuator used has enough torque or load to create such response. The actuator limitations must always be included because they are some of the most important design constraints.

Frequency-Domain Design

Now let us carry out the design of the PD controller in the frequency domain using the following performance criteria:

> Steady-state error due to a unit-ramp input ≤ 0.00443
>
> Phase margin $\geq 80°$
>
> Resonant peak $M_r \leq 1.05$
>
> BW ≤ 2000 rad/sec

To start designing the frequency-domain design process, click the Bode button in the Controller Design main window (Fig. 10-85) to get the MATLAB SISO Design Tool, as shown in Fig. 10-94.

As in the root-locus design approach, a PD controller with a zero at $z = -1/0.001772$ may be used for this purpose. Recall that this number was obtained from the steady-state error criterion. Similar to the root-locus approach, in order to enter zero for the controller, click the C(s) block in the block diagram in the top right corner of the screen shown in Fig. 10-94. The new open-loop Bode diagram of the system appears in Fig. 10-95 for the system with a PD controller $G_c(s) = C(s) = 181.2 (1 + 0.001772s)$. The gain margin in this case is infinity, while the phase margin is $83°$. As a result, the system has also fully complied with all the design criteria in the frequency domain. Review the effect of adding a zero to a closed-loop

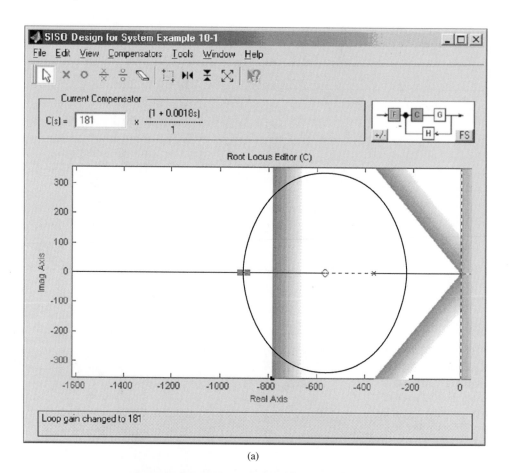

(a)

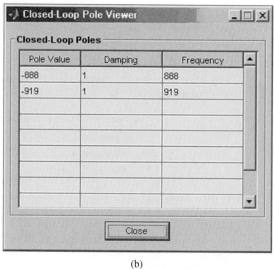

(b)

Figure 10-92 (a) The root-locus diagram, (b) the closed-loop poles for Example 10-22 after incorporating a zero in the PD controller at $-1/0.001772$ and using a gain of 181.

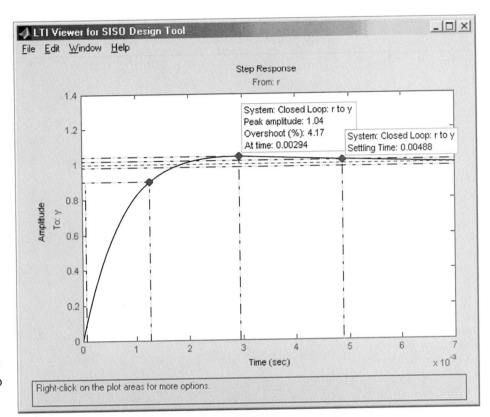

Figure 10-93 Step response of the system of Example 10-22 with a PD controller, $C(s) = 181$ $(s + 1/0.001772)$.

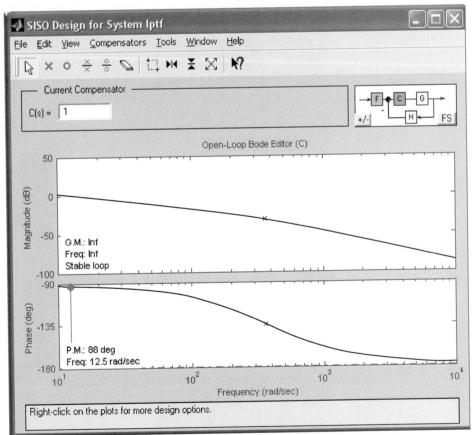

Figure 10-94 The loop magnitude and phase diagrams in the frequency-domain SISO Design Tool for Example 10-22.

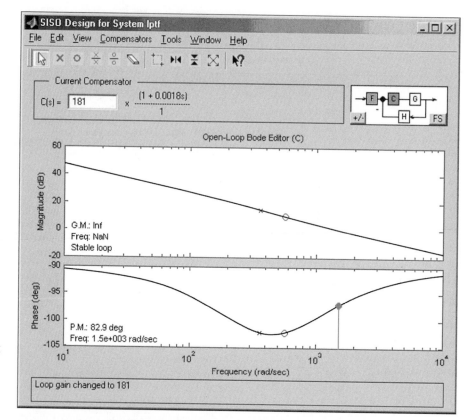

Figure 10-95 The loop magnitude and phase diagrams for Example 10-22, after incorporating a zero in the PD controller at $-1/0.001772$ and a gain of $K = 181.2$.

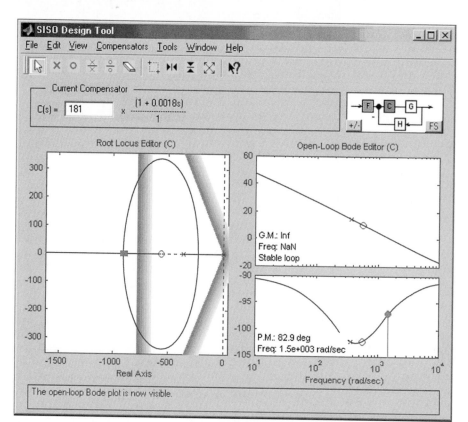

Figure 10-96 The loop magnitude and phase diagrams and the root locus for Example 10-22, after incorporating a zero in the PD controller at $-1/0.001772$ and a gain $K = 181.2$.

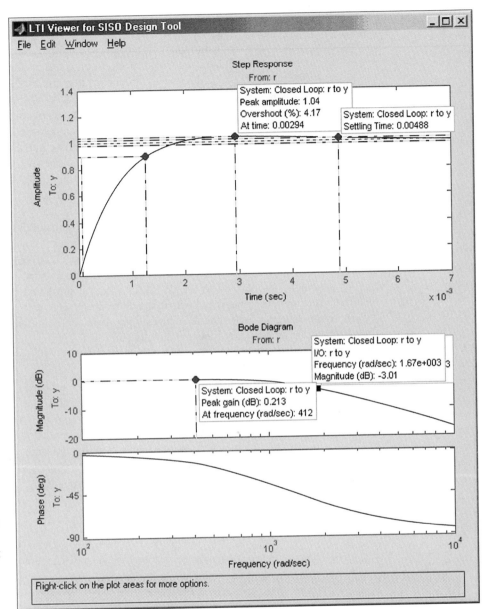

Figure 10-97 The closed-loop, unit-step, and magnitude and phase diagrams for Example 10-22, after incorporating a zero in the PD controller at $-1/0.001772$ and a gain $K = 181.2$.

transfer function, which was discussed in Section 9-3 (see also Section 9-4 for addition of a pole) to appreciate this behavior.

   Alternatively, you may select the Open-Loop Bode option from the View menu in the root-locus diagram in Fig. 10-91 to obtain a concurrent root-locus and frequency-response representations of the system as shown in Fig. 10-96. Finally, closed loop responses of the system appear in Fig. 10.97. ◄

• Refer to Chapter 5 to learn how to use the statetool.

Example 10-22 covered most aspects associated with the functionality of the **controls** tool. You may use this design tool for all examples and homework problems in this chapter that have a block diagram representation as shown in Figure 10-83. You may also use the State Space Tool (statetool) to solve problems involving state-feedback controller design (refer to Chapter 5 for a detailed discussion of the functionality of statetool).

►10-16 SUMMARY

This chapter was devoted to the design of linear continuous-data control systems. It began by giving some fundamental considerations on system design and the reviewed the specifications in the time domain and frequency domain. Fixed configurations of compensation schemes used in practice, such as series, forward and feedforward, feedback and minor loop, and state feedback, were illustrated. The types of controllers considered were the PD, PI, PID, phase-lead, phase-lag, lead–lag, pole-zero cancellation, and notch filters. Designs were carried out in the time-domain (s-domain) as well as the frequency domain. The time-domain design was characterized by specifications such as the relative damping ratio, maximum overshoot, rise time, delay time, settling time, or simply the location of the characteristic-equation roots, keeping in mind that the zeros of the system transfer function also affects the transient response. The performance was generally measured by the step response and the steady-state error. Frequency-domain designs were generally carried out using Bode diagrams or gain-phase plots. The performance specifications in the frequency domain were the phase margin, gain margin, M_r, BW, and the like.

The effect of feedforward control on noise and disturbance reduction was demonstrated. A section was devoted to the design of robust control systems. State feedback through constant-feedback gains and with dynamic feedback were discussed. The state-feedback design is more versatile than the conventional fixed-configuration controller design, since the characteristic-equation roots are directly controlled. An unstable system that is controllable can always be stabilized by state-feedback control. The disadvantage with state feedback is that all the states must be sensed and fed back for control, which may not be practical.

While the design techniques covered in this chapter were outlined with analytical procedures, the text promotes the use of computers. The Controller Design Tool (controls) and the State Space Tool (statetool) have been developed by the authors for this purpose. Through the GUI approach, these programs were intended to create a user-friendly environment to reduce the complexity of control systems design.

► REVIEW QUESTIONS

1. What is a PD controller? Write its input-output transfer function.

2. A PD controller has the constants K_D and K_P. Give the effects of these constants on the steady-state error of the system. Does the PD control change the type of a system?

3. Give the effects of the PD control on rise time and settling time of a control system.

4. How does the PD controller affect the bandwidth of a control system?

5. Once the value of K_D of a PD controller is fixed, increasing the value of K_P will increase the phase margin monotonically. (T) (F)

6. If a PD controller is designed so that the characteristic-equation roots have better damping than the original system, then the maximum overshoot of the system is always reduced. (T) (F)

7. What does it mean when a control system is described as being robust?

8. A system compensated with a PD controller is usually more robust than the system compensated with a PI controller. (T) (F)

9. What is a PI controller? Write its input-output transfer function.

10. A PI controller has the constants K_P and K_I. Give the effects of the PI controller on the steady-state error of the system. Does the PI control change the system type?

11. Give the effects of the PI control on the rise time and settling time of a control system.

12. How does the PI controller affect the bandwidth of a control system?

13. What is a PID controller? Write its input-output transfer function.

14. Give the limitations of the phase-lead controller.

15. How does the phase-lead controller affect the bandwidth of a control system?

16. Give the general effects of the phase-lead controller on rise time and settling time.

17. For the phase-lead controller, $G_c(s) = (1 + aTs)/(1 + Ts)$, $a > 1$, what is the effect of the controller on the steady-state performance of the system?

18. The phase-lead controller is generally less effective if the uncompensated system is very unstable to begin with.　　　　　　　　　　　**(T)**　　**(F)**

19. The maximum phase that is available from a single-stage phase-lead controller is 90°.　　　　　　　　　　　　　　　　　　　　　　**(T)**　　**(F)**

20. The design objective of the phase-lead controller is to place the maximum phase lead at the new gain-crossover frequency.　　　　　　　　**(T)**　　**(F)**

21. The design objective of the phase-lead controller is to place the maximum phase lead at the frequency where the magnitude of the uncompensated $G_p(j\omega)$ is $-10\log_{10}a$, where a is the gain of the phase-lead controller.　**(T)**　　**(F)**

22. The phase-lead controller may not be effective if the negative slope of the uncompensated process transfer function is too steep near the gain-crossover frequency.　**(T)**　　**(F)**

23. For the phase-lag controller, $G_c(s) = (1 + aTs)/(1 + Ts)$, $a < 1$, what is the effect of the controller on the steady-state performance of the system?

24. Give the general effects of the phase-lag controller on rise time and settling time.

25. How does the phase-lag controller affect the bandwidth?

26. For a phase-lag controller, if the value of T is large and the value of a is small, it is equivalent to adding a pure attenuation of a to the original uncompensated system at low frequencies.　　　　　　　　　　　**(T)**　　**(F)**

27. The principle of design of the phase-lag controller is to utilize the zero-frequency attenuation property of the controller.　　　　　　　　**(T)**　　**(F)**

28. The corner frequencies of the phase-lag controller should not be too low or else the bandwidth of the system will be too low.　　　　　　　**(T)**　　**(F)**

29. Give the limitations of the pole-zero-cancellation control scheme.

30. How does the sensitivity function relate to the bandwidth of a system?

Answers to the true-and-false questions are given following the Problems section.

► REFERENCES

1. J. C. Willems and S. K. Mitter, "Controllability, Observability, Pole Allocation, and State Reconstruction," *IEEE Trans. Automatic Control*, Vol. AC-16, pp. 582–595, Dec. 1971.

2. H. W. Smith and E. J. Davison, "Design of Industrial Regulators," *Proc. IEE (London)*, Vol. 119, pp. 1210–1216, Aug. 1972.

3. F. N. Bailey and S. Meshkat, "Root Locus Design of a Robust Speed Control," *Proc. Incremental Motion Control Symposium*, pp. 49–54, June 1983.

► PROBLEMS

Most of the following problems can be solved using a computer program. This is highly recommended if the reader has access to such a program.

• Design of series controller to satisfy steady-state-error requirement

10-1. The block diagram of a control system with a series controller is shown in Fig. 10P-1. Find the transfer function of the controller $G_c(s)$ so that the following specifications are satisfied:

　The ramp-error constant K_v is 5.

　The closed-loop transfer function is of the form

$$M(s) = \frac{Y(s)}{R(s)} = \frac{K}{(s^2 + 20s + 200)(s + a)}$$

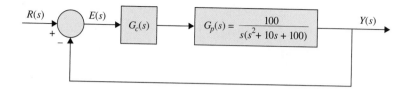

Figure 10P-1

where K and a are real constants. Find the values of K and a.

The design strategy is to place the closed-loop poles at $-10 + j10$ and $-10 - j10$, and then adjust the values of K and a to satisfy the steady-state requirement. The value of a is large so that it will not affect the transient response appreciably. Find the maximum overshoot of the designed system.

• Design of series
controller to satisfy steady-
state-error requirement

10-2. Repeat Problem 10-1 if the ramp-error constant is to be 9. What is the maximum value of K_v that can be realized? Comment on the difficulties that may arise in attempting to realize a very large K_v.

• Design of system with
PD controller

10-3. A control system with a PD controller is shown in Fig. 10P-3.

(a) Find the values of K_P and K_D so that the ramp-error constant K_v is 1000 and the damping ratio is 0.5.

(b) Find the values of K_P and K_D so that the ramp-error constant K_v is 1000 and the damping ratio is 0.707.

(c) Find the values of K_P and K_D so that the ramp-error constant K_v is 1000 and the damping ratio is 1.0.

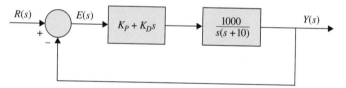

Figure 10P-3

• Design of system with
PD controller

10-4. For the control system shown in Fig. 10P-3, set the value of K_P so that the ramp-error constant is 1000.

(a) Vary the value of K_D from 0.2 to 1.0 in increments of 0.2 and determine the values of phase margin, gain margin, M_r, and BW of the system. Find the value of K_D so that the phase margin is maximum.

(b) Vary the value of K_D from 0.2 to 1.0 in increments of 0.2 and find the value of K_D so that the maximum overshoot is minimum.

• Design of aircraft
attitude control system with
PD controller

10-5. Consider the second-order model of the aircraft attitude control system shown in Fig. 7-25. The transfer function of the process is

$$G_p(s) = \frac{4500K}{s(s + 361.2)}$$

(a) Design a series PD controller with the transfer function $G_c(s) = K_D + K_Ps$ so that the following performance specifications are satisfied:

Steady-state error due to a unit-ramp input ≤ 0.001

Maximum overshoot ≤ 5 percent

Rise time $t_r \leq 0.005$ sec

Settling time $t_s \leq 0.005$ sec

(b) Repeat part (a) for all the specifications listed, and, in addition, the bandwidth of the system must be less than 850 rad/sec.

• Design of liquid-level control system with PD controller

10-6. Figure 10P-6 shows the block diagram of the liquid-level control system described in Problems 6-13 and 7-29. The number of inlets is denoted by N. Set $N = 20$. Design the PD controller so that with a unit-step input the tank is filled to within 5 percent of the reference level in less than 3 sec without overshoot.

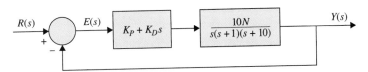

Figure 10P-6

• Design of liquid-level control system with PD controller

10-7. For the liquid-level control system described in Problem 10-6,

(a) Set K_P so that the ramp-error constant is 1. Vary K_D from 0 to 0.5 and find the value K_D that gives the maximum phase margin. Record the gain margin, M_r, and BW.

(b) Plot the sensitivity functions $|S_G^M(j\omega)|$ of the uncompensated system and the compensated system with the values of K_D and K_P determined in part (a). How does the PD controller affect the sensitivity?

• Design of a system with PI controller

10-8. A control system with a type 0 process $G_p(s)$ and a PI controller is shown in Fig. 10P-8.

(a) Find the value of K_I so that the ramp-error constant K_v is 10.

(b) Find the value of K_P so that the magnitude of the imaginary parts of the complex roots of the characteristic equation of the system is 15 rad/sec. Find the roots of the characteristic equation.

(c) Sketch the root contours of the characteristic equation with the value of K_I as determined in part (a) and for $0 \le K_P < \infty$.

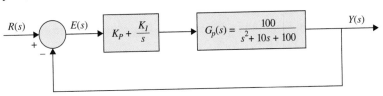

Figure 10P-8

• Design of a system with PI controller

10-9. For the control system described in Problem 10-8,

(a) Set K_I so that the ramp-error constant is 10. Find the value of K_P so that the phase margin is minimum. Record the values of the phase margin, gain margin, M_r, and BW. How does this optimal K_P relate to the root contours constructed in Problem 10-8(c)?

(b) Plot the sensitivity functions $|S_G^M(j\omega)|$ of the uncompensated and the compensated systems with the values of K_I and K_P selected in part (a). Comment on the effect of the PI control on sensitivity.

• Design of system with PI controller

10-10. For the control system shown in Fig. 10P-8, perform the following:

(a) Find the value of K_I so that the ramp-error constant K_v is 100.

(b) With the value of K_I found in part (a), find the critical value of K_P so that the system is stable. Sketch the root contours of the characteristic equation for $0 \le K_P < \infty$.

(c) Show that the maximum overshoot is high for both large and small values of K_P. Use the value of K_I found in part (a). Find the value of K_P when the maximum overshoot is a minimum. What is the value of this maximum overshoot?

• Design of system with PI controller

10-11. Repeat Problem 10-10 for $K_v = 10$.

• Design of system with PID controller

10-12. A control system with a type 0 process and a PID controller is shown in Fig. 10P-12. Design the controller parameters so that the following specifications are satisfied:

$$\text{Ramp-error constant } K_v = 100$$
$$\text{Rise time } t_r < 0.01 \text{ sec.}$$
$$\text{Maximum overshoot} < 2 \text{ percent}$$

Plot the unit-step response of the designed system.

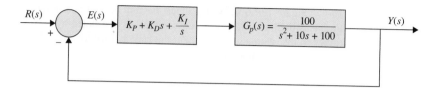

Figure 10P-12

Design of auto engine
control system. Time
domain

10-13. A considerable amount of effort is being spent by automobile manufacturers to meet the exhaust-emission-performance standards set by the government. Modern automotive-power-plant systems consist of an internal combustion engine that has an internal cleanup device called the catalytic converter. Such a system requires control of the engine air-fuel ratio (A/F), the ignition-spark timing, exhaust-gas recirculation, and injection air. The control system problem considered in this exercise deals with the control of the air-fuel ratio. In general, depending on fuel composition and other factors, a typical stoichiometric A/F is 14.7:1, that is, 14.7 grams of air to each gram of fuel. An A/F greater or less than stoichiometry will cause high hydrocarbons, carbon monoxide, and nitrous oxides in the tailpipe emission. The control system whose block diagram is shown in Fig. 10P-13 is devised to control the air-fuel ratio so that a desired output variable is maintained for a given command signal. Figure 10P-13 shows that the sensor senses the composition of the exhaust-gas mixture entering the catalytic converter. The electronic controller detects the difference or the error between the command and the sensor signals and computes the control signal necessary to achieve the desired exhaust-gas composition. The output variable $y(t)$ denotes the effective air-fuel ratio. The transfer function of the engine is given by

$$\frac{Y(s)}{U(s)} = G_p(s) = \frac{e^{-T_d s}}{1 + \tau s}$$

where T_d is the time delay and is 0.2 sec. The time constant τ is 0.25 sec. Approximate the time delay by a power series:

$$e^{-T_d s} \cong \frac{1}{1 + T_d s + T_d^2 s^2/2}$$

(a) Let the controller be a PI controller so that

$$G_c(s) = \frac{U(s)}{E(s)} = K_P + \frac{K_I}{s}$$

Find the value of K_I so that the ramp-error constant K_v is 2. Determine the value of K_P so that the maximum overshoot of the unit-step response is a minimum and the settling time is a minimum. Give the values of the maximum overshoot and the settling time. Plot the unit-step response of $y(t)$. Find the marginal value of K_P for system stability.

(b) Can the system performance be further improved by using a PID controller?

$$G_c(s) = \frac{U(s)}{E(s)} = K_P + K_D s + \frac{K_I}{s}$$

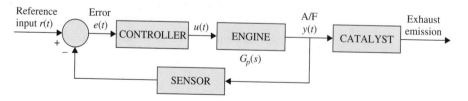

Figure 10P-13

10-14. One of the advantages of the frequency-domain analysis and design methods is that systems with pure time delays can be treated without approximation. Consider the automobile-engine control system treated in Problem 10-13.

The process has the transfer function

$$G_p(s) = \frac{e^{-0.2s}}{1 + 0.25s}$$

Design of auto engine
control system. Frequency
domain

Let the controller be of the PI type so that $G_c(s) = K_P + K_I/s$. Set the value of K_I so that the ramp-error constant K_v is 2. Find the value of K_P so that the phase margin is a maximum. How does this "optimal" K_P compare with the value of K_P found in Problem 10-13(a)? Find the critical value of K_P for system stability. How does this value of K_P compare with the critical value of K_P found in Problem 10-13?

• Design of space
telescope control system

10-15. The telescope for tracking stars and asteroids on the space shuttle may be modeled as a pure mass M. It is suspended by magnetic bearings so that there is no friction, and its attitude is controlled by magnetic actuators located at the base of the payload. The dynamic model for the control of the z-axis motion is shown in Fig. 10P-15(a). Since there are electrical components on the telescope, electric power must be brought to the telescope through a cable. The spring shown is used to model the wire-cable attachment, which exerts a spring force on the mass. The force produced by the magnetic actuators is denoted by $f(t)$. The force equation of motion in the z direction is

$$f(t) - K_s z(t) = M\frac{d^2 z(t)}{dt^2}$$

where $K_s = 1$ lb/ft, and $M = 150$ lb (mass), $f(t)$ is in pounds, and $z(t)$ is measured in feet.

(a) Show that the natural response of the system output $z(t)$ is oscillatory without damping. Find the natural undamped frequency of the open-loop space-shuttle system.

(b) Design the PID controller:

$$G_c(s) = K_P + K_D s + \frac{K_I}{s}$$

shown in Fig. 10-15(b) so that the following performance specifications are satisfied:

Ramp-error constant $K_v = 100$

The complex characteristic equation roots correspond to a relative
damping ratio of 0.707 and a natural undamped frequency of 1 rad/sec

Compute and plot the until-step response of the designed system. Find the maximum overshoot. Comment on the design results.

(c) Design the PID controller so that the following specifications are satisfied:

Ramp-error constant $K_v = 100$
Maximum overshoot < 5 percent

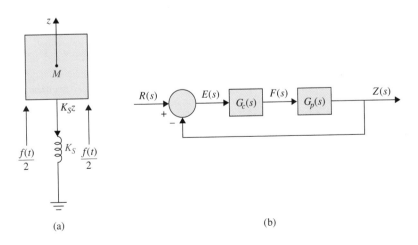

Figure 10P-15   (a)   (b)

Compute and plot the unit-step response of the designed system. Find the roots of the characteristic equation of the designed system.

10-16. Repeat Problem 10-15(b) with the following specifications:

Ramp-error constant $K_v = 100$

The complex characteristic equation roots correspond to a relative damping ratio of 1.0 and a natural undamped frequency of 1 rad/sec

10-17. An inventory control system is modeled by the following state equations:

$$\frac{dx_1(t)}{dt} = -2x_2(t)$$

$$\frac{dx_2(t)}{dt} = -2u(t)$$

where $x_1(t)$ = level of inventory, $x_2(t)$ = rate of sales of product, and $u(t)$ = production rate. The output equation is $y(t) = x_1(t)$. One unit of time is one day. Figure 10P-17 shows the block diagram of the closed-loop inventory control system with a series controller. Let the controller be a PD controller, $G_c(s) = K_P + K_D s$.

(a) Find the parameters of the PD controller, K_P and K_D, so that the roots of the characteristic equation correspond to a relative damping ratio of 0.707 and $\omega_n = 1$ rad/sec. Plot the unit-step response of $y(t)$ and find the maximum overshoot.

(b) Find the values of K_P and K_D so that the overshoot is zero and the rise time is less than 0.06 sec.

(c) Design the PD controller so that $M_r = 1$ and BW ≤ 40 rad/sec.

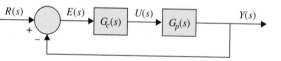

Figure 10P-17

10-18. The block diagram of a type 2 control system with a series controller $G_c(s)$ is shown in Fig. 10P-18.

The objective is to design a PD controller so that the following specifications are satisfied:

Maximum overshoot < 10 percent

Rise time < 0.5 sec

(a) Obtain the characteristic equation of the closed-loop system, and determine the ranges of the values of K_P and K_D for stability. Show the region of stability in the K_D-versus-K_P plane.

(b) Construct the root loci of the characteristic equation with $K_D = 0$ and $0 \leq K_P < \infty$. Then construct the root contours for $0 \leq K_D < \infty$ and several fixed values of K_P ranging from 0.001 to 0.01.

(c) Design the PD controller to satisfy the performance specifications given. Use the information on the root contours to help your design. Plot the unit-step response of $y(t)$.

(d) Check the design results obtained in part (c) in the frequency domain. Determine the phase margin, gain margin, M_r and BW of the designed system.

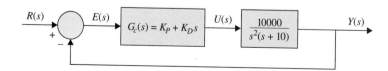

Figure 10P-18

• Design of PD controller for broom-balancing system

10-19. Consider the broom-balancing control system described in Problems 4-22 and 5-41. The $\mathbf{A}^*$ and $\mathbf{B}^*$ matrices are given in Problem 5-41 for the small-signal linearized model.

$$\Delta\dot{\mathbf{x}}(t) = \mathbf{A}^*\Delta\mathbf{x}(t) + \mathbf{B}^*\Delta r(t)$$
$$\Delta y(t) = \mathbf{C}\Delta\mathbf{x}(t)$$
$$\mathbf{D}^* = \begin{bmatrix} 0 & 0 & 1 & 0 \end{bmatrix}$$

Figure 10P-19 shows the block diagram of the system with a series PD controller. Determine if the PD controller can stabilize the system; if so, find the values of K_P and K_D. If the PD controller cannot stabilize the system, explain why not.

Figure 10P-19

• Design of series controller

10-20. The process of a unity-feedback control system has the transfer function

$$G_p(s) = \frac{100}{s^2 + 10s + 100}$$

Design a series controller (PD, PI, or PID) so that the following performance specifications are satisfied:

Steady-state error due to a step input $= 0$

Maximum overshoot < 2 percent

Rise time < 0.02 sec

Carry out the design in the frequency domain and check the design in the time domain.

• Design of phase-lead controller for inventory-control system

10-21. For the inventory control system shown in Fig. 10P-17, let the controller be of the phase-lead type:

$$G_c(s) = \frac{1 + aTs}{1 + Ts} \qquad a > 1$$

Determine the values of a and T so that the following performance specifications are satisfied:

Steady-state error due to a step input $= 0$

Maximum overshoot < 5 percent

(a) Design the controller using the root contours with T and a as variable parameters. Plot the unit-step response of the designed system. Plot the Bode diagram of $G(s) = G_c(s)G_p(s)$, and find PM, GM, M_r, and BW of the designed system.

(b) Design the phase-lead controller so that the following performance specifications are satisfied:

Steady-state error due to a step input $= 0$

Phase margin $> 75°$

$M_r < 1.1$

Construct the Bode diagram of $G(s)$ and carry out the design in the frequency domain. Find the attributes of the time response of the designed system.

• Design of phase-lead controller

10-22. Consider that the process of a unity-feedback control system is

$$G_p(s) = \frac{1000}{s(s + 10)}$$

Let the series controller be a single-stage phase-lead controller:

$$G_c(s) = \frac{1 + aTs}{1 + Ts} \qquad a > 1$$

(a) Determine the values of a and T so that the zero of $G_c(s)$ cancels the pole of $G_p(s)$ at $s = -10$.

The damping ratio of the designed system should be unity. Find the attributes of the unit-step response of the designed system.

(b) Carry out the design in the frequency domain using the Bode plot. The design specifications are

$$\text{Phase margin} > 75° \qquad M_r < 1.1$$

Find the attributes of the unit-step response of the designed system.

• Design of phase-lead controller for liquid-level control system

10-23. Consider that the controller in the liquid-level control system shown in Fig. 10P-6 is a phase-lead controller:

$$G_c(s) = \frac{1 + aTs}{1 + Ts} \qquad a > 1$$

(a) For $N = 20$, select the values of a and T so that the maximum overshoot is barely 0 percent. The value of a must not exceed 1000. Find the attributes of the unit-step response of the designed system. Plot the unit-step response.

(b) For $N = 20$, design the phase-lead controller in the frequency domain. Find the values of a and T so that the phase margin is maximized subject to the condition that $BW > 100$. The value of a must not exceed 1000.

• Design of single-stage and two-stage phase-lead controllers

10-24. The transfer function of the process of a unity-feedback control system is

$$G_p(s) = \frac{6}{s(1 + 0.2s)(1 + 0.5s)}$$

(a) Construct the Bode diagram of $G_p(j\omega)$ and determine the PM, GM, M_r, and BW of the system.

(b) Design a series single-stage series phase-lead controller with the transfer function

$$G_c(s) = \left(\frac{1 + aTs}{1 + Ts}\right) \qquad a > 1$$

so that the phase margin is maximum. The value of a must not be greater than 1000. Determine PM and M_r of the designed system. Determine the attributes of the unit-step response.

(c) Using the system designed in part (b) as a basis, design a two-stage phase-lead controller so that the phase margin is at least 85°. The transfer function of the two-stage phase-lead controller is

$$G_c(s) = \left(\frac{1 + aT_1s}{1 + T_1s}\right)\left(\frac{1 + bT_2s}{1 + T_2s}\right) \qquad a > 1 \qquad b > 1$$

where a and T_1 are determined in part (b). The value of T_2 should not exceed 1000. Find the values of PM and M_r of the designed system. Find the attributes of the unit-step response.

(d) Plot the unit-step responses of the output in parts (a), (b), and (c).

• Design of phase-lock-loop system, phase-lead controller

10-25 A phase-lock-loop, dc-motor-speed-control system is described in Problem 4-19. The block diagram of the system is shown in Fig. 10P-25. The system parameters and transfer functions are given as follows:

$$\text{Reference speed command, } \omega_r = 120 \text{ pulse/sec}$$
$$\text{Phase-detector gain, } K_p = 0.06 \text{ V/pulse/sec}$$
$$\text{Amplifier gain, } K_a = 20$$
$$\text{Encoder gain, } K_e = 5.73 \text{ pulse/rad}$$
$$\text{Counter gain, } N = 1$$
$$\text{Motor transfer function,}$$
$$\frac{\Omega_m(s)}{E_a(s)} = \frac{10}{s(1 + 0.05s)}$$

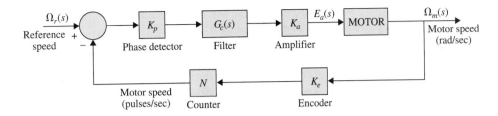

Figure 10P-25

(a) Let the filter (controller) transfer function be of the form:

$$G_c(s) = \frac{E_o(s)}{E_i(s)} = \frac{1 + R_2 Cs}{R_1 Cs}$$

where $R_1 = 2 \times 10^6 \, \Omega$ and $C = 1 \, \mu F$. Determine the value of R_2 so that the complex roots of the closed-loop characteristic equation have a maximum relative damping ratio. Sketch the root loci of the characteristic equation for $0 \leq R_2 < \infty$. Compute and plot the unit-step responses of the motor speed $f_\omega(t)$ (pulse/sec) with the values of R_2 found, when the input is 120 pulse/sec. Convert the speed in pulse/sec to rpm.

(b) Let the filter transfer function be

$$G_c(s) = \frac{1 + aTs}{1 + Ts} \qquad a > 1$$

where $T = 0.01$. Find a so that the complex roots of the characteristic equation have a maximum relative damping ratio. Compute and plot the unit-step response of the motor speed $f_\omega(t)$ (pulse/sec) when the input is 120 pulse/sec.

(c) Design the phase-lead controller in the frequency domain so that the phase margin is at least 60°.

• Liquid-level control
system design. Phase-lag
control

10-26. Consider that the controller in the liquid-level control system shown in Fig. 10P-6 is a single-stage phase-lag controller:

$$G_c(s) = \frac{1 + aTs}{1 + Ts} \qquad a < 1$$

(a) For $N = 20$, select the values of a and T so that the two complex roots of the characteristic equation correspond to a relative damping ratio of approximately 0.707. Plot the unit-step response of the output $y(t)$. Find the attributes of the unit-step response. Plot the Bode plot of $G_c(s)G_p(s)$ and determine the phase margin of the designed system.

(b) For $N = 20$, design the phase-lag controller in the frequency domain so that the phase margin is approximately 60°. Plot the unit-step response of the output $y(t)$, and find the attributes of the unit-step response.

• Phase-lead and phase-lag
controller design

10-27. The controlled process of a unity-feedback control system is

$$G_p(s) = \frac{K}{s(s + 5)^2}$$

The series controller has the transfer function

$$G_c(s) = \frac{1 + aTs}{1 + Ts}$$

(a) Design a phase-lead controller ($a > 1$) so that the following performance specifications are satisfied:

Ramp-error constant $K_v = 10$ Maximum overshoot is near minimum

The value of a must not exceed 1000. Plot the unit-step response and give its attributes.

(b) Design a phase-lead controller in the frequency domain so that the following performance specifications are satisfied:

$$\text{Ramp-error constant } K_v = 10$$
$$\text{Phase margin is near maximum}$$

The value of a must not exceed 1000.

(c) Design a phase-lag controller ($a < 1$) so that the following performance specifications are satisfied:

$$\text{Ramp-error constant } K_v = 10$$
$$\text{Maximum overshoot} < 1 \text{ percent}$$
$$\text{Rise time } t_r < 2 \text{ sec}$$
$$\text{Settling time } t_s < 2.5 \text{ sec}$$

Find the PM, GM, M_r, and BW of the designed system.

(d) Design the phase-lag controller in the frequency domain so that the following performance specifications are satisfied:

$$\text{Ramp-error constant } K_v = 10$$
$$\text{Phase margin} \geq 70°$$

Check the unit-step response attributes of the designed system and compare with those obtained in part (c).

• Design of dc-motor control system with shaft compliance

10-28. The controlled process of a dc-motor control system with unity feedback has the transfer function

$$G_p(s) = \frac{6.087 \times 10^{10}}{s(s^3 + 423.42s^2 + 2.6667 \times 10^6 s + 4.2342 \times 10^8)}$$

Due to the compliance in the motor shaft, the process transfer function contains two lightly damped poles, which will cause oscillations in the output response. The following performance criteria are to be satisfied:

$$\text{Maximum overshoot} < 1 \text{ percent}$$
$$\text{Rise time } t_r < 0.15 \text{ sec}$$
$$\text{Settling time } t_s < 0.15 \text{ sec}$$
$$\text{Output response should not have oscillations}$$
$$\text{Ramp-error constant is not affected}$$

(a) Design a series phase-lead controller,

$$G_c(s) = \frac{1 + aTs}{1 + Ts} \qquad a > 1$$

so that all the step-response attributes (except for the oscillations) are satisfied.

(b) To eliminate the oscillations due to the motor shaft compliance, add another stage to the controller with the transfer function

$$G_{c1}(s) = \frac{s^2 + 2\zeta_z \omega_n s + \omega_n^2}{s^2 + 2\zeta_p \omega_n s + \omega_n^2}$$

so that the zeros of $G_{c1}(s)$ will cancel the two complex poles of $G_p(s)$. Set the value of ζ_p so that the two poles of $G_{c1}(s)$ will not have an appreciable affect on the system response. Determine the attributes of the unit-step response to see if all the requirements are satisfied. Plot the unit-step responses of the uncompensated system, the compensated system with the phase-lead controller designed in part (b).

• Design of computer-tape-drive system. PI controller

10-29.   A computer-tape-drive system utilizing a permanent-magnet dc-motor is shown in Fig. 10P-29(a). The closed-loop system is modeled by the block diagram in Fig. 10P-29(b). The constant K_L represents the spring constant of the elastic tape, and B_L denotes the viscous-friction coefficient between the tape and the capstans. The system parameters are as follows:

$$K_i = \text{motor torque constant} = 10 \text{ oz-in./A}$$
$$K_b = \text{motor back-emf constant} = 0.0706 \text{ V/rad/sec}$$
$$B_m = \text{motor friction coefficent} = 3 \text{ oz-in./rad/sec}$$
$$R_a = 0.25 \ \Omega \qquad\qquad L_a \cong 0 \text{ H}$$
$$K_L = 3000 \text{ oz-in./rad} \qquad B_L = 10 \text{ oz-in./rad/sec}$$
$$J_L = 6 \text{ oz-in./rad/sec}^2 \qquad K_f = 1 \text{ V/rad/sec}$$
$$J_m = 0.05 \text{ oz-in./rad/sec}^2$$

(a) Write the state equations of the system between e_a and θ_L using θ_L, ω_L, θ_m, and ω_m as state variables and e_a as input. Draw a state diagram using the state equations. Derive the transfer functions:

$$\frac{\Omega_m(s)}{E_a(s)} \quad \text{and} \quad \frac{\Omega_L(s)}{E_a(s)}$$

(b) The objective of the system is to control the speed of the tape, ω_L, accurately. Consider that PI controller with the transfer function $G_c(s) = K_P + K_I/s$ is to be used. Find the values of K_P and K_I so that the following specifications are satisfied:

$$\text{Ramp-error constant } K_v = 100$$
$$\text{Rise time} < 0.02 \text{ sec}$$
$$\text{Settling time} < 0.02 \text{ sec}$$
$$\text{Maximum overshoot} < 1 \text{ percent or at minimum}$$

Plot the unit-step response of $\omega_L(t)$ of the system.

(c) Design the PI controller in the frequency domain. The value of K_I is to be selected as in part (b). Vary the value of K_P and compute the values of PM, GM, M_r, and BW.

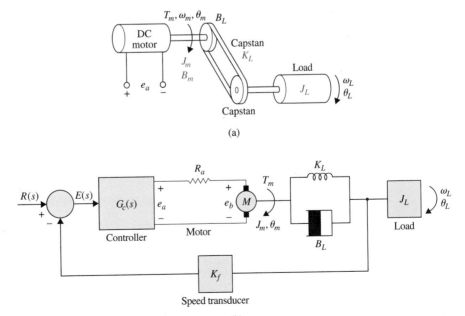

(a)

(b)

Figure 10P-29

Find the value of K_P so that PM is maximum. How does this value of K_P compare with the result obtained in part (b)?

• Motor-control system with flexible shaft, notch controller

10-30. Figure 10P-30 shows the block diagram of a motor-control system that has a flexible shaft between the motor and the load. The transfer function between the motor torque and motor displacement is

$$G_p(s) = \frac{\Theta_m(s)}{T_m(s)} = \frac{J_L s^2 + B_L s + K_L}{s[J_m J_L s^3 + (B_m J_L + B_L J_m)s^2 + (K_L J_m + K_L J_L + B_m B_L)s + B_m K_L]}$$

where $J_L = 0.01$, $B_L = 0.1$, $K_L = 10$, $J_m = 0.01$, $B_m = 0.1$, and $K = 100$.

(a) Compute and plot the unit-step response of $\theta_m(t)$. Find the attributes of the unit-step response.

(b) Design a second-order notch controller with the transfer function

$$G_c(s) = \frac{s^2 + 2\zeta_z \omega_n s + \omega_n^2}{s^2 + 2\zeta_p \omega_n s + \omega_n^2}$$

so that its zeros cancel the complex poles of $G_p(s)$. The two poles of $G_c(s)$ should be selected so that they do not affect the steady-state response of the system, and the maximum overshoot is a minimum. Compute the attributes of the unit-step response and plot the response.

(d) Carry out design of the second-order controller in the frequency domain. Plot the Bode diagram of the uncompensated $G_p(s)$, and find the values of PM, GM, M_r, and BW. Set the two zeros of $G_c(s)$ to cancel the two complex poles of $G_p(s)$. Determine the value of ζ_p by determining the amount of attenuation required from the second-order notch controller and using Eq. (10-155). Find the PM, GM, M_r, and BW of the compensated system. How do the frequency-domain design results compare with the results in part (b)?

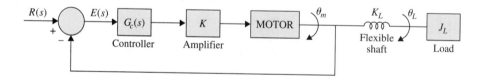

$R(s)$ $E(s)$ $G_c(s)$ K MOTOR θ_m K_L θ_L J_L

Controller   Amplifier   Flexible shaft   Load

Figure 10P-30

• Design of notch controller

10-31. The transfer function of the process of a unity-feedback control system is

$$G_p(s) = \frac{500(s + 10)}{s(s^2 + 10s + 1000)}$$

(a) Plot the Bode diagram of $G_p(s)$ and determine the PM, GM, M_r, and BW of the uncompensated system. Compute and plot the unit-step response of the system.

(b) Design a series second-order notch controller with the transfer function

$$G_c(s) = \frac{s^2 + 2\zeta_z \omega_n s + \omega_n^2}{s^2 + 2\zeta_p \omega_n s + \omega_n^2}$$

so that its zeros cancel the complex poles of $G_p(s)$. Determine the value of ζ_p using the method outlined in Section 10-8-2. Find the PM, GM, M_r, and BW of the designed system. Compute and plot the unit-step response.

(c) Design the series second-order notch controller so that its zeros cancel the complex poles of $G_p(s)$. Determine the value of ζ_p so that the following specifications are satisfied:

$$\text{Maximum overshoot} < 1 \text{ percent}$$
$$\text{Rise time} < 0.4 \text{ sec}$$
$$\text{Settling time} < 0.5 \text{ sec}$$

• Forward controller design

10-32. Design the controllers $G_{cf}(s)$ and $G_c(s)$ for the system shown in Fig. 10P-32 so that the following specifications are satisfied:

Ramp-error constant $K_v = 50$

Dominant roots of the characteristic equation at $-5 \pm j5$ approximately

Rise time < 0.1 sec

System must be robust when K varies ± 20 percent from the nominal value, with the rise time and overshoot staying within specifications

Compute and plot the unit-step responses to check the design.

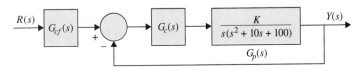

Figure 10P-32

• Forward controller design, motor-control system

10-33. Figure 10P-33 shows the block diagram of a motor-control system. The transfer function of the controlled process is

$$G_p(s) = \frac{1000K}{s(s + a)}$$

where K denotes the aggregate of the amplifier gain and motor torque constant, and a is the inverse of the motor time constant. Design the controllers $G_{cf}(s)$ and $G_c(s)$ so that the following performance specifications are satisfied.

Ramp-error constant $K_v = 100$ when $a = 10$

Rise time < 0.3 sec

Maximum overshoot < 8 percent

Dominant characteristic equation roots $= -5 \pm j5$

System must be robust when a varies between 8 and 12

Compute and plot the unit-step responses to verify the design.

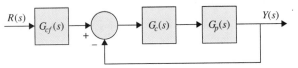

Figure 10P-33

• Design of system with tachometer feedback

10-34. Figure 10P-34 shows the block diagram of a dc-motor control system with tachometer feedback. Find the values of K and K_t so that the following specifications are satisfied:

Ramp-error constant $K_v = 1$

Dominant characteristic equation roots correspond to a damping ratio of approximately 0.707; if there are two solutions, select the larger value of K

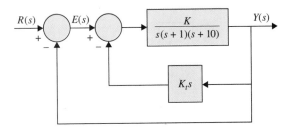

Figure 10P-34

• Design of system with
tachometer feedback

10-35. Carry out the design with the specifications given in Problem 10-34 for the system shown in Fig. 10P-35.

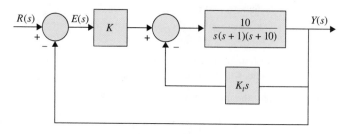

Figure 10P-35

• Design of system with
tachometer feedback

10-36. The block diagram of a control system with a type 2 process is shown in Fig. 10P-36. The system is to be compensated by tachometer feedback and a series controller. Find the values of a, T, K, and K_t so that the following performance specifications are satisfied:

Ramp-error constant $K_v = 100$

Dominant characteristic equation roots correspond to a damping ratio of 0.707

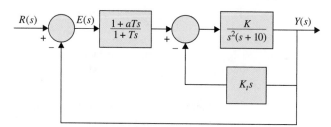

Figure 10P-36

• Design of aircraft-
attitude control system,
tachometer feedback

10-37. The aircraft-attitude control system described in Section 7-7 is modeled by the block diagram shown in Fig. 10P-37. The system parameters are

$K = $ variable $K_s = 1$ $K_1 = 10$ $K_2 = 0.5$ $K_t = $ variable $R_a = 5$

$L_a = 0.003$ $K_i = 9.0$ $K_b = 0.0636$ $J_m = 0.0001$ $J_L = 0.01$

$B_m = 0.005$ $B_L = 1.0$ $N = 0.1$

Find the values of K and K_t so that the following specifications are satisfied:

Ramp-error constant $K_v = 100$

Relative damping ratio of the complex roots of the characteristic equation is approximately 0.707

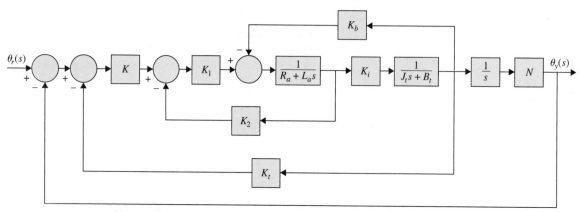

Figure 10P-37

Plot the unit-step response of the designed system. Show that the system performance is extremely insensitive to the value of K. Explain why this is so.

• Design of position-feedback control system, Single- and two-stage phase-lead controllers

10-38. Figure 10P-38 shows the block diagram of a position-control system with a series controller $G_c(s)$.

(a) Determine the minimum value of the amplifier gain K so that the steady-state value of the output $y(t)$ due to a unit-step torque disturbance is ≤ 0.01.

(b) Show that the uncompensated system is unstable with the minimum value of K determined in part (a). Construct the Bode diagram for the open-loop transfer function $G(s) = Y(s)/E(s)$, and find the values of PM and GM.

(c) Design a single-stage phase-lead controller with the transfer function

$$G_c(s) = \frac{1 + aTs}{1 + Ts} \qquad a > 1$$

so that the phase margin is 30°. Show that this is nearly the highest phase margin that can be achieved with a single-stage phase-lead controller. Find GM, M_r, and BW of the compensated system.

(d) Design a two-stage phase-lead controller using the system arrived at in part (c) as a basis so that the phase margin is 55°. Show that this is the best PM that can be obtained for this system with a two-stage phase-lead controller. Find GM, M_r, and BW of the compensated system.

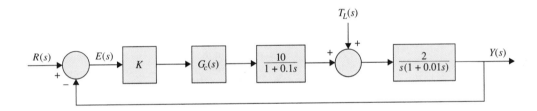

Figure 10P-38

• Design of two-stage phase-lead controller, phase-lag controller, lag-lead controller

10-39. The transfer function of the process of a unity-feedback control system is

$$G_c(s) = \frac{60}{s(1 + 0.2s)(1 + 0.5s)}$$

Show that due to the relative high gain, the uncompensated system is unstable.

(a) Design a two-stage phase-lead controller with

$$G_c(s) = \left(\frac{1 + aT_1s}{1 + T_1s}\right)\left(\frac{1 + bT_2s}{1 + T_2s}\right) \qquad a > 1, \quad b > 1 \tag{1}$$

so that the phase margin is greater than 60°. Conduct the design by first determining the values of a and T_1 to realize a maximum phase margin that can be achieved with a single-stage phase-lead controller. The second stage of the controller is then designed to realize the balance of the 60° phase margin. Determine GM, M_r, and BW of the compensated system. Compute and plot the unit-step response of the compensated system.

(b) Design a single-stage phase-lag controller with

$$G_c(s) = \frac{1 + aTs}{1 + Ts} \qquad a < 1$$

so that the phase margin of the compensated system is greater than 60°. Determine GM, M_r, and BW of the compensated system. Compute and plot the unit-step response of the compensated system.

(c) Design a lag-lead controller with $G_c(s)$ as in Eq. (1) in part (a). Design the phase-lag portion first by setting the phase margin at 40°. The resulting system is then compensated by the phase-lead portion to achieve a total of 60° of phase margin. Determine GM, M_r, and BW of the compensated system. Compute and plot the unit-step response of the compensated system.

• Design of steel-rolling
control system

10-40. The block diagram of the steel-rolling system described in Problem 4-25 is shown in Fig. 10P-40. The transfer function of the process is

$$G_p(s) = \frac{Y(s)}{E(s)} = \frac{5e^{-0.1s}}{s(1 + 0.1s)(1 + 0.5s)} \qquad K_s = 1$$

(a) Approximate the time delay by

$$e^{-0.1s} \cong \frac{1 - 0.05s}{1 + 0.05s}$$

Design a series controller of your choice so that the phase margin of the compensated system is at least 60°. Determine GM, M_r, and BW of the compensated system. Compute and plot the unit-step responses of the compensated and the uncompensated systems.

(b) Repeat part (a) without using the approximation of the time delay.

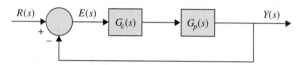

Figure 10P-40

• Design of respiratory
control system, PI
controller

10-41. Human beings breathe in order to provide for gas exchange for the entire body. A respiratory control system is needed to ensure that the body's needs for this gas exchange are adequately met. The criterion of control is adequate ventilation, which ensures satisfactory levels of both oxygen and carbon dioxide in the arterial blood. Respiration is controlled by neural impulses that originate within the lower brain and are transmitted to the chest cavity and diaphragm to govern the rate and tidal volume. One source of signals consists of the chemoreceptors located near the respiratory center, which are sensitive to carbon dioxide and oxygen concentrations. Figure 10P-41 shows the block diagram of a simplified model of the human respiratory control system. The objective is to control the effective ventilation of the lungs so that a satisfactory balance of concentrations of carbon dioxide and oxygen is maintained in the blood circulated at the chemoreceptor.

(a) Plot the Bode diagram of the transfer function $G(s) = Y(s)/E(s)$ when $G_c(s) = 1$. Find the PM and GM. Determine the stability of the system.

(b) Design a PI controller, $G_c(s) = K_P + K_I/s$, so that the following specifications are satisfied:

$$\text{Ramp-error constant } K_v = 1 \qquad \text{Phase margin is maximized.}$$

Plot the unit-step response of the system. Find the attributes of the unit-step response.

(c) Design a PI controller so that the following specifications are satisfied:

$$\text{Ramp-error constant } K_v = 1 \qquad \text{Maximum overshoot is minimized.}$$

Plot the unit-step response of the system. Find the attributes of the unit-step response. Compare the design results in parts (b) and (c).

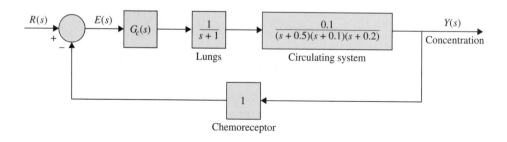

Figure 10P-41

• State-feedback control

10-42. The block diagram of a control system with state feedback is shown in Fig. 10P-42. Find the real feedback gains, k_1, k_2, and k_3 so that:

The steady-state error e_{ss} [$e(t)$ is the error signal] due to a step input is zero.

The complex roots of the characteristic equation are at $-1 + j$ and $-1 - j$.

Find the third root. Can all three roots be arbitrarily assigned while still meeting the steady-state requirement?

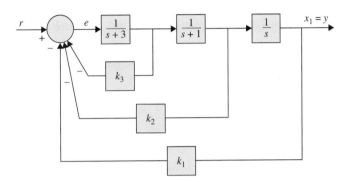

Figure 10P-42

State-feedback control

10-43. The block diagram of a control system with state feedback is shown in Fig. 10P-43(a). The feedback gains k_1, k_2, and k_3 are real constants.

(a) Find the values of the feedback gains so that:

The steady-state error e_{ss} [$e(t)$ is the error signal] due to a step input is zero.

The characteristic equation roots are at $-1 + j$, $-1 - j$, and -10.

(b) Instead of using state feedback, a series controller is implemented, as shown in Fig. 10P-43(b). Find the transfer function of the controller $G_c(s)$ in terms of k_1, k_2, and k_3 found in part (a) and the other system parameters.

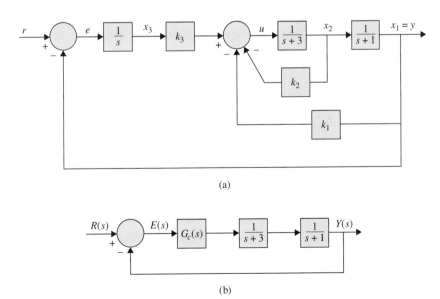

Figure 10P-43

(b)

• Broom-balancing system with state-feedback control

10-44. Problem 10-19 has revealed that it is impossible to stabilize the broom-balancing control system described in Problems 4-22 and 5-41 with a series PD controller. Consider that the system is now controlled by state feedback with $\Delta r(t) = -\mathbf{K}\mathbf{x}(t)$, where

$$\mathbf{K} = \begin{bmatrix} k_1 & k_2 & k_3 & k_4 \end{bmatrix}$$

(a) Find the feedback gains k_1, k_2, k_3, and k_4 so that the eigenvalues of $\mathbf{A}^* - \mathbf{B}^*\mathbf{K}$ are at $-1 + j$, $-1 - j$, -10, and -10. Compute and plot the responses of $\Delta x_1(t)$, $\Delta x_2(t)$, $\Delta x_3(t)$, and $\Delta x_4(t)$ for the initial condition, $\Delta x_1(0) = 0.1$, $\Delta\theta(0) = 0.1$, and all other initial conditions are zero.

(b) Repeat part (a) for the eigenvalues at $-2 + j2$, $-2 - j2$, -20, and -20. Comment on the difference between the two systems.

• Ball-suspension system with state-feedback control

10-45. The linearized state equations of the ball-suspension control system described in Problem 4-24 are expressed as

$$\Delta\dot{\mathbf{x}}(t) = \mathbf{A}^*\Delta\mathbf{x}(t) + \mathbf{B}^*\Delta i(t)$$

where

$$\mathbf{A}^* = \begin{bmatrix} 0 & 1 & 0 & 0 \\ 115.2 & -0.05 & -18.6 & 0 \\ 0 & 0 & 0 & 1 \\ -37.2 & 0 & 37.2 & -0.1 \end{bmatrix} \quad \mathbf{B}^* = \begin{bmatrix} 0 \\ -6.55 \\ 0 \\ -6.55 \end{bmatrix}$$

Let the control current $\Delta i(t)$ be derived from the state feedback $\Delta i(t) = -\mathbf{K}\Delta\mathbf{x}(t)$, where

$$\mathbf{K} = \begin{bmatrix} k_1 & k_2 & k_3 & k_4 \end{bmatrix}$$

(a) Find the elements of $\mathbf{K}$ so that the eigenvalues of $\mathbf{A}^* - \mathbf{B}^*\mathbf{K}$ are at $-1 + j$, $-1 - j$, -10, and -10.

(b) Plot the responses of $\Delta x_1(t) = \Delta y_1(t)$ (magnet displacement) and $\Delta x_3(t) = \Delta y_2(t)$ (ball displacement) with the initial condition:

$$\Delta\mathbf{x}(0) = \begin{bmatrix} 0.1 \\ 0 \\ 0 \\ 0 \end{bmatrix}$$

(c) Repeat part (b) with the initial condition:

$$\Delta\mathbf{x}(0) = \begin{bmatrix} 0 \\ 0 \\ 0.1 \\ 0 \end{bmatrix}$$

Comment on the responses of the closed-loop system with the two sets of initial conditions used in (b) and (c).

• Temperature control of electric furnace, state feedback, PI control

10-46. The temperature $x(t)$ in the electric furnace shown in Fig. 10P-46 is described be the differential equation

$$\frac{dx(t)}{dt} = -2x(t) + u(t) + n(t)$$

where $u(t)$ is the control signal, and $n(t)$ the constant disturbance of unknown magnitude due to heat loss. It is desired that the temperature $x(t)$ follows a reference input r that is a constant.

(a) Design a control system with state and integral control so that the following specifications are satisfied:

$$\lim_{t\to\infty} x(t) = r = \text{constant}$$

The eigenvalues of the closed-loop system are at -10 and -10.

Plot the responses of $x(t)$ for $t \geq 0$ with $r = 1$ and $n(t) = -1$, and then with $r = 1$ and $n(t) = 0$, all with $x(0) = 0$.

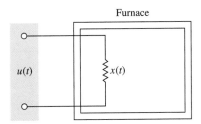

Furnace

$u(t)$ $x(t)$

Figure 10P-46

(b) Design a PI controller so that

$$G_c(s) = \frac{U(s)}{E(s)} = K_P + \frac{K_I}{s}$$

$$E(s) = R(s) - X(s) \qquad \text{where } R(s) = R/s$$

Find K_P and K_I so that the characteristic equation roots are at -10 and -10. Plot the responses of $x(t)$ for $t \geq 0$ with $r = 1$ and $n(t) = -1$, and then with $r = 1$ and $n(t) = 0$, all $x(0) = 0$.

Answers to True-and-False Review Questions
5. (F)   6. (F)   8. (T)   18. (T)   19. (T)   20. (T)   21. (T)   22. (T)   26. (T)
27. (T)   28. (T)

CHAPTER 11

The Virtual Lab

To enhance their understanding of control systems, students must participate in labs and projects involving practical problems. In this chapter we have created a so-called virtual lab with problems mimicking actual practical systems. A real system may exhibit properties that are not discussed in most theoretical textbooks but, ultimately, affect the controller design.

In a realistic system, including an actuator (e.g., a dc motor) and mechanical (gears) and electrical components (amplifiers), issues such as saturation of the amplifier, friction in the motor, or backlash in the gears will seriously affect the controller design. Students always become disenchanted after spending many hours designing an impractical controller that would ultimately not work in reality.

It was a challenging task to develop software that provides the reader with practical appreciation and understanding of dc motors including modeling, system identification, and controller design. However, through the use of MATLAB and Simulink, we managed to create a virtual lab representing some of the behaviors observed in an actual system. All the experiments presented here were compared to real systems in the lab environment, and their accuracy has been verified. These virtual labs include experiments on speed and position control of dc motors followed by a controller design project involving control of a simple robotic system. The focus on dc motors in these experiments was intentional, because of their simplicity and wide usage in numerous industrial applications.

The main objectives of this chapter are as follows:

1. To provide an in-depth description of dc motor speed response, speed control, and position control.

2. To provide preliminary instruction on how to identify the parameters of a system.

3. To show how different parameters and nonlinear effects such as friction and saturation affect the response of the motor.

4. To give a better feel for controller design through realistic examples.

5. To gain insight into the SIMLab and Virtual Lab software.

Before starting the lab, we need to provide a comprehensive summary of the theoretical development for dc motors.

▶ 11-2 IMPORTANT ASPECTS IN THE RESPONSE OF A DC MOTOR

Servomechanisms are probably the most frequently encountered electromechanical control systems. Applications include robots (each joint in a robot requires a position servo), numerical control (NC) machines, and laser printers, to name but a few. The common characteristic of all such systems is that the variable to be controlled (usually position or velocity) is fed back to modify the command signal. The servomechanism that will be used in the experiments in this chapter comprises a dc motor and amplifier that are fed back the motor speed and position values.

One of the key challenges in the design and implementation of a successful controller is obtaining an accurate model of the system components, particularly the actuator. In Chapters 4 and 7 we discussed various issues associated with modeling of dc motors. We will briefly revisit the modeling aspects in this section.

11-2-1 Speed Response and the Effects of Inductance and Disturbance-Open Loop Response

Consider the armature-controlled dc motor shown in Fig. 11-1, where the field current is held constant in this system. The system parameters include

R_a = armature resistance, ohm

L_a = armature inductance, henry

v_a = applied armature voltage, volt

v_b = back emf, volt

θ = angular displacement of the motor shaft, radian

T = torque developed by the motor, N-m

J_L = moment of inertia of the load, kg-m²

T_L = any external load torque considered as a disturbance, N-m

J_m = moment of inertia of the motor (motor shaft), kg-m²

J = equivalent moment of inertia of the motor and load connected to the motor-shaft, $J = J_L/n^2 + J_m$, kg-m² (refer to Chapter 4 for more details)

n = gear ratio

B = equivalent viscous-friction coefficient of the motor and load referred to the motor shaft, N-m/rad/sec (in the presence of gear ratio, B must be scaled by n; refer to Chapter 4 for more details)

K_t = speed sensor (usually a tachometer) gain

As shown in Fig. 11-2, the armature-controlled dc motor is itself a feedback system, where back-emf voltage is proportional to the speed of the motor. In Fig. 11-2, we have included the effect of any possible external load (e.g., the load applied to a juice machine

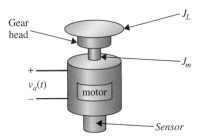

Figure 11-1 An armature-controlled dc motor with a gear head and a load inertia J_L.

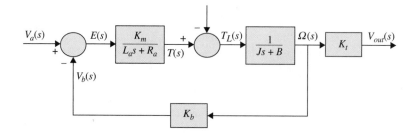

Figure 11-2 Block diagram of an armature-controlled dc motor.

by the operator pushing in the fruit) as a disturbance torque T_L. The system may be arranged in input-output form such that $V_a(s)$ is the input and $\Omega(s)$ is the output.

$$\Omega(s) = \frac{\dfrac{K_m}{R_a J_m}}{\left(\dfrac{L_a}{R_a}\right)s^2 + \left(J_m + B\dfrac{L_a}{R_a}\right)s + \dfrac{K_m K_b + R_a B}{R_a J_m}} V_a(s)$$

$$- \frac{\left\{1 + s\left(\dfrac{L_a}{R_a}\right)\right\}/J_m}{\left(\dfrac{L_a}{R_a}\right)s^2 + \left(J_m + B\dfrac{L_a}{R_a}\right)s + \dfrac{K_m K_b + R_a B}{R_a J_m}} T_L(s)$$

(11-1)

The ratio L_a/R_a is called the *motor electric-time constant*, which makes the system speed-response transfer function second order, and is denoted by τ_e. Also, it introduces a zero to the disturbance-output transfer function. However, as discussed in Chapter 7, since L_a in the armature circuit is very small, τ_e is neglected, resulting in the simplified transfer functions and the block diagram of the system. Thus, the speed of the motor shaft may be simplified to

- If the electric time is small, it may be neglected.

$$\Omega(s) = \frac{\dfrac{K_m}{R_a J_m}}{s + \dfrac{K_m K_b + R_a B}{R_a J_m}} V_a(s) - \frac{\dfrac{1}{J_m}}{s + \dfrac{K_m K_b + R_a B}{R_a J_m}} T_L(s)$$

(11-2)

or

$$\Omega(s) = \frac{K_{eff}}{\tau_m s + 1} V_a(s) - \frac{\dfrac{\tau_m}{J_m}}{\tau_m s + 1} T_L(s)$$

(11-3)

where $K_{eff} = K_m /(R_a B + K_m K_b)$ is the motor gain constant, and $\tau_m = R_a J_m /(R_a B + K_m K_b)$ is the motor mechanical time constant. If the load inertia and the gear ratio are incorporated into the system model, the inertia J_m in Eqs. (11-1) through (11-3) is replaced with J (total inertia).

Using superposition, we get

$$\Omega(s) = \Omega(s)|_{T_L(s)=0} + \Omega(s)|_{V_a(s)=0}$$

(11-4)

To find the response $\omega(t)$, we use superposition and find the response due to the individual inputs. For $T_L = 0$ (no disturbance and $B = 0$) and an applied voltage $V_a(t) = A$, such that $V_a(s) = A/s$,

$$\omega(t) = \frac{A}{K_b}(1 - e^{-t/\tau_m})$$

(11-5)

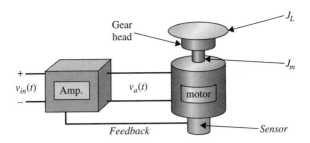

Figure 11-3 Feedback control of an armature-controlled dc motor with a load inertia J_L.

In this case, note that the motor mechanical time constant τ_m is reflective of how fast the motor is capable of overcoming its own inertia J_m to reach a steady state or constant speed dictated by voltage V_a. From Eq. (11-5) the speed final value is $\omega(t) = A/K_b$. As τ_m increases, the approach to steady state takes longer.

• Steady-state speed response of a dc motor is adversely affected by a disturbance load.

If we apply a constant load torque of magnitude D to the system (i.e., $T_L = D/s$), the speed response (11-5) will change to

$$\omega(t) = \frac{1}{K_b}\left(A - \frac{R_a D}{K_m}\right)(1 - e^{-t/\tau_m}) \tag{11-6}$$

which clearly indicates that the disturbance T_L affects the final speed of the motor. From Eq. (11-6), at steady state, the speed of the motor is $\omega_{fv} = \frac{1}{K_b}\left(A - \frac{R_a D}{K_m}\right)$. Here the final value of $\omega(t)$ is reduced by $R_a D/K_m K_b$. A practical note is that the value of $T_L = D$ may never exceed the motor stall torque, and hence for the motor to turn, from Eq. (11-6), $AK_m/R_a > D$, which sets a limit on the magnitude of the torque T_L. For a given motor, the value of the stall torque can be found in the manufacturer's catalog.

If the load inertia is incorporated into the system model, the final speed value becomes $\omega_{fv} = A/K_b$. Does the stall torque of the motor affect the response and the steady-state response? In a realistic scenario, you must measure motor speed using a sensor. How would the sensor affect the equations of the system (see Fig. 11-2)?

11-2-2 Speed Control of DC Motors: Closed-Loop Response

As seen previously, the output speed of the motor is highly dependant on the value of torque T_L. We can improve the speed performance of the motor by using a proportional feedback controller. The controller is composed of a sensor (usually a tachometer for speed applications) to sense the speed and an amplifier with gain K (proportional control) in the configuration shown in Fig. 11-3. The block diagram of the system is also shown in Fig. 11-4.

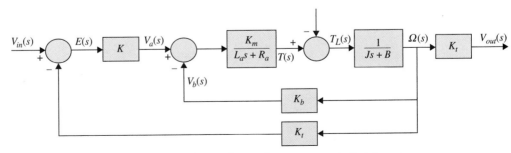

Figure 11-4 Block diagram of a speed-control, armature-controlled dc motor.

Note that the speed at the motor shaft is sensed by the tachometer with a gain K_t. For ease in comparison of input and output, the input to the control system is converted from voltage V_{in} to speed Ω_{in} using the tachometer gain K_t. Hence, assuming $L_a = 0$, we have

$$\Omega(s) = \frac{\dfrac{K_t K_m K}{R_a J_m}}{s + \left(\dfrac{K_m K_b + R_a B + K_t K_m K}{R_a J_m}\right)} \Omega_{in}(s)$$

$$- \frac{\dfrac{1}{J_m}}{s + \left(\dfrac{K_m K_b + R_a B + K_t K_m K}{R_a J_m}\right)} T_L(s)$$

(11-7)

For a step input $\Omega_{in} = A/s$ and disturbance torque value $T_L = D/s$, the output becomes

$$\omega(t) = \frac{A K K_m K_t}{R_a J_m} \tau_c (1 - e^{-t/\tau_c}) - \frac{\tau_c D}{J_m}(1 - e^{-t/\tau_c})$$

(11-8)

where $\tau_c = \dfrac{R_a J_m}{K_m K_b + R_a B + K_t K_m K}$ is the system mechanical-time constant. The steady-state response in this case is

- Speed control reduces the motor sensitivity to a disturbance load.

$$\omega_{fv} = \left(\frac{A K K_m K_t}{K_m K_b + R_a B + K_t K_m K} - \frac{R_a D}{K_m K_b + R_a B + K_t K_m K}\right)$$

(11-9)

where $\omega_{fv} \to A$ as $K \to \infty$. So, speed control may reduce the effect of disturbance. As in Section 11-2-1, the reader should investigate what happens if the inertia J_L is included in this model. If the load inertia J_L is too large, will the motor be able to turn? Again, as in Section 11-2-1, you will have to read the speed-sensor voltage to measure speed. How will that affect your equations?

11-2-3 Position Control

The position response in the open-loop case may be obtained by integrating the speed response. Then, considering Fig. 11-2, we have $\Theta(s) = \Omega(s)/s$. The open-loop transfer function is therefore

$$\frac{\Theta(s)}{V_a(s)} = \frac{K_m}{s(L_a J_m s^2 + (L_a B + R_a J_m)s + R_a B + K_m K_b)}$$

(11-10)

where we have used the total inertia J. For small L_a, the time response in this case is

$$\theta(t) = \frac{A}{K_b}(t + \tau_m e^{-t/\tau_m} - \tau_m)$$

(11-11)

which implies that the motor shaft is turning without stop at a constant steady-state speed. To control the position of the motor shaft, the simplest strategy is to use a proportional controller with gain K. The block diagram of the closed-loop system is shown in Fig. 11-5. The system is composed of an angular position sensor (usually an encoder or a potentiometer for position applications). Note that for simplicity the input voltage can be scaled to a position input $\Theta_{in}(s)$ so that the input and output have the same units

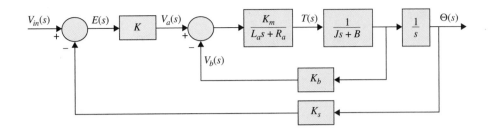

Figure 11-5 Block diagram of a position-control, armature-controlled dc motor.

and scale. Alternatively, the output can be converted into voltage using the sensor gain value.

The closed-loop transfer function in this case becomes

$$\frac{\Theta_m(s)}{\Theta_{in}(s)} = \frac{\dfrac{KK_mK_s}{R_a}}{(\tau_e s + 1)\left\{J_m s^2 + \left(B + \dfrac{K_bK_m}{R_a}\right)s + \dfrac{KK_mK_s}{R_a}\right\}} \tag{11-12}$$

• If the electric time is small, it may be neglected.

where K_s is the sensor gain, and, as before, $\tau_e = (L_a/R_a)$ may be neglected for small L_a.

$$\frac{\Theta_m(s)}{\Theta_{in}(s)} = \frac{\dfrac{KK_mK_s}{R_aJ}}{s^2 + \left(\dfrac{R_aB + K_mK_b}{R_aJ_m}\right)s + \dfrac{KK_mK_s}{R_aJ_m}} \tag{11-13}$$

We next set up an experimental system to test and verify the preceding concepts and learn more about other practical issues.

▷ 11-3 DESCRIPTION OF THE VIRTUAL EXPERIMENTAL SYSTEM[1]

The experiments that you will perform are intended to give you hands-on (virtually!) experience in analyzing the system components and experimenting with various feedback control schemes. To study the speed and position response of a dc motor, a typical experimental test bed is shown in Fig. 11-3. The setup components are as follows:

- A dc motor with a speed sensor (normally a tachometer) or a position sensor (usually an encoder with incremental rotation measurement)
- A power supply and amplifier to power the motor
- Interface cards to monitor the sensor and provide a command voltage to the amplifier input and a PC running MATLAB and Simulink to control the system and to record the response (alternatively, the controller may be composed of an analog circuit system)

The components are described in the next sections.

[1] The authors acknowledge the assistance of Professor Jan Huissoon, Department of Mechanical Engineering at University of Waterloo Ontario, Canada, in the development of some of the experiments discussed in this chapter. Some of the experiments were inspired by the third-year undergraduate courses ME 360 and ME 380 at the University of Waterloo. (See Reference 1.)

11-3-1 Motor

The motor used is a permanent magnet dc motor, with the following parameters (as given by the manufacturer):

K_m	Motor (torque) constant	0.10 Nm/A
K_b	Speed Constant	0.10 V/rad/sec
R_a	Armature resistance	1.35 ohm
L_a	Armature inductance	0.56 mH
J_m	Armature moment of inertia	0.0019 E-4 kg-m^2
τ_m	Motor mechanical time constant	2.3172 E-005 sec

A reduction gear head may be attached to the output disk of the motor shaft. If the motor shaft's angular rotation is considered as the output, the gear head will scale the inertia of the load by $1/n^2$ in the system model, where n is the gear ratio (refer to Chapter 4 and verify).

11-3-2 Position Sensor or Speed Sensor

• Sensor-shaft inertia and damping are too small to be included in the system model.

For position-control applications, an incremental encoder or a potentiometer may be attached directly to the motor shaft to measure the rotation of the armature. In speed control, it is customary to connect a tachometer to the motor shaft. *Normally sensor-shaft inertia and damping are too small to be included in the system model.* The output from each sensor is proportional to the variable it is measuring. We will assume a proportionality gain of 1, that is, $K_t = 1$ *(speed control)*, and $K_s = 1$ *(position control)*.

11-3-3 Power Amplifier

The purpose of the amplifier is to increase the current capacity of the voltage signal from the analog output interface card. The output current from the interface should normally be limited, whereas the motor can draw many times this current. The details of the amplifier design are somewhat complex and will not be discussed here. But we should note two important points regarding the amplifier:

1. The maximum voltage that can be output by the amplifier is effectively limited to 20 V.

2. The maximum current that the amplifier can provide to the motor is limited to 8 A. Therefore,

Amp gain	2 V/V
Amplifier input saturation limits	±10 V
Current saturation limits	±4 A

11-3-4 Interface

In a real experiment, interfacing is an important issue. You would be required to attach all the experimental components and to connect the motor sensor and the amplifier to the computer equipped with MATLAB and Simulink. Simulink would then provide a voltage output function that would be passed on to the amplifier via a digital-to-analog (D/A) interface card. The sensor output would also have to go through an analog-to-digital (A/D) card to reach the computer. Alternatively, you could avoid using a computer and an A/D or D/A card by using an analog circuit for control.

▶ 11-4 DESCRIPTION OF SIMLAB AND VIRTUAL LAB SOFTWARE

SIMLab and Virtual Lab are series of MATLAB and Simulink files that make up an educational tool for students learning about dc motors and control systems. SIMLab was created to allow students to understand the basic simulation model of a dc motor. The parameters of the motor can be adjusted to see how they affect the system. The Virtual Lab was designed to exhibit some of the key behaviors of real dc motor systems. Real motors have issues such as gear backlash and saturation, which may cause the motor response to deviate from expected behavior. Users should be able to cope with these problems. The motor parameters cannot be modified in the Virtual Lab because in a realistic scenario, a motor may not be modified but must be replaced by a new one!

In both the SIMLab and the Virtual Lab, there are five experiments. In the first two experiments, feedback speed control and position control are explored. Open-loop step response of the motor appears in the third experiment. In the fourth experiment, the frequency response of the open-loop system can be examined by applying a sinusoidal input. A controller design project is the last experiment.

The SIMLab or Virtual Lab main menu windows can be called from the Automatic Control Systems launch applet (**ACSYS**) by clicking on the appropriate button. The main SIMLab and Virtual Lab menu windows are shown in Fig. 11-6. The experiments are selected from these windows. You may directly call SIMLab or Virtual Lab from the MATLAB Command window by typing simlab or virtuallab respectively.

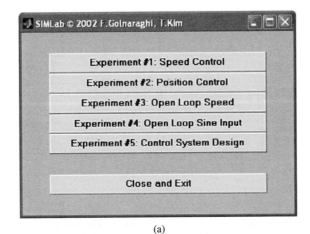

(a)

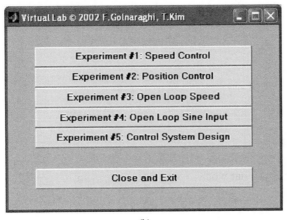

(b)

Figure 11-6 The menu windows for (a) SIMLab and (b) Virtual Lab.

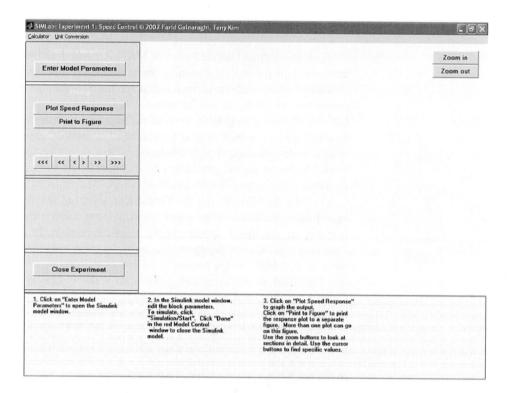

Figure 11-7 Typical experiment control window.

An experiment control window pops up after selecting one of the experiment buttons in Fig. 11-6 (e.g., Fig. 11-7 represents the Speed Control experiment in the SIMLab). The deep blue panel on the left contains the control buttons for the experiment. Every experiment has a button to enter model parameters, plot the response curve, and move a cursor across the response curve. The plots or animations that the experiment supports appear in

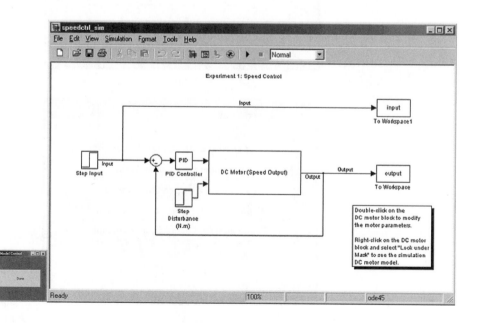

Figure 11-8 Typical SIMLab experiment model.

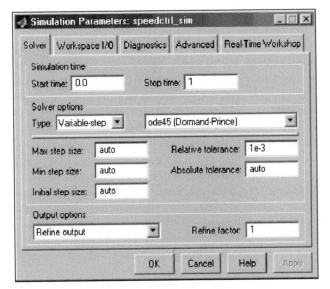

Figure 11-9 Simulation parameters: solver tab.

• Refer to MATLAB Help on Simulink to learn more on its functionality.

the right panel. There are Zoom buttons in the top right corner to zoom in on plots. The blue panel skirting the bottom of the main window gives step-by-step instructions on how to use the experiment window. The standard Microsoft Windows 9x calculator and a unit conversion tool can be selected from the top menu.

The model parameters must be set first in any experiment. By selecting the Enter Model Parameters button, a Simulink (.mdl) window containing the experiment model pops up, along with a red window with a button marked Done. The model shown in Fig. 11-8 contains a simple closed-loop system using PID speed control.

All the simulation parameters for the Simulink model should be set previously. Selecting Simulation from the Simulink menu, and next choosing Simulation Parameters can access these settings. The settings are shown in Figs. 11-9 and 11-10. The Start Time and Stop Time settings in the Solver tab are most important as far as SIMLab and Virtual Lab

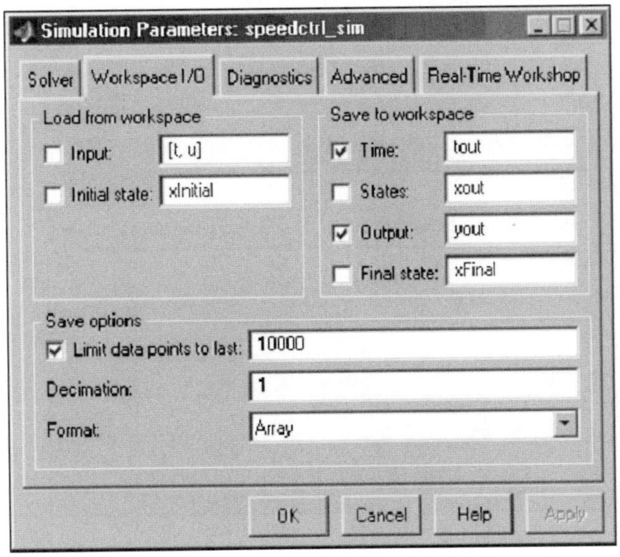

Figure 11-10 Simulation Parameters: Workspace I/O tab.

examples are concerned, and they could be manipulated in order to modify the simulation running time.

Double-clicking on the appropriate model block can modify model parameters such as the PID values. For SIMLab, double-clicking on the motor block brings up a window containing a list of adjustable motor parameters (see Fig. 11-11). All motor parameters, such as the resistance, back-emf constant, load inertia, and damping coefficient, may be modified. Right-clicking on a SIMLab motor block and selecting Look under Mask makes the dc motor model available. However, the Virtual Lab motor blocks are completely opaque to the user since they model actual dc motors. One other feature that SIMLab has, which Virtual Lab does not, is a torque-disturbance input into the motor. This can be used to investigate the stall torque and the effect of an integral controller.

• SIMLab motor blocks can be adjusted by the user.

Clicking Start from the Simulation menu runs the simulation. Clicking the Done button in the red window closes the Simulink window and returns to the experiment control window. To view the time response of the system, click Plot [Speed or Position] Response in the experiment control window (e.g., Fig. 11-7). The graph can be zoomed in and out using the Zoom buttons. The cursor buttons allow the graphed values to be displayed corresponding to the position of a cursor dot. The Print to Figure button allows the current response plot to be sent to a separate MATLAB figure. Several plots can be stored in this figure, which is useful to compare the system response after changing a parameter in the Simulink model. This figure can also be saved as a (.fig) file for later reference. Again, in the Virtual Lab you cannot change the system parameters. This is because of the black-box approach used. Each new scenario must be recompiled.

• Virtual Lab motor blocks are completely opaque to the user.

Some of the experiments have additional features, such as animation and calculation tools. These are discussed in the following sections. Selecting Close Experiment in the control window exits the experiment.

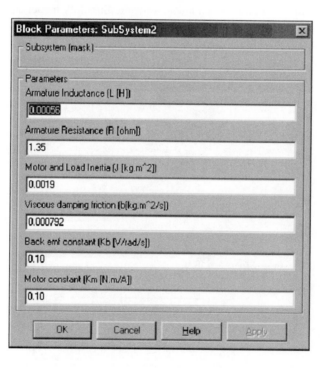

Figure 11-11 Adjustable parameters for the SIMLab motor blocks.

▶ 11-5 SIMULATION AND VIRTUAL EXPERIMENTS

It is desired to design and test a controller offline by evaluating the system performance in the safety of the simulation environment. The simulation model can be based on available system parameters, or they may be identified experimentally. Since most of the system parameters are available (see motor specifications in Section 11-3-1), it will be useful to build a model using these values and to simulate the dynamic response for a step input. The response of the actual system (in this case, the virtual system) to the same test input will then verify the validity of the model. Should the actual response to the test input be significantly different from the predicted response, certain model parameter values would have to be revised or the model structure refined to reflect more closely the observed system behavior. Once satisfactory model performance has been achieved, various control schemes can be implemented.

• Use SIMLab as a simulation tool to test your controllers, and later test your design on the Virtual Lab to check the design against practical implications.

In this chapter, SIMLab represents the simulation model with adjustable parameters, and Virtual Lab represents the actual (virtual) system. Once the model of the Virtual Lab system is identified and confirmed, the controller that was originally designed using SIMLab should be tested on the Virtual Lab model.

11-5-1 Open-Loop Speed

The first step is to model the motor. Using the parameter values in Section 11-3-1 for the model of the motor in Fig. 11-1, simulate the open-loop velocity response of the motor to a step voltage applied to the armature. You should run the following tests using SIMLab:

1. Apply step inputs of $+5$ V, $+15$ V, and -10 V. Note that the steady-state speed should be approximately the applied armature voltage divided by K_b as in Eq. (11-5) (try dc motor alone with no gear head or load applied in this case).

2. Study the effect of viscous friction on the steady-state motor speed. First set $B = 0$ in the Simulink motor parameter window. Then gradually increase its value and check the speed response.

3. Repeat Step 2 and connect the gear head with a gear ratio of 5.2:1, using additional load inertia at the output shaft of the gear head, of 0.05 kg-m^2 (requires modification of J in the Simulink motor parameters).

4. Determine the viscous friction required at output shaft to reduce the motor speed by 50 percent from the speed it would rotate at if there were no viscous friction.

5. Introduce a step-disturbance input T_L and study its effect on the system in Step 3.

6. Assuming that you do not know the overall inertia J for the system in Step 3, can you use the speed-response plot to estimate its value? If so, confirm the values of motor and load inertia. How about the viscous-damping coefficient? Can you use the time response to find other system parameters?

In this experiment, we use the open-loop model represented in Experiment #3: Open Loop Speed button. The Simulink system model is shown in Fig. 11-12, representing a simple open-loop model with a motor speed output.

In a realistic scenario, the motor is connected to an amplifier that can output a voltage only in a certain range. Outside of this range, the voltage saturates. The current within the motor may also be saturated. To create these effects in software, right-click the

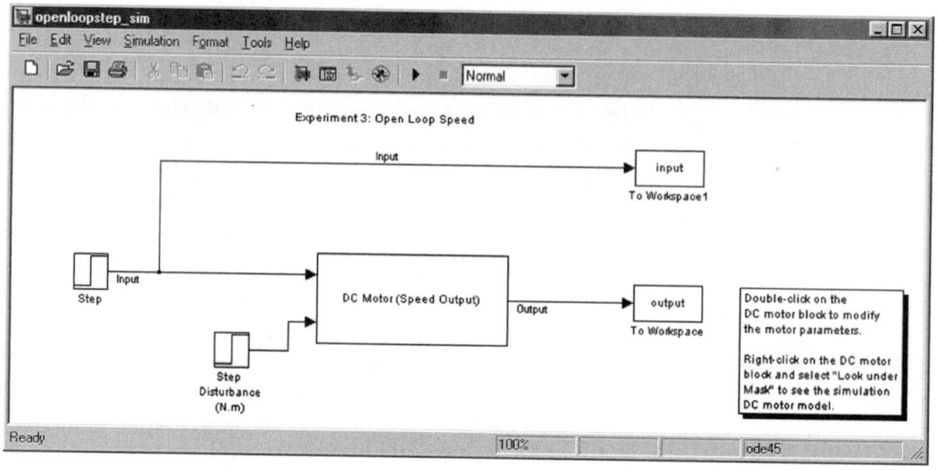

Figure 11-12 SIMLab open-loop speed response of dc motor experiment.

dc motor block and select the Look under Mask to obtain the motor model shown in Fig. 11-13. Double-click both the voltage and current blocks and adjust their values (default values of ±10 volts and ±4 amps have already been set). If you do not wish to include saturation, set the limits very large (or delete these blocks altogether). Run the above experiments again and compare the results.

Assuming a small electric-time constant, we may model the dc motor as a first-order system. As a result, the motor inertia and the viscous damping friction could be calculated with measurements of the mechanical-time constant using different input magnitudes. For a unit-step input, the open-loop speed response is shown in Fig. 11-14. After measuring the mechanical-time constant of the system τ_m, you can find the inertia J, assuming all other parameters are known. Recall that for a first-order system, the time constant is the time to reach 63.2 percent of final value for a step input [verify using Eq. (11-5) or (11-6)]. A typical open-loop speed response is shown in Fig. 11-14. The steady-state velocity and the time constant τ_m can be found by using the cursor.

In SIMLab, the disturbance torque default value is set to zero. To change an input value, simply change its final value.

Now that you have gained insight into the motor speed response, it is time to apply your knowledge to test the virtual experiment. Here you have no access to the system parameter values. Use the Virtual Lab to test the following:

7. Apply step inputs of $+5$ V, $+15$ V, and -10 V. How different are the results from Step 1?

• Use the Simulink Library Browser to add blocks or change your Simulink model.

• Use the cursor buttons to move along the response and read the values at a given location.

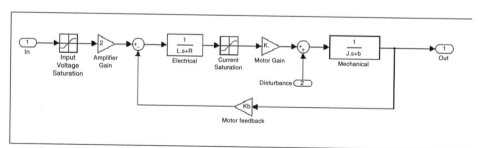

Figure 11-13 DC motor model including voltage and current saturation.

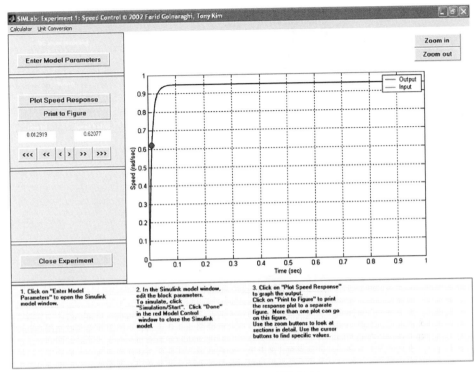

Figure 11-14 Speed response of the open-loop system (SIMLab).

8. From the transient and steady-state responses, identify the system model as closely as possible.

Recall that the motor and amplifier have built-in nonlinear effects due to noise, friction, and saturation. So in Step 8, your model may vary for different input values. Incorrect values may be obtained if the input to the motor is excessive and saturates. Caution must be taken to ensure that the motor input is low enough such that this does not happen. Use the mechanical time constant and final value of the response in this case to confirm the system parameters defined in Sec. 11-3-1. These parameters are needed to conduct the speed-and position-control tasks. Figure 11-15 shows the Virtual Lab motor speed response to a unit-step input. The friction effect is observed when the motor starts. The noise at steady state may also be observed. For higher input magnitudes, the response will saturate.

• The Virtual Lab response exhibits effects including friction, saturation, and noise.

11-5-2 Open-Loop Sine Input

The objective of using open-loop sine input is to investigate the frequency response of the motor using both SIMLab and Virtual Lab.

9. For both SIMLab and Virtual Lab, apply a sine wave with a frequency of 1 rad/sec and amplitude of 1 V to the amplifier input, and record both the motor velocity and sine wave input signals. Repeat this experiment for frequencies of 0.2, 0.5, 2, 5, 10, and 50 rad/sec (keeping the sine wave amplitude at 1 V).

10. Change the input magnitude to 20 V and repeat Step 9).

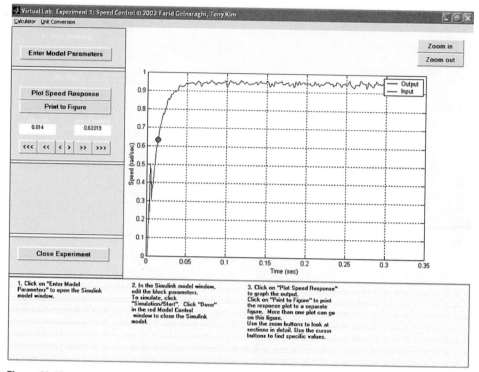

Figure 11-15   Speed response of the open-loop system (Virtual Lab).

Open up Experiment #4: Open-Loop Sine Input from the SIMLab or Virtual Lab menu windows, as shown in Fig. 11-16. The input and disturbance blocks and the motor parameters are adjustable in the SIMLab model. For the Virtual Lab version, the amplitude should be low to avoid amplifier or armature current saturation. Amplitude of 1 is a low enough value to avoid saturation in this example. In the SIMLab version, the saturation values are adjustable to allow you practice with their effect. The SIMLab response for sine input of

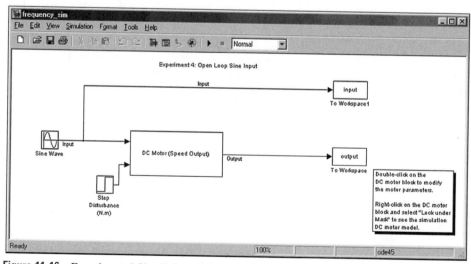

Figure 11-16   Experiment 4 Simulink model.

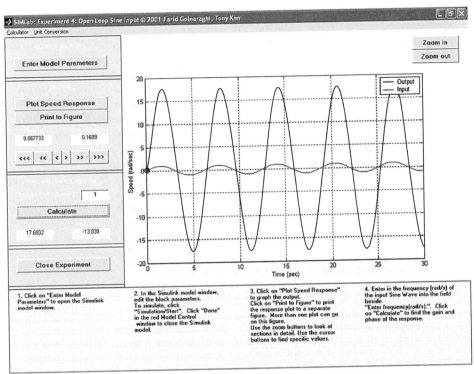

Figure 11-17 SIMLab time response and gain and phase calculation for input = sin(*t*).

magnitude and frequency of 1 is shown in Fig. 11-17. You may also try adding dead zone and backlash to your motor block to test their effects (these functions are available in the Simulink Library Browser, briefly discussed at the end of Section 11-6). For a sine input of magnitude 20 V, the Virtual Lab system exhibits saturation as shown in Fig. 11-18.

• Use the Simulink Library Browser to add blocks to your Simulink model. (See the example in the end of the chapter.)

The frequency of the sine wave will dictate the gain and phase of the response curve. There is a Gain and Phase Calculator in the Experiment 4 control window. To measure the magnitude and phase of the steady-state response, enter a frequency of 1 rad/sec in the edit block. Entering the input frequency and clicking on Calculate displays the gain and phase of the system (see Fig. 11-16). Using the Gain and Phase Calculator, record the gain and phase of the response. Repeat with other input frequencies, and discuss any trends.

11-5-3   Speed Control

Having simulated the open-loop motor characteristics in previous sections, we can now extend the model to include velocity feedback from the motor and use a proportional controller. Assume that the motor velocity is measured using a sensor that provides 1 V/rad/sec. The block diagram that you should be modeling is shown in Fig. 11-4. For proportional gains of 1, 10, and 100, perform the following tests using SIMLab:

11.   Apply step inputs of +5 V, +15 V, and −10 V (try dc motor alone; no gear head or load applied in this case).

12.   Repeat Step 11 using additional load inertia at the output shaft of the gear head (gear ratio 5.2:1) of 0.05 kg-m^2 (requires adjustment of the *J* value in SIMLab motor parameter block).

• The Virtual Lab response exhibits effects including friction, saturation and noise.

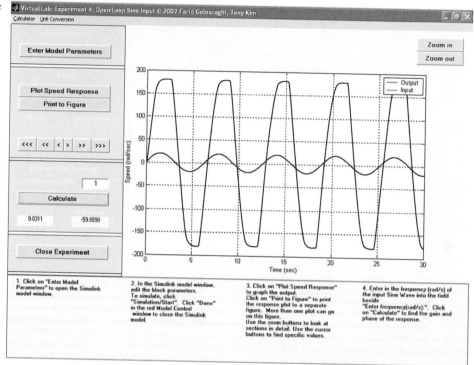

Figure 11-18 Virtual Lab time response and gain and phase calculation for input = 20 sin(t).

13. Apply the same viscous friction to the output shaft as obtained in Step 4 in Section 11-5-1, and observe the effect of the closed-loop control. By how much does the speed change?

14. Repeat Step 5 in Section 11-5-1, and compare the results.

Open up Experiment 1: Speed Control from the SIMLab menu window. A screen similar to Fig. 11-7 will be displayed. Next, select the Enter Model Parameters to get the system Simulink model, as shown in Fig. 11-19. This figure is a simple PID speed-control model. Double-clicking on the PID block displays the editable PID values. The values of the step-input and the disturbance-torque blocks may also be adjusted. The disturbance-torque default value is set to zero. To change an input value, simply change its final value.

Increasing the proportional gain in the PID block will decrease the rise time. For an unsaturated model, the SIMLab version of this experiment could exhibit extremely fast rise times at very high proportional gains, because the dc motor can utilize unlimited voltage and current levels. To create this effect in software, right-click the dc motor block, and select the Look under Mask to obtain the motor model similar to Fig. 11-13. Double-click both the voltage and current blocks and adjust their values to very large (or delete their blocks). Recall from Section 11-5-1 that the default saturation limits are ±10 V and ±4 A, respectively. Fig. 11-20 displays a typical speed response from the SIMLab.

For a given input to change the proportional gain values, enter in the following sets of PID values and print all three plots on the same figure (use the Print to Figure button in experiment main window).

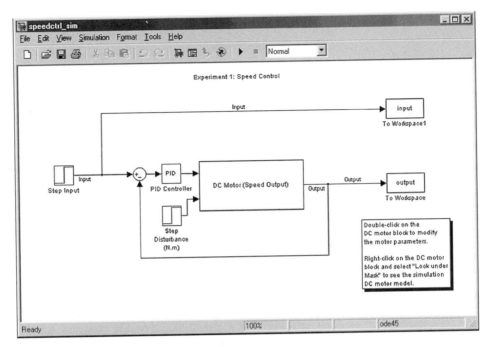

Figure 11-19 Experiment 1: Simulink model.

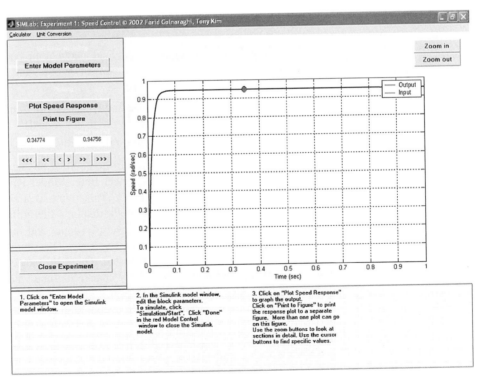

Figure 11-20 Speed-control response in the SIMLab control window.

$$P = 1 \qquad I = 0 \quad D = 0$$
$$P = 10 \qquad I = 0 \quad D = 0$$
$$P = 100 \quad I = 0 \quad D = 0$$

The input units used in these simulations are specified in volts, while the feedback units at the motor are in radians per second. This was done intentionally to illustrate the scaling (or conversion) that is performed by the sensor. Had the velocity been specified in volts per radians per second, a different response would have been obtained. To check the effect of the velocity feedback scaling, repeat the preceding experiments using a proportional gain of 10, but assume that the velocity feedback signal is a voltage generated by a sensor with a conversion factor of 0.2 V/rad/sec. (Note: in commonly used industry standards the tachometer gain is in volts per RPM.)

Next, for the Virtual Lab test the following:

15. Apply step inputs of +5 V, +15 V, and −10 V. How different are the results from the SIMLab?

You may again confirm the system parameters obtained in Section 11-5-1.

11-5-4 Position Control

Next, investigate the closed-loop position response; modify your model to position feedback. For proportional gains of 1, 10, and 100 (requires modification of PID block parameters), perform the following tests using SIMLab:

16. For the motor alone, apply a 160° step input.
17. Apply a step disturbance torque (−0.1) and repeat Step 16.
18. Examine the effect of integral control in Step 17 by modifying the Simulink PID block.
19. Repeat Step 16, using additional load inertia at the output shaft of 0.05 kg-m² and the gear ratio 5.2 : 1 (requires modification of J_m and B in the motor parameters).
20. Set $B = 0$ and repeat Step 19.
21. Examine the effect of voltage and current saturation blocks (requires modification of the saturation blocks in the motor model).
22. In all above cases, comment on the validity of Eq. (11-13).

Open up Experiment 2: Position Control from the SIMLab menu window. A screen similar to Fig. 11-7 will be displayed. Next select Enter Model Parameters to get the system Simulink model, as shown in Fig. 11-21. This figure is a simple PID position-control model. Double-clicking on the PID block displays the editable PID values. The Deg to Rad and Rad to Deg gain blocks convert the input and the output such that the user enters inputs and receives outputs in degrees only. The values of the step-input and the disturbance-torque blocks may also adjusted. The disturbance-torque default value is set to zero. To change an input value, simply change its final value. Figure 11-22 displays a typical position response from the SIMLab.

The position time response can also be animated. The Animate Position Response button initiates the animation, and the Stop Animation button halts it. This is a very useful tool that gives the user a physical sense of how a real motor turns. The time, input-angle, and output-angle values are displayed on the animation field, as shown in Fig. 11-23.

The nonlinearities due to voltage and current limits cause the time response to saturate at a high enough proportional gain. In the SIMLab version, the saturation values may

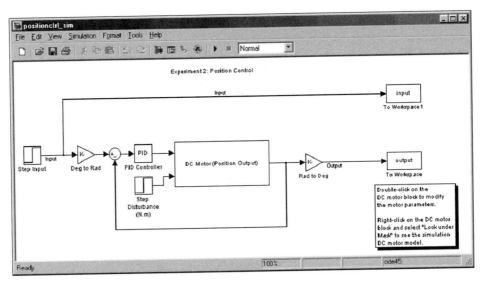

Figure 11-21   Experiment 2: Position Control Simulink model.

be adjusted as discussed in Section 11-5-1. Try to examine the saturation effect for different proportional gain values.

23.   For proportional gains of 1, 10, and 100 (requires modification of PID block parameters), repeat Steps 16 and 18 tests using Virtual Lab.

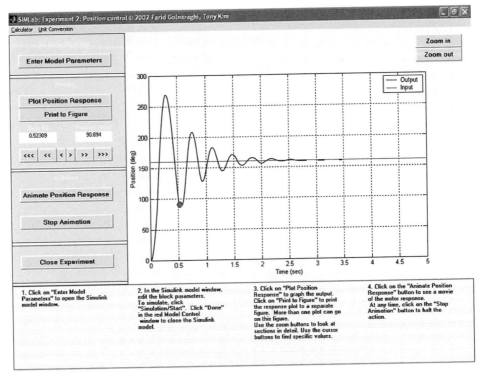

Figure 11-22   Position response in the Experiment 2 control window.

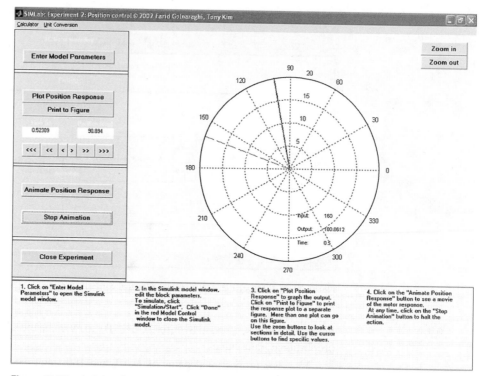

Figure 11-23 Animated position response in the Experiment 2 control window.

▶ 11-6 DESIGN PROJECT

The primary goal of this section is to help you to gain experience in applying your control knowledge to a practical problem. You are encouraged to apply the methods that you have learned throughout this book, particularly in Chapter 10, to design a controller for your system. The animation tools provided make this experience very realistic. The project may be more exciting if it is conducted by teams on a competitive basis. The SIMLab and Virtual Lab software are designed to provide enough flexibility to test various scenarios. The SIMLab, in particular, allows introduction of a disturbance function or changes of the system parameters if necessary.

Description of the Project: Consider the system in Fig. 11-24. The system is composed of the dc motor used throughout this chapter. We connect a rigid beam to the motor shaft to create a simple robotic system conducting a pick and place operation. A solid disk is attached to the end of the beam through a magnetic device (e.g., a solenoid). If the magnet is on, the disk will stick to the beam, and when the magnet is turned off, the disk is released.

Objective: to drop the disk in to a hole as fast as possible. The hole is 1 in. (25.4 mm) below the disk (see Fig. 11-25).

Design Criteria: The arm is required to move in only one direction from the initial position. The hole location may be anywhere within an angular range of 20° to 180° from the initial position. The arm may not overshoot the desired position by more than 5°. A tolerance of ±2% is acceptable (settling time). These criteria may easily be altered to create a new scenario.

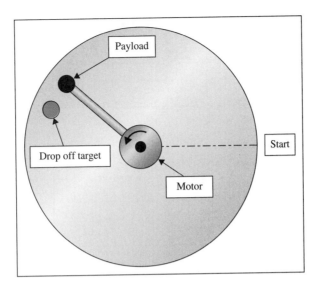

Figure 11-24 Control of a simple robotic arm and a payload.

The objective may be met by looking at the settling time as a key design criterion. However, you may make the problem far more interesting by introducing other design constraints such as the percent overshoot and rise time. In SIMLab, you can also introduce a disturbance torque to alter the final value properties of the system. The Virtual Lab system contains nonlinear effects that make the controller design more challenging. You may try to confirm the system model parameters first, from earlier experiments. It is highly recommended that you do the design project only after fully appreciating the earlier experiments in this chapter and after understanding Chapter 10. Have fun!

This experiment is very similar to the position-control experiment. The idea of this experiment is to get a metal object attached to a robot arm by an electromagnet from position 0° to a specified angular position with a specified overshoot and minimum overall time.

Open up Experiment 5: Control System Design from the SIMLab menu window. A Screen similar to Fig. 11-7 will be displayed. Next, select Enter Model Parameters to get the system Simulink model, as shown in Fig. 11-26. As in Section 11-5-4, this figure represents a simple PID position-control model with the same functionalities. The added feature in this model is the electromagnetic control. By double-clicking the Electromagnet Control block, a parameter window pops up, as in Fig. 11-27, which allows the user to

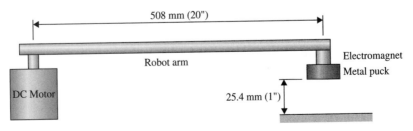

Figure 11-25 Side view of the robot arm.

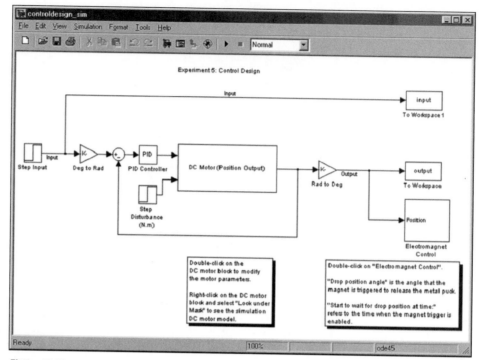

Figure 11-26   Experiment 5 Simulink model.

adjust the drop-off payload location and the time delay (in seconds) to turn the magnet off after reaching the target. This feature is particularly useful if the response overshoots and passes through the target more than once. So, in Fig. 11-27, the "Drop position angle" is the angle where the electromagnet turns off, dropping the payload. "Start to wait for drop position at time" refers to the time where the position trigger starts to wait for the position specified by "Drop position angle."

An important note to remember is that in the Virtual Lab the electromagnet will never drop the object exactly where it is specified. Since any electromagnet has residual magnetism even after the current stops flowing, the magnet holds on for a short time after the trigger is tripped. A time response of the system for proportional gain and derivative gain of 4 and 0.3, respectively, is shown in Fig. 11-28.

The system response can also be animated, as shown in Fig. 11-29. This feature makes the problem more realistic. The red circular object represents the payload, which in this

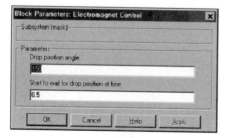

Figure 11-27   Parameter window for the electromagnet control.

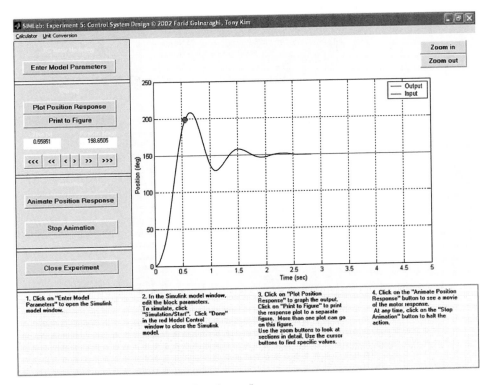

Figure 11-28 Position response for Experiment 5.

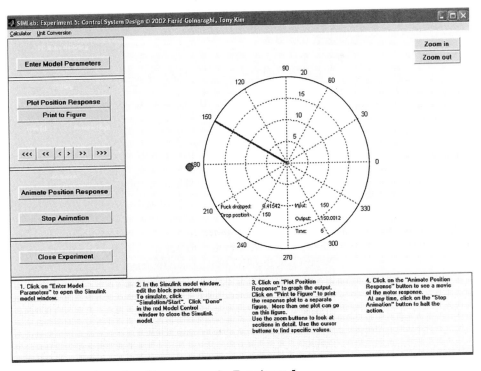

Figure 11-29 Animated position response for Experiment 5.

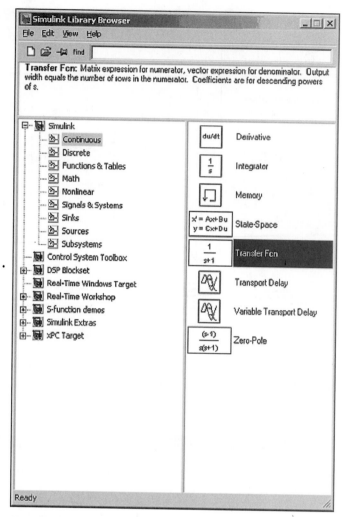

Figure 11-30 Selecting a Transfer Function block in Simulink Library Browser.

case, has been overthrown. The drop angle and drop time are displayed on the animation field. Note that in this case, the magnet drop off takes place prematurely. As a result, the magnet has been released earlier and is not on target!

- Other controllers may also be tried.

You can apply other controllers such as lead or lag. To do so, first click the Enter Model Parameters button to get the system Simulink window, as shown in Fig. 11-26. Next, click the PID block once to select it, and then press the delete button to clear the block. Go to the View menu and choose Show Library Browser. In the resulting Simulink Library Browser window (as shown in Fig. 11-30), select the Transfer Function block and drag it to your Simulink window, while depressing the left mouse button. Place the block

- Use the Simulink Library Browser to add blocks to your Simulink model.

in the controller location as shown in Fig. 11-31. You must ensure that all blocks are fully connected, as shown in Fig. 11-31. You may have to move the comparator block a bit to the left to create more room for the new Transfer Function block. Next, double-click the block and enter the desired values.

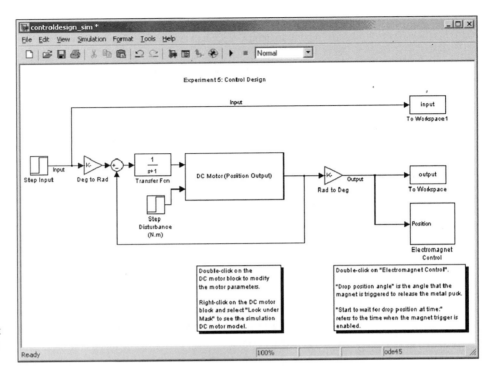

Figure 11-31 Experiment 5: Simulink model with a new controller block.

▷ 11-7 SUMMARY

In this chapter we described the SIMLab and Virtual Lab software to improve your understanding of control and to provide a better practical appreciation of the subject. We discussed that in a realistic system including an actuator (e.g., a dc motor) and mechanical (gears) and electrical components (amplifiers), issues such as saturation of the amplifier, friction in the motor, or backlash in gears will seriously affect the controller design. This chapter focused on problems involving dc motors including modeling, system identification, and controller design. We presented experiments on speed and position control of dc motors, followed by a controller design project involving control of a simple robotic system. The focus on dc motors in these experiments was intentional, because of their simplicity and wide use in industrial applications.

▷ REFERENCE

1.   J. P. Huissoon and F. Golnaraghi, "ME360 Laboratory Experiment," University of Waterloo Ontario, Canada, *ME 3B Lab Manual,* Fall 1999.

▷ PROBLEMS

11-1.   Create a model of the motor as shown in Fig. 11-2. Use the following parameter values: $J_m = 0.0004$ kg-m^2; $B = 0.001$ Nm/rad/sec, $R_a = 2\ \Omega$, $L_a = 0.008$ H, $K_m = 0.1$ Nm/A, and $K_b = 0.1$ V/rad/sec. Assume that the load torque T_L is zero.

• Speed response: SIMLab

Apply a 5-V step input to the motor, and record the motor speed and the current drawn by the motor (requires modification of SIMLab blocks by making current the output) for 10 sec following the step input.

a. What is the steady-state speed?

b. How long does it take the motor to reach 63% of its steady-state speed?

c. What is the maximum current drawn by the motor?

11-2. Set the viscous friction B to zero in Problem 11-1. Apply a 5-V step input to the motor, and record the motor speed and current for 10 sec following the step input. What is the steady-state speed?

a. How long does it take the motor to reach 63% of its steady-state speed?

b. What is the maximum current drawn by the motor?

c. What is the steady-state speed when the applied voltage is 10 V?

11-3. Set the armature inductance L_a to zero in Problem 11-2. Apply a 5-V step input to the motor, and record the motor speed and current drawn by the motor for 10 sec following the step input.

a. What is the steady-state speed?

b. How long does it take the motor to reach 63% of its steady-state speed?

c. What is the maximum current drawn by the motor?

d. If J_m is increased by a factor of 2, how long does it take the motor to reach 63% of its steady-state speed following a 5-V step voltage input?

11-4. Repeat Problems 11-1 through 11-3, and assume the load torque $T_L = -0.1$ Nm (don't forget the minus sign) starting after 5 sec (requires change of the disturbance block parameters in SIMLab).

a. How does the steady-state speed change once T_L is added?

b. How long does it take the motor to reach 63% of its new steady-state speed?

c. What is the maximum current drawn by the motor?

d. Increase T_L and further discuss its effect on the speed response.

• Speed control: SIMLab

11-5. For the system in Fig. 11-3, use the parameters for Problem 11-1 (but set $L_a = 0$) and an amplifier gain of 2 to drive the motor (ignore the amplifier voltage and current limitations for the time being). What is the steady-state speed when the amplifier input voltage is 5 V?

11-6. Modify the model in Problem 11-5 by adding a proportional controller with a gain of $K_p = 0.1$, apply a 10 rad/sec step input, and record the motor speed and current for 2 sec following the step input.

a. What is the steady-state speed?

b. How long does it take the motor to reach 63% of its steady-state speed?

c. What is the maximum current drawn by the motor?

11-7. Change K_p to 1.0 in Problem 11-6, apply a 10 rad/sec step input, and record the motor speed and current for 2 sec following the step input.

a. What is the steady-state speed?

b. How long does it take for the motor to reach 63% of its steady-state speed?

c. What is the maximum current drawn by the motor?

d. How does increasing K_p affect the response (with and without saturation effect in the SIMLab model)?

11-8. Repeat Problem 11-6, and assume the load torque $T_L = -0.1$ N-m starting after 0.5 sec (requires change of the disturbance block parameters in SIMLab).

a. How does the steady-state speed change once T_L is added?

b. How long does it take the motor to reach 63% of its new steady-state speed?

• Position control: SIMLab

11-9. Insert a velocity sensor transfer function K_s in the feedback loop, where $K_s = 0.2$ V/rad/sec (requires adjustment of the SIMLab model). Apply a 2-rad/sec step input, and record the motor speed and current for 0.5 sec following the step input. Find the value of K_p that gives the same result as in Problem 11-6.

11-10. For the system in Fig. 11-5, select $K_p = 1.0$, apply a 1-rad step input, and record the motor position for 1 sec. Use the same motor parameters as in Problem 11-1.

a. What is the steady-state position?

b. What is the maximum rotation?

c. At what time after the step does the maximum occur?

11-11. Change K_p to 2.0 in Problem 11-10, apply a 1-rad step input, and record the motor position for 1 sec.

a. At what time after the step does the maximum occur?

b. What is the maximum rotation?

11-12. Using the SIMLab, investigate the closed-loop position response using a proportional controller. For a position control case, use proportional controller gains of 0.1, 0.2, 0.5, 1 and 2, record the step response for a 1 rad change at the output shaft, and estimate what you consider to be the best value for the proportional gain. Use the same motor parameters as in Problem 11-1.

11-13. Using the SIMLab, investigate the closed-loop position response using a PD controller. Modify the controller used in Problem 11-12 by adding derivative action to the proportional controller. Using the best value you obtained for K_p, try various values for K_D and record the step response in each case.

11-14. Repeat Problem 11-13 and assume a disturbance torque $T_D = -0.1$ N-m in addition to the step input of 1 rad (requires change of the disturbance block parameters in SIMLab).

11-15. Use the SIMLab and parameter values of Problem 11-1 to design a PID controller that eliminates the effect of the disturbance torque, with a percent overshoot of 4.3.

• Virtual Lab experiments

11-16. Investigate the frequency response of the motor using the Virtual Lab Tool. Apply a sine wave with a frequency of 0.1 Hz (don't forget: 1Hz = 2π rad/sec) and amplitude of 1 V the amplifier input, and record both the motor velocity and sine wave input signals. Repeat this experiment for frequencies of 0.2, 0.5, 1, 2, 5, 10, and 50 Hz (keeping the sine wave amplitude at 1 V).

11-17. Using the Virtual Lab Tool, investigate the closed-loop motor speed response using a proportional controller. Record the closed-loop response of the motor velocity to a step input of 2 rad/sec for proportional gains of 0.1, 0.2, 0.4, and 0.8. What is the effect of the gain on the steady-state velocity?

11-18. Using the Virtual Lab Tool, investigate the closed-loop position response using a proportional controller. For a position-control case, use proportional controller gains of 0.1, 0.2, 0.5, 1, and 2, record the step response for a 1-rad change at the output shaft, and estimate what you consider to be the best value for the proportional gain.

11-19. Using the Virtual Lab Tool, investigate the closed-loop position response using a PD controller. Modify the controller used in Problem 11-13 by adding derivative action to the proportional controller. Using the best value you obtained for K_p, try various values for K_D, and record the step response in each case.

Index

Laplace Transform Table

Laplace Transform $F(s)$	Time Function $f(t)$
1	Unit-impulse function $\delta(t)$
$\dfrac{1}{s}$	Unit-step function $u_s(t)$
$\dfrac{1}{s^2}$	Unit-ramp function t
$\dfrac{n!}{s^{n+1}}$	$t^n (n = \text{positive integer})$
$\dfrac{1}{s + \alpha}$	$e^{-\alpha t}$
$\dfrac{1}{(s + \alpha)^2}$	$te^{-\alpha t}$
$\dfrac{n!}{(s + \alpha)^{n+1}}$	$t^n e^{-\alpha t} (n = \text{positive integer})$
$\dfrac{1}{(s + \alpha)(s + \beta)}$	$\dfrac{1}{\beta - \alpha}(e^{-\alpha t} - e^{-\beta t})(\alpha \neq \beta)$
$\dfrac{s}{(s + \alpha)(s + \beta)}$	$\dfrac{1}{\beta - \alpha}(\beta e^{-\beta t} - \alpha e^{-\alpha t})(\alpha \neq \beta)$
$\dfrac{1}{s(s + \alpha)}$	$\dfrac{1}{\alpha}(1 - e^{-\alpha t})$
$\dfrac{1}{s(s + \alpha)^2}$	$\dfrac{1}{\alpha^2}(1 - e^{-\alpha t} - \alpha te^{-\alpha t})$
$\dfrac{1}{s^2(s + \alpha)}$	$\dfrac{1}{\alpha^2}(\alpha t - 1 + e^{-\alpha t})$
$\dfrac{1}{s^2(s + \alpha)^2}$	$\dfrac{1}{\alpha^2}\left[t - \dfrac{2}{\alpha} + \left(t + \dfrac{2}{\alpha}\right)e^{-\alpha t}\right]$
$\dfrac{s}{(s + \alpha)^2}$	$(1 - \alpha t)e^{-\alpha t}$